SACRED CAUSES

ALSO BY MICHAEL BURLEIGH

Prussian Society and the German Order

Germany turns Eastwards

The Racial State: Germany 1933–1945

Death and Deliverance

(ed.) *Confronting the Nazi Past*

Ethics and Extermination: Reflections on Nazi Genocide

The Third Reich: A New History

*Earthly Powers: Religion and Politics in Europe
from the French Revolution to the Great War*

SACRED CAUSES

THE CLASH OF RELIGION AND POLITICS,
FROM THE GREAT WAR TO THE WAR ON TERROR

Michael Burleigh

HarperCollins*Publishers*

HarperCollins books may be purchased for educational, business,
or sales promotional use. For information please write: Special
Markets Department, HarperCollins Publishers, 10 East 53rd
Street, New York, NY 10022.

T. S. Eliot *Choruses from 'The Rock'* from *Collected Poems
1909–1962* by kind permission of Faber and Faber.

Lyrics from John Lennon's 'Imagine' by kind permission
from Lenono Music.

First printed in Great Britain in 2006 by Harper Press,
an imprint of HarperCollins Publishers.

FIRST U.S. EDITION

Library of Congress Cataloging-in-Publication Data
is available upon request.

ISBN: 978-0-06-058095-7
ISBN-10: 0-06-058095-X

07 08 09 10 11 RRD 10 9 8 7 6 5 4 3 2 1

O weariness of men who turn from GOD
To the grandeur of your mind and the glory of your action,
To arts and inventions and daring enterprises,
To schemes of human greatness thoroughly discredited . . .
Plotting of happiness and flinging empty bottles,
Turning from your vacancy to fevered enthusiasm
For nation or race or what you call humanity . . .

Thomas Stearns Eliot, *Choruses from 'The Rock'*

Imagine there's no heaven – It's easy if you try.
No hell below us,
Above us only sky.
Imagine all the people
Living for today.

Imagine there's no countries – It isn't hard to do.
Nothing to kill or die for,
And no religion too.
Imagine all the people
Living life in peace.

John Lennon, 'Imagine'

the incorruptible Professor walked, too, averting his eyes
from the odious multitude of mankind. He had no future.
He disdained it. He was a force. His thoughts caressed the
images of ruin and destruction. He walked frail, insignificant, shabby, miserable – and terrible in the simplicity of his
idea calling madness and despair to the regeneration of the
world. Nobody looked at him. He passed on unsuspected
and deadly, like a pest in the street full of men.

Joseph Conrad, *The Secret Agent*

Be not afraid.   John Paul II

CONTENTS

For Linden, Martin Ivens and Adolf Wood

PREFACE

This is not a history of Christianity, of which there are many, nor a history of modern times, of which Paul Johnson has already written an outstanding example. Rather, the book operates in the middle ground between them, where culture, ideas, politics and religious faith meet in a space for which I cannot find a satisfactory label. Perhaps one should not try. Establishing that space has been one of the major challenges in writing this book. It is easy to recognise what one wants to avoid, for below my rope bridge snap such crocodiles as 'ecclesiastical history', 'the history of ideas' and 'theology'. The general ambition has been to write a coherent history of modern Europe primarily organised around issues of mind and spirit rather than the merely material, although in no sense do I discount the material as an important factor in history, being as I am inordinately credulous towards simple displays of production statistics.

A previous book, *Earthly Powers*, began with the 'political religion' created during the Jacobin phase of the French Revolution with its Cults of Reason or the Supreme Being. These were not simply cynical usurpations of religious forms, but were what the Italian thinker Luigi Sturzo in the mid-1920s referred to as 'the abusive exploitation of the human religious sentiment'. Like much earlier attempts to realise heaven on earth – vividly described in Norman Cohn's classic account of medieval heresies *The Pursuit of the Millennium* – these resulted in hell for many people, as anyone who walks around the sites of Jacobin massacres in the bleak and depopulated Vendée can readily establish. This dystopian strain recurred in various guises throughout the nineteenth century, whether in the crackbrained schemes of Auguste Comte or Charles Fourier, the moral insanity of Russian nihilists, or the scientific socialism of Marx and Engels, which was morally insane in other ways. Although Christianity was an integral aspect of many early socialist movements – and in Britain remains so to this day – in general the

Churches arranged themselves on the side of conservatism, partly as a result of their traumatic experiences at the hands of democratic mobs in revolutionary France and elsewhere.

This alliance of throne and altar duly broke down as the temporal power of the Churches was challenged by nation states which vied for ultimate human loyalties. A succession of popes, more or less gifted in public diplomacy, doggedly tried to shore up their powers in the face of this assault, whether from the combination of liberals and the reactionary conservative Bismarck in Germany, or from the anticlerical zealots of the French Third Republic. Meanwhile many of the Protestant Churches feebly accommodated themselves to the latest secular ideologies such as nationalism and scientism. These conflicts took place in conjunction with a broader series of changes – for which the label secularisation is unsatisfactory – whereby 'science', 'progress', 'morality', 'money', 'culture', 'humanity' and even 'sport' became objects of devotion and refocused religiosity. By the end of the century, when God was invoked by all sides in a catastrophic world war, the 'strange gods' of Bolshevism, Fascism and Nazism were already discernible as alternative objects of religious devotion, those political religions being the initial focus of this book.

Sacred Causes begins amid the terrible trauma of the Great War, the shock that reverberated throughout the first half of the twentieth century. These were strange times. One of the assassins of the Weimar foreign minister Walter Rathenau, who was slain in 1922, claimed that he had been (spiritually) dead since Armistice Day (9 November 1918). Another extreme right-winger, depicted in a post-war play, says: 'What does it matter whether I die of a bullet at twenty, or of cancer at forty, or of apoplexy at sixty. The people need priests who have the courage to sacrifice the best – priests who slaughter.' There were many self-appointed priests (and prophets) in the 1920s, ranging from the strange individuals who briefly cropped up in Weimar Germany (the most successful of whom was Adolf Hitler) to the puritanical sectarians of Bolshevism. Rather than retell the over-familiar story of Fascism, Nazism and Communism, I have tried to evoke their pseudo-religious pathologies, ranging from the Nazis' skilful manipulation of such notions as 'rebirth' and 'awakening' to the Bolsheviks' bizarre resort to perpetual confession and remorseless search for heretics. Although there were important differences between these totalitarian regimes, they drew from a common well of enthusiasm, and shared such heretical goals (or rather temptations) as fashioning a 'new man' or establishing heaven on earth.

They metabolised the religious instinct. The thinkers who first identified and conceptualised these worrying developments lead on to the next part of the story, for many of the most insightful critics of totalitarian political religions came from a religious background, whether the Catholics Luigi Sturzo and Eric Voegelin, the Orthodox Nikolai Berdyaev, or the Protestants Frederick Voigt and Adolf Keller.

The complex responses of the Churches to these challenges are a major concern of this book. While how a national Church reacted certainly requires comment, it is also the case that these were international institutions, so that whenever one writes that 'the' Catholic Church did this or that, this generalisation does not hold, for example, for Britain, the US, Africa or the whole of Central and Latin America. Indeed, international events are indispensable for understanding this subject. The general predisposition of the Churches towards authoritarian (rather than totalitarian) regimes in the inter-war period is inexplicable without reference to the anticlerical atrocities that took place in Russia, Spain and Mexico – what Pius XI called the 'terrible triangle' in direct anticipation of contemporary talk of 'axes of evil'. If one wants a sense of the sort of polity the inter-war Church supported, then it is a matter of looking at Austria, Ireland, and Portugal, rather than Fascist Italy or Nazi Germany, although again British or US Catholics were perfectly at home in their respective democracies regardless of their external sympathies in particular conflicts. Moving on to the period of the Second World War, I have tried to treat Pius XII in a historical way, which means giving him credit for one of the most penetrating intellectual demolitions of Nazism – in the 1937 encyclical *Mit brennender Sorge* – and by trying to evoke his personality and world, and hence the options that were realistically open to him as the Church grappled with a continent-wide conspiracy to murder Europe's Jews. Very little of the cruder – Soviet inspired – 'black legend' survives close analysis, although legitimate questions remain about his hesitations and tone.

The intervention of the Churches in post-war politics – for their 'good war' facilitated this amid the collapse of other authorities – is an important part of the book, notably regarding the extraordinary success of European Christian Democrats in ensuring that Stalin's surrogates did not achieve power in the western half of the continent. It is fashionable, on the left, to decry those aged French, German or Italian leaders, including Pius XII, as well as Adenauer, Bidault and de Gasperi; this is a view I do not share in view of the dizzy alternative prospect of rule by a Marxist nomenklatura, a secret police, and trades union hacks. Turning eastwards, the book charts the state imposition of atheism on the intensely religious societies of eastern Europe, and the extraordinary heroism of persecuted churchmen in Hungary and Poland, who ensured the survival of a heavily restricted form of civil society amid the ambient corruption and darkness of Communism. That theme is taken up in connection with the role of Pope John Paul II (himself a protégé of cardinal Stefan Wyszyński) and the Catholic Church in Poland in the implosion of European Communism in the late 1980s, a role whose importance has been independently recognised by such leading historians of the Cold War as John Lewis Gaddis, and an Italian parliamentary commission unravelling the 1981 KGB/Bulgarian plot to kill the pope.

Three chapters of *Sacred Causes* deal with Europe's present and possible futures. I cast a rather dyspeptic eye over the 1960s, which in many ways were the chief motor of what then seemed like a highly secularised future, with Churches scrambling to articulate every evanescent secular gospel in a manner trenchantly analysed by Edward Norman. The politicisation of religion is as important in this story as the 'sacralisation' of politics. So are the forces that seemed to be turning Europe into a post-Christian desert, in which 'wisdom' would be represented by the lyrics of John Lennon.

There was one regional exception, that along with Franco's Spain seemed immune not just to the 1960s – although it certainly had its barricades – but to the European Enlightenment. No discussion of religion and politics would be complete without reference to the long war in Northern Ireland. Initially, I regarded this as an almost inexplicable, atavistic, tribal struggle fitfully audible as distant bombs rattled the windows of various places I've lived in London. However, in the long term, this squalid little conflict also anticipated the sinister surrender of power to so-called 'moderate' community leaders (and the creation of exceptional pockets where the law does not appear to apply) that is becoming evident in the responses of European governments to the much wider threat of Islamic radicalism. The spectrum of such responses ranges from the appeasement practised by the Spanish socialists – with their vain dialogue about a common 'Mediterranean' culture with people who think 'Andalus' belongs in a revived Caliphate – to the harder line of the Netherlands with its threats of compulsory Dutch and the banning of the burqa – an understandable reaction to the murder of Theo van Gogh, the prominent film-maker, and to the fact that some of its MPs, notably the redoubtable 'Infidel' Ayaan Hirsi Ali, now have to sleep on army bases surrounded by bodyguards. Those Americans who disparage what they see as an emerging 'Eurabia' might bear a thought for the many Europeans who not only dread that prospect but are doing their best to avert it, sometimes risking their lives.

There are a few grounds for hope in this present 'age of anxiety'. Most obviously, Islamist terrorism is not the same order of threat as that of the thermonuclear destruction that overshadowed the planet during the Cold War. Furthermore, whether in Britain or once-liberal Holland, there are definite signs that the worm has turned, suggesting that ordinary people – as opposed to politicians with inner-city Muslim constituents – are not ready to tolerate indefinitely those who wish to eradicate homosexuals, reduce women to second-class citizens, or openly call for the murder of Danish cartoonists, Dutch politicians or Jews and Israelis, activities that may be acceptable in Saudi Arabia or Iran, but which are not all right here. Anyone with those views is irreconcilable with our civilisation and should take the opportunity to leave before Europe's history repeats itself. There are encouraging signs that the Churches – and in particular the Catholic Church of Benedict XVI – are ready to make certain non-negotiable positions clear rather than to mouth the platitudes of a discredited multiculturalism that only exists in the Left university and within local government, neither of them at the cutting edge of European thinking.

Finally, what of the long-term relationships between religion and politics? Atheists and anticlericals (many regarding themselves as 'liberals') like to rehearse the rote of Crusades and Inquisition, wars of religion and US evangelical Christians to extrude the Churches from any involvement in politics. Insofar as there is a debate, this is conducted on the level of alarm aroused when a British prime minister casually mentions that he is accountable to God, a rather unremarkable admission in a broad sweep of European history from Louis the Pious to Gladstone. Historically, of course, as has been pointed out by such thinkers as Marcel Gauchet and George Weigel, Christianity had much to do with the notion of the autonomous, sacrosanct individual, with the preservation of a sphere beyond the state that anticipated civil society, with the notion of elected leadership, and with holding rulers accountable to higher powers. It is almost superfluous to add that Christianity played an integral part in Europe's high culture, and in such campaigns (or crusades) as abolishing the slave trade or ameliorating the social evils of industrialisation. How many atheistic liberals run soup kitchens for homeless drug addicts? Is the culture of guns and gangster rap, which thrills progressive cultural commentators, a better alternative to the thriving black Pentecostal churches? More controversially, the Churches upheld necessary inhibitions and taboos, without which we seem degraded, judging by much of what TV commissioning editors regularly inflict upon us in an obsession with sex that they share with some clergy. Christianity's historical achievements deserve more notice than they customarily receive. Interestingly, it is increasingly secular intellectuals, like Régis Debray or Umberto Eco, who are mounting the defence of Christianity against silly politically correct attempts to deny or marginalise it.

There also seems no rational reason to exclude Christians – to range no further – from political debate, any more than there is to deny the vote to people with blue eyes or red hair. That is particularly so where they speak with authority, namely regarding the aged, imprisoned, sick and disadvantaged whom bureaucratised welfare has done little or nothing to help. Whether they have anything relevant to contribute to, for example, foreign policy seems more dubious, especially when they simply replicate the predictable views of the progressive intelligentsia regarding, say, Israel and Palestine. Matters become more complex regarding such issues as the creation or expansion of faith schools, with all their potentialities for consolidating antagonistic ghettos through what amounts, in the worst scenarios, to monocultural indoctrination, whatever lip-service is tactically paid to a self-serving multiculturalism. That a cardinal archbishop of France, of Jewish extraction, has become one of the main defenders of the separation of Church and state or that Bavaria has banned Muslim head-scarves while making crucifixes mandatory on school walls, illustrates the complexity of current developments that radical Islam has been largely responsible for.

A number of people have helped in the writing of this book and it is a pleasure to thank them. My friend Andrew Wylie has been a great 'pit-stop boss', of a team that includes Katherine Marino and Maggie Evans. HarperCollins in New York and London have been amazingly sympathetic publishers, notably Tim Duggan, Arabella Pike, Kate Hyde and Helen Ellis, who have all brought a great deal of thought to bear on the entire project. Peter James deserves my special thanks for his careful work on what is now his third manuscript by an author who can almost anticipate his learned queries.

Several people have helped with specific subjects, some of which I was unfamiliar with when I started. Hermann Tertsch and Miguelangelo Bastinar of *El Pais* have helped deepen my knowledge of their remarkable country whenever I surface in Madrid. Detective Chief Superintendent Janice McClean was kind enough to facilitate meetings with retired RUC and current PSNI officers, and to show me Belfast. My wife's relative Andrew Robathan MP kindly set aside time in the Opposition Whips' lair to explain the army view of the conflict in Northern Ireland, while Sean O'Callaghan provided insights into armed republicanism from the former practitioner's point of view. Dean Godson and Paul Bew extended my perceptions of a conflict they both know so well. Hazhir Temourian has been a tremendous help with anything to do with the Middle East. I was also privileged to meet Norman Cohn whose work stimulated my own.

William Doino was generous with his knowledge of Pius XII, sharing the latest archival findings and his own publications. Rabbi David Dalin, Karol Gadge and Ronald Rychlak also kept me abreast of their work. In Rome, fathers Peter Gumpel SJ and Giovanni Sale SJ gave encouragement and advice, while in London father James Campbell SJ explained an especially opaque biblical prophecy that made more sense to Max Weber than it initially did to me. John Cornwell, who reanimated the debate about Pius, kindly commented on the entire manuscript, which helped clarify the few remaining areas where we may disagree. Professor Gerhard Besier kept me supplied with his stream of books on the Churches in the former German 'Democratic' Republic and on cognate subjects, while Professor Hans Maier has been a constant source of wisdom and encouragement as a leading historian and philosopher of religion. I am also grateful to Denys Blakeway and James Burge for helping turn some of these ideas into the programme *Dark Enlightenment*, and for such memorable experiences as sheltering from a mini-tornado while filming in Mussolini's Foro Italico. The editors of the *Sunday Times*, *The Times*, *Daily Telegraph* and *Evening Standard*, as well as Nancy Sladek at the *Literary Review*, encouraged me to write about Islamist terrorism after 9/11, thereby liberating me from the ghastly prospect of writing about Nazis for the next twenty years.

The book's dedication is divided three ways. My wife Linden has been a constant source of love and encouragement despite health problems not made any easier by Islamist bombers striking near her workplace on two occasions in 2005. Martin Ivens is both a fund of knowledge – on anything ranging from St Augustine to City churches – and someone who thinks deeply about contemporary issues. Finally, Adolf Wood has been a true and wise friend for twenty years now, reading every page of my work when I suspect he'd rather be in the company of Conrad, Dickens, James or Eliot. He has always been ready with a point of style or literary allusion, all delivered with his characteristic reticent firmness. None of them are responsible for my conclusions – the chief of which is that clearly identifying a problem takes one halfway to its resolution, the viewpoint that accounts for the qualified optimism with which I end the book.

Michael Burleigh
London *January 2006*

CHAPTER 1

'Distress of Nations and Perplexity': Europe after the Great War

I 'HAVE YOU NEWS OF MY BOY JACK?'

Some future archaeologist, should all written records vanish, may speculate that early-twentieth-century Europe witnessed a regression to the age of megaliths and funerary barrows before it succumbed to a more general primitive fury. The extent of this commemorative enterprise can be gauged from the fact that each of France's 35,000 communes erected a war memorial, mainly between 1919 and 1924, as did most of the parish churches, with a special chapel, plaque or stained-glass window dedicated to local representatives of the two million French war dead.[1] Such memorials proliferated across the continent and beyond, with memorial arches, cenotaphs, obelisks, ossuaries and crosses, and plinths peopled by eyeless poilus and tommies in bronze or stone. At Douaumont, Hartmanwillersdorf or Lorette, imposing memorials marked these vast necropolises for the dead. The continent's culture was more generally permeated by the loss of nine million men in a conflict that had become maniacal in its relentless destructiveness. There were a further twenty-eight million wounded and millions too who had experienced captivity. The dead left three million widows, not including women they might have married, and, on one calculation, six million fatherless children, not to speak of tens of millions of grieving parents and grandparents, for the war burned its way up and down the generations with heedless ferocity. Total war also struck directly at civilians, whether in the form of burned villages, reprisal shootings and the sinking of merchant ships, or as naval blockades gradually decimated entire populations through calculated starvation.

Myriad individual griefs welled into a greater sense of public loss, in some quarters sentimentalised as a culturally significant 'lost generation' – although plenty of butcher's boys and postmen were 'lost' as well as minor painters and poets. The querulous homosexual Oxford don A. L. Rowse remembered an encounter from his schooldays during the unveiling of a war memorial:

> A little man came up to me and started talking in a rambling way about his son who was killed. I think the poor fellow was for the moment carried away with sorrow. He said 'Sidney Herbert – Sidney Herbert – you know they called him Sidney Herbert, but really he was called Sidney Hubert: he was my boy. He was killed in the War – yes: I thought you would like to know.' And he went on like that till I dared not stay any longer with him.[2]

Rudyard Kipling lost his son John, a subaltern in the Irish Guards, at the battle of Loos in 1915. John's (or 'Jack's') body was never found; it was presumed to have disappeared during a German bombardment, along with half of the British war dead, whose bodies remained unrecovered. Kipling wrote 'My boy Jack' to express his desolation:

> 'Have you news of my boy Jack?'
> Not this tide.
> 'When d'you think that he'll come back?'
> Not with this wind blowing, and this tide . . .
> 'Oh, dear, what comfort can I find?'
> None this tide,
> Nor any tide,
> Except he did not shame his kind –
> Not even with that wind blowing, and that tide.

Possessed even in old age of indefatigable energy, fuelled by implacable hatreds not exclusively exhausted by the Germans, Kipling became a leading member of the Imperial War Graves Commission, overseeing the creation of decorous cemeteries and memorials to John and his kind. They include the Tyne Cot Memorial, where twenty-one-year-old Lieutenant James Emil Burleigh MC of the 12th Battalion Argyle and Sutherland Highlanders is remembered 'with honour', while my other uncle, Lieutenant Robert Burleigh, twenty-three years old, of the Royal

Flying Corps, lies in Knightsbridge Cemetery at Mesnil-Martinsart. For others the war left no mortal remains to bury.

Powerful emotions once accompanied monuments experienced nowadays in a blur of traffic – such as the Artillery memorial in London's busy Marble Arch or the Arc de Triomphe in Paris. Others are too modest to attract a second glance unless one consciously seeks them out, or they have disappeared into the uncertainty that eventually disperses the material effects of even the most scrupulous. For years after the war, reminders of this colossal tragedy lay in drawers or were displayed on mantelpieces and sideboards: photographs of sons, brothers, husbands, uncles in uniform; bundles of letters and field postcards; civilian clothes and juvenilia, augmented by fragments of the soldier's life – perhaps a ring, wristwatch or lucky charm that had brought no luck – if relatives were so fortunate.

The final British war memorial was unveiled in July 1939 at the seaside resort of Mumbles in Wales, the last summer before Europe's civil war resumed on a larger scale. Memorials included simple stone markers in obscure villages; plaques in Oxford college chapels and public schools (at Repton alone 355 alumni had perished) or on the walls of metropolitan stations, recalling 19,000 dead railwaymen; and last, but not least, the two and a half thousand cemeteries that transformed French hectares into permanent corners of England and its dominions.[3]

Memorialising the dead evolved from practices that initially accompanied armies of the willing. In Britain, rolls of honour, recording the names of pre-1916 volunteers, mutated into lists of the dead, whose names appeared on separate tablets, or proliferated below an ominous black line separating them from men still alive. Primitive street shrines were created in the East End of London, often at the prompting of the same Anglo-Catholic clergy who had introduced settlements into those dismal areas. These were simple affairs of names, illustrative kitsch clipped from the newspapers, and arrangements of wilting flowers, to which more puritanically minded clerics would object at their peril, for the shrines protected men at the front. Permanent memorials, intended to focus mass bereavement, superseded these impromptu shrines, although resort to spiritualists, to which modern technologies had given an enormous fillip since the late nineteenth century, suggests a reluctance to accept that the dead were beyond human contact regardless of disapproval by the Church of England.

In purely artistic terms, the greatest of these shrines was Edward

Lutyens's Cenotaph in London's Whitehall, a stark 'empty tomb' that re-placed a plaster and timber affair erected to focus the marching veterans' salute on Peace Day in July 1919. The randomly selected Unknown Warrior was interred in Westminster Abbey, an interior already so cluttered with illustrious dead that it could provide no clear focus as the Arc de Triomphe did in Paris, in a ceremony involving the king walking to Westminster from Whitehall. A representative war widow, a father who had lost a son, and a child who had lost its father accompanied the French unknown soldier to his final resting place. The Cenotaph reaches its considerable affective power not only through its emptiness, but by inviting the spectator to project his or her thoughts and emotions on to its largely unadorned surfaces. Quiet reflection was encouraged by the accompanying Great Silence – the culminating point of Armistice Day – although surrounding the Cenotaph with a section of rubber road to enhance the silence did not prove a success. Well into the 1930s men doffed their hats as they passed. Respect was something owed to other people, not something on demand. Remembrance Sunday was, and remains, one of the few occasions when the Church of England – in the form of the Bishop of London – is at the centre of national affairs, addressing matters of import to most citizens.

The Cenotaph, copied up and down the country where people did not opt for chapels, crosses or non-denominational obelisks, became the focus for a very British, reticent form of public grief, in which, as the newspapers reported, sobs were muted, voices cracked, and tears flowed silently. Some places opted for more utilitarian reminders of the war, in the form of memorial bowling greens and hospital wings, a solution much favoured in the US too. In Paris, enterprising clergy constructed a memorial housing estate, where the children of the war dead would be raised surrounded by their memory. War memorials, which were the outcome of discussions involving more than the customary range of local worthies, reflected a collective sense of what the war had been for, a con-sensual minimum beyond which lay more contentious expectations in the new mass democracies where sacrifice brought a sense of entitlement. The overwhelming majority of these memorials drew on traditional classical or romantic imagery, although Catholic countries employed a greater range of religious exemplars such as a grieving mother cradling a dead son. Parallel with this public art, artists of considerable distinction brought their talents to bear on the greatest event of those times. Quite possibly the finest example of this tradition was the cycle of etchings

Miserere produced by the intensely religious Georges Rouault between 1916 and 1928, but made public only in 1948, in which works of small compass achieve the monumentality of images on a medieval cathedral, while encapsulating something essential about the war from a Christian perspective.[4]

War memorials were not simply constructed to focus grief, but often carried a moral message to the future. In bronze or stone, at least, the dead became pensive paragons of service and sacrifice, no longer caked with mud, crawling with lice or numbed by serial percussive detonations, but hidden beneath sculpted helmets and the stone folds of trench coats. An acceptable narrative was imposed on an experience that defied most imaginations except those of the men who had been to hell and back.

Many writers, whether consciously creating art or not, chose to transpose hell into the made-to-measure clothing of received literary traditions in which birdsong, poppies, roses and warriors prettified the reality of industrial-scale slaughter involving barbed wire, bombardment, gas and machine gun that in some respects prefigured the Holocaust. Everyday speech was contaminated by terms only explicable from that era, although nowadays it comes more readily to those who write for tabloid newspapers than it does in normal intercourse, where it strikes a false note.[5] While grief remained a presence at commemorations – and does so every 11 November – so participants were encouraged to see themselves as guardians of the unfinished legacy of the dead, whether fulfilling some real yet inchoate vision of a better world, or by imagining that blood spilled had ended bloodlust, a theme reflected in a naive enthusiasm for the inter-war League of Nations.

Individuals in Britain or France may have relished the war experience, but this did not translate into a 'political religion' that subsumed the myth of the Great War into an apocalyptic and redemptive politics. The war temporarily shook these societies, but it did not destabilise their institutions or shatter their forms of government. For that we have to turn to Germany and Italy. The German empire was one of four major European empires not to survive the war. Its first democratic republican experiment lasted a mere fifteen years before conflicts that the war exacerbated and which peace did not resolve resulted in a totalitarian tyranny. Italy's liberal regime barely survived the war, to be blamed for a 'mutilated peace' and was hijacked by Mussolini's Fascists a mere four years after the war ended.[6]

Although Germany had its memorial to the Unknown Soldier –

installed in Berlin's Neue Wache – there was no single national monument to the dead equivalent to the Cenotaph or the French necropolis at Douaumont, commemorating huge losses incurred during the victory at Verdun. The Weimar Republic eventually managed to construct the Tannenberg memorial in East Prussia, an ugly series of squat towers enclosing an immense space, which was opened in 1927 in the presence of President von Hindenburg, but no agreement was reached regarding where to site a single memorial to Germany's war dead, and Tannenberg commemorated two mythically connected victories, defeat by the Poles in 1410 and victory over the Russians in 1915. Put slightly differently, one could argue that the Republic failed to capture the symbolic representations that are essential if any regime is to survive.[7] Since the experience of grief was universal, local memorials served German mourners in the same ways as their British or French equivalents. But in the circumstances of what to many seemed inexplicable defeat and post-war chaos, they were overshadowed by the war as part of a nationalist myth, in which the dead were restless rather than deeply sleeping, waiting to join Germany's self-appointed political saviours. Vivid myths were stronger than the quotidian complexities of operating a democratic regime in unpropitious circumstances.

British and French veterans may have hoped that this terrible conflict had been the war to end all wars, but in both Italy and Germany such resolve was often trumped by the rival view that the war was the prelude to the triumphal resurrection of the fatherland.[8] Writing in 1925, Ernst Jünger exclaimed that 'this war is not the end, but the chord that heralds new power. It is the anvil on which the world will be hammered into new boundaries and new communities. New forms will be filled with blood, and might will be hammered into them with a hard fist. War is a great school, and the new man will be of our cut.' In the space left vacant by a stridently pacifist left, the political right successfully represented its own fighting formations, whose first incarnation were the Freikorps bands of demobilised veterans and radicalised students, as the apostolic successors of the men who had fought and died in the trenches. These units of paramilitary freebooters evolved from the elite units that general Erich Ludendorff had created to break the tactical deadlock created by the clash of conscript armies whose training was almost designed to stifle individual initiative. Men were ordered and trained to attack in waves, since to duck, weave and zigzag was deemed beyond their limited capabilities and intelligence. By contrast, the stormtroopers were armed for close-quarter combat and were expected to range opportunistically around the battlefield so as to identify weak points in massed positions. These units were relatively democratic, in the sense that distinctions between officers and men were based on ability rather than convention or class, and they consisted of men who went about carnage with excitement as well as grim determination: 'gathering men about us and playing soldiers with them; brawling and drinking, roaring and smashing windows – destroying and shattering what needs to be destroyed. Ruthless and inexorably hard. The abscess on the sick body of the nation must be cut open and squeezed until clear red blood flows. And the blood must be left to flow for a good long time till the body is purified.'[9]

After the war, the Socialist-dominated republican government unleashed these marauders upon Bolsheviks in the Baltic States, upon Poles in Upper Silesia and upon the revolutionary left throughout post-war Germany. This was rather like sowing the dragon's teeth, since Freikorps veterans subsequently flooded into anti-republican conspiratorial organisations or the paramilitary arms of the Nazis. Inevitably the literary imagination – for left-wing writers have no monopoly of glorifying political violence – was drawn to these gaunt figures, many of whom, like Ernst von Salomon, were themselves passable writers. Salomon described these armed bohemians in idealised terms: 'We were cut off from the world of bourgeois norms . . . the bonds were broken and we were freed . . . We were a band of fighters drunk with all the passion of the world; full of lust, exultant in action.' These men had overcome human sympathy, which was routinely dismissed in such circles as insipid sentimentality. This overcoming gave the stormtroopers the narcissistic delusion, common among psychopaths, that they themselves were a new predatory type of being in whom hardness trumped humanity. According to Ernst Jünger, a former stormtrooper himself, they were 'magnificent beasts of prey', for whom war was not sporting, and whose soldierly contempt for civilian existence tipped over into murderous rage towards republicans and revolutionaries. These were fierce figures. As Arnold Zweig wrote in 1925: 'We have become a wrathful people / committed to the waging of war / as a bloodied and enraged knighthood of men / we have sworn with our blood to attain victory.'[10]

Values engendered by total war – notably the inward-focused camaraderie of what the British called 'bands of brothers' – were perpetuated and turned outwards in what became a murderous war against the weak Weimar Republic, political parties that camouflaged vested interests, Jews and socialists, overlooking the fact that large numbers of Jews and men of all political persuasions had made their own patriotic sacrifices. If in Britain local worthies worried about whether having statues of men armed with rifles and bayonets conjured up a killer instinct that many wanted to forget, in both Italy and Germany elite fighting units (the Italian *arditi*) who had brought fanatical courage and tenacity to the wartime battlefields, provided the prototypical 'new man' who, despite his self-professed dehumanisation, was supposed to be the nation's future redeemer. The brutality that total war had engendered, and which in Armenia, Belgium, the Balkans, northern France and East Prussia had spilled over into violence towards civilians, became a permanent condition, in the sense that political *opponents* were regarded as deadly *enemies*.[11] In Italy people who revelled in violence for political purposes acquired a political label earlier than elsewhere: that of Fascists, the very symbol – of axes tightly bound in lictorial rods – conveying the closed community of the exultantly thuggish better than the mystic iron octopus of the Nazi swastika. But this is to anticipate; there were states of mind that we must first visit.

II THE LAST DAYS OF MANKIND

The Great War cast a very long shadow over the creative literature dedicated to warfare, inspiring novelists to this day – the obvious contemporary analogy being the imaginative writing, good, bad and indifferent, generated by the Holocaust. The pre-war apocalyptic imaginings of the artist Ludwig Meidner became wartime apocalyptic facts as even cathedrals were blown to oblivion on the grounds they were used as artillery observation points. The conflict destroyed a world that combined ordered social relations with a degree of cultural experimentation in what the US novelist Scott Fitzgerald called 'a gust of high explosive love'.[12]

As the historian of memory Jay Winter has argued, regardless of whether they were personally religious, imaginative writers often drew on religious traditions – broadly conceived – as they tried to capture the essence of the war experience, leaving the matter of causes for historians to discover in the pre-war diplomatic traffic. The apocalyptic mode dominated fiction, as if the war signified divine judgement upon the civilisation of the pre-war era or mankind as a whole. Literary prophets abounded. The French Socialist Henri Barbusse spent seventeen months on active duty in the trenches of the Western Front. He was cited for bravery on two occasions and then invalided out, exhausted and suffering from dysentery and damage to his lungs. In early 1917 he published *Le Feu*, which within a year had sold two hundred thousand copies and had won the author the prestigious Prix Goncourt, although some critics thought the novel lacked verisimilitude to their own war experiences. In terms of character, the novel does not amount to much – a socially and regionally heterogeneous French band of brothers, gone astray from any novel by Zola, is rapidly thinned out through the random impact of battle – all interspersed with vague socialist yearnings for a better tomorrow that seems questionable to anyone unfortunate enough to have experienced even a simulacrum of it.

But, despite its romantic political predictability, Barbusse's book succeeds in depicting war as an additional natural element, alongside fire and earth, air and, above all, water. The action alternates between miles of trenches and villages and towns that have been smashed to smithereens but there is another, much more pervasive presence even than the smells of death. Water is the novel's dominant element, as rain found its way through even the most carefully buttoned tunic and spread upwards along trouser legs from boots swollen with damp mud, or trickled down the waders that were essential in water-logged trenches. Everywhere there was an ocean of deep mud, regularly churned up by shelling to reveal new layers of corpses in varying states of decomposition, or, bizarrely, springing open the coffins in bombarded cemeteries. The battlefields were submerged by a flood of biblical proportions, leading Barbusse to announce 'hell is water' – rather than other people.

> Where are the trenches?
> We see lakes, and between the lakes there are lines of milky and motionless water. There is more water even than we had thought. It has taken everything and spread everywhere, and the prophecy of the men in the night [that the trenches were disappearing] has come true. There are no more trenches; these canals are the trenches enshrouded. It is a universal flood. The battlefield is not sleeping; it is dead.

The flood provokes Barbusse's surviving soldiers into angry denunciations of a war:

> that is about appalling, superhuman exhaustion, about water up to your belly and about mud, dung and repulsive filth. It is about moulding faces and shredded flesh and corpses that do not even look like corpses anymore, floating on the greedy earth. It is this infinite monotony of miseries, interrupted by sharp, sudden dramas. This is what it is – not the bayonet glittering like silver or the bugle's call in the sunlight!

The men cry 'no more war' and argue such banalities as 'When all men have become equal we'll be forced to unite,' while denouncing such comic-strip villains as bankers, priests, lawyers, economists and historians. The novel concludes with the remark, 'if this present war had advanced progress by a single step, its miseries and its massacres will count for little', at which, as if on cue, 'a tranquil ray shines out and this line of light, so tightly enclosed, so edged with black, so meagre that it seems to be merely a thought, brings proof none the less that the sun exists'.[13]

At roughly the time Barbusse was converting the war into socialist prophecy, the Austrian satirist Karl Kraus was training his larger talents upon the enthusiasts who welcomed war in 1914. Kraus was an intriguing figure. Paper factories owned by his family meant that he was rich enough not to have to earn a living. He edited *Die Fackel*, one of the most successful journals in central Europe, and was easy in the company of the beautiful young aristocrat Sidonie Nadherny, with whom he fell in love. Bespectacled, bookish and slight with a curvature of the spine, he took up horse riding so as to fit in with the aristocratic country set he admired. Although Jewish by birth, he frequently gave vent to wounding antisemitism, particularly against the liberal Jewish bourgeoisie of his home city who furnished a number of his hate figures. He detested the superficial Positivism of the times, with its belief in Enlightenment, Progress and Science, and its credulity towards journalism, sociology, psychiatry and eugenics. The liberal *Neue Freie Presse* became the *Neue feile Presse* (best rendered as 'New Presstitute') in his hands. Kraus converted to Roman Catholicism, while drifting politically, during the First World War, from a conservative anarchism to republicanism and socialism.[14] As a leading journalist, he was inclined to exaggerate the power of the press and words in general.[15] His published talk 'In these great times' delivered in November 1914 was an attack upon the vicarious heroics of the editorial bench and war profiteers, as well as other writers who in August so eagerly prostituted their pens. He was scathing about the role of the press in bringing about its own sanguinary fantasies by engendering the vicious mass enthusiasms that had propelled Europe into war. His technique relied upon absurd details, of 'patriotic' Viennese restaurants that renamed macaroni 'Treubruchnudeln' ('perfidy noodles') to condemn what seemed treachery by the Italians, to identify some symptom of the age, his other obsession being an advertisement for Berson's shoe rubbers which 'progress' sought to inflict even on babies – the advertisement had appeared opposite the Austrian proclamation of war:

> May the times grow great enough not to fall prey to a victor who places his heel on the intellect and the economy, great enough to overcome the nightmare of the opportunity to have a victory redound to the credit of those uninvolved in it, the opportunity for wrongheaded chasers after decorations in peacetime to divest themselves of what honour they have left, for utter stupidity to discard foreign words and names of dishes and for slaves whose ultimate goal all their lives has been the 'mastery' of languages henceforth to desire to get around in the world with the ability not to master them! What do you who are in the war know about the war?! You are fighting! You have not remained behind! Even those who have sacrificed their ideals to life will some day have the privilege of sacrificing life itself. May the times grow so great that they measure up to these sacrifices and never so great that they transcend their memory as they grow into life![16]

In 1915 Kraus commenced work on a documentary drama called *The Last Days of Mankind*, which took seven years to complete and ten hours to perform on the stage with a cast of hundreds. According to his greatest biographer, the documentary form was partly inspired by Georg Büchner's *Danton's Death* which Kraus saw in Berlin in 1902, but the influence of Shakespeare is also apparent, including vengeful ghosts and juxtapositions of high and low conversation, although Kraus regards the gravediggers as more important than Hamlet. Kraus claimed that even the most outré utterances in the play were grounded in documentary fact; that he had to defend two libel actions related to people caricatured in the play suggests that his satire hit home. One of his key dramatic devices is the 'gruesome contrast'. The belligerent crudity of the Viennese mob is transformed into purple prose by the no less belligerent mob of journalists who reported it. Pope Benedict XV prays in the Vatican imploring God to stop the mindless bloodshed; his name-sake, the Jewish newspaper editor Benedikt, dictates a gruesome piece involving the Adriatic's fish and lobsters dining better than before on the bodies of Italian seamen whose ships have been sunk by the Austrians.[17] In a Protestant church 'Pastor Buzzard' assures his congregation:

> Let us acknowledge clearly and unequivocally that Jesus' commandment 'Love thy enemies' applies only to individuals and not between nations. In the struggle of the nations there is no room for loving one's enemies. Here the individual soldier need have no scruples! In the heat of battle, Jesus' command of love is suspended! In combat, killing is no sin but a service to the Fatherland, a Christian duty – indeed, even a service to God![18]

Kraus also repeatedly uses the device of reducing and ridiculing such 'world historical' figures as Berchtold, Conrad, Hindenburg and the German and Austrian emperors, while inflating nonentities, such as the typical reader of the *Neue Freie Presse*, into embodiments of the age. Although the drama does not develop in any conventional sense, Kraus employs a running commentary shared between an Optimist and a Grumbler to register his sense of moral outrage, aroused not only by the home front but also by a war that had degenerated into summary killings of prisoners and the wounded, or the execution of deserters and shirkers by brutal NCOs and officers. The play's epilogue uses Shakespearean spectres to accuse those Kraus held responsible for the war – including the soldiers who allowed themselves to be abused – before order is restored as God defeats the Antichrist. The play ends with a series of nightmare apparitions, of children drowned in the *Lusitania*; of an elderly Serb digging his own grave; of a bomb landing on a school-room; of civilians and prisoners of war being shot and so forth until Kraus plunges the stage world into darkness, as a wall of fire rises on the horizon and God says: 'I never wanted this.'

> What is that sound high in the air
> Murmur of maternal lamentation
> Who are those hooded hordes swarming
> Over endless plains, stumbling in cracked earth
> Ringed by the flat horizon only
> What is the city over the mountains
> Cracks and reforms and bursts in the violet air
> Falling towers
> Jerusalem Athens Alexandria
> Vienna London
> Unreal

If many writers and artists turned to compelling Christian idioms to interpret the Great War, others subsumed traditional elements within a deliberately 'fragmented' vision that seemed to reflect the condition of the post-war years. Actually, the ensuing fragmentation was evident well before the war. In a series of lectures on 'Civilisation at the Cross-roads', delivered at Harvard in 1911, the Anglican monk John Neville Figgis said:

> amid the Babel of the world's religions and moralities, it is not possible to state what are the governing ideals of the triumphant classes at the moment, and it is ten to one that if you met two dozen at dinner, you would hear a dozen different faiths asserted, with all that voluble enthusiasm that befits 'the light half-believers of our casual creeds' ... if we judge by their conduct, we may ask with Archbishop Benson, when he arrived in London, 'What do these people believe?'[19]

The decades before the war were almost as rich in devotees of occult practices as the 'New Age' is now. The war makes many oblique appearances in a work which, despite its saturation with traditional images, is regarded as a waypost of artistic modernism because of its fashionable anthropological references, jazz-like rhythms and random snatches of the pulsing city's polyphonic argot. But the war is there all right, in the references to the archduke, to rats crawling along alleys, to dead men's bones, to fear and dust, in the demobbing of Albert, maternal lamentations, Madame Sosostris, hooded hordes and the marching dead of commuters through the dingy London air.

T. S. Eliot began working on *The Waste Land* – although the original title was 'He Do the Police in Different Voices' – in 1921, completing the poem the following year, after a supervening convalescence in Margate. It was to be a large modernist statement, reminiscent of Joyce's *Ulysses* or Stravinsky's *Rite of Spring*, although Eliot would later claim that he had just reassembled a few fragments – fashionable references to primitive and Eastern religions, Jacobean drama, jazz syncopations and pseudo footnotes – while denying that the poem had any major point to make. He sounds an ironist embarrassed by the credulity of admirers and disciples, such as the undergraduate aesthete Anthony Blanche who declaims the poem to shock Oxford hearties in Evelyn Waugh's *Brideshead Revisited* or those who, without irony, dubbed themselves 'The Waste Landers'. The desire to make a cult of a poem in which cryptic and eclectic allusions to a variety of religions abound was *in itself* symptomatic of the spiritual appetency of the post-war wasteland it evoked, and which Eliot would mock in his later *Four Quartets* after he had turned to Anglo-Catholicism.[20] According to Eliot the poem was variously 'just a piece of rhythmical grumbling', or as he later admitted, 'I wasn't even bothering whether I understood what I was saying.'[21]

IV AGE OF ANXIETY, TIME OF THE PROPHETS

Modern sociologists of religion tend to relate the strength of religion in the contemporary world to existential anxiety. While this argument involves leaving aside the US as an 'inexplicable' anomaly, it does seem to account for the increasing purchase of religion in what used to be called the Third World.[22] It holds good not only for the transcendental monotheisms, but also for the cults, fads and sacralised mundanities that accompanied, if not secularisation, then de-Christianisation and the remorseless atomisation of life in the modern world. In 'The Dry Salvages', the third of his *Four Quartets*, T. S. Eliot captured this vapid spiritual experimentation:

> To communicate with Mars, converse with spirits,
> To report the behaviour of the sea monster,

Describe the horoscope, haruspicate or scry,
Observe disease in signatures, evoke
Biography from the wrinkles of the palm
And tragedy from fingers; release omens
By sortilege, or tea leaves, riddle the inevitable
With playing cards, fiddle with pentagrams
Or barbituric acids, or dissect
The recurrent image into pre-conscious terrors –
To explore the womb, or tomb, or dreams; all these are usual
Pastimes and drugs, and features of the press:
And always will be, some of them especially
When there is distress of nations and perplexity
Whether on the shores of Asia, or in the Edgware Road.[23]

More disturbing than these more or less harmless pastimes were the political manifestations of what could be called mass spiritual need in deranged times. As Langmead Casserley argued long ago: 'Totalitarianism is founded not only on the will to power of autocratic statesmen, but also on the will to security, and the impulse to adore and propitiate, of the mass of citizens ... The pseudo-divinity of the modern state is perhaps not so much a divinity which it has arrogantly usurped as a divinity thrust upon it by masses of insecure and frustrated people, insistently demanding some powerful and venerable object of faith and trust.'[24]

A bare recital of what Germans underwent from 1918 onwards reveals the magnitude of their existential crisis at a time when intellectual doubts had already undermined belief in science and progress as well as revealed religion.[25] We begin with the series of external events before moving on to the parallel world of the mind and spirit, which are poorly handled in most accounts of the Third Reich.[26] It was one of those times of what Emile Durkheim called 'effervescence' in which, like the night of 4 August 1789 when feudal privilege was renounced, men and women experienced life with an intensity that is hard to evoke except in terms of religion.[27] The German armed forces, whose triumphs were so integral to national identity, and which wartime propaganda had presented as invincible, had been defeated, despite Russia having been knocked out of the war by revolution, and following a vast final push that promised to break years of stalemate on the Western Front. Defeat seemed inexplicable. The German-Jewish philosopher Karl Löwith had served under Ritter von Epp on the Austro-Italian front, until he was shot in the chest and captured by the Italians. After being repatriated after two years through a prisoner exchange, he recalled his father, a respected painter, at home in Munich during the later stages of the war: 'The pinning of miniature flags on the wall-map of theatres of war I left to my patriotic father, who was saddened by his son's indifference. He never took any notice of the retreat of German troops when he was engaged in this. The miniature flags always stayed in the most advanced positions, and when the Western front collapsed, the war on the map seemed almost won.'[28]

Since there were no enemy soldiers on German soil – although the effects of Allied blockade were palpable to starving civilians – defeat seemed to many the result of domestic treason or of a more malign conspiracy involving interconnected 'racial' actors. A wartime hunt for Jews who had allegedly shirked their patriotic duty became a post-war hunt to identify Jewish preponderance in such areas as banking, the arts and journalism. The Versailles peace settlement blithely blamed Germany for a war whose causes are still debated and criminalised commanders recently deified as heroes. Being neither generous nor punitive, its ambiguities increased the sense of having lost control of one's destinies, especially since the economy seemed to have been put in hock to foreigners in a perpetuity whose horizon was an improbable 1988. Venerable institutions collapsed, with many people already having lost faith in them, the Hohenzollern dynasty being a major case in point. Once capable of inspiring awe in every carbuncular young clerk, as well as in the obsequious monster conjured forth in a controversial novel by Heinrich Mann, Wilhelm II became a forgotten figure in Dutch exile.

Revolutionaries, who were readily conflated with the 'Asiatic hordes' on the loose with their firebrands in Russia, brought chaos to the streets of German and other central European cities. These Bolsheviks acquired a racial aspect since many of the leaders of evanescent socialist republics in Budapest, Berlin and Munich were radicalised Jews tantalised by Marxism's messianic vision. In 1923 the Reichsmark went into freefall, upsetting a moral order based on constant values. Karl Löwith experienced the havoc this played on the finances of his own family. In four decades, his father had worked his way up from being a penniless immigrant Moravian Jew to being a pillar of Munich society. Now, the sale of a villa on nearby Lake Starnberg brought nothing. His wife's dowry was rendered valueless. He could not pay the life-insurance premiums on his wife's life. His patriotic investments in war loans were worthless. He kept a packet of thirty thousand Mark notes; when his son tried to sell them, they fetched ten Pfennigs as collectors' items. Karl had inherited shares worth 30,000 Marks from his grandfather; they were worth three Marks at the height of the inflation. His monthly salary as a tutor in Mecklenburg was the equivalent of a hundredweight of rye or five small cigars. Löwith was no more nor less outraged than the rest of the bourgeoisie as the scum rose:

> Old and well-situated families were impoverished overnight, while young have-nots acquired great wealth through bank speculation. The buyers of my father's paintings were no longer the rich distinguished businessmen of the Wilhelmine era but major industrialists, speculators and shoe manufacturers who wanted to invest their money in material assets. Even the four year war did less to loosen morality and the whole fabric of social life than this raging turmoil, which eroded people's foundations anew every day, and instilled a desperate daring and unscrupulousness in the younger generation. It was only this grotesque occurrence that laid bare the true significance of the war: the total overspending and destruction resulting in the zeros of the inflationary period and the Thousand Year Reich. The virtues of the German bourgeoisie were swept away then, and this dirty brown torrent bore the movement which formed around Hitler.

As Löwith sensed, 'Germany was undergoing universal devaluation – not only of money, but of all values – and the National Socialist "revaluation" was a result of that.'[29]

A sense of moral order was further outraged by the pockets of artistic nihilism and pseudo-radicalism and sexual self-advertisement in the major cities, cities surrounded by rural seas of conservative traditionalism. The myth of the modern artist would prove to have tragic consequences when he became a model for a new generation of 'artist–politicians' whose egoism dwarfed that of the denizens of Bloomsbury, Montmartre or Schwabing.[30] Creative artists, the majority belonging to the left by way of gesture, contributed to undermining the Republic. While some glorified conmen and criminals – such moral relativism being a sure sign of cultural decadence – others, like Kurt Tucholsky, failed to discriminate between such worthy statesmen as Gustav Stresemann and the paramilitary Stahlhelm. The left's desire to see every opponent to the right of them as a 'Fascist' duly led them massively to underestimate the genuine phenomenon. Foreign armies, including the nightmare of 'Black' French colonial troops pushing around white Europeans, were ensconced in the occupied western regions of Germany, which indigenous separatists threatened to detach permanently.

No wonder the apocalyptic mode of thought that the war had encouraged intensified during a 'peacetime' that had many of the characteristics of a civil war as well as a material and moral catastrophe. Prophets of the end of days abounded, whether on the political left or right. Oskar Jaszi, a Hungarian government minister who witnessed Béla Kun's orgy of violence in post-war Soviet Budapest, described the latterday possessed:

> Now for the first time, in circumstances most agreeable, the demonic spark lurking behind Marxism has caught fire. Indeed, like every true mass movement, it ignited firstly with powers of religious character ... Constantly we would witness excited discussions in the streets and coffee houses, in theatres and lectures, in which people with feverish eyes and fierce gesticulations prophesied and discussed the nearing of a new world order ... The days of Capitalism were numbered, the world revolution is loudly nearing, Lenin will soon unify the labour force of all Europe in one single revolutionary union ... In the brains of these people the new deity was alive: the belief in the unavoidable dialectic of said economic development which will bring to fall the evil Capitalism and with the irresistibility of the laws of nature – divine laws – will bring to life the new society, dreamed of by all prophets, the land of peace, equality, brotherhood – the Communist society.[31]

A messianic mood was abroad in Germany, which invariably took the form of expectations of a leader to redeem the German chosen people from the Egypt of Allied captivity. Such hopes had a long tradition in Germany, with figures such as Emperor Frederick Barbarossa, Bismarck or, on the left, Ferdinand Lassalle, indicative of the exceptionally gifted individuals who would come to the nation's rescue. If such longings partly represented the recasting of messianism in secular form, so it also reflected a democratisation of the traditional relationship between monarch and subject who became, respectively, the 'leader' and his 'following', although the German word Gefolgschaft continued to reflect feudal origins.³² Historians, like many academics deeply inimical to the Weimar Republic, contributed their anti-democratic pennyworth by encouraging their credulous students to inhabit a mental universe consisting of supposedly ordered past societies dominated by genial leader-figures, whom they contrasted with the dull pragmatic politicians who were making a hash of the present. The theologians were not much better. Although the Lutheran Paul Althaus deplored the fact that even pastors were not immune to 'political messianism' as a substitute for belief in redemption through Jesus Christ, in the same breath he argued that the Old Testament's conflation of the history of a people with salvation was ample precedent for 'political preaching' about events in Germany in the present. Did Lutherans owe the Weimar Republic the loyalty prescribed in Romans 13? Only in a heavily qualified way, since the 'temporary structure' of Weimar was 'the expression and means of German degradation and apathy'.³³ This betrayal of professional objectivity was so pervasive that the sociologist Max Weber devoted a talk in a Munich bookstore to these 'tenured prophets' of a future Führer. Bearded and tired, Weber spoke without notes, although his words were taken down. After his second lecture, which became 'Politics as a Vocation', he concluded with these delphic verses from Isaiah 21: 11–12:

> The burden of Dumah. He calleth to me out of Seir, Watchman, what of the night? Watchman, what of the night?
>   The watchman said, The morning cometh, and also the night: if ye will enquire, enquire ye: return, come.

Night was a metaphor for the lordship of the Babylonians over Dumah, an oasis in Arabia; Seir was a mountain in Edom sometimes used as a metonym for it. From there comes the question to the Watchman, another name for prophet: 'Watchman, what of the night?' The answer suggested only temporary relief, since the signs were unsure – it was neither night nor day – for the prophet refused to raise false hopes. Indeed, according to Weber, it was his duty to lower expectations until matters became more transparent. Weber used this passage about an unusually equivocating prophet to urge his students to reject those who claimed to divine the course of events, while retaining their focus on the pragmatic issues of the day.³⁴ One of the students who heard Weber speak was Karl Löwith:

At the end of his two lectures Max Weber had prophesied what was soon to happen: that those who could not endure the tough fate of the times would be returning into the arms of the old churches, and that the 'conviction politicians', who intoxicated themselves with the Revolution of 1919, would become the victims of the reaction whose onset he anticipated within ten years. Because before us lay not a blossoming spring but a night of impenetrable darkness, and it was therefore pointless to wait for prophets to tell us what we should be doing in our disenchanted world. From this Weber drew his lesson: we should set to work and meet 'the demands of the day'; this is plain and simple.[35]

Such counsels of caution had virtually no effect as young people, including large numbers of students, rebelled against conventional political parties and threw themselves into bizarre cults, orders and sects, or into political parties that stressed absolute obedience and practised military drill. The rebellion took entirely predictable forms: naive prostration before any convincing charlatan or the retreat from the chaos of modern life into communes and rural settlements, on a scale that would not be repeated until the 1970s. During both the period of hyperinflation from 1919 to 1923 and then the Depression between 1929 and 1933, Germany also witnessed the phenomenon of wandering 'prophets', who went about barefoot, bearded and long-haired, charging people considerable sums to attend meetings at which they prophesied the end of the world and called for moral renewal and a new type of man to create a new type of society before it was too late. According to a journalist on a Cologne newspaper who attended such a meeting in Berlin:

> Today the public flocks to the meeting halls of these fantasists because in its monumental mental confusion it seeks any kind of prop to console itself. Already, shortly after the end of the war, as the fruitlessness of so many efforts became apparent, a mood of limitless disappointment set in. On top of that, in recent months, people's minds have been totally deranged by the ever increasing material distress, the hopeless struggle against inflation . . . Everyone, and especially weaker natures, flocks to these contemporary redeemers with their long hair and mad fantasies, because they cannot do without such support. Prophecy is a dangerous symptom of the spiritual condition of Germany at

the moment. One should not underestimate it; it will become more pervasive in the crises to come. The time is out of joint! As Hamlet said.[36]

Interestingly, one of the most astute observers of those times had difficulty distinguishing between the barefooted beardie-weirdies and the future German Führer, whose ideas sounded as insane as theirs. Writing in British exile on the eve of the Second World War, Sebastian Haffner recalled these prophets of the early 1920s, with one significant addition:

> Gradually the mood had even become apocalyptic. Hundreds of saviours were running around Berlin, people with long hair, wearing hairshirts, claiming that they had been sent by God to save the world ... The most successful of them was a certain Haeusser, who advertised on advertising pillars and staged mass gatherings and had many followers. According to the newspapers, his Munich counterpart was a certain Hitler ... Whereas Hitler wanted to bring about the thousand-year Reich by the mass murder of all Jews, in Thuringia a certain Lamberty wanted to bring it about by having everyone do folk dancing, singing, and leaping about.[37]

Who were these people? As it happens, we know quite a lot about them, even if this involves studying the files of psychiatric institutions and courts where many of the prophets and their followers washed up. Stuttgart in Württemberg was the epicentre of the movement, the home-town of Bosch and Daimler-Benz being an unlikely location to choose for the renewal of the world. In reality, the area had a strong tradition of peasant pietism, which seeped back into the countryside as the workers forsook the factories for the wooded hills that ringed the productive twentieth-century cauldron below. The town and surrounding heights attracted a wide range of mystics from the pedagogue Rudolf Steiner to the 'Vagrant King' Gregor Gog, the name being indicative of the madness. The war and the ensuing hyperinflation greatly contributed to the phenomenon of life on the open road, setting hundreds of thousands of indigents in motion, as vagrancy became as epidemic as it would be in the US Depression. The prophets catered to a very Teutonic sense of 'longing' (Sehnsucht) for a big idea expressed by a charismatic leader, who would give meaning to the lives of humble workers as well as Viktor Emil von Gebsattel, who would subsequently discover his life's purpose as Germany's first professor of psychotherapy. Many of them espoused nudity and relaxed sexual relations. This must have been convenient, since several of these bearded satyrs went through a remarkable number of sexually voracious young women, with the startlingly successful claim that they had been chosen to give birth to the new redeemer, a variant of which promise leaders of the 1968 student movement (not to speak of an entire generation of slightly sordid academics) would also routinely exploit to confuse sex and Sartre.

The movement completely rejected most of the fundamental principles the Enlightenment has given to the modern world, notably the separation of politics and religion. Several of these prophets made sallies into politics, offering a novel ideological synthesis, or the transcendence of politics as such. Although a few of the prophets gravitated to the *völkisch* right, the majority were attracted to the anarchist or radical Communist left, in either its socialist or its nationalist variants. Many other boundaries were also fluid, since the prophets were sometimes welcomed by Protestant clergy, who admired their followers' enthusiasm, and by the artistic avant-garde, of Bauhaus and Dada, who themselves were attuned to provocative 'happenings'. In most cases, the failure of hopes of revolution in 1918–19 led the prophets and their followers to refocus their enthusiasm away from the prospect of radical socio-economic change to the world of consciousness and personal development. This was truly the 'Ich' generation. The private and the personal were then politicised and generalised in the form of a moral–political revolution, of which Hitler was merely a mutant and successful manifestation, for in some respects these prophets were like a parody of the much more politically astute future Führer, sometimes saying what he was clearly thinking and employing similar means of mobilisation on a more modest scale.

Ludwig Christian Haeusser, who styled himself 'President of the United States of Europe', was born in 1881, the son of a brutal and ill-tempered farmer who beat him every time he showed any interest in learning. Eventually, Haeusser managed to escape this grim environment, learning commerce in London and Paris. After various scams, involving close calls with the law, he established an apparently successful champagne-exporting business, the elegant clothes, top hats, rich wife and house on the Champs-Elysées being some of the external fruits of his enterprise. In 1912 on a business trip to Frankfurt, he seems to have rediscovered religion, although not the conventional pieties that he had imbibed as a boy. The outbreak of war in 1914 – the French confiscated his local assets – intensified his belief that mankind was on the verge of a rebirth, and that he had the duty to help bring this about. He began to write endless, unpublished articles as well as a book called 'The Coming Superman', which posterity was similarly spared. His business went to pieces as customers who came to buy champagne were treated to hours of prophetic ramblings.

In the summer of 1918 he turned up, dressed in a top hat and starched shirt, at the Italian mecca for drop-outs and redeemers at Ascona. Returning to neutral Switzerland, after his forty days in the Italian wilderness, he exchanged his starched shirts for the bearded, long-haired look, and forsook swanky restaurants for the local soup kitchen. Sometimes he ate leaves and slept in ditches. He took up nudism so as to personify the 'naked truth': in his case it was never a pretty sight. His newfound role of prophet took him back to his native Württemberg, where he honed such rhetorical devices as referring to his audiences as 'apes, donkeys and swine' to get their attention. He attracted a following. A number of his acolytes were young women, many drawn by his wish to sire the mother of God through them. Although Haeusser took every opportunity to speak of sexual purity, he was addicted to cunnilingus and sado-masochism, as seem to have been some of the ladies in suits and ties with short-cropped hair in his permanent entourage. When admirers gave him gifts, their choice invariably fell upon silver-encrusted whips. The entourage was blindly devoted to the prophet. On one occasion a very drunk Haeusser leaned over the lectern in the middle of a public lecture and vomited over the audience, whereupon several young women rushed to get mops and buckets to preserve the stomach contents of 'the saviour'. Beginning in 1922, Haeusser turned to a political vocation, hoping that a morally purified German 'master race' would lead Europe. His programme included the closure of all asylums and prisons and the pardoning of their inmates; the universal abolition of property; a ten-day general strike; and a reformed officialdom, whose watchword would be to be nice to the disadvantaged. The guillotine awaited anyone who resisted the 'spirit of truth' revealed by the 'peoples' Kaiser'. In a sense, Haeusser spoke as Hitler's holy fool:

Blood! Blood! Blood! Blood!
    Blue blood! Black blood! Red blood! Blood in every colour!
Even white blood! Only blood! Nothing but blood! Blood again!

Once more blood! Cold blood! Flowing blood! Hot blood! Blood! A very special taste! Blood is the universal panacea. Blood is healthy! Blood is a sign! With this sign you will conquer! With German blood and broom the world will soon recover. I am the true blood-wind! Bloodhound! Blood storm! Blood-Blood-Blood-Blood shall flow. Blood must flow!

Although others, equally obsessed with 'blood' would put into effect Haeusser's psychopathic fantasies with greater thoroughness, the prophet himself attempted to found a political movement. By now calling himself 'Louis the Christ, King of Germany and Emperor of the World', he founded a Christian Radical People's Party in 1922, together with a journal immodestly called *Haeusser*. In an odd prefiguring of what Hitler would subsequently announce in 1927, he called this a 'partyless party', consisting of 'men of the deed' who would resist all compromise. He wanted it to attract all the extremes. Hence he referred to it as the 'Swastika-Communists'. While the extreme-right German Racial Defence and Offence League was cool, Haeusser did attract support from the self-styled National Communists, who were drawn by his promise of an 'enabling law' in which 'millions of superfluous, parasitic, unproductive officials' would be forced Pol Pot style to work. This platform attracted about 25,000 votes in March 1924, although his party's share of the poll plummeted in successive elections. By this time, Haeusser was embarked on a personal Golgotha. His intemperate letters to various authorities – 'I shit on your lazy, mindless laws – yes, I shit a great big heap of shit!' – and his publishing of lists of the names of judges with whom he had had contretemps, whom he said he would execute within three days of establishing his dictatorship, inevitably attracted the concerted malice of the law. From 1919 onwards, the authorities in each federal state shunted him hither and thither and he was held under protective arrest during the right-wing Kapp putsch in 1920 as a menace to public order.

Haeusser eventually washed up in north-western Oldenburg, where in 1922 he became engaged to Hetty von Pohl, the niece of a wealthy aristocratic landowner and daughter of the former chief of the imperial navy. He relinquished his waist-length beard and acquired decent suits and shoes so as to fit in with this smart company. Although his future uncle had initially been won over to his teachings, things turned sour when Haeusser moved his entourage into the baron's home and the family silver found its way to a pawnbrokers in Hamburg. Appalled by the prospect of this mésalliance with a maniac, Hetty's mother had her daughter confined in a secure psychiatric hospital, to break the spell of a fiancé who more properly belonged in such a setting. Haeusser lashed out at the authorities in Oldenburg, 'small, fly-blown, rancid, lousy' being the printable outbursts, which resulted in a storm of prosecutions against the prophet. He was sentenced to twenty-one months' imprisonment and a one million Mark fine. During what proved to be three years inside, he wrote a 2,413-page diary on wrapping and lavatory paper, which a team of female devotees duly transcribed for future publication. His chances of early release diminished when the authorities came across articles in *Haeusser* which appeared to threaten the life of the prosecutor at his trial. His mental and physical health worsened, until in 1923 he almost died. It seemed that every other political jailbird was being pardoned, including the anarchist Erich Mühsam and Adolf Hitler, while the prophet stayed behind bars. In July 1925 he was released and settled in Hamburg. After one final spurt of political prophecy, in June 1927 the forty-five-year-old Haeusser died.[38]

Meanwhile, in deepest Thuringia, Friedrich Muck-Lamberty was leading his followers on a merry dance. Muck was born in 1891 into a family with fourteen children. He was christened Muck because, like the character in the fairy tale, he had a large head. However Muck grew up to be a man with the clean-cut good looks of a Robin Hood in a 1950s film, an image he actively promoted with his medieval jerkins and hunting horn. The young Muck was raised as a Catholic, but his faith took a knock when he surprised the priest with his housekeeper and the priest bribed him with chocolate to ensure his silence. At thirteen he left home. By sixteen, he had become a 'lifestyle reformer' and vegetarian, managing to turn the former passion into a successful business by helping to design orthopaedic footwear. By eighteen he had gravitated to the wandering youth movement and sought to found his own utopian rural settlements consisting of skilled craftsmen who would sell their goods through co-operatives. During the Great War, Muck was stationed by the navy on the island of Heligoland in the North Sea, where he was appalled by the arrogance of officers and the vulgarity of the sailors. The revolutionary events at the end of the war convinced him that all those involved were locked into outmoded forms of thought. In a programmatic statement that he published in January 1919, he called for a supra-political 'German national community' based on abstinence, lifestyle reform and a reversion to craftsmanship. Above all he thought that 'without religion there would be no nation' and called for popular solstice festivals to bring this about. Muck's Catholic background was also evident in the enormous importance he attached to woman as mother, being especially exercised by the appalling way in which working-class men treated their wives.

During the warm summer of 1920, Muck led a throng of youthful middle-class followers through Franconia and Thuringia. They set out from Kronach and marched via Coburg, Jena and Weimar to Eisenach. They were dressed in blue and white coloured fabrics and went either barefoot or in self-made 'Jesus sandals'. Each evening Muck blew a horn to summon his followers to a 'Thing' where he heard and resolved their complaints and concerns. One would confess to having eaten meat, and agree to a three-day fast. Another would ask to sleep alone, since the sound of snoring kept him awake when he slept in the close-packed row. In Eisenach the group refused to stand for the national anthem at a nationalist youth festival since the ambient fug of tobacco and beer detracted from the dignity of the occasion. Eventually some thousands of people joined Muck's merry moralising band. They were quickly encouraged by some enterprising Protestant pastors into their churches, notwithstanding Muck's pronounced Mariolatry, impressed as they were by the group's charismatic enthusiasms. There was more. In each village and town they journeyed through, the group took over the square and used it to form concentric circles of dancers, which the incredulous inhabitants were invited to join. The dances were more reminiscent of a Kirmesse by Breughel than of the era of the Foxtrot, Rouli Rouli and Tipsy Step in which these people were living. There was much 'swinging' to induce a quasi-religious ecstasy, although many young men and women took the opportunity to 'swing' in a less innocent sense. In the autumn of 1920 the 'throng' retraced their route, wintering on the Leuchtenberg near Kahla where Muck established a crafts settlement whose products they exchanged for apples and potatoes from local farmers.

Initially the local authorities were sympathetic to the settlement, but there was a snake in paradise. A disgruntled female member of Muck's entourage wrote to the authorities accusing the prophet of running a 'household harem'. Investigations proved that, although he already had a child with a married woman, a girl in the throng was also pregnant, with one of the many children he would name after trees, even as he was carrying on with another of his young disciples. All the mothers of his children were themselves members of the extreme racialist Mittgart League. A year earlier he had also tried to seduce various blonde girls with the prospect that they might give birth to a German Christ. A court subsequently established the, invariably compelling, nature of his pitch: 'A personality like Christ, a redeemer, is necessary in the present time. Every female being, whether married or unmarried, but physically and spiritually sound, is entitled to help bring about the birth of this redeemer.'

This accusation resulted in an embarrassing examination by the authorities in Altenburg. Muck and his throng were ordered out of town and he became the object of sly commentaries. The throng did not desert the disgraced prophet. He set up a new settlement in a villa at Naumburg, where they concentrated on wood-turning and produced such knickknacks as nutcrackers and sewing boxes. The scandal was over. During the Depression, Muck made a brief reappearance as a public prophet, when he organised a 'religious week' in Hildburghausen. Among the young Bolsheviks, young Catholics, young Nordicists and a sprinkling of Russians, he established contact with representatives of the more socialist wing of the Nazis. Although he was no more a friend of the Republic than they were, he found some aspects of Nazism unsympathetic – they put power before the spirit, and deceived themselves with their prodigious feats of organisation. While the Nazis subsequently forbade Muck to use the term 'throng' – their own German Labour Front appropriated it – autarchy and an emphasis on craftsmanship meant that his workshops boomed in the 1930s. He spent the Second World War training naval cadets, and then returned to Naumburg where the scrap from a huge weapons plant enabled him to re-establish his communal business. Relocated to Königswinter in 1949, this settled into a conventional family firm, which exists to this day.

Although sophisticates mocked the sudden appearance of deranged figures on village squares and the streets of modern cities, other commentators detected a stirring of ancient spirits, as if deep currents had erupted through the surface of modern life. In 1924 a Franciscan called Erhard Schlund wrote:

> The war of Christianity against Teutonic paganism was not over when Bonifatius felled the sacred oak. Even after the general victory of Christianity and the Christianisation of the German tribes, the battle continued as a guerrilla war in the souls and in the beliefs and religious customs, even in certain individuals and

there were always men who preferred Wotan to Christ. Today it seems as though this century-old skirmish will again become an open battle.[39]

Early in the same year, D. H. Lawrence wrote his perspicacious 'Letter from Germany', although it was not published for another decade. Lawrence felt that 'the great leaning of the Germanic spirit is once more eastwards, towards Russia, towards Tartary. The strange vortex of Tartary has become the positive centre again, the positivity of western Europe is broken. The positivity of our civilisation has broken. The influences that come, come invisibly out of Tartary. So that all Germany reads *Beasts, Men and Gods* with a kind of fascination. Returning again to the fascination of the destructive East, that produced Attila.' In Heidelberg Lawrence encountered hordes of students, including 'queer gangs of Young Socialists, youths and girls, with their non-materialistic professions, their half-mystic assertions'. The country was 'whirling to the ghost of the old Middle Ages of Germany, then to the Roman days, then to the days of the silent forest and the dangerous, lurking barbarians'.[40] In an essay entitled 'The Longing for a World View' written two years later, the novelist Hermann Hesse described the almost frenetic quest for the stable beliefs and morality that had once accompanied rural and small-town society:

> Making itself felt with particular urgency, however, is the need for a replacement for the values of the vanishing culture, for new forms of religiosity and community. That there is no shortage of tasteless, silly, even dangerous and bad substitute candidates is obvious. We are teeming with seers and founders; charlatans and quacks are mistaken for saints; vanity and greed leap at this new, promising area . . . In itself this awakening of the soul, this burning resurgence of longings for the divine, this fever heightened by war and distress, is a phenomenon of marvellous power and intensity that cannot be taken seriously enough.[41]

In July of that year, Joseph Goebbels attended a series of meetings in and around Berchtesgaden. One of his diary entries recorded his gushing impression of Hitler, whose star Goebbels had begun to follow. He added: 'In the afternoon he [Hitler] spoke about the conquest of the state and the meaning of political revolution. Thoughts, that I had certainly had myself, but which I had never articulated. After supper we sat for a long time in the garden of the Naval Home, and he preached on the subject of the new state and how we would fight for it. It sounded like prophecy. In the heavens above a white cloud took on the form of a swastika. There was a flickering light in the sky, which could not be a star. – A sign of destiny?'[42]

Nineteen-twenty-four saw the publication of one of the most remarkable and least-known books of the twentieth century, Christoph Bry's *Verkappte Religionen* or 'Hidden Religions'. It was reprinted in Germany in 1964, but has never been translated, although it certainly deserves to be since it speaks to the twenty-first century as much as to Bry's own time. Bry was fascinated by books, and acutely conscious that so much of what was being published was inconsequential rubbish that skilfully concealed its absence of thought behind various adopted manners. The facts of his life are sparse. He was born in Pomeranian Stralsund on the Baltic coast. His father ran a sausage shop. He had two brothers. One went missing in the First World War, the other died of a heart attack as the Gestapo arrested him. Bry, born so lame on his left side that his foot dragged behind him, died in 1926 at the age of thirty-three. The best pupil in his local grammar school, between 1911 and 1916 he studied history, political economy, jurisprudence, German, philosophy and theatre at Munich and Heidelberg. From 1916 onwards he worked for the publisher Ullstein, before establishing his own small house in Munich. Despite his difficulty in walking, he was well known on the Munich artistic scene. In order to boost his income, he wrote (he actually dictated them to his wife) newspaper articles and reviews on books, film, theatre, mass meetings and trials, including that of Ludendorff and Hitler after the failed putsch. Hitler struck him as a cross between a holy-roller and a provincial prima donna, whose audiences loved the predictability of the hectic hysteria, the sweeping arm gestures, and above all those southern German 'rolling-RRRs'. As Bry argued, Hitler belonged in the company of Rudolf Steiner, Haeusser and the other 'miracle workers'.[43] He was spared the worst ravages of the inflation. A student friend owned a major Argentine German newspaper, which meant that Bry was paid in US dollars, enabling him to write his extraordinary book. Unfortunately, his health deteriorated to the point where he had to visit Davos in Switzerland to recuperate. He died there in 1926, although not before telling a poet friend how he had written his book, how the public responded, and how he envisaged its themes developing. He died an optimist.

*   *   *

Verkappte Religionen is both brilliantly written and bold in tackling several modern spiritualised monomanias, regarding which, even today, in the case of gender politics which he also included, many people might pull their punches with if they wanted to get on in life. Bry savoured the fact that in Argentina a newspaper advertisement for his book appeared next to one selling pesticides for household vermin. His acerbic, mocking tone would doubtless upset many US academics, as he believed that to adopt a 'solemn pomposity' was inadvertently to compound the earnestness he detected in the ecstasies he wrote about, thereby perpetuating discussion of them to infinitude. Bry was what Americans call mean – in other words, without spurious civility or collegiality. Writing of a monomaniacal American book of meaningless statistics about the 'success' of Prohibition, he defended the right not just 'to get pissed' – as he put it – but to resist the wider vision of the stone-cold sober 'new man' that lurked behind the walls of numbers and percentages. A 'homunculus' would result, who consumed nothing but the air he breathed, since unlike religions that seek to enhance and refine human drives, 'all forms of lifestyle reform – whether sophisticated or crude – constitute a form of spiritual suicide' as deadly as smoking or drinking oneself to death.[44]

Bry was concerned with many total explanations of the world, whether these were numerological or political – notably Communism and Fascism – or pseudo-religious – Freudian psychoanalysis and the 'department store' anthroposophy of Rudolf Steiner, as well as with those lifestyle choices that become all-consuming vocations, such as abstinence from alcohol, yoga, vegetarianism, feminism and homosexuality. Bry was especially concerned to show how the adoption of specific identities did not 'liberate' an individual, but imprisoned them within such a narrow carapace that a caricature (or stereotype) automatically resulted. His other major concern was with the more familiar theme of the 'hidden' logic – the world *behind* rather than *beyond* – that animated many of the resulting crazes, fads, sects and movements. In the case of antisemitism, which many of his cults and sects shared, this involved a salt-cellar never being just a thing used to deposit seasoning on the side of a plate, but being also physical 'evidence' of Jewish control of the ancient salt trade or 'their' majority stock holdings in modern salination works. If racism was the hidden logic of the extreme right, then it was Marx's 'achievement' to transform the inchoate utopian enthusiasms of early socialism into a hidden religion supported by what passed for science but which was a form of prophecy involving his own chosen people – the industrial proletariat.

Interestingly, in passages which he omitted from his book, Bry claimed that the moral cowardice of the Churches (rather than of Christianity) was responsible for the elephantic growth of hidden religions, a surrender that had already been evident in their defensive response to the growth of science. Instead of seeking the widest possible position within society, they had fallen back on their own narrowing ramparts, where they duly fell into fighting among themselves. The bien-pensants had successfully caricatured Christianity as deceptive, stupid and reactionary. It was something so retrograde that there was not even any point in fighting it, an argument not lost on Friedrich Engels. Bry felt that instead of routinely blaming religious indifference upon 'anti-religious powers', Christians should spend more time laughing at the risible beliefs of the 'modern' indifferent that he analysed so sharply in his book. At about the same time, James Joyce came to a remarkably similar conclusion, when in *Portrait of the Artist as a Young Man* his Catholic hero remarked: 'What kind of liberation would that be to forsake an absurdity which is logical and coherent and to embrace one which is illogical and incoherent?'[45]

V POLITICS AND RELIGION IN THE 1920S

It would be wrong to imagine the post-war period in overly dramatic terms, or to view it exclusively through the apocalyptic or dyspeptic optic of many of its fashionable writers and artists. During the 1920s, Britain witnessed a remarkable efflorescence of Anglo-Catholic social endeavour, the emphasis on collective responsibility tending to accompany the more corporately conscious, medievalising wing of the Anglican Church rather than its Evangelical individualists.[46] Its most prominent advocate was William Temple, successively bishop of Manchester and York and, for two years before his death in 1944, archbishop of Canterbury. Hugely fat and smug in appearance, Temple was the son of a former archbishop of Canterbury, who at the age of seven had wept on learning that the servants in the family's Lake District hotel were prohibited from eating chicken. The guilty moralism of the privileged members of an established church would hang over much of the Church of England's social endeavours.

In 1924 – the year the first Labour government briefly came to power – Temple organised the Conference on Christian Politics, Economics and Citizenship (or COPEC) which met in Birmingham in that year. Temple was not calling for the Church of England to put itself behind Labour, although he belonged to the Party for a couple of years, nor for a new, continental-style Christian-democratic party; his vision of establishment combined the belief that Christianity should be at the heart of the nation with the view that it should also encourage a radical, reformist social agenda through the enunciation of general principles. The COPEC conference was massively prepared with detailed expert reports on such subjects as industry and property; the treatment of crime; and politics and citizenship. The conference called:

> on all Christian people to do all in their power to find and apply
> the remedy for recurrent unemployment, to press vigorously for
> the launching of efficient housing schemes, whether centrally or
> locally, and to secure an immediate extension of educational
> facilities, especially for the unemployed adolescents, whose case
> is perhaps the most deplorable of all the deplorable features of
> our social life today ... we urge the immediate raising of the
> school leaving age to sixteen, and the diminution as rapidly as
> possible of the maximum size of classes.[47]

While many of these ideas, already familiar in spirit from the Christian Social Union of the pre-war era, would bear fruit in the form of the post-1945 welfare state, at the time they were unwelcome, not least to the elderly archbishop of Canterbury, Randall Davidson, who thought that the Church would be better advised to eschew partisan politics. The archbishop discovered this himself when in May 1926 he made what seems a reasonable proposal to the nation as to how to resolve the General Strike that had polarised the country: the strikers should return to work; the government should restore limited subsidies to the coal industry; and the mine-owners should withdraw their reduced wage scales. The BBC refused to broadcast the archbishop's appeal, although it did transmit the Catholic cardinal Bourne's denunciation of the strike as 'a sin against the obedience which we owe to God', sentiments which won the approval of the minority of right-wing Anglicans. As for Temple, he was ill and abroad when the strike occurred, although he lent his support to the Standing Committee of the Christian Churches on the Coal Dispute, which vainly tried to mediate between the miners and the coal-owners. These unqualified interventions prompted prime minister Stanley Baldwin to remark that perhaps the Federation of British Industry should seek to revise the Athanasian Creed.[48]

The years after the war saw a remarkable burgeoning of Catholic politics in Europe, which seemed to promise a 'third way' between Marxist socialism, whether in its democratic or totalitarian guises, and the atomised individualism of liberal capitalism. Since Catholic politics encompassed left-wing trades unionists as well as clericalist authoritarians, urban as well as rural voters, it straddled the more familiar ideological divides of the modern continent. It was hostile to the power of the modern state, while its cure for the social injustices of liberalism involved reviving autonomous associations rather than multiplying faceless bureaucracies. Political Catholicism stood in a sometimes uneasy relationship with the social and political vision of the papacy, which was dominated by the goal of 'the re-establishment of the Kingdom of Christ by peace in Christ', a goal that could be realised by a variety of means other than through dedicated confessional parties. Thus, while some Catholics continued to work through political parties, others – and in particular many of the young – regarded such vehicles as Catholic Action as the better way of achieving spiritual goals that were imperfectly addressed by the parties.[49]

The war had a profound impact upon European Catholicism. In most countries there was a brief upsurge in church attendance and an increase in diffuse religiosity, albeit much of it bent on self-preservation or protecting the lives of combatant relatives. Most countries proclaimed a civic truce, or what in France was called a 'sacred union', designed not only to suspend class conflict, but clashes between rival confessions or, notably in France, recent clashes between Church and state. While in some Mediterranean countries the power of the Catholic Church continued as a source of resentment to militant anticlericals, elsewhere there was a marked diminution of the passions this issue had incited during the pre-war period. This partly reflected the sacrifices that Catholics, including the clergy, had made on the battlefields. Although some Catholic political parties originated in the decades before the war, such as the Belgian Catholic Party or the German Centre Party, many were created to represent Catholic interests in a new age of mass politics and parliamentary regimes. In Italy, pope Benedict XV reluctantly gave his consent to the formation of the Partito Popolare Italiano which was founded in January 1919. To appease the Church, its leader, a remarkable priest called Luigi Sturzo, insisted that the Party was 'aconfessional' but broadly based on Christian principles. These were reflected in a concern for the well-being of the family and small farmers, the transference of power to subsidiary associations and the regions, and, since Sturzo was a Sicilian, for the development of the backward South. In the general elections in late 1919, the Party won 20 per cent of the vote and a fifth of the seats in parliament, becoming the second largest party after the Socialists. It did especially well in the traditionally White areas in north-eastern Italy – especially Lombardy and the Veneto – and conspicuously badly in the South, where in some regions it scraped a mere 5 per cent. Since the Socialists refused to participate in 'bourgeois politics', the Popolari joined each of the six cabinets that attempted to rule Italy between 1919 and Mussolini's 'March on Rome' in October 1922.

Essentially a broad coalition of potentially opposed interests, the Popolari disintegrated amid the tensions that were endemic in Italy in the first years of the 1920s. Sturzo's clerical status proved to be the Party's Achilles heel, particularly after Pius XI succeeded Benedict in February 1922. The new pope was not convinced that political parties were the best means of Catholic self-assertion, while taking the view that perhaps Mussolini could finally resolve the status of Rome. The Vatican forced Sturzo to resign the leadership in July 1923. In elections held in 1924, the Party's share of the poll slumped to 9 per cent, as Catholic voters turned to the Fascists. Shortly afterwards the Party dissolved itself.

The advent of the Weimar Republic affected the German Churches in different ways. The soft, and only partial, separation of Church and state dismayed Protestants, who were used to the external legitimisation that had come with ecclesiastical establishment in the Empire. Since Catholics had never enjoyed such privileges, they were less affected by their disappearance. In addition to lacking a single political vehicle to defend their interests – until, that is, the Nazis appeared to address that deficit – Protestants watched with trepidation the ascendancy of the Catholic Centre Party, which, in addition to dominating most Weimar and Prussian coalition governments, supplied chancellors Fehrenbach, Wirth, Marx and Brüning. Other prominent Catholics included Matthias Erzberger, who signed the armistice at Compiègne, and was subsequently assassinated in 1921 by two right-wing extremists while out hiking during a holiday, and Munich's archbishop Michael von Faulhaber, who became an influential voice in the land. Although Catholics had been as patriotic as the next man in the recent war, some, ignoring the dissolution of the Habsburg Empire and the role of Protestant Britain and the US, regarded the victory of, inter alia, France, Belgium and Italy as a delayed triumph for Catholicism over the Lutheran Reformation.[50]

The Weimar Constitution adopted confessional and ideological neutrality, rejecting an established state Church, but nonetheless included generous support for the Churches within a framework of religious toleration. The Churches enjoyed special legal protection and continued to be subsidised through taxes, while God was defended with blasphemy laws. Sunday and Church holidays continued as days of rest, while chaplains were still attached to asylums and the army. Religion remained a compulsory subject in schools, its content to be determined by the clergy, although no solution was reached to satisfy both Catholics and liberals on the wider balance between parental choice vis-à-vis a system designed to transcend confessional divisions.

Both Churches were affected by the general climate of the times, whether in terms of Allied impositions, domestic political conflict, economic dislocation or a no less tangible sense of moral disintegration, a perennial subject of clerical rumination. Protestant clergy responded indignantly to the war-guilt clause in the Treaty of Versailles by holding a day of national mourning, and tolling their church bells on the day the treaty was signed. Many Catholic priests would have done so too, had they not had to respect Benedict XV's diplomatic efforts to soften the terms of the treaty in response to entreaties from the Catholic bishops. Both Churches were affected by the subtraction of German territory stipulated by the treaty, as well as by the drastic curtailment of missionary activity in Germany's former colonies. While some liberal Protestants realistically adhered to the new Constitution and tried to win over their ecumenical friends for treaty revision, rather than supporting those actively seeking to subvert it, others continued to trumpet the 'war theology' that had resulted in one catastrophe and would contribute to another once it had mutated into the so-called German Christians.

The Catholic Centre Party was involved in all thirteen coalition governments between 1919 and 1930. Its strategy was to preserve denominational interests, while mediating between the various ideological camps that otherwise dominated the Republic. It was symptomatic of a time when being in government did parties no favours that the Centre Party regarded being in power as a form of sacrifice for the troubled fatherland. The role of mediator meant that the Centre Party was a classic party of compromise, rather than one capable of setting forth a bold vision for the future. In the early 1920s there were attempts, associated with the trades unionist Adam Stegerwald, and then with Wilhelm Marx and Konrad Adenauer, to broaden the Party's appeal to encompass Christians in general.[51] Not even every Catholic, however, voted for the Party, its share of the vote being much higher among regular churchgoers than among nominal Catholics. Conservative nationalist Catholics supported the German National People's Party (DNVP), while those on the left backed the evanescent Christian-Social Reich Party. In Bavaria, the Catholic vote went to the particularist Bavarian People's Party. The combined vote for the two major Catholic parties fell from 19.7 per cent in 1919 to 13.9 per cent in March 1933, a decline that would have been much worse had newly enfranchised women voters not supported these parties in impressive numbers. Whereas almost 63 per cent of Catholics voted for the Centre in 1919, by 1930 this had fallen to 47 per cent.

The politics of Germany's Protestants were more dispiriting. A prominent few, such as Adolf Harnack or Ernst Troeltsch, were *Vernunftrepublikaner*, that is, people who thought that there was no going back to a non-existent imperial utopia, and that they had to work within the framework of present political realities. Initially many of those Protestants who were involved with their Churches gave their backing to the DNVP as the only party that promised to defend Protestant interests. In 1925 the Evangelical League for the Defence of German-Protestant Interests played a part in the rejection of Wilhelm Marx and the election of field marshal von Hindenburg as the president of the Republic. Thereafter, the votes of churchgoing Protestants migrated to a plethora of evanescent splinter parties, which were the religious analogue of the narrow moralising interest-based parties – for sound money, creditors' interests and so forth – that massively fragmented the middle-class electorate without being capable of sustaining it, in the middle years of the decade. Like many of their secular analogues, such oddities as the German Reformation Party, formed in 1928 to defend Protestantism against political Catholicism, Marxism and liberalism, failed to secure a single seat in the Reichstag. The same fate met the prophet–politicians such as Louis Haeusser. The political homelessness of Protestant Germany would be resolved after 1930 when its citizens gave their votes in increasing numbers to a party that promised authority, order and respect for religion: the Nazis successfully presented themselves as the sword of an awakening semi-religious German spirit. That in turn was part of a much broader challenge from totalitarian movements with a more or less conscious mimetic relationship to the Churches, not least the Bolsheviks in Russia, the first illegitimate brother of religion to assume political power and to demonstrate the horrors of applied rationality.

CHAPTER 2

The Totalitarian Political Religions

I STORMING THE HEAVENS

In 1920 the British philosopher Bertrand Russell spent five weeks in Bolshevik Russia as a member of a Labour Party delegation. The group hoped to discover a promised land, breaking into spontaneous choruses of the Internationale and Red Flag on spying the first Red banners across the border. After twenty-four hours Russell realised that there was not much to sing about. What he mistook for a gaggle of vagrants turned out to be a party of distinguished mathematicians keen to pay homage. This did not augur well. One wonders what Russell would have made of the 'intellectual' Lenin's deportation of two hundred prominent scientists and thinkers in that very year. Sharing at least two of his nastier prejudices with an (aristocratic) correspondent, Russell called the Bolshevik leaders arrogant and flashy, 'an aristocracy as insolent and unfeeling (as the tsar's) composed of Americanised Jews'.[1]

Lenin granted Russell a side-audience as he sat for a portrait sculptor. This may have irked. Russell found Lenin's laugh especially ghoulish since it went with gloating accounts of poor peasants hanging their richer fellows from trees. To escape the oppressive attentions of the Bolsheviks in Moscow, the delegation took a steamer south through the darkly desolate countryside. Boldly venturing ashore, from a barque that was soon stricken by disease, Russell encountered beings to whom it was less easy to relate than to 'a dog or a cat or a horse'. He meant the starving peasantry. After ten days on the boat, his party hastened to Saratov, from where the express train returned them to civilisation in Estonia.[2]

Russell recycled the bits and pieces he had published in the *New Republic* into an instant book called *The Practice and Theory of Bolshev-*

ism. The message was clear enough: 'I felt that everything I valued in human life was being destroyed in the interests of a glib and narrow philosophy, and that in the process untold misery was being inflicted upon many millions of people.'

He racked his brain for an apt analogy: the French Directory? Cromwell's Puritans? Plato's guardians? Perhaps the followers of Mohammed, an 'orientalist' analogy that Alexis de Tocqueville had already applied to the Jacobins? Russell finally alighted on a Christianity whose Sermon on the Mount had not inhibited inquisitions or obscurantism: 'The hopes which inspire Communism are, in the main, as admirable as those instilled by the Sermon on the Mount, but they are held as fanatically, and are likely to do as much harm ... The war has left throughout Europe a mood of disillusionment and despair which calls aloud for a new religion, as the only force capable of giving men the energy to live vigorously. Bolshevism has supplied the new religion. It promises glorious things ... .'[3]

The Russian religious philosopher Semyon Frank had given such comparisons between revolutionaries and religious fanatics far more profound expression decades earlier, even though he never achieved Russell's celebrity. His fate was that of an émigré and a grave in North London's Hendon. Born in 1877 into a Russian-Jewish family, he had quickly outgrown a juvenile Marxism, converting to Orthodox Christianity in 1912, on the ground that the Jewish God was as remote from the world as the utopia of socialist imaginings. He rejected a Russian-Jewish sympathy for messianic radicalism in favour of liberal conservatism too. Long before the October Revolution he devoted a remarkable essay in *Landmarks* to the nihilistic moralism of the Russian intelligentsia. Of the socialists' infatuation with the idea, he presciently declared:

> Sacrificing himself for the sake of this idea, he does not hesitate to sacrifice other people for it. Among his contemporaries he sees either merely the victims of the world's evil he dreams of eradicating or the perpetrators of that evil ... This feeling of hatred for the enemies of the people forms the concrete and active psychological foundation of his life. Thus the great love of mankind of the future gives birth to a great hatred for people; the passion for organizing an earthly paradise becomes a passion for destruction.

The revolutionaries were 'militant monks of the nihilistic religion of earthly contentment'. The monk–revolutionary, Frank continued, 'Shuns reality, avoids the world, and lives outside genuine, historical, everyday life, in a world of phantoms, daydreams and pious faith . . . The content of this faith is an idolatry founded on religious unbelief, of earthly material contentment . . . A handful of monks, alien to and contemptuous of the world, declare war on the world in order to forcibly do it a great favour and gratify its earthly, material needs.'[4]

A witness to the famine on the Volga, Frank was one of two hundred leading intellectuals who were deported by the Bolsheviks in 1922. Courses on religion and philosophy he had organised at the University of Moscow for students weary of atheism proved too popular for the new masters to tolerate. The final page of his (cancelled) passport was stamped with a warning that he would be shot if he ever re-entered the Soviet Union.

The new Bolshevik religion arose amid the ruins of the old, but it was never free of its imprint, including that of the egalitarian millenarianism of the sects that had broken with Orthodoxy in the late seventeenth century. The Orthodox Church was integral to the tsarist autocracy and essential to the lives of millions of Russians. Most people believed in supernatural intervention in human affairs. Religious rites accompanied births, weddings and funerals, and were integral to folk medicine. There were 150–200 Orthodox holidays a year, of which fifty – in addition to Sundays – were opportunities for days off and booze-ups of a herculean nature. Each region had four or so patron saints, whose feast days were holidays too. So to strike at the Orthodox Church was not simply a matter of taking on its clerical hierarchy, but of assaulting the traditional beliefs of much of the population.

On the eve of the Bolshevik coup d'état, the Orthodox Church claimed a hundred million adherents, two hundred thousand priests and monks, seventy-five thousand churches and chapels, over eleven hundred monasteries, thirty-seven thousand primary schools, fifty-seven seminaries and four university-level academies, not to speak of thousands of hospitals, old people's homes and orphanages. Within a few years, the institutional structures were swept away, the churches were desolated, vandalised or put to secular use. Many of the clergy were imprisoned or shot; appropriately enough the first concentration camp of the gulag was opened in a monastery in Arctic regions. Religiosity itself remained, disappearing underground, or diverted into shallower affective channels, and focused on false gods, the mightiest of whom gave socialism one omnipresent, pock-marked, smiling face.

The future Bolshevik leader Lenin had lost the Orthodox faith of his parents at sixteen, although he and Krupskaya were married in an Orthodox wedding ceremony.[5] In common with much of the secular intelligentsia, his faith was replaced by an ideological creed which professed that all religion would atrophy once the material conditions that had engendered it had been abolished. Since the Orthodox Church was so integral to tsarism, Russian radicals, including the Bolsheviks, were correspondingly militantly atheist, although their own surrogate cults were not without a certain religiosity. The philosopher Nikolai Berdyaev, an acquaintance of Semyon Frank, caught this when he wrote: 'Just as pious mystics once strove to make themselves into an image of God, and finally to become absorbed in Him, so now the modern ecstatics of rationalism labour to become like the machine and finally to be absorbed into bliss in a structure of driving belts, pistons, valves and fly-wheels.'[6]

Aggressive atheism did not entirely preclude tacking to whatever gust of wind that promised to bring the revolutionary ship closest to the peasant masses' distant shore. In the early 1900s Lenin encouraged Vladimir Bonch-Bruevich to find common ground with collectivist, millenarian Sectarian peasants through a new journal called *Dawn*, while in 1905 Lenin himself briefly flirted with the 'little father' George Gapon, the police-spy turned demagogue who had led the march on 'Bloody Sunday' to the Winter Palace.[7] After the Revolution, the occasional renegade Christian tried to invest it with religious significance, notably the leading Symbolist poet Alexandr Blok (1880–1921) in his 1918 poem 'The Twelve'. The poem's conclusion must have perplexed the Bolshevik leadership.

> So they march with sovereign tread . . .
> behind them limps the hungry dog,
> and wrapped in wild snow at their head
> carrying a blood-red flag –
> soft-footed where the blizzard swirls,
> invulnerable where bullets crossed –
> crowned with a crown of snowflake pearls,
> a flowery diadem of frost,
> ahead of them goes Jesus Christ.[8]

For a time Blok made a pittance from reciting this poem to large audiences before cold and hunger killed him when he had no more books to sell and no more furniture to chop up for firewood.

The idiosyncratic enthusiasms of literary collaborators such as Blok apart, the Bolsheviks had two potential tactical means of extinguishing religion. One was to combine repression with ridicule, destroying the institutional fabric of the Orthodox Church; the other was to use crude scientific materialism to discredit its personnel by mocking their most deep-seated beliefs. But there was a rival variant which, while acknowledging a general religious instinct, described as a 'necessary illusion', sought to divert it from a transcendental God to worship of mankind and science. The essence of this faith was 'Man does not need God, he himself is God. Man is a God to man.'

This faith was called 'God-building' and was associated with Alexander Bogdanov, Leonid Krasin and above all Anatoly Lunacharsky, who in 1909 founded a 'God-building' summer school at Maxim Gorky's villa on Capri, a step which resulted in Bogdanov and Lunacharsky being temporarily expelled from the Party. Lenin agreed with Plekhanov's view that the 'God-builders' 'start out by declaring God a fiction, and end by proclaiming man a god. But since humanity is not a fiction, why call it god?'[9]

While these two strategies were evident in the subsequent Bolshevik approach towards religious faith, there was no disagreement about how to deal with institutionalised religion. The Provisional Government had pursued policies that sought to disestablish the Orthodox Church. The Bolsheviks regarded this relative restraint as a form of 'bourgeois' inhibition, and resolved to eradicate Christianity as such. In 1918 the Churches were deprived of their legal personality and their lands and properties were nationalised. Ten Orthodox hierarchs were summarily shot, with the explanation: 'Soviet power will keep shooting these lords until we smash and crush the criminal counter-revolutionary activity of Church leaders.' Registration of the rites of passage passed to a civil authority which also assumed control of education, whether or not it was subsidised by the state. A liberal-sounding 'Decree on the Freedom of Conscience and on Church and Religious Associations' separated Church and state and appeared to guarantee freedom of both religion and irreligion, although the power of the state was unevenly massed behind atheism. While individual religious communities were allowed to lease houses of worship, the Orthodox Church was deprived of any income, and its hierarchy was divested of authority on the ground that it was assumed to be counter-revolutionary.[10] Religious believers used the right to petition the authorities to lease back their own churches as a form of resistance to these measures.[11] Children were deprived of any religious education outside the home, and the married secular clergy were stripped of the right to vote, of adequate rations, and of the ability to put their own children through higher education, while being subject to a vindictively onerous tax burden. In a letter to the Council of People's Commissars in October 1918 the patriarch Tikhon, who had scrupulously refrained from commenting on politics, used the anniversary of the Bolshevik coup d'état to communicate his misgivings about the regime. He denounced both the Treaty of Brest-Litovsk – a necessity for the Bolsheviks rather than the Russian people – and the Red Terror, enjoining the regime 'to celebrate the anniversary of taking power by releasing the imprisoned, by ceasing bloodshed, violence, ruin, constraints on the faith. Turn not to destruction but establish order and legality.'[12]

The many crises facing the new regime meant that a temporary truce prevailed between the Bolsheviks and the Church between late 1918 and early 1922. Widespread famine in 1921–2, for which the government's confiscation of food and seed was largely responsible, provided the pretext for a renewal of hostilities. With 25 million people projected to starve in the Volga region, the Bolsheviks began inventorising Church property, while simultaneously banning any voluntary assistance on the part of churchmen or others to the starving. Trotsky was put in charge, behind the scenes, of both the general campaign against religion and the confiscation of Church valuables, there being no mention of the famine in his dual remits. A campaign of oblique agitation, consisting of letters to newspapers, was designed to connect confiscation with relief of the famine.

On 6 February 1922 patriarch Tikhon offered non-consecrated Church valuables to alleviate the plight of millions. However, Lenin insisted that the Church also surrender those valuables that were intrinsic to celebration of the eucharist. Tikhon warned those contemplating even voluntary surrender of such objects that the penalty for sacrilege was excommunication. By early March, the quantity of foreign relief piled up on quaysides was so great that the transport system could not handle it; sale of Church valuables abroad to buy more grain would have been pointless. Nonetheless, violent confiscations of valuables went ahead. In some places, notably Rostov-on-Don, Smolensk, Shuia and

Staraia Russa, there were spontaneous confrontations between crowds and church robbers. On 12 March, Trotsky told the Politburo that 'our entire strategy at the present moment must be calculated to provoke a schism among the clergy on the concrete issue of the seizure of church valuables'.[13] The famine would be exploited to divide and rule the Church.

On 19 March Lenin dictated a memorandum that was to guide the next day's Politburo meeting on how it should respond to events like those in Shuia. The second paragraph is a textbook example of how to demonise opponents, in this case by eliding the moderate Tikhon with the antisemitic and reactionary Black Hundreds, while projecting one's own habitual conspiratorial mode on to how these opponents allegedly operated:

> Connected with what we know of the illegal appeal of Patriarch Tikhon, it becomes crystal clear that the Black Hundreds clergy, headed by its leader, quite deliberately implements a plan to give us decisive battle precisely at this moment. Apparently at secret consultations of the most influential groups of the Black Hundred clergy this plan had been thought through and quite firmly adopted. The events in Shuia are but one manifestation of the fulfillment of this plan.

The millions of starving peasants were merely a tactical opportunity to smash an opponent:

> This precise moment is not only uniquely favourable, but offers us a 99 per cent chance of shattering the enemy and ensuring for ourselves for many decades the required positions. *It is now and only now, when in the regions afflicted by the famine there is cannibalism and the roads are littered with hundreds if not thousands of corpses, that we can (and therefore must) pursue the acquisition of church valuables with the most ferocious and merciless energy, stopping at nothing in suppressing all resistance.*

And so to the grim conclusion: 'For this reason I have come to the unequivocal conclusion that we must now give the most decisive and merciless battle to the Black Hundreds clergy and subdue resistance with such brutality that they will not forget it for decades to come.' Agents were to be despatched to Shuia with a remit to arrest 'no fewer than a dozen representatives of the local clergy, local burghers, and local bourgeois on suspicion of direct or indirect involvement in violent resistance'. The Politburo was then to instruct the court 'that the trial of the Shuia rebels who oppose help to the starving should be conducted with maximum swiftness and end with the execution of a very large number of the most influential and dangerous Black Hundreds of Shuia and, insofar as possible, not only of that city but also of Moscow and several other church centres'.[14]

Since the American Relief Administration had more food piled up in Russia's ports than could be distributed, the confiscation of Church valuables had little or nothing to do with ameliorating the plight of the starving. A well-publicised disbursement of one million rubles realised from Church valuables derived from a confiscation campaign itself funded to the tune of ten times that amount. Apart from the fact that the regime also had the tsar's crown jewels as an alternative, it also rejected a generous offer from the Vatican to cover a sum equivalent to the confiscated Church valuables. Most of these valuables found their way into museums, where experts began assessing their market value – which, as it happened, was in the low millions rather than the 'billions' of Lenin's imagination. By late 1922 the regime was exporting nearly a million tons of grain, which again suggests that the confiscations of Church valuables had had nothing to do with famine relief.

Resistance to confiscation provided an opportunity to calumniate the clergy as crypto-Fascists, or as agents of 'the Rothschilds and international capital', thereby allowing the organisation of show trials throughout European Russia against people who were not even clerics. Of course, some Orthodox clergy were indeed involved in drumming up support for White armies. Be that as it may, the outcome of these clerical trials was assured even before the defendants had been notified of any charges. At these trials clerics appeared as witnesses to a putative counter-revolutionary conspiracy within the Orthodox Church. Ironically, there was no mention of the Russian Orthodox Church Abroad, which in 1922 had indeed called for the overthrow of the Bolshevik regime from its Yugoslav exile, in these trumped-up charges. These witnesses were drawn from a minority of reforming clergy opposed to the Orthodox hierarchy who under archpriest Vvedensky coalesced into the Renovationist or Living Church in March 1922. These conciliarist idealists, leftists, malcontents and opportunists were just the schismatic entity that the Bolsheviks needed for an ecclesial coup. This group, the ancestor of various stooge Churches that subsequently proliferated throughout the post-1944 Communist bloc, proclaimed the October Revolution as a 'Christian creation' and its atheist leader the 'tribune of social truth'. The Living Church declared that the separation of Church and state was beneficial to religion: 'Freedom of religious propaganda (in addition to freedom to propagate antireligious ideas) enables believers to defend the value of their purely religious convictions in ideal circumstances. Therefore adherents of the Church cannot regard Soviet power as the realm of Antichrist. On the contrary, the Council [of the Living Church] draws attention to the fact that the Soviet power is the sole entity in the world that is in a position to realize the Kingdom of God,' this being their pathetic accommodation to the spirits of those times.[15]

Such concessions to unreality ensured that the Living Church soon had sixteen thousand functioning churches, seventeen thousand priests and two hundred bishops, although in Bolshevik eyes they were always just a tool that could be put back in the box after it had been used. Meanwhile, on 9 May 1922, eleven of the fifty-four defendants in the Moscow clerical show trial were sentenced to death, most of them parish priests and laymen. When the Petrograd metropolitan, Veniamin, excommunicated the schismatic Renovationists, he was tried for counter-revolutionary conspiracy, along with eighty-eight other defendants, with local newspapers calling for 'No Mercy for Black Hundred Clergy'. Ten of the defendants received the death sentence, with Veniamin being secretly shot on the night of 12 August 1922 at the Porokhovye railway station. In total, between May 1922 and early 1923, over seven hundred people were tried for obstructing the confiscations of valuables, with forty-four of them executed and 346 receiving long prison sentences.

The patriarch Tikhon had been put under house arrest early in this process. In May 1923 he was deposed and his title usurped by the Renovationists. Faced with the alternative of being tried for counter-revolution or capitulating to Bolshevik power, he chose the latter course, and was restored the year before he died. Under his more amenable successor Sergei an uneasy accommodation with the Soviet authorities ensued, albeit punctuated by further bouts of savage persecution, notably in the years 1928–32 and 1937–41. In May 1929 a nominal constitutional right to 'religious propaganda' was replaced by a 'right of religious confession' which was interpreted in a highly restrictive fashion, while such devices as raising fire-insurance premiums on churches ensured further closures. By that time Stalin's plans for crash industrialisation and forced collectivisation of agriculture were under way, the latter presaged by the 1927 Violations of the Regulations Concerning the Separation of the Church from the State incorporated into the new Criminal Code. Giving religious instruction or 'inspiring superstition in the masses of the population' was penalised by up to a year's 'corrective labour', while the section entitled 'State Crimes' stipulated that propaganda or agitation calling for the 'overthrow, undermining or weakening' of Soviet power which exploited the 'religious prejudices of the masses' would warrant execution or no less than three years' imprisonment.[16] These measures, together with the removal of church bells and the closure of churches, were designed to uproot one of the obvious poles around which opposition to collectivisation could gather. In fact, it led angry and terrified peasants to identify the regime's agents and the state farms themselves as manifestations of the Antichrist, a view derived from the only conceptual template they had to explain this devastating incursion into their lives and to inspire resistance towards the outsiders responsible for it. Rumours went abroad to the effect that on these farms women would have to share a 'common blanket' – in other words, become rural groupies – or that children would be exported elsewhere. Those who wanted to enter such farms were told that they faced massacres of the order of St Bartholomew's Night.[17]

Another wave of persecution set in after 1937. The adoption of the 1936 Constitution seems to have encouraged clergy in the delusion that they could reach a modus vivendi with the Soviet regime, by claiming biblical authority for such slogans as 'He who does not work shall not eat' or declaring 'Stalin – we respect him, because he was put in place by the Lord God'. Religious believers began to use the Constitution to reassert lost religious rights. The 1937 census, conducted during the Orthodox Christmas, included information on the religious affiliations of the population which proved so disconcerting for the regime that the results failed to appear.[18] Finally, the clergy 'misinterpreted' the Constitution, under which they received full citizenship rights, to mean that their representatives could stand for election to the Supreme Soviet alongside candidates from other legally established organisations. They were wrong. Before the elections, clerics were arrested throughout the country and charged with organising espionage and sabotage. By 1938 eighty bishops had lost their lives, while thousands of clerics were sent to the Solovetsky labour camp set up in a former monastery on an island in the White Sea. By 1939, when a pre-war thaw set in that two years later would result in a cynical and desperate resort to religion, the situation of the Orthodox Church was near catastrophic, with a ragged remnant on the surface, and a dedicated but endangered underground Church, with its itinerant lay priests and clandestine 'house churches'. Some extreme Sectarians fled to the northern wildernesses to escape the maw of the Antichrist.[19]

These repressive measures were accompanied by anti-religious policies that oscillated between the two variant approaches to religion that we mentioned at the start of this section. Newsreels show the militant godless permanently on the march in a dusty flurry of banners and placards. Bolshevik raiding parties were sent into the churches (and synagogues, but not into the mosques) to pillage sacred paraphernalia, while icons and relics were subjected to mocking 'scientific' scrutiny worthy of the most militant early modern European Protestantism. Icons that allegedly glistened and glowed in the dark were exposed as so much hocus-pocus reliant on luminous gold paint – one group of painters responsible for the clandestine production of such images were shot. Since in Orthodox tradition – as opposed to doctrine – the bodies of saints were supposedly immune to decomposition, the Bolsheviks exultantly opened coffins and tombs to reveal bones collapsed within dusty rags, or, if the uncorrupted figures turned out to be waxen, demonstrated that they were rigged with devices that 'miraculously' induced tears. Newsreel footage shows Bolsheviks gleefully exhibiting decomposing skulls. The bodies of saints were juxtaposed with the accidentally preserved corpses of a counterfeiter and mummified frogs and rats that had been preserved by dry air in ventilation shafts. Doctors and scientists were on hand to explain these phenomena in the required anti-miraculous manner.[20] When George Bernard Shaw was shown the perfectly preserved bodies of two peasants to give the lie to the indestructibility of saints, he characteristically inquired how anyone could know that the two peasants were not saints themselves. The first of forty-four anti-religious museums was opened in 1924, the biggest being the Museum of the History of Religion and Atheism established within Leningrad's Kazan cathedral eight years later.[21]

Individual irreligious enthusiasts, puffed up with dull scientific certitudes, periodically appeared in village streets to challenge God to punish their blasphemies with lightning bolts, or declaimed bits of wisdom about the isosceles triangles or the like from an encyclopaedia outside the porch of a church. Since schoolteachers, the vehicle for militant secularism in many western European countries, were often the sons of priests in Russia, and hence useless as propagators of irreligion, the campaign against religion lacked a substantial group to advance the new tidings. This soon changed.

The League of the Militant Godless was founded in 1925 under the leadership of the veteran atheist Emelyan Yaroslavsky, founder editor since 1922 of a weekly called 'The Godless'. The league consisted of Party members, hooligans from the Komsomol youth movement, immature workers and army veterans.[22] Its members fanned out in cell-like groups, although their propaganda was occasionally more dramatically re-inforced by aircraft deliberately buzzing those churches still in use, or by the arrival of a train called 'The Godless Express' bringing light through puffs of engine smoke to the dark vastness of the Russian countryside.

Because of the pervasiveness of rural illiteracy, atheist propaganda was reliant upon the spoken word and visual images as well as an avalanche of print aimed at activists. Crude dramatics in the style of agitprop took the customary Bolshevik form of a Punch and Judy-like contest between the powers of good and evil, light and darkness, a Manichean concept of the world that was itself ironically much indebted to a religious view of things. Debates were organised between atheists and priests, whose outcome sometimes included the latter admitting their 'deception' and dramatically throwing off their clerical costume. More often than not, the peasant audiences for these charades took the side of religion against atheism. Priests won debates with their poorly educated Bolshevik interlocutors, which contributed to the replacement in 1928 of debates by lectures where there was no opportunity to contest the message. Propaganda about priestly parasitism, which dovetailed felicitously with conventional peasant anticlericalism, was gradually superseded as the 1920s progressed by more sinister accusations against 'kulak-priestly terror', a formula aimed at eradicating the last spiritual refuge for farmers dragooned into collective farms in the early Stalin era. Churches were closed and vandalised, or turned over to secular uses, for example as cinemas. Their bells were taken away and smelted and their crosses hauled from the roofs with grappling irons and ropes. Some of the biggest churches were blown up. The attack on the large number of religious festivals was moralised. Many festivals required weeks of fasting as a prelude to day after day of inebriation. The Bolsheviks claimed that people were too weak to fast – they had of course created famine conditions – while arguing that mass drunkenness was ruining productivity.

Counter-festivals were organised to obliterate saints' days and other religious feasts, with the godless going into overdrive each Easter and Christmas as they tugged on their phoney tiaras and mitres. The celebrity theatre director Vsevelod Meyerhold was co-opted to choreograph these events – the same Meyerhold who would subsequently write to the Soviet prison authorities with his left hand, since his torturers had broken his right arm just after urinating down his throat. He was then shot.

The anti-Christmas celebrated in Moscow and over four hundred other towns from 25 December 1922 to 6 January 1923 was a particular nadir, with clowns mocking God, a figure of God embracing a naked woman, and mock-priests and rabbis chanting indecent liturgies. This culminated in images of Buddha, Christ, Mohammed and Osiris being burned on a bonfire. Komsomol 'carol singers' went from house to house singing an adapted version of the Christmas Troparion of the Orthodox Church: 'Thy Komsomol Christmas / Restoring to the world the light of reason / Serving the workers' revolution / Blooming under the five-pointed star / We greet thee, sun of the Commune / We see thee on the heights of the Future / Russian Komsomol, glory to thee!'

As this carol indicates, the carnivalesque, allegedly playful aspects of Bolshevik cultural utopias had an intolerant, sinister aspect that was as inherent in the socialist project as the coercion and repression that were coeval with the regime, and integral to its revolutionary iconoclasm and Manichean, Red-and-White worldview. To detach utopian dreams from terror or to regard them as a colourful 'if-only' before the onset of Stalin's grey 'Thermidor' is to indulge in vicarious utopianism from the safety of the modern Western campus. Bolshevik utopianism, it has been argued, oscillated between an innate and pervasive peasant desire for dignity, equality and justice and attempts to create militarised oases of order that tantalised aristocrats infatuated with nineteenth-century Prussia. The latter's enthusiasm was then adopted by technocrats much taken with a Fordist or Taylorist fantasy world, in which robot men had no names but numbers, a vision brilliantly satirised by the Russian novelist Zamyatin.[23] Actually, Bolshevism was also the legatee of a wider left-wing mythology that stretched back to the Jacobins, and that was incorporated into Russia's own sectarian and conspiratorial traditions. While one should be careful not to mistake the choreographed images of Bolshevik films about 1917 as a faithful reflection of the chaotic reality, there was a mythic plot-line in the story as it unfolded, consisting of storming certain key buildings, whether palaces or prisons, the renaming of ships, streets, and squares, and the importance of signs, songs and symbols. In other words, the plot had largely been scripted in Paris in 1789. There was nothing especially 'playful' about the sentiments expressed in the old/new utopian anthem, Lavrov's 1875 'the Workers' Marseillaise':

> To the parasites, to the dogs, to the rich!
> Yes and to the evil vampire-Tsar!
> Kill and destroy them, the villainous swine!
> Light up the dawn of a new and better life![24]

Many of the phenomena encountered in this book's predecessor *Earthly Powers* in the discussion of revolutionary France were repeated in revolutionary Russia, as part of a similar drive to make a permanent psychological and cultural break with the old order. There was a similar onomastic revolution involving the renaming of squares, streets, ships and so forth. The quintessential Communist emblem, the hammer and sickle, was widely used during the February Revolution before it was appropriated by the Bolsheviks. People began exchanging names that had stigmatised them, such as 'Lackey', 'Idiot' or 'Romanov', in favour of personalised statements of ideological fervour, such as 'Citizen', 'Democrat' or 'Freedom'.[25] This process was institutionalised in new 'Red' rites of passage, that is pseudo-christenings in which babies were 'Octobered' as 'Avangarda', 'Octobrina', or 'Spartak' with 'Giotin' (guillotine) and 'Robesper' (Robespierre) being direct references to two notorious names from the French Revolution. One such ceremony to mark the birth of a girl in Nadezhdinsk in 1923 included the following declaration by the participants:

> We cover thee not with a cross, not with water and prayer – the
> inheritance of slavery and darkness – but with our Red banner of
> struggle and labour, pierced by bullets and torn by bayonets . . .
> We bid the parents of the newborn child: bring up thy child to
> be a devoted fighter for the liberation of the toilers of the entire
> world, an advocate of science and labor, an enemy of darkness
> and ignorance.[26]

There were corresponding efforts to institute Red weddings and funerals, the latter taking the form of cremation, which was clean and scientific – the first dedicated installation being provocatively installed in Moscow's Donskoi monastery in 1926. Since such secular civil ceremonies lacked a transcendental dimension, the customary etiquette and days of ambient drunkenness that characterised the traditional ceremonies, they were not a success. Mass revolutionary festivals were an attempt to impose one narrative on the chaotic events of the Revolution, which could then be permanently conjured up ad infinitum, in a forlorn attempt to relive that liberating moment while the enthusiasm that had driven it died as choreographed rigor mortis set in. These festivals began with the May Day celebrations in Petrograd in 1918, events that managed to incorporate some of the enthusiasm of a carnival, with dancing and fireworks as well as more choreographed march-pasts. Within six months, this loose-knit affair had been superseded by the more planned arrangements in Moscow when on 7 November the regime celebrated the first anniversary of the Bolshevik coup, with highly organised parades and rituals whose function was to put Lenin at the centre of the proceedings, however much he may have disdained such developments. Soon it would clear vast open spaces and wide boulevards in the major cities to celebrate armies of marching men and the muscular frames of a latter-day pagan body-cult, the two things being interlinked in this song to 'physical culture' (Fizkul'tura):[27]

> So your body and soul can be young,
> Can be young, can be young –
> Don't shrink away from the cold or the heat,
> Temper yourself, like steel!
> Fizkul'tura, Hurrah!
> Fizkul'tura Hurrah!
> Fizkul'tura Hurrah! Hurrah!
> When the hour comes to bash all our enemies,
> To drive them from our borders, be prepared!
> Left! Right! Don't hang back! Don't be slack![28]

Utopian cults of Promethean man and his machines, of electricity, tractors and speeding trains, lacked the affective, focused power of a single God. Lenin commented sarcastically on the adulation he received from immediate colleagues, particularly at his fiftieth-birthday celebrations, but he was powerless to stop its growth, not least because such a cult fulfilled a popular demand based on historical and psychological expectations. The combination of failed assassination attempts and strokes, together with 'miraculous' recoveries, Lenin's own undoubted tactical political skill as leader of what was at once a beleaguered and global revolutionary movement, and the ways in which simple people in remote villages regarded him as either demonic or a miracle-worker meant that by May Day 1918 he was being referred to as *vozhd'* or supreme leader, or simply (capitalised) Leader by his own closest colleagues.[29] After a failed assassination bid, Lenin's colleagues talked about him as if he were a mortal god: 'Lenin's long years in emigration was the trial of an ascetic ... and he came to be the apostle of world communism ... Lenin became a leader of cosmic stature, a mover of worlds ... He is really the chosen one of millions. He is the leader by the Grace of God. He is the authentic figure of a leader such as is born once in 500 years in the life of mankind,' wrote his colleague Zinoviev. Busts of the leader were despatched to twenty-nine cities. Posters showed him larger than the sun, with his arm outstretched delivering a benediction.

This God-in-the making woke for the last time at 10.30 a.m. on 21 January 1924 in his country retreat at Gorky. Even for a chronic invalid, Lenin felt terrible, and spent a listless day in bed, until his condition became highly unstable late in the afternoon. After a massive stroke, he was confirmed dead shortly before 7 p.m. The coffin was taken to Moscow. Lenin's funeral took place six days later on the coldest day of the year with hundreds of thousands of people focusing the myriad sufferings of the past years in cathartic solemnities before the body of the man largely responsible for them. Trotsky led the cortège, while Stalin trudged manfully beside the coffin. Stalin spoke on the eve of the funeral in tones he had learned at Tiflis theological seminary, where he had gone to receive a cheap education – rather than to join the priesthood – since he had lost whatever religious faith he had at thirteen:

> Leaving us, Comrade Lenin ordered us to hold high and keep pure the great calling of member of the party. We vow to thee, Comrade Lenin, that we will honour this, thy commandment.
>
> Leaving us, Comrade Lenin enjoined us to keep the unity of the party like the apple of our eye. We vow to thee, Comrade Lenin, that we will with honour fulfil this, thy commandment.
>
> Leaving us, Comrade Lenin enjoined us to keep and strengthen the dictatorship of the proletariat. We vow to thee, Comrade Lenin, that we will with honour fulfil this, thy commandment . . .[30]

Krupskaya's wish that her husband be interred with other old comrades was ignored in favour of mummifying his corpse, a step apparently inspired by worldwide fascination with the contemporary excavation of Luxor and discovery of the tomb of the pharaoh Tutankhamen, although the intention was to preserve for eternity what Robert Service has dubbed 'Saint Vladimir of the October Revolution'.[31] Lenin's mummified corpse was displayed in a temporary timber mausoleum in the Wall of the Kremlin before this was replaced in 1930 by a permanent stone structure. The design reminded one Russian commentator of the tomb of King Cyrus near Murgaba in Persia, although the model was actually the mausoleum of Tamerlane.[32] The prime movers in the preservation of Lenin's body were Bonch-Bruevich, Leonid Krasin and Lunacharsky, ironically all erstwhile God-builders who had clashed with Lenin on this very issue. They formed an 'Immortalisation Commission'. The reasons for Lenin's mummification were several. His early death, probably brought about by chronic bureaucratic overwork that he had been unaccustomed to in the earlier decades of his life, was a metaphor for the years of revolutionary élan and enthusiasm that were ineluctably passing away. Mummification meant that the moment would exist in this curious symbolic form throughout time. His spirit would also endure in the Party: 'Lenin lives in the heart of every member of our Party. Every member of our Party is a small part of Lenin. Our whole communist family is a collective embodiment of Lenin.'[33] The aura of this dead St Vladimir would spread to his lesser successors, who henceforth were in control of what he had or had not said or written during his lifetime. Significantly, Stalin managed to gain influence over the fledgling Lenin Institute at the Party's Sverdlov university, and through *The Foundations of Leninism*, in which *he* explained Lenin's ideology to the new Party intake, thereby establishing himself as guardian of the canonical texts.[34]

And what was the net result of this vicious campaign against religion? The Party-state could certainly deploy more force, and did so against the Orthodox clergy. But the ranks of the militant godless waned as quickly as they had waxed, and they were usually filled with the intellectually low grade in the first place. Peasants, whether on the land or newly transplanted to the cities, found ways of resisting this assault on their beliefs, perhaps by sending grannies to obstruct four-eyed student atheists or using loopholes in the law to retain use of a church. Committed religious believers became more entrenched in their faith, while the more casually secure fell away, probably without turning to the dominant secular creed.

The Marxist Benito Mussolini was a rising star on the revolutionary left-wing of the Italian Socialist Party. From 1912 he was a member of its Executive Committee and editor of *Avanti*, the Socialists' principal newspaper, journalism being Mussolini's true métier. He hailed from the Romagna, one of the most anticlerical regions of Italy, a trait fully reflected by his anarchist farrier father, even though Rosa Maltoni, his schoolmistress mother, was rather pious. Her story would come in useful later when Mussolini required a more conventional background. Unlike Hitler, Mussolini was widely read in modern thought, as we know from the books he borrowed in 1902–4 from Geneva university's library. He read three foreign languages, French, German and English, although he spoke only the first two fluently. Much of this reading was in aid of a public debate with an Italian Protestant minister on the proposition 'God does not exist: science proves that religion is an absurdity, is actually immoral and a disease among men'. Mussolini would remain vociferously anticlerical – his first publication was a book called *God does Not Exist* – but he would also develop a regard for the political utility of religion.

Although during his Swiss sojourn Mussolini began to acquaint himself with Marxism, his imagination was stirred by social Darwinism, Friedrich Nietzsche and Georges Sorel, a combination which led him to believe that the real revolution would be in the realms of culture and values. He remained an increasingly maverick socialist until the Great War, but his reading of Nietzsche led to intellectual tensions that Fascism would ultimately resolve, since how exactly was the superior New Man of will to be reconciled with a philosophy based on egalitarianism and the masses? His concern with culture and values led Mussolini to a quasi-religious conception of politics, in which a dedicated elite would help regenerate mankind from the social and spiritual ills that were commonly held to debilitate it in the closing decades of the nineteenth century. The problem was that no such elite seemed to exist.[35]

By 1912, Mussolini had become critical of the reformist pragmatism of the Socialist majority, regarding the act of violent revolution with the sort of limitless expectation with which syndicalists regarded the general strike or extreme Italian nationalists viewed a major international conflict.[36] Both his volatile temperament and his voracious reading led him to stray further and further from the Marxist worldview which he thought was too preoccupied with external forms and not enough with the moral content of the inner man. In 1914 he founded his own paper, *Popolo d'Italia*, returning from war service to establish five years later a new political grouplet called the Fasci di Combattimento. For the war had created the elite that Mussolini had hitherto sought in vain. His highly eclectic political philosophy alighted upon a viable vehicle to express an anthropological revolution from which would emerge a new Fascist man.

As well as destroying the myth of international proletarian solidarity, the Great War created the affective conditions in which the Fascist credo might resonate, especially since some of its values were transpositions of the experiences of the wartime 'trenchocracy', the holy warrior band that lived and perished together on some bleak mountainside fighting the Austrians. Fascism began life as a rag-bag militia amid the bohemians, ex-soldiers, schoolboys and students of urban north Italy; it only discovered its true vocation in the squads formed to terrorise the left in the 'Red' rural provinces of the Po valley and central Italy. The militancy of rural labourers, railwaymen, and urban proletarians ensured that such people as foremen, gang masters, ticket sellers and station masters flocked to the Party of Order.[37] While bankrupt liberal governments seemed impotent in the face of what was actually a divided socialist movement, whose threat never constituted more than an exasperating and often gestural nuisance, the Fascists went about matters with castor oil, clubs, petrol bombs and explosives. In this fashion, provincial Fascist bosses achieved notoriety and power, while Mussolini – whose leadership went rarely uncontested in these circles – simultaneously smoothed Fascism's entrée to the ruling political and business and banking elites of Italy. Hitler would work the same dual strategy in Germany by periodically exchanging his revolutionary's animalistic leather jacket for a frock coat. In 1921 thirty-five Fascists – half of them under forty years of age – entered parliament, as part of a National Coalition under the veteran liberal statesman Giovanni Giolitti, who imagined that he could assimilate Fascism, much as he had successfully assimilated other challenges in the past, through compromise and clientelism. Giolitti confidently predicted: 'You will see. The Fascist candidates will be like fireworks. They will make a lot of noise but will leave nothing behind except smoke.'[38]

'Fascism', reminisced Giuseppe Bottai in 1922, 'was, for my comrades or myself, nothing more than a way of continuing the war, of transforming its values into a civil religion.'[39] Mussolini agreed when four years later he proclaimed: 'Fascism is not only a party, it is a regime, it is not only a regime, but a faith, it is not only a faith, but a religion that is conquering the labouring masses of the Italian people.'[40] Membership of the Fascist militias involved the swearing of oaths, the constant affirmation of the sacrificial community, the consecration of holy symbols, and the veneration of both the war dead and the victims of their own rampages in such a manner that the two categories became indistinct. Each crime against opponents tightened the bonds of moral complicity between members of the gang, although their nefarious activities were rarely disturbed by the forces of law and order, with whom the Fascists elided themselves under the banner of a common patriotism.

Much of the evolving public manner – the theatricalisation of the piazza – derived from the operatic regime established in the Adriatic port of Fiume by the capricious and colourful nationalist poet Gabriele D'Annunzio. This self-styled 'Comandante' usurped power there for a year in 1919–20, with the aid of deserters and the connivance of the military authorities. The so-called Constitution of Carnaro promulgated by D'Annunzio included plans for a public political cult, whose centrepiece was to be a circular auditorium capable of containing ten thousand. In the absence of such a venue, D'Annunzio addressed crowds of supporters from a modest balcony, who responded to the question 'A chi l'Italia?' with a thunderous 'A noi!' while his blackshirts emitted a barbaric war cry – 'eia eia alalà' – in a demonstrative break with a bourgeois political class which wore frock coats and winged collars.[41] Things degenerated after that – apparently 'Holocaust City', as the poet dubbed it, witnessed orgies of cocaine and wild sex. After a year, the Italian government decided to oust the poet. A few shells from the battleship *Andrea Doria* sent D'Annunzio packing and cleared the public square. Ungenerously, the Fascists proceeded to treat him as a non-person, although a good deal of their style derived from him.[42]

There was much more to Fascism than political aesthetics, the staged aspect that most tantalises postmodern historians of 'culture' who fight shy of visceral conflicts over social and political power in precisely the manner that the Fascists themselves encouraged. The postmodern Left university is ostentatiously 'apolitical', preferring talk of frontiers, trees and mountains. Fascism itself was an attempt to transcend the narrow horizons of conventional class or interest politics, whether of the left or right, in favour of an all-embracing anti-politics based on a series of potent myths whose veneration was taken to religious heights.

Although the uniform social profile of many Fascists suggested otherwise, they justified their rejection of parliamentary politics on the ground that they, rather than a greying liberal gerontocracy, truly represented the Italian people. The conspicuous youth of Fascist leaders enabled them to posture as the coming wave of Italy's future. As Mussolini had it in August 1922: 'Democracy has done its work. The century of democracy is over. Democratic ideologies have been liquidated.'[43] In a strikingly tasteless metaphor he announced his desire to trample on the 'more or less decomposed body of the Goddess of Liberty'.[44] In place of democracy, Fascism offered a militarised hierarchy, and the abolition of any distinction between the political and the private, the essential totalitarian aspiration, albeit like most aspirations rarely totally realised. Possession of a PNF card became the key to advancement in virtually every walk of life from inspecting fish to awarding literary prizes; the meaning and value of an individual life was weighed in terms of how it advanced the greatness of the state, a form of state-worship that such Catholic opponents as Luigi Sturzo dubbed 'statolatria'. Enemies, real or imagined, would be dragooned by the organs of state power, themselves subject to creeping Fascist control, or by the informal violence of Fascist thugs who continued to operate under licence. A Party that said, 'The fist is the synthesis of our theory,' replaced reasoned argument with violence. Fascism also espoused an anthropological revolution, sometimes pretentiously called 'palingenesis', whose goal was the creation of a new Fascist man, and a species of economic corporatism, which in superficial respects chimed with Social Catholicism.[45]

The liberal Italian state had devoted few resources to the invention of national traditions among a people whose primary loyalties were to the family, their region, and the Roman Catholic Church. Such efforts as the giant wedding-cake monument to the first king seemed ludicrous. The Fascists worked with what lay to hand in a country whose cityscapes were almost designed for public spectacle and which provided a powerful architectural backdrop. In the capital, venerable buildings were smashed down to create marching routes to show off the new Roman goose-step. Virtually every city had a piazza, reached by an avenue from the railway station, which because of the relative proximity of Italian towns could be used to import hordes of semi-professional activists, after careful scheduling and with heavily discounted tickets. The Fascist Party liaised with central and local government agencies who were legally responsible for public holidays and public festivals, and with the Church which commenced some of the ceremonies with a mass.[46]

After 1922 the Fascists reconsecrated and built upon the traditional rites and symbols of patriotism that they inherited, into which they merged much of their own limited political repertory. The national flag became omnipresent, to suggest that the triumph of Fascism had brought about national rebirth, while both the dates of Italian intervention and victory in the Great War became Fascist state occasions, commemorating the dead of the Fascist Revolution – notionally put at three thousand – as well as those of the war, an elision that must have seemed grotesque to surviving veterans of other political persuasions.

In 1926 the regime instituted a new calendar, with October 1922 declared to be the advent of 'Year I' in obvious echo of the Jacobins. It also devoted much effort to remodelling the cycle of public holidays, abolishing not only the socialist May Day, but also the Statuto which commemorated the liberation of Rome from French troops. Instead of this contentious holiday in late September, the government decided to celebrate the Conciliation between state and Church in late February. It would be tedious to review each Fascist spectacle, most of which went through several evolutions to reflect the regime's current requirements. Apart from the hyperbolic accounts in the government's own newspapers, we are in the dark as to how ordinary people reacted to these events, whether they were enthused by them or found them a tedious nuisance.

The anniversary of the March on Rome in late October 1922 was carefully stage-managed to transform what had been an exercise in political bluff into an event with multiple symbolisms. In Rome itself, ancient churches and houses were demolished to make way for the triumphal Via dell'Impero, linking the Colosseum with the Piazza Venezia. Key cities were used to exemplify different stages in the Fascist version of recent history: Milan (the birthplace of the movement); Cremona (where the largest number of Fascist martyrdoms had occurred); Bologna (scene of pitched battles with the left); Perugia (where the March had been co-ordinated) and Rome (site of the memorial to the Unknown Soldier and home of national government). The ceremonies incorporated the Church by commencing with an early-morning mass; they involved the armed forces (as well as Fascist militias) with fly-pasts and military parades, while women were honoured as the mothers and widows of

Fascist victims of 'Red' aggression and of the war dead, who were also represented by surviving veterans. Eventually, the commemoration of the March fused with the celebration of the battle of Vittorio Veneto which replaced armistice day on 4 November expanding into a week of commemorative festivals.

Successive Fascist Party secretaries, Roberto Farinacci, Augusto Turati, Giovanni Giurati and Achille Starace, elaborated the cultic elements of the Fascist faith, with eternal flames, votive woods and rituals for handling the holy banners, flags and pennants from the epic period of struggle. The talismanic image became the Roman lictorial (and Jacobin) Fasces, consisting of an axe bound in a bundle of rods, which began to crop up on the sides of public buildings and the walls of motorways. Commemorative meetings generated a range of Fascist memorabilia, such as medals, plaques and ribbons, while commercial firms cashed in with such offerings as a perfume called Fascio. Buildings for the innumerable formations that comprised the Fascist Party proliferated, each opening being the occasion for a solemn ceremony of dedication, as were the opening of dams, highways, public buildings and factories. In 1932 Starace decreed that each Fascist headquarters should have a tower and bells, which would summon the faithful to special Party occasions.

Fascism is associated with a visual culture derived from Roman antiquity and modern Futurism, as manifested in the rather interesting buildings – such as the square pyramid – at EUR, the Universal Roman Exhibition created in a suburb of the capital. Nevertheless the contribution of a vulgarised understanding of Church history to the spirit of Fascism should not be neglected. Commemoration of Fascist martyrs freely confused Fascism with Christianity, which the presence of so many clerics at such rituals did little to dispel, while Fascist memorabilia owed much to pious kitsch. But while one should not force these parallels, the history of Fascism was also congruent with a crudely anticlerical version of the history of the Church, to which any north European Protestant would happily have subscribed. It was a grotesque parody of what the Church was, in its sinister way reminiscent of Signorelli's Antichrist in Orvieto cathedral. What they celebrated in the Church provides clues to the desired Fascist temperament.

The Fascist squads were the totalitarian community in embryo, a dedicated masculine band motivated against impossible odds by the intensity of their faith and loyalty to one another. Thugs who managed to get themselves killed brawling with socialists became political martyrs, whose lengthening list of names resonated at ever more elaborate Fascist ceremonies, at which they were solemnly registered as 'Present'. Commemoration of such martyrs would be a permanent feature of future Fascist ceremonial, with votive woods and sacred parks dedicated to their memory, which in turn were watched over by guards of honour.[47]

While still a grimly dedicated minority, Fascists regarded themselves as missionaries: 'scattered in the unexplored regions of the world among savages and idolatrous tribes'. In this version, Mussolini became a Messiah figure 'who began speaking to fifty people and ended up evangelizing a million', although that was only clear in retrospect. The missionaries metamorphosed into crusaders, liberating Italians from the infidel socialists who had temporarily occupied the patria, with the aid of such weapons as the 'holy Manganello', the wooden club which 'brightened every brain' into a glistening bloody pulp. In their wake they left the citadels of the infidels (Socialist offices) in flames, with everything that could be smashed broken.

Expansion was a product of disciplined ruthlessness. Intelligent opponents of Fascism, such as the liberal journalist Giovanni Amendola, recognised that Fascism differed in intensity and ambition from traditional political movements: 'Fascism wants to own the private conscience of every citizen, it wants the "conversion" of Italians . . . Fascism has pretensions to being a religion . . . the overweening intransigence of a religious crusade. It does not promise happiness to those who convert; it allows no escape to those who refuse baptism.' The Fascists gloried in the alleged intolerance of the medieval preaching orders, notably the Dominican friars, turning purblind fanaticism into a Fascist virtue. Notoriously, in 1926 Roberto Davanzati proudly announced: 'When our opponents tell us we are totalitarian, Dominicans, implacable, tyrannical, we don't recoil from these epithets in fright. Accept them with honour and pride . . . Don't reject any of it! Yes indeed, we are totalitarians! We want to be from morning to evening, without distracting thoughts.'[48] The Church's destruction of unrepentant heretics became the model for Fascist treatment of political dissidence: 'Fascism is a closed political party, not politically but religiously. It can accept only those who believe in the truth of its faith . . . As the Church has its own religious dogmas, so Fascism has its own dogmas of national faith.'

Alfredo Rocco made the totalitarian analogy between the Church and Fascism explicit:

One of the basic innovations of the Fascist State is that in some respects, like another centuries-old institution, the Catholic Church, it too has, parallel to the normal organization of its public powers, another organization with an infinity of institutions whose purpose is to bring the State nearer to the masses, to penetrate them, organize them, to look after their economic and spiritual well-being at a more intimate level, to be the channel and interpreter of their needs and aspirations.[49]

From here it was a relatively short step to lauding the more sanguinary episodes in the history of the Catholic Church as they have settled in vulgar memory. Fascism had learned 'from those great and imperishable pillars of the Church, its great saints, its pontiffs, bishops and missionaries: political and warrior spirits who wielded both sword and cross, and used without distinction the stake and excommunication, torture and poison – not of course in pursuit of temporal or personal power, but on behalf of the Church's power and glory'.

The militant orders of the Counter-Reformation became paradigmatic as the Fascists attempted to settle down the thuggish squadrisiti into state-controlled paramilitary formations, a task subverted by the desire of the provincial Fascist bosses to retain a measure of autonomy vis-à-vis the central government and the prefectoral regional administration. The Fascist youth organisation would be modelled after the Society of Jesus, with the operating credo 'Believe, Obey, Fight', while Fascism's protean and pretentious doctrine would be condensed into a simple catechism for schoolchildren.

Official statements of Fascist doctrine were routinely characterised by a pretentiously woolly religiosity, whose opacity (in any language) faithfully reflected the philosophical tone of the times. In 1932 Mussolini himself claimed that 'Fascism is a religious conception in which man in his immanent relationship with a superior law and with an objective Will that transcends the particular individual and raises him to conscious membership of a spiritual society.' He was careful, however, to eschew the vaulting ambitions of either the Jacobins or Bolsheviks: 'The Fascist State does not create a "God" of its own, as Robespierre once, at the height of the Convention's foolishness, wished to do; nor does it vainly seek, like Bolshevism, to expel religion from the minds of men; Fascism respects the God of the ascetics, of the saints, of the heroes, and also God as seen and prayed to by the simple and primitive heart of the people.'[50]

Mussolini's first statement, 'The Fascist State does not create a God,' was tactically astute but also unduly modest. Mussolini may have had to share power with the monarchy, and never entirely mastered either state institutions like the army or the bigger personalities in what his PNF opponents regarded as *their* party, but there were more ways than one to skin a cat. Like his contemporary, Edward, Prince of Wales, Mussolini was astute enough to see the advantages of using the same media that made Hollywood filmstars into demi-gods: trashy biographies, flashy magazines and newsreel films. In contrast to earlier Italian politicians, who had the charisma of aged lawyers, Mussolini was a virile and omnipresent figure: fencing, riding, skiing or wrestling submissive lions and tigers in the zoo. Since, unlike Hitler, Mussolini had learned to drive and could pilot aircraft, he was perpetually seen rushing about, at the controls of planes or speeding by on motorbikes or in racing cars, this activist haste being essential to the image of any self-respecting dictator in the 1930s. Like Hitler, Mussolini was also a 'workerist', although in common with the Führer he had successfully avoided honest toil most of his life. Film of Mussolini stripped down to his bronzed barrel-chest grounded him among sweaty peasants in the Fascist 'Battle for Grain', for Mussolini was probably the first Italian leader to venture so close to ordinary Italians on their own home patch. As a French journalist noted, they responded by waiting for hours at crossings or stations as he sped through obscure places. The evanescent nature of modern celebrity was countered by associating the regime and its leader with Roman antiquity, the most grandiose setting imaginable. The Duce became the DUX. Rome also provided an almost unparalleled model of creative imperialism, and of the complete subordination of everyone and every-thing – including religion – to the higher interests of the state. The allure of Rome, eternal and universal, was therefore irresistible, especially after the regime launched its imperial ventures in the Horn of Africa in the mid-1930s. It was then that it simply craved mass adulation, which it got in the form of the 'adunate nazionali', the four mega-rallies held between 1935 and 1937 to indicate the nation's defiance of the League of Nations.[51]

Mussolini's personal charisma antedated his involvements with Fascism – he was known as Duce during his long socialist apprenticeship – and increased as his sole claim to leadership was contested by rival Fascist barons. Although they curtly told him, 'Fascism is not summed up in you,' the provincial bosses found him indispensable as a broker in their own intrigues against rivals. The Fascist Party ensured that Mussolini occupied centre-stage in the emerging political cult. Intellectual sycophants and propagandists characterised him as a prodigy of genius in terms that would not have embarrassed Stalin: messiah, saviour, man of destiny, latterday Caesar, Napoleon and so forth. His brother Arnaldo became head of a new School of Fascist Mysticism exclusively devoted to the man and his thought. If this constituted the rarefied heights of the personality cult, the popular base consisted of the usual ways in which people project their hopes and longings on to one charismatic figure. The Fascist Party could orchestrate those sentiments, through such devices as the judicious use of dictatorial bad timing, in a country where the trains allegedly ran on time, but they did not create them. Like Hitler, and for that matter medieval monarchs, people dissociated the infallible dictator from a party whose corruptions and oppressions they increasingly detested.[52]

So far we have deliberately postponed discussion of how the political religion of Fascism related to the Catholicism of the Church. Mussolini ascribed inordinate power to Catholicism, and acknowledged that an outright clash with the Church would be disastrous for the Fascist regime. That was why he was so keen to resolve the Roman Question, and why he eschewed both the atheist Bolsheviks and the civic cultism of the Jacobins, to which he added – in a brief moment of enlightenment at a time of estrangement from Nazi Germany in 1934 – the 'writing [of] a new gospel or other dogmas . . . overthrowing old gods and substituting them with others, called "blood", "race", "Nordic", and things of the kind'. But having lanced the boil of the Roman Question, a party that deified the state, made explicitly totalitarian claims over the minds and morals of the young and that was so profligate in its own use of religious metaphors, vocabulary and sentiments was unlikely to settle for mere cohabitation with the Catholic Church. However much dewy-eyed Catholic Fascists (and there were many of them) may have seen their values embodied in Mussolini's regime, relations between Church and state were marked by multiple tensions, which no amount of flattering references by Fascist intellectuals to the Latinity and universality of the Roman Church could conceal. In fairness, it also should be added that, whether on the issues of abortion, contraception, the role of women or which side to support in Spain, there were also areas of broad agreement.

The rapprochement or Conciliation between the (Fascist) Italian state and the Vatican had a history that antedated the advent of Mussolini.

We need to know something of this to understand how the Church, and Catholics more generally, responded to Fascism. Catholic politics in the late nineteenth century consisted of an intransigent strain, which, true to the spirit of Pius IX, abstained from any contaminating involvements with national politics and with those known as 'conciliatorists' or 'clerico-moderates', who sought an accommodation with the less anticlerical elements in the ruling liberal establishment. Although the papacy was hostile to the creation of a Catholic political party, on the ground that it might slip out of clerical harness, the growing menace of socialism encouraged a more emollient response to moderate liberals who felt equally threatened by the socialists, anarchists and a rowdy artistic avant-garde, notably the Futurists.

In 1909 the liberal government itself sought Catholic support to defeat a challenge from a left-wing bloc of radicals, republicans and Socialists. Pius X responded by relaxing the Holy See's blanket ban on voting in national elections so as to strenghen the liberal vote in Catholic northern Italy. In 1909 thirty-eight clerico-moderates entered parliament. Following the introduction of universal suffrage in 1912, the so-called Gentiloni Pact delivered Catholic votes to a couple of hundred government candidates in northern Italy, who privately promised to respect Catholic interests.

Catholic opinion was represented across the left–right ideological spectrum; the material concerns of Catholic bankers and landowners were very different from those of landless labourers. On the left, the Christian Democrats were situated well to the left of advocates of intransigent theocratic corporatism, but to the right of socialism, although some 'Red Catholics' acknowledged the need for both labour unions and strikes to attract a mass following. On the right, some Catholics listened to the siren voice of a nationalism that spoke of an 'ethical state' which recognised the importance (and Romanity) of Catholicism; whose quest for social order chimed with the doctrine of corporatism; and whose desire for empire in Africa could be construed as a crusade on behalf of Catholicism in the minds of the gullible. In fact, the Nationalists were hostile towards what they regarded as a pro-Habsburg papacy, and at most thought of the Catholic masses as biddable footsoldiers, rather in the delusional way that some neo-conservatives regard the Christian right in the contemporary US. This did not stop influential members of the hierarchy, the Catholic press and the Bank of Rome from supporting the 1911–12 war in Libya.[53]

In 1914 the Nationalist lawyer Alfredo Rocco wrote presciently in a pamphlet:

> The Nationalists do not believe that the State should be an instrument of the Church; instead they believe that the State must assert its sovereignty also in regard to the Church. Since, however, they recognize that the Catholic religion and Church are most important factors of national life, they wish to watch over Catholic interests as far as possible, always safeguarding the sovereignty of the State. And at this stage of Italian life, such protection should take the form of respect for the freedom of conscience of Italian Catholics, against the antireligious persecutions of anticlerical democrats. In the future it will perhaps be possible to go farther and establish an agreement with the Catholic Church, even if only tacit, by which the Catholic organization could serve the Italian nation for its expansion in the world.

Rocco would lead the Fascist regime's negotiations of the 1929 Lateran Treaties, the pamphlet already anticipating what both sides thought they would be gaining.

The crisis over Italy's intervention in the First World War deepened the ideological cleavages within Catholicism, at a time when the political system had to weather the displacement of an elite, liberal-dominated politics, whose venial sins at least guaranteed continuity and stability, with much more volatile mass arrangements. Catholic opinion was divided by the question of war. The papacy had respectable reasons for neutrality, since the war crisis had been sparked by the murder of the heir to the most important Catholic throne in Europe, while that Empire's largely Protestant German ally had then invaded overwhelmingly Catholic Belgium. Pope Benedict XV also feared that a cataclysmic conflict would result in a vast social revolution, of which parts of Italy had a brief foretaste in 1914. The pope's neutralism was not shared by either left-wing Christian Democrats or clerico-moderates and Nationalists, who supported prime minister Salandra's fateful decision to take Italy into the war on the side of the Entente.

Catholic support for, and participation in, the war removed the final obstacle to their direct involvement in Italian politics. The new Partito Popolare Italiano (PPI) was founded in January 1919 and led by the Sicilian priest don Luigi Sturzo. An enormously attractive and intelligent man, Sturzo was handicapped by the fact that his clerical status disbarred him from parliament, while ensuring that he was subjected to an ecclesiastical discipline that reflected the Church's serpentine political manoeuvres. The PPI was a self-styled non-confessional party (necessarily of Catholics) rather than one dominated by priests. In its first electoral outing in 1919, it won 20 per cent of the votes and a fifth of the seats in parliament, support being especially strong in the traditionally White northern Catholic heartlands of Lombardy and the Veneto.[54] Since the largest party in parliament – the Socialists – refused to enter coalitions with 'bourgeois' parties, the PPI participated in six such coalition governments formed between July 1919 and October 1922. It must have been an increasingly demoralising experience, since in those 'Red Years' sections of the armed forces and the police ceased to be reliable agents for dealing with mounting Fascist violence, which began to affect the apparatus of the Partito Popolare as well as that of the left. Entire regions of Italy simply slipped out of the government's control and into that of local Fascist bosses. A further source of disillusionment was that Pius XI, elected in early 1922, disapproved of the PPI's strategic neglect of the Roman Question and its concentration on secular political issues, which in his eyes made it 'no better than the liberals'.

The PPI had a right and a left wing. The former was associated with Stefano Cavazzoni, count Giovanni Groscoli and father Agostino Gemelli, the latter with Guido Miglioli. The more right-wing members of the PPI began to contemplate a government of National Concentration that would include Fascists; some effortlessly metamorphosed from 'clerico-moderates' into 'clerico-Fascists'. Other prominent PPI leaders were at pains to distance themselves from Fascism. As its rising star Alcide de Gasperi explained:

> As opposed to this Right, the PPI are a party of the Left. In the politics of everyday, they want common law protected, and they do not admit the legitimacy of reprisals and punitive expeditions ... They recognize the value of labor unions and of the co-operative movement, and in fact cooperate with their greater development. The Fascists instead too often lend themselves to the support of the proprietary class ... Finally the Right, blind worshipper of the unitary State, is opposed to all political and administrative decentralization and renounces all local autonomy, considering it destructive to the national framework.

Can we, proponents of decentralization, of autonomy . . . orient ourselves toward the Right?[55]

One striking abdication of responsibility was when Filippo Meda, the leader of its parliamentary caucus, thrice declined forming a government on the ground that it would interfere with his law practice.[56] While Mussolini intrigued with the leaders of the liberal elite, his paramilitary forces took over ever larger swathes of the country. Summoning him to form a government probably seemed to many of that elite something of a relief. The Vatican pressurised the Popolari to support the government he formed in November 1922; two PPI figures became ministers in the coalition government, although they were dismissed by Mussolini the following spring when the PPI congress in Turin reaffirmed constitutional values. In the summer, the papacy used a tame newspaper to encourage Sturzo to resign as PPI general secretary, after Fascist threats to exploit his anomalous position for an attack on the clergy as a whole. The parliamentary PPI then split over Mussolini's controversial Acerbo law whereby the party gaining a narrow plurality of votes cast would be rewarded with an absolute majority in the chamber. The former 'clerico-moderates' on the PPI right broke away to form a pro-Fascist Unione Nazionale. Don Sturzo went into exile in England, while de Gasperi became leader of the rump PPI that continued to espouse the politics of liberty, advocating co-operation with the Socialists in the wake of the regime's murder of Giacomo Matteotti. This strategy was actively blocked by Pius XI, who felt that such an arrangement would benefit only the Socialists. The PPI deputies joined the so-called Aventine Secession of 150 parliamentarians who for sixteen months boycotted the Fascist-dominated chamber in protest at Matteotti's murder. When the PPI deputies eventually returned to parliament, the Fascists forced them to withdraw. De Gasperi had resigned as Party secretary a month earlier. The PPI was dissolved in November 1926 and de Gasperi went into internal exile.

The road to the 1929 Concordat and Lateran Treaties was paved by small but significant gestures whose ulterior motive was to render the PPI irrelevant long before it was abolished. The librarian pope was presented with the Chigi collection of books and manuscripts, purchased by the Italian government in 1918. The Vatican removed its interdict upon a chapel in the Quirinal Palace, enabling the king's eldest daughter to marry there a few days later. Crucifixes reappeared on the walls of classrooms and lecture theatres, with an imposing wooden cross in the middle of the pagan Colosseum. Holy Week in 1925 went smoothly, due in no small part, as Pius XI acknowledged, to the co-operation of the Fascist government. Since not even Mussolini had the effrontery to grace the seven centuries' anniversary of the death of St Francis of Assisi, secretary of state Merry del Val had to make do with the education minister. But in 1925 Mussolini made a point of marrying Donna Rachele in church, a decade after their civil union. Totally ignoring their own Party programme, the Fascists restored properties once confiscated from religious orders, bailed out the ailing Bank of Rome, increased clerical salaries and modified the law in directions that benefited the Church. The regime closed fifty-three brothels and suppressed the freemasons – widely regarded within the Church as the dark power behind liberal anticlericalism – notwithstanding the fact that the masons had contributed generously to Fascist Party coffers, while several Fascist hierarchs, including Acerbo, Balbo, Farinacci and Rossi, were of the apron-and-trowel persuasion. In 1931 the regime banned abortion and beauty contests, measures that were welcomed by the Church.[57]

The first formal initiative in solving the perennial Roman Question began in 1925 with the appointment of a commission designed to soothe certain neuralgic sensitivities in relations between Church and state. Despite the fact that Pius XI disowned the commission, changes in the government – the dismissal of the anticlerical Roberto Farinacci as Party secretary and the appointment of the Nationalist lawyer Alfredo Rocco as minister of justice – facilitated contacts. Two lawyers handled the talks, Francesco Pacelli, brother of Eugenio, at that point nuncio to Germany, and Domenico Barone, a senior civil servant in Rocco's Justice Ministry. These men resolved such issues as the sovereign status of the Vatican City and the extraterritoriality of papal basilicas and palaces; a compensation package that the papacy was to receive in lieu of its lost revenues from the former Papal States; and guarantees of unimpeded communications between the Vatican and the wider Catholic world. These measures formed the basis of the 1929 Lateran Treaties. Thenceforth the temporal patrimony of the papacy has consisted of a 109-acre territory, roughly comparable in size with London's St James's Park or about a tenth of the area of New York's Central Park. It had its own coinage, garage, postal system, radio transmitter, newspaper and printing press, a jail and a school, a mini-railway line and, of course, separate diplomatic accreditation and the famed Swiss Guard. Vatican Radio (whose transmitter rather than broadcasting station is within the enclave) was intended to underline the Church's role in the wider world.

The miniscule size of the Vatican State was designed to contrast advantageously with the limitlessness of the claim to spiritual power. The wealth of the Vatican was also mythic, as can be seen from the related financial convention. The grant of 750 million lire in cash and a billion in consolidated government stock was urgently needed, even though the papacy agreed to take the cash in instalments and not to sell the stock. During the First World War, pope Benedict XV had given away his own fortune and then the Holy See's ordinary revenue to repatriate prisoners of war and to afford succour to civilian refugees, so that by 1922 the Vatican Treasury consisted of the lire equivalent of £10,000 or roughly US$19,000. Unable to pawn a Bernini, Michelangelo or Raphael, his successor managed to deplete the financial resources still further, with generous donations to those ruined by inflation in Weimar Germany and gifts to the starving multitudes in the Soviet Union. Only the generosity and financial acumen of North American Catholics, who contributed half the papacy's income in the 1920s, staved off financial ruination.

Unlike the Treaty, the Concordat between the Vatican and the Italian state took two years to negotiate. For Pius XI it was a significant step in the re-Christianisation of Italian society, in the re-establishment of a 'Res publica Christiana'. It ended the unified Italian state's usurpation of the right of defunct Italian principalities to veto nominations to bishoprics and many other ecclesiastical offices and to appropriate the revenues of vacant benefices. The state now accorded civil recognition to the sacrament of marriage, which remained indissoluble as it had been under the civil code. The Roman Segnatura, the supreme ecclesiastical court, would henceforth deal with dispensations or nullifications. In other respects, the Church's antipathy to artificial birth-control harmonised with the Fascist state's militant quest for births. Fascism also wanted women on the maternity bed or in the kitchen in ways that conformed with Catholic models. Religious instruction was reintroduced into secondary as well as primary schools, thus negating the wish of the first Fascist education minister to teach older children philosophy rather than religion. The state also agreed to recognise diplomas awarded by pontifical universities. Most importantly, in article 43, the state conceded an autonomous space to Catholic Action: 'The Italian state recognises the organisations affiliated to the Italian Catholic Action in so far as these shall, as has been laid down by the Holy See, develop their activities outside all political parties and in immediate dependence on the hierarchy of the Church for the diffusion and realisation of Catholic principles.' In other words, a state that in May 1929 formally styled itself 'totalitarian' had conceded the Church's right to operate a variety of associations independently of such Fascist organisations as the Balilla youth movement, which had to desist from scheduling its activities to subvert Catholic holidays. Of course, the general climate created by Fascism stealthily leached into the Italian Church itself through something resembling osmosis. Even as it resisted Fascism, the Church tried to keep up with its heroic version of modernity. Under a regime that was ostentatiously virile, the Church endeavoured to 'de-feminise' its own image in favour of a more muscular tone. Clerical novels celebrated priests who were war veterans and athletically built devotees of 'extreme sports' – Pius XI himself being a keen climber.[58]

III SOCIALISM WITH ONE HUMAN FACE

In September 1936, an NKVD secret police agent codenamed 'Volgin' within the Soviet Academy of Sciences recorded a conversation he had overheard between four academics about the future role of the Communist Party. One of these men, an orientalist called Krachkovsky, made the following comments:

> I am almost sure that the president will be Stalin, who will that way be transformed into Joseph the First, the new all-Russian emperor. It's not a question of intentions, but of the general course of history. Communism is becoming the national religion of Russia, just as fascism is becoming the national religion of Germany and Italy, and Kemalism the national religion of Turkey. With all these movements what is characteristic on the one hand is hatred for the pre-existing religions – Orthodoxy, Catholicism, Lutheranism, Islam – and on the other – a cult of the *vozhd'*. For when Stalin is publicly called the father and *vozhd'* of the peoples, then the last line between him and the Führer Hitler is eliminated.[59]

In the Soviet Union, the early 1930s witnessed the replacement of an anonymous collective leadership with the cult of the paramount *vozhd'* –

Stalin – a cult which may have owed something to the leadership's keen appreciation of the role of the contemporaneous cult of the Führer in Germany.[60] Quite independently of that, Stalin felt the psychological need to progress from being 'boss', the bureaucrats' bureaucrat, to 'leader' (*vozhd*) – never an acknowledged position, but all the more charismatically potent for that. The first ominous sign of what was coming occurred in December 1929 with ten days of celebrations to mark Stalin's fiftieth birthday. Three hundred and fifty expressions of joy – including from such non-existent entities as 'the women collective farm workers of Armenia' – were published in newspapers, with the choicest examples of oleaginous sycophancy then being anthologised. In an article entitled 'Stalin and the Red Army', Voroshilov rewrote the history of the Civil War substituting Stalin for Trotsky as the heroic roving troubleshooter who had guaranteed Bolshevik victory.[61] A further wave of adulation coincided with the Sixteenth Party Congress the following summer, leading the American correspondent Louis Fischer to observe:

> A good friend might also advise Stalin to put a stop to the orgy of personal glorification which has been permitted to sweep the country ... Daily, hundreds of telegrams pour in on him brimming over with Oriental super-compliments: 'Thou art the greatest leader ... the most devoted disciple of Lenin' and the like. Three cities, innumerable villages, collectives, schools, factories, and institutions have been named after him, and now somebody has started a movement to christen the Turksib the 'Stalin Railway'.

A Soviet press officer let Fischer know that Stalin's comment on this piece was 'The bastard!'

While modestly disclaiming any intention of creating a personality cult, Stalin took several steps to ensure that one came into being. In 1931 he lured the socialist writer Maxim Gorky back to the Soviet Union from Italy, with a view to Gorky writing his biography. In 1932 the town of Nizhny Novgorod was named in honour of a writer whose juvenilia Stalin compared to Goethe's *Faust*. Despite such flattery, nothing came of the biography, but it spoke chapters regarding Stalin's intent, as did the fact that from 1933 onwards Gorky was forbidden to leave Russia and acquired a secretary cum NKVD agent. Failure here – although biographies there would be aplenty – was paralleled by success on the philosophical front. Stalin was best known as a practical operator rather than as a subtle dialectician although this may be an injustice. Indeed, in the mid-1920s, the director of the Marx–Engels Institute, after hearing him mangle the doctrine of socialism in one country, had said to him: 'Stop it, Koba, don't make a fool of yourself. Everybody knows that theory is not your strong point.'[62] Stalin went about establishing his 'unrivalled' credentials as a theorist with characteristic native cunning. Lenin was deliberately built up as a canonical authority, at the expense of both Plekhanov, the doyen of Russian Marxist theory, and Bukharin, its most adept living exponent, the subtext being that no one should be deceived by the current general secretary's protestations of intellectual modesty. Soon Stalin was part of an illustrious philosophical quartet: Marx, Engels, Lenin and Stalin, from which vantage point he went on to be an avatar of every humanistic and scientific discipline, the master of all he surveyed.

Historians of the Party were the next to feel Stalin's hand on their collar. An article in the journal *Proletarian Revolution* had made some minor criticisms of Lenin's analysis of pre-1914 German Social Democracy, namely that he had been slow to recognise the dangers of inertia in the 'centrism' represented by Bebel and Kautsky, continuing to see them as the great white hope of any future German revolution. Stalin denounced the author of this piece of 'Trotskyist contraband' in a withering letter to the editor that illustrated his concern to be the arbiter of all historical questions, his dismissal of facts as the concern of mere 'archival rats', and, as this already indicates, his automatic resort to abusive generic labels to negate any discussion. If every other Bolshevik leader had to be diminished in the telling of the Party's history, Stalin was initially content to hyphenate Lenin, who, in death as in life, found himself shadowed by his younger alter ego. When *Pravda* celebrated its twentieth anniversary in 1932, Stalin was 'found' to have ghostwritten many of Lenin's contributions, and it was his photograph, and not Lenin's, that accompanied *his* recollections of the paper's early history.[63]

The Stalin cult took off in earnest that year, with Gerasimov's portrait of Stalin addressing the Sixteenth Party Congress and Voroshilov's May Day speech in Red Square. Voroshilov concluded with a rousing 'Long live its [the Party's] Leader, the leader of the workers of our countryside and the whole world, our glorious, valorous Red Army man, fighter for the world proletarian revolution COMRADE STALIN!' References to Stalin proliferated in the news media, increasingly accompanied by such epithets as 'the great leader', 'father of the people', 'the great helmsman', the 'genius of our epoch' and 'titan of the world revolution'.[64] So-called poets strained to depict Stalin as the sun, an eagle, a panther and so forth: 'O Thou mighty one, chief of the peoples, Who callest man to life, Who awakest the earth to fruitfulness, Who summonest the centuries to youth ... O sun, Who art reflected by millions of human hearts' being among the choicest examples of this extensive genre.[65] By 1934 it was possible to anthologise the visual images devoted to Stalin in a book entitled *Stalin. Paintings, Posters, Graphics, Sculpture.* To foster the image of father-figure and to counteract the impression of remoteness, Stalin was frequently photographed with adoring children, notably in January 1936 when he was shown with a dark-eyed, high-cheek-boned Gelya Markinova, an image replicated on millions of posters. This poster gave pleasure to millions, ignorant of the fact that Gelya's father had been shot as an enemy of the people and that her mother was arrested and later killed herself.[66] Stalin's own family life made the Macbeths seem functional – alcoholism, divorce and suicide being the lot of his own children, and insanity or the labour camps for many of his own side of the family. By the late 1930s, Lenin had become a sort of St John the Baptist prophesying Stalin, or least an abstract presence in the background to the man of the moment:

> Lenin died. But stronger than steel,
> Firmer than the flinty mountain races
> Came his pupil – splendid Stalin.
> He is leading us to victories and happiness.[67]

'He' slipped easily into the tsars' role of genial father-figure, to whose justice desperate people turned when they sought to outflank un-responsive officialdom. This was qualitatively little different from the contemporary German insistence that 'if only the Führer knew' he would make short work of corrupt or unfeeling petty Party bureaucrats, in itself an almost classical trope derived from medieval kingship, in which everything maleficent was the work of wicked underlings. This belief in Stalin's good-natured blindness sat oddly with the repeated claim to omniscience – the essence of the 'fantasy state' based on the interaction of the inner workings of a dictatorial mind and the wider society (including its institutions) as a whole. The following Stalin-era poem reflected the Orwellian spirit:

And so – everywhere. In the workshops, in the mines
In the Red Army, the kindergarten
He is watching . . .
You look at his portrait and it's as if he knows
Your work – and weighs it
You've worked badly – his brows lower
But when you've worked well, he smiles in his moustache.[68]

If the popular tropes of divine-right kingship structured how totalitarian rulers interacted with wider society, the totalitarian parties reproduced an evolution from sectarian adepts or *virtuosi* to established Church, while their outlook was essentially Manichean, dividing the world into good and evil, light and darkness, old and new, a view which led to the demonisation of their enemies, especially heretics within their own party.

As we have seen, contemporary observers were often struck by the similarities between the Bolsheviks and religious communities. René Fülöp-Miller compared the Bolsheviks and the Society of Jesus:

> Bolshevism, therefore, is the result of the transference of Jesuit maxims to revolutionary tactics; its spirit is the same as that of the *ecclesia militans* of Ignatius Loyola. In both we find the principle that the end justifies the means . . . Man, therefore, if he is to be happy in the Bolshevik sense, must obey not the inner truth of conscience, but the commands of a number of authorities who claim to be able, as being cleverer, to weigh soberly what is best and most useful for the community.

This is precisely how some of the most knowledgeable and sophisticated contemporary historians of Stalinism, such as Marc Lazar, Stephen Kotkin, Klaus-Georg Riegel, Robert Service and Robert Tucker, describe the functioning of the Communist Party within their wider discussion of Stalinist civilisation. One merit of this approach is to get away from sterile debates, which have their analogues in the less imaginative literature on Nazism, about whence – top down or bottom up – the impetus to persecute and destroy emanated.

Semyon Frank was one of the first to draw attention to the sectarian characteristics of the Russian revolutionary intelligentsia – 'monk–revolutionaries' practising ascetic self-discipline, who persecuted unbelievers with hate, intolerance and annihilation, and proclaimed infallible doctrines of salvation. Max Weber similarly noted a stratum of déclassé Russian intellectuals who espoused 'an almost superstitious veneration of science as the possible creator or at least prophet of social revolution, violent or peaceful, in the sense of salvation from class rule'. Lenin thought that the working class was merely capable of trade union consciousness, and that there was a potentially much larger 'counter-community of the estranged' who could be revolutionised under the tutelage of a missionary–Marxist party.[69] The Bolsheviks consisted of individuals prepared to sacrifice their entire lives to the socialist eschatology, identifying and obeying its dogmas, and surrendering themselves to the Party's disciplinary norms and amoral values in the manner of members of a sect of *virtuosi*. The Party was like holy water in which deracinated intellectuals and the occasional worker would be baptised into the proletarian vanguard that was destined to reforge mankind and society through apocalyptic revolutionary violence. The sinless proletariat would be the vehicle of redemption, even if this might entail its imposition of a sinful dictatorship to ensure that the forces of evil did not regroup and rally.

While the sect routinely practised amorality, conspiracy and deception towards the world without, within the sect the transparency of a panopticon prison was to prevail, with each member of the sect open to collective scrutiny of his or her revolutionary soul through confession, purification and purge, practices which drifted from the Christian Church into the milieu of ostensibly atheistic revolutionaries. In power, Bolshevism replicated the traditional dualism of Church and state, but with the Communist nomenklatura paralleling and penetrating state structures, which – in a society where the state included culture, education, health, agriculture and industry – meant what Kotkin calls 'a kind of theocracy', where the state was responsible for technical administration and the Party for ideological orthodoxy and the overall sense of direction towards building a socialist society. The point of the (post-revolutionary) Party became to goad those with mere technical competence and expertise towards the achievement of revolutionary consciousness. The Party would imbue (or infect) these relatively inert and unimaginative forces with 'Party spirit'. Nothing lay beyond the Party's reach, including thoughts.[70]

For the sect had mutated into a hierocratic Church. It was structured like a Church, with the hierarchy ascending from the humblest cells (or parishes), upwards via the urban or regional *gorkoms* and *obkoms* (the bishoprics), and onwards to the Olympian figures in the Kremlin. Party meetings were highly ritualised services, held under the gaze of the icons Marx, Engels, Lenin and Stalin, or their local surrogates, and replete with the symbolic paraphernalia of busts and banners, meetings that were punctuated with extravagant professions of faith and loyalty. Admission was a complicated procedure, beginning with confession of biographical suitability, which, if successful, brought candidate status and then full membership. After all, one was joining an elect, with a separate legal status and privileges, an identifiable form of dress – jack-boots, leather jackets, flat or peaked caps – and a gang-like common tone and vocabulary. Victor Kravchenko described the sensation when he joined the Party in 1929:

> It seemed to me the greatest event in my life. It made me one of the elite of the new Russia. I was no longer an individual with a free choice of friends, interests, views. I was dedicated forever to an idea and a cause. I was a soldier in a highly disciplined army in which obedience to the centre was the first and almost sole virtue. To meet the wrong people, to listen to the wrong words, thereafter would be inadmissible.[71]

That initial confession of class suitability (and every subsequent 'incident' or covert denunciation added to the individual's files) formed the basis for the verifications of membership and purges which swept through the Party from its inception, but which reached heights of surreality in the Stalin era. Periodically the Party sought to expand its mass base through crash recruitment drives, such as the 1924 Lenin Enrolment or the October Enrolment three years later. Membership rose from 625,000 in 1921 to 1,678,000 nine years later.[72] Expansion was invariably followed by a corrective weeding out of the delinquent or unsuitable who had slipped in through lack of revolutionary vigilance. Popular perceptions that the Party consisted of self-important fat cats would be countered by the restoration of an appropriate degree of neurotic tension among the privileged who witnessed their errant comrades fall from grace. Stalin explained this in his closing speech to the Thirteenth Party Congress: 'The basic idea in the purging is the fact that people of this kind feel that there is a master who may call them to account for their transgressions against the Party. I believe that sometimes, from time to time, the master must without fail go through the ranks of the Party with a broom in his hands.'[73] Finally, at a time when the Party's policies had wrought human havoc through a combination of crash industrialisation and forced agricultural collectivisation, a purge would stifle any murmured criticisms of the leadership at a time when its international postures – 'social fascists' – seemed perverse. Purges were public degradation rituals, which began when the Party member laid his Party card on the table of the Purge Commission. Kravchenko underwent such an ordeal in late 1933, shortly after he had returned from enforcing collection of the harvest in a Ukrainian countryside where people were reduced to eating animal manure in order to find the odd grain, an ordeal which began for him with dread anticipation and raking of memory:

> Didn't you talk too much one night three years ago under the influence of good fellowship? Perhaps one of the good fellows reported your unguarded remarks ... One of your uncles had been an officer under the Tsars. True, you had never met him. But what if someone has dug up that ghost and you're accused of 'hiding' him from the Party? A woman who was your lover was later arrested as a Right deviationist. What if this relationship with a class enemy was suddenly thrown up to you? Pavlov is likely to be expelled – how shall I disassociate myself from him before he drags me with him to ruin? Save your own skin – somehow, anyhow – for the stakes are life itself.

Kravchenko watched as his fellow Party members performed what he called 'a political and spiritual strip act'. An engineer called Dukhovtsev was doing fine, confidently batting off a flurry of questions from the moral vantage point of his impeccably proletarian background. Then things deteriorated for him:

> 'Comrade Dukhovtsev, are you married?' Galembo [the prosecutor and judge] inquires, almost casually.
> 'Yes, I am.'
> 'When were you married and who is your wife?'
> 'I was married last year. My wife is the daughter of a bookkeeper and is now a nurse in a hospital.'
> 'Tell me, did you register your marriage or not? In other words, how was your marriage consecrated?'
> Dukhovtsev turns red. He fidgets with embarrassment. Suddenly he recognizes the import of this line of inquiry. The

audience becomes tense, expectant. There is not a sound in the hall. Finally the purgee, in a low voice, admits the awful truth:

'I was married in church,' he says dejectedly.

The tension is broken. The audience rocks with laughter.

'I know, comrades, that it sounds funny,' Dukhovtsev raises his voice above the laughter. 'It's ridiculous and I admit it. A church ceremony means nothing to me, believe me. But I was in love with my wife and her parents just wouldn't let her marry me unless I agreed to a church comedy. They're backward people. My wife doesn't follow superstitions any more than I do but she is an only daughter and didn't want to hurt her old people. I argued with her and begged her and warned her it would lead to no good. But she wouldn't budge, and on the other hand I couldn't live without her. So in the end we married secretly in a distant village church. On the way back I hid the veil and flowers in my briefcase . . . We are not believers, I can assure you. My wife is working, I am studying, we have a child. I beg you, comrades, to forgive my mistake. I confess I'm guilty for having hidden this crime from the Party.'

Dukhovtsev was expelled from the Party. Kravchenko survived this experience, partly because being examined late he studied how the Commission had dealt with those who preceded him, partly because he had carefully documented his actions in the Ukraine which enabled him to practise a form of 'the best defence is attack' by denouncing the delators.[74]

If purges brought expulsion from the Party, and hence denial of access to the privileges that went with it, charges of 'sabotage' or 'wrecking', or failure to maintain vigilance in combating this, not to speak of consorting with foreign agents, involved that sword of the righteous, the NKVD, and either a public show trial (the first of which was held in 1922) or disappearance during the night, imprisonment or a bullet in the back of the neck.

Trials for 'wrecking' – a means of putting a sinister spin on the accidents and wastage that accompanied reckless industrialisation – began with the Shakhty trial in 1928. Over fifty engineers and technicians were tried for both espionage and sabotage in the Don Basin coalmines, whose higher purpose was to indicate the price that would be paid by those who failed to keep pace with Stalin's plans for crash industrialisation. The trial was a travesty of legal procedure, under the presiding genius of Andrei Vyshinsky. Two of the accused failed to appear, it being announced that one had gone insane, the other had committed suicide, and more recanted earlier confessions. One elderly figure consistently outwitted Krylenko, the loutish prosecutor. Such glitches would not be allowed to recur.[75]

In the show trials that commenced that autumn with the trial of Lev Kamenev and Grigori Zinoviev, there was no evidence – circumstantial or otherwise – connecting the accused to vast conspiracies, which were projections of how the Communists themselves viewed the world; rather there was the bizarre spectacle of lifelong Bolsheviks making abject public confessions. Virtually every aspect of these trials was rigged, with Stalin using Vyshinsky to update the charges, or the lists of accused, and with the NKVD hastening to secure the corresponding additional confessions via their dextrous use of boots and chair legs. Vyshinsky also received direct instructions on how to conduct the proceedings: 'Don't let the accused speak too much ... Shut them up ... Don't let them babble.' Defence counsel was indistinguishable from the prosecution: 'Comrade judges,' said a distinguished defence lawyer in the trial of Arnold, Pushin and another long-standing Bolshevik Knyazev,

> the picture of treachery and betrayal which has unfolded before you in the course of these few days is monstrous. The gravity of the defendants' guilt is immense. The wrath of the popular masses of our Union is understandable. Both the work itself of the Trotskyite organization and the methods it used to entice people into its midst have been revealed here in court with the utmost cogency and clarity ... The range of arguments which have been brought to your attention, the range of debates which may be produced as factors extenuating the guilt of one or other accused in this case is becoming extremely limited.

Sometimes the defendants did not bother to conceal that they were literally reading from a script, as is illustrated by the following exchanges between defendant Sharangovich and Vyshinsky, with the judge Ulrikh lending a hand as impromptu coach when Sharangovich fluffed his lines:

> V. Let us briefly sum up what you plead guilty to in the present case.
> S. Firstly, to being a traitor to the Motherland.

V. An old Polish spy.

S. Secondly, to being a conspirator. Thirdly to being directly involved in wrecking.

V. No, thirdly, to being one of the main leaders of the National Fascist Group in Byelorussia and one of the active participants in the 'Rightist Trotskyite Anti-Soviet Bloc'.

S. Correct. Then to being personally involved in wrecking.

V. Acts of sabotage.

S. Correct.

U. To being the organizer of terrorist acts against the leaders of the Party and Government.

S. That is right.

U. And all this was done with a view to . . .

S. And all this was done with a view to overthrowing the Soviet regime, with a view to Fascism triumphing, with a view to defeating the Soviet Union in the event of war against the Fascist states.

U. Directed at the division of the USSR, the separation of Byelorussia, its transformation . . .

S. Its transformation into a capitalist state under the yoke of Polish landowners and capitalists.[76]

The medieval and early modern Inquisition played a part in the Communists' demonisation of the Christian Church, with Young Pioneers chasing inquisitors from the stage in a celebrated Bolshevik play. In fact, the modus operandi of the Communist Party itself bore a marked similarity to the Spanish Inquisition, an arm of the Spanish monarchy rather than the Church, with the important differences that torture was an acknowledged and legal part of the latter's proceedings, whose overarching objective was to induce heretics to seek forgiveness for the sake of their souls. Only unrepentant heretics were ceremonially burned. In the Soviet cover version, torture was frequently used but never publicly acknowledged, and confession did not bring forgiveness, but rather either a swift death or disappearance into the camps.

The point of these confessions was various. They would demonstrate the legality and professionalism of the authorities both to themselves and to the outside world, with distinguished observers from such august bodies as the International Association of Lawyers on hand to testify that 'the accused were sentenced quite lawfully'.[77] Confession would give substance to the chimera of ramifying conspiracies, dramatising the existence of an evil against which the NKVD was fighting the good fight. The Party's own sectarian culture met the interrogators halfway, for as we have seen confession was integral to the cleansing of the cadres. Another favourite metaphor was that of vomiting, the human body's own most dramatic way of cleansing itself of impurities. The Party dictated that everyone should play out their role, regardless of mere matters of guilt or innocence. As prosecutor Krylenko explained: 'I have no doubt that you are personally not guilty of anything. We are performing our duty to the Party – I have considered and consider you a Communist. I will be the prosecutor at the trial, you will confirm the testimony given during the investigation. This is our duty to the Party, yours and mine.'

As a community of faith, of self-proclaimed 'miracle men', the Bolsheviks had long accustomed themselves to believing that 'that black was white, and white black, if the Party required it'. The same person, Grigory Pyatakov, continued: 'In order to become one with this great Party he would fuse himself with it, abandon his own personality, so that there was no particle left inside him which was not at one with the Party, did not belong to it.' Better to confess, which had so routinely entailed expulsion and chastened readmission, than to risk being cast out permanently into the cold and darkness, although this was the fate of hundreds of thousands, including Pyatakov. These men's capacity to resist making false confessions was permanently damaged by the alacrity with which they had believed the false confessions of others, managing even to suggest their own criminal negligence as they bayed for the blood of such criminals as Kamenev and Zinoviev. Here is Pyatakov himself calling in *Pravda* for the death of Zinoviev:

> One cannot find the words fully to express one's indignation and disgust. These people have lost the last semblance of humanity. They must be destroyed like carrion which is polluting the pure, bracing air of the land of the Soviets: dangerous carrion which may cause the death of our leaders, and has already caused the death of one of the best people in our land – that wonderful comrade and leader S. M. Kirov . . . Many of us, including myself, by our heedlessness, our complacency and lack of vigilance towards those around us, unconsciously helped these bandits to commit their black deeds . . . It is a good thing

that the People's Commissariat of Internal Affairs had exposed this gang ... It is a good thing that it can be exterminated ... Honour and glory to the workers of the People's Commissariat of Internal Affairs.[78]

Pyatakov, one of the tyros of Stalinist industrialisation, was arrested in the autumn of 1936 as an alleged member of a passive 'Reserve Centre' to the active 'Troskyite–Zinovievite Terrorist Centre's' alleged conspiracy to murder senior Bolshevik leaders. His estranged wife was induced to testify against him by the simple expedient of threatening their young son. Pyatakov duly confessed. He was tried along with Radek, Sokolnikov and Serebryakov in January 1937 in the cold gloom of the October Hall. The defendants were charged with industrial sabotage (the absurd excuse for the inevitable disasters of rapid industrialisation) and spying on behalf of the Germans and Japanese. Stalin personally inserted Trotsky into the conspiracy as a sort of hidden, but omnipresent, demonic presence. After circumspectly admitting chronic industrial inefficiencies, Pyatakov claimed that in December 1935 he had flown from Berlin, where he was on official business, to Oslo for a clandestine meeting with Trotsky – who, he claimed, had been in contact with the Nazi leader Rudolf Hess. When the Norwegian press complained that no aircraft had actually landed at Oslo's Kjeller airfield between September 1935 and May 1936, Vyshinsky was reduced to citing 'corroborative' evidence about the possibility of such winter landings from the Soviet Union's Oslo consulate. And so things dragged to their ineluctable conclusion. Vyshinsky's final paroxysm of abuse was doubtless un-connected with the fact that he was already in the process of acquiring the dacha of one of the accused, which he had taken a fancy to when accompanying his erstwhile host on enchanting woodland walks. Now, in altered circumstances, he railed: 'they [the accused] sank lower than the worst Denikinites or Kolchakites ... The Denikinites, Kolchakites, Milyukovites, did not sink as low as these Trotskyite Judases.' Reaching unplumbed depths of victimology, Vyshinsky summoned forth the Stakhanovites and Young Communist League members who had per-ished, not in industrial accidents, but as a result of sabotage and terrorist atrocities: 'I do not stand here alone! The victims may be in their graves, but I feel that they are standing beside me, pointing at the dock, at you, accused, with their mutilated arms, which have mouldered in the graves to which you sent them!' The defence lawyers readily concurred in this diatribe. Pyatakov addressed the court with downcast eyes: 'In a few hours you will pass sentence. And here I stand before you in filth, crushed by my own crimes, bereft of everything through my own fault, a man who has lost his Party, who has no friends, who has lost his family, who has lost his very self.' Pyatakov, whom Lenin in his Testament had tipped along with Bukharin as the ablest of the younger leaders, was shot shortly after the verdict.[79]

These people were shot or sent to an empire of camps that stretched across the vastness of the Russian countryside partly because they had been dehumanised and demonised by propaganda. The fact that these victims were leading Bolsheviks was all that distinguished them from the thousands of people who had met a brutal end under Lenin's terror, about which he jested, even to strangers like Bertrand Russell, with sardonic, vicious directness. What conceivable 'upside' could there have been in talking in this fashion to one of the West's most influential (and not automatically unsympathetic) intellectuals? Here is Lenin in December 1917 calling for a 'war to the death against the rich, the idlers, and the parasites' in which every village and town should find a way of:

> cleansing the Russian land of all vermin, of scoundrels and fleas, the bedbug rich and so on. In one place they will put in prison a dozen rich men, a dozen scoundrels, half a dozen workers who shirk on the job . . . In another place they will be put to cleaning latrines. In a third they will be given yellow tickets [such as prostitutes are given] after a term in prison, so that everyone knows they are harmful and can keep an eye on them. In a fourth one out of every ten idlers will be shot. The more variety the better . . . for only practice can devise the best methods of struggle.[80]

Or Lenin on the kulaks, the relatively prosperous farmers:

> The most beastly, the coarsest, the most savage exploiters . . . These bloodsuckers have waxed rich during the war on the people's want . . . These spiders have grown fat at the expense of the peasants, impoverished by the war, of hungry workers. These leeches have drunk the blood of toilers, growing the richer the more the worker starved in the cities and factories. These vampires have gathered and continue to gather in their hands the lands of the landlords, enslaving, time and again, the poor peasants. Merciless war against these kulaks! Death to the

kulaks! Hate and contempt to the parties defending them; the rightist Social Revolutionaries, the Mensheviks and today's left Social Revolutionaries.[81]

It is a platitude in the study of the Nazis that their wartime annihilatory rampage was preceded by the relentless stereotyping of their victims – as onrushing vermin who were the advance guard of shape-shifting quasi-satanic forces. It is also commonplace that these stereotypes drew on older folkloric beliefs and prejudices (some of which originated in Christianity and persisted long after the Church had disowned them) as well as the more 'objective' language of modern medical pathology.

The demonisation of the class enemy in the Soviet Union – as asocial deviants, insects and vermin, or tools of such foreign 'devils' as the French Prime Minister Poincaré or Uncle Sam – similarly conflated historic hatreds – against the 'idle rich' with their white collars and smooth hands – with the hygienic obsession evident in what Lenin had said in the passages quoted. The Hungarian screenwriter René Fülöp-Miller provided an astute portrait of the psychological processes at work during these sessions of organized hate. May Day on Red Square was like the democratic child 'mass man's' birthday, since it was bedecked with the equivalent of toys: giant papier-mâché dolls, guns, trains and so on.

Sometimes he suddenly stops, looks round, considers one by one the enormous figures made of cardboard or cloth stuffed with straw; all at once he notices that the dolls have the faces of foreign statesmen and capitalists, that is to say of people against whom he has a grudge at the moment. In a mad rage, he hurls himself against them, furiously tears out their stuffing, holds them in his many outstretched hands, and gloats in the intoxication of victory. Often the figures are hanged on a rope; the raging 'mass' sticks a long tongue of red ribbon in their mouths, or burns them ceremoniously. All this is done with the naïve cruelty of savages or children, with the primitive joy in smashing toys which is natural to both. Like a child the collective man, in his games, avenges himself on his enemies. He amuses himself in this way on the Red Square till late in the evening; if he finally gets tired, the megaphone from the platform above sounds the signal for 'closing', and the mass man goes off and lies down obediently to sleep in his ten thousand beds.[82]

Like the Nazis, the Bolsheviks were implacable wielders of brooms, an image that crops up repeatedly in the propaganda and rhetoric of both groups. Soviet posters abound with crows, dogs, pigs, rats, snakes, spiders, whose function was to strip real people of their humanity to make it easier to disfranchise, incarcerate or kill them. Some of these noxious images came from the wider European left – for example Wilhelm Liebknecht's 1917 pamphlet *The Spider and the Flies* – whose blood-sucking bourgeois spiders bore an uncanny resemblance to how antisemites depicted the Jews and which was similarly informed by pseudo-Darwinian zoomorphism. The argument that the discourses of class and race were somehow distinct does not find universal assent, however much it suits some to regard these notions as mutually exclusive, thereby disregarding the 'class profile' of some notional 'race'. As the leading authority on Soviet political posters remarks: 'The authorities made no distinction between individuals in this category and their families. In fact, the official approach to this group of people was genetic, since the class defect could not be removed by repentance or good deeds and family members were likewise considered unredeemable.'[83]

All the atrocities of the Bolsheviks were notionally related to the idea of realising a perfect society on earth in the here and now. If the eradication of anyone or anything thought to obstruct that objective was one side of the project, its corollary was the occasional glimpse of what regenerated mankind could be, for without the vision of the new society and the new beings who would comprise it, there would be no hope and the suffering would seem as meaningless as it does to anyone looking at it, dispassionately and without nostalgia, and with the benefit of nearly a century's hindsight.

Lenin was introduced to the gist of Tommaso Campanella's utopian tract, *The City of the Sun*, while he was visiting Maxim Gorky on Capri before the First World War. Campanella was an early-seventeenth-century Dominican friar who spent twenty-seven years of his life in Neapolitan dungeons, periods punctuated with bouts on the rack for his heretical and seditious opinions, although he would end his days as a propagandist of universal papal monarchy. Lenin was clearly impressed by what he heard since he later wished Campanella's name to be inscribed on the refashioned Romanov Tricentennial Obelisk in Moscow. In Campanella's tract, a Genoese mariner tells of an ideal city that would be ruled according to 'scientific' principles. These would be gathered together in one book, with knowledge translated into images decorating both sides of the city's seven concentric rings of walls. These walls were interspersed with statues of outstanding figures. The family, money and private property were to be abolished, and reproduction would be controlled by eugenic intervention. People were brought up in communal dormitories and lived in what amounted to a unisex society where love was focused on the whole society. They all wore white clothes. Industriousness was the highest virtue; idleness was despised and punished. Solarian society was authoritarian and hierarchical, with the omniscient Metafisico or Sole at the top, although lesser officials were elected, and replaced when someone more competent appeared.[84]

The communication of ideology through striking images was the feature of this tract that captured Lenin's imagination. Having obliterated the monuments of the tsarist past, he wanted to fill the streets of cities with inspiring inscriptions engraved on giant stone tablets and images of Russian and European revolutionary figures – Chernyshevsky, Lavrov, Spartacus, Brutus, Babeuf, Blanqui, Danton and Marat, as well as Marx, Engels, Liebknecht and Luxemburg. Between 1918 and 1921 some fifty such statues were erected, most of which crumbled and disintegrated because of the inclement weather and shoddy materials. The utopian musings of countless architects remained similarly unrealised because money was short and the country in chaos, science fiction being a cheaper surrogate, in the sense that Bolshevised cities, where the Red Star glowed in the darkness, could more easily be erected, in the imagination at least, on the planet Mars in the twenty-third century.

When conditions – which were only ever relatively normal – made it feasible to construct the new socialist civilisation, its realities did resemble life on another planet. Between 1929 and 1936 a gigantic smudge began to take shape in a cold white landscape on the River Ural: a massive steel works, with blast furnaces and rolling mills, amid barracks, a prison camp, tents and mud huts for its inhabitants. Since in the past compasses had been disoriented by the rich iron-ore deposits in the hills, this place was called 'Magnetic Mountain' or Magnitogorsk. As a sort of afterthought, this chaos of muddy tracks, rail lines and rusting machinery was fashioned into a city in typical fits and starts, despite the toxic artificial lake and with chemical fumes wafting over its forlorn public buildings and inadequate housing projects. There was no sanitation, scant public transport, no street-lighting or ways of distinguishing one bleak barracks settlement from another. Housing was calculated in terms of square metres of living space, thus rendering unnecessary anything so luxurious as one person per room, or indeed, exclusive occupancy of a bug-ridden bed. There was not, and never would be, a church, although Magnitogorsk did run to a cinema and circus, the latter a capacity venue for the larger local show trials.

People teemed into this improbable environment, whether in the form of enthusiasts, the curious and expectant, who, disabused, quickly moved elsewhere, former farmers undergoing compulsory 'dekulakisation', or convicts, who had their own barbed-wire encampment. Collectively, these people disembarked in the middle of nowhere after a journey of a week or so on railway lines so poorly constructed that trains slowed to such a speed that it would have been quicker to walk, finding to their surprise that there was not even a station to indicate where they had alighted. By 1932 Magnitogorsk had 215,000 inhabitants, but these people lived in what amounted to Third World shanty towns, with housing and an urban infrastructure only being put in place by the end of the decade. Whether they knew it or not, they were in a socialist crucible, where a new type of human being was to be forged alongside the ingots and girders (many of them faulty) produced in the blast furnaces and rolling mills. The primary purpose and identity of such human beings was derived from work in the accident-ridden and poorly constructed edifices by which the city was overshadowed – work being both the core identity of the vanguard class and their contribution to the deadly battle between the (depressed) capitalist world and socialism in the making. The regime decided that these human beings did not need a family, for both the cramped living conditions and the communal baths, laundries and kitchens were designed to make such narrow, old-fashioned attachments redundant, until the line changed to engender greater social stability.[85]

Crash industrialisation had the hubristic goal of catching up with capitalism, not in fifty or a hundred years, but in ten, with the aid of centralised planning. To that end some nine million farmers were moved into industrial cities and zones during the course of the First Five Year Plan. In a culture that dramatised and militarised production and much else, crash industrialisation relied heavily on 'shock work', a term that had already been used to denote performance of especially arduous tasks during the Civil War. Shock brigades were supposed to lift the performance of the generality of workers through example in return for enhanced privileges. In 1929 shock work was elaborated by intra-factory, group or individual 'socialist competition'. Since up to 40 per cent of the workforce were eventually classified as shock-workers, by anxious managers worried about losing disgruntled workers and political brownie-points, the concept of shock work became a debased currency, devalued by the industrial equivalent of grade inflation.[86] With the introduction, from 1931, of individuated output norms and differential pay scales, the anonymous mass aspect of shock work no longer corresponded to the desired reality. The search was on for extraordinary heroes. In 1924 Trotsky had typically characterised such beings with a rhetoric that was so high flown as to be ludicrous in a Chernyshevskian pamphlet entitled *Literature and Revolution*: 'Man will be incomparably stronger, more intelligent and finer: his body will be more harmonized, his movements more rhythmic and his voice will become more musical. The forms of everyday life will take on a dynamic theatricality. The average human type will be raised up the level of Aristotle, Goethe and Marx. And over this mountain chain new peaks will come into view.' The reality of the exemplary new man was quite different.[87]

The Soviet Union was not immune to what was emerging as a global cult of celebrity, or notoriety, focused on athletes, aviators, boxers, film-stars, gangsters, mountaineers and, as we have seen, dictators. Already, the commissar for heavy industry, Sergo Ordzhonikidze, had launched the search for 'new people', saying, 'In capitalist countries, nothing can compare with the popularity of gangsters like Al Capone. In our country, under socialism, heroes of labour, our Izotovites, must become the most famous,' a reference to Nikita Izotov, a miner whom colleagues described rather sourly as 'the human cutting machine'. But Izotov was destined to be eclipsed, along with the new hybrid Marx, Aristotle and Goethe.

In 1931 *Pravda* ran features under the slogan 'The Country Needs to Know its Heroes', consisting of photographs of aviators, collective farmers, shock-workers and the like. The concept of the exemplary elite was primarily associated with Aleksei Stakhanov, a thirty-year-old Donbass coalminer, who in August 1935 managed to cut 102 tons of coal (or fourteen times his norm) in a single shift – moreover, with the aid of a trusty Soviet-produced pneumatic pick. Stakhanov had migrated from a village in Orel, working his way up from pony-brakeman to manual pick operative, before getting his hands on the air-powered pick that brought him fame and fortune. Of course the work was done at night, enabling Stakhanov to maximise his labours as compressed air went to his pick alone, and his six-hour continuous stint was facilitated by a lengthy logistical chain beginning with the men installing timber props behind him. Nonetheless, the anonymous battalions of shock-workers were thenceforth superseded by a Soviet Hercules with a human face.[88] 'Recordmania' spread like a feverish sickness, with managers and foremen sweating too lest they be denounced as 'bigwigs', 'windbags', 'routiners', 'wreckers'; or 'saboteurs' for failing to make these 'Stakhanovite' feats feasible, rendering them liable to what the Kremlin's own Al Capone sinisterly called 'straightening out' or 'a tap on the jaw'. It mattered not that these epic episodes tended to deplete machinery and leave 'Stakhanovites' spent, or that some workers resented the diversion of resources, the subsequent lifting of their own norms, or the rich rewards such Promethean heroics brought. Schadenfreude best describes those who said of a young female Stakhanovite, who had been rewarded (one hopes she was grateful) with the selected works of Lenin: 'That's what the whore deserves!' Resentment towards Stakhanovites bestriding the factory floors 'like gods' was compounded when they became fixtures of the factory 'production courts'.

Much of the time of stellar Stakhanovites was increasingly spent on tour, whether visiting the Kremlin, addressing other workers or venturing confidently into places – such as the opera or theatre – where workers already did not comfortably go. Even society pages in the newspapers included such gems as 'The brigadier-welder Vl. Baranov (28), the best Stakhanovite at Elektrozavod, glided across the floor in a slow tango with Shura Ovchinnovka (20), the best Stakhanovite at TsAGI. He was dressed in a black Boston suit that fully accentuated his solidly built figure; she was in a crêpe de chine dress and black shoes with white trimming.'[89]

In other words, although they talked incessantly about work, Stakhanovites did less and less of it, recalling it, like millionaire footballers or pop stars from humble origins, as something that took on roseate hues in memory of things past. Of course, Stakhanovites had a role to play within a wider myth-in-the-making. As an explicitly hierarchical society replaced one allegedly based on fraternity, they had to acknowledge the crucial guiding role of the nation's father-figure, whose speeches had allegedly originally inspired them to break through artificial barriers while using technology almost as an extension of their own brain.[90] Stakhanovites, who were often not members of the Party, were also model citizens in respects other than dutiful sons and daughters of the ultimate patriarch. Their lifestyle was supposed to exemplify the theme that 'life is joyous, comrades', and since they were showered with official munificence while simultaneously enjoying very high wages, the joyous life seemed like an idyllic shopping spree, for clothes, clocks, furniture, motorbikes, perfume, phonographs and so forth. Thus adorned and kitted out, Stakhanovites appeared having their leisurely breakfasts, reading the papers, lunching with friends, playing a little volleyball, tea and a game of checkers, while their wives undertook charitable work as 'housewife–activists' and their children were exhorted to their own heroics at school.

Like the other totalitarian dictatorships, the Soviet Union was especially interested in the moulding of the coming generations, who were impressed into youth organisations such as the Young Pioneers, and simultaneously formally educated and politically indoctrinated at kindergartens and schools. In addition to the promotion of literacy and numeracy, the Party saw to it that children were exposed to a 'new morality' or, rather, to new forms of social behaviour, since there was not much that was moral about it. That this happened can be seen from the Shakhty trial, when the son of one of the alleged 'wreckers' in the coal industry wrote to *Pravda* calling for condign punishment for his father:

> As the son of one of the conspirators, Andrei Kolodub, and at the same time a Young Communist . . . I cannot react calmly to the treacherous deeds of my father . . . Knowing my father as a confirmed enemy and hater of the working people I add my voice to the demand of all the workers that the counter-revolutionaries should be severely punished . . . Since I consider it shameful any longer to bear the name Kolodub I am changing it to Shakhtin.[91]

A painting by Nikolai Chebakov celebrates one of the key exemplars used to illustrate the penetration of the family by the Party. An upright blond youth in the uniform of a Pioneer looks accusingly at two shifty-looking men disporting themselves around a table in a cottage in the Urals, with their wives as shadowy as the icon display in the background.[92] The youth was Pavlik Morozov, who in September 1932 as a member of a Pioneer group acting as auxiliaries in the collectivisation campaign 'unmasked' his own father, the former president of the Gerasimovka village soviet, for 'falling under the influence of kulak relations'. Shortly after the father was shot, his grandfather and uncle murdered Morozov by way of revenge. They and their peasant accomplices were shot too. For good measure, the grandmother was sent to a prison camp, leaving Pavlik's mother alone to tend the flame of his memory. Maxim Gorky was on hand to spell out the new moral tidings: 'If a "blood" relative turns out to be an enemy of the people, then he is no longer a relative but simply an enemy and there is no longer any reason to spare him.' Worryingly, the Pioneers who attended the first Writers' Congress to laud these 'engineers of the human soul' announced proudly that they 'had thousands like Pavlik'.[93]

The goal of a morally rearranged new Soviet man or woman was not confined to hoping people would behave or work like Morozov or Stakhanov. Recently discovered personal diaries from that era show how people sought to replace their 'old' self with a 'new' Soviet personality, an activity already prefigured in the *Spiritual Exercises* of St Ignatius Loyola. Their diaries enable us to follow this process of self-reconstruction – the diary itself was part of this auto-political 'therapy', rather than a record of a private world or the random musings of an individual. Stepan Podlubnyi was born into a well-off farming family in Vinnitsa in the Ukraine on the eve of the Great War. Although most of the family's property was confiscated during the Revolution, memories lingered and they were regarded as residual 'kulaks' by their resentful fellow villagers. In 1929 the family were 'dekulakised', that is everything was taken and the father was deported for three years' administrative exile. Armed with forged documents that described them as 'workers', the boy and his mother found work in a printing plant. He joined the Komsomol youth movement and after middle school in 1935 went to the Moscow Medical Institute as a student. He was on the way up, in a modest sort of way.

The diary probably started as a Komsomol task, an objective method of gauging people's inner consciousness. Podlubnyi's covers the years 1931–9. He regarded it as part of the process of re-educating himself, a 'rubbish heap' on which to jettison the dirt that the 'kulak' past he was concealing had left in his soul. His 'alien' class origins led him to construe his own being as a battleground between the old and the new, as he tried to slough off what he called his 'sick psychology'. He sloughed off his unreconstructed father too: 'A halfway old man, of no use to anybody and completely superfluous . . . This old man's weak will can destroy us as well as him. We have to help him with many things. We must force him to work on himself . . . I look at him as an acquaintance.

Coldly. I can see in him only qualities negative for me . . . His character is one of a wretched old man. Actually he's not really an old man.' This division of humanity into the reconstructed and the reprobate spread to acquaintances, who were deemed to be 'cultured' in a new Soviet sense rather than like the dancing, drunken hogs of the old order. The comprehensive social transformation that Stalin was undertaking was literally replicated in the way Podlubnyi viewed the world: there were 'old' beings and 'new', 'old' ways of behaving and 'new'.

How this battle between his past and his present was resolved was no academic matter since people with the wrong class pedigrees were being ejected from the cities, with the aid of an internal passport system, and returned to the rural collectives, where they might starve. Podlubnyi's reconstructed self eventually secured him a passport in 1933, for the reconstructed self – utterly synthetic though it might be – had more substance than his 'kulak' origins. He began to read Marx and Lenin in an attempt to rearrange his consciousness. In late 1932 he became a secret informer for the GPU, forerunner of the NKVD, reporting on his classmates and workmates. Clearly oppressed by the thought of his origins being 'found out' or by the terror he felt for the GPU, Podlubnyi occasionally mused about the world beyond Moscow: 'I want to be free! I'll live at the end of the world! In Arkangel'sk! In the tundra! I don't care, I just want to be free, so that nobody can reproach me any longer: ah, so you are one of those? We know who you are etc.' Gradually he moved from careerism to belief. When his mother reported famine conditions in the Ukraine in mid-1933, he wrote:

> By the way, about the news that Mama reported; an incredible famine is going on over there. Half of the people have died of hunger. Now they are eating cooked beet tops. There are plenty of cases of cannibalism . . . All in all it's a terrifying thing. I don't know why, but I don't have any pity for this. It has to be this way, because then it will be easier to remake the peasants' smallholder psychology into the proletarian psychology that we need. And those who die of hunger, let them die. If they can't defend themselves against death from starvation, it means that they are weak-willed, and what can they give to society?

In March 1933 he visited a graphologist to have his handwriting analysed. This was a lifelong ambition fulfilled. The report pleased him:

A personality full of initiative, who easily grasps the essence of a matter. Materialistic worldview. Politically oriented. At an early stage escaped the ideological influence of his family. Has a gift for observation. Can distinguish lies from sincerity in the voice of another. Sociable and pleasant; soft, even good-natured, in the company of others; but when decisive action is called for, or when an obligation or a strong desire has to be fulfilled, neither his close friends nor any other temptations can distract him from the goal he has set himself ... Shows little trust and is suspicious, has developed professional caution. Leans toward formal and logical reasoning, shows talent for treating issues with a scientific methodology, suited for activities in law and administration, is also mechanically talented.

In fact, as these values suggest, Podlubnyi was a new Soviet man. Even when things happened to disturb that mindset, like the purges after Kirov's murder, he could not construe these events in anything other than prefabricated Soviet categories. Even when his mother was arrested in 1937 as an alleged Trotskyite, he regarded this as a mistake – 'to number Mama, a half-literate woman, among the Trotskyites, that would never have occurred to me' – in an essentially rational economy of terror. By that Orwellian construction 'that would never have occurred to me', one can gauge how far the Soviets had achieved their goal.[94]

IV   THE MAN FROM NAZIRETH

Hitler was a lazy, dilettantish autodidact rather than a systematic thinker, so one should not strain to discover coherence or consistency in his views on religion or much else. In fact there is something faintly ridiculous about the weight of learning brought to bear in the last six decades on this less than fascinating figure, a cavernous blank behind the impassioned postures. 'Hitler brings nothing to my mind,' as Karl Krauss memorably had it. Hitler commented, off the cuff, on every religion: Buddhism, Christianity, Islam, Shinto and Judaism, without knowing much about them beyond the wisdom of everyman. He thought that belief in higher powers was a value in itself, for without that capacity for belief mankind would be unable to believe in nation, race or the future

Führer. The young Goebbels, another lapsed Catholic, came to the same conclusion, when he wrote in his quasi-autobiographical novel *Michael*: 'It is almost immaterial what we believe in, so long as we believe in something.' Of the Christian injunctions to faith, hope and charity, only faith was all-important, although charity in a corrupted form played a major role later on.[95] Hitler thought that a people needed a common faith, whether religious or otherwise. He argued that, with the exception of the Communists, the political parties of the Weimar Republic were uninspiring, lacking as they did 'the fanatically religious' ingredient of 'blind faith'. Belief was a dormant constant; the trick was how to activate it through a compelling political creed, like putting a match to a trail of dry straw.[96]

Hitler's understanding of Christianity was as vulgarly confined to the externals of Church history as that of the Italian Fascists. But it went beyond common recognition of the suggestibility of twilight and darkness for inculcating belief. Hitler thought that the virtue of 'fanaticism' was a characteristic of religious belief that was also indispensable in politics. The word crops up again and again in his conversation, speeches and writings. Like the Fascists, he admired the implacability of the Roman Catholic Church towards the pagan altars: 'It was only as a result of this fanatic intolerance that absolute faith could have been established.' This was the model for the political faith of Nazism, which also brooked no dissent or opposition.[97] Similarly Catholicism's dogmatic impermeability to the fashionable creeds of the moment was something Hitler esteemed.[98] The meritocrat in him found good words for clerical celibacy, since it ensured a constant flow of young men from the mass of the population, contributing to 'the amazing youthfulness of this gigantic organism, its spiritual suppleness and iron will-power'.[99] Finally, whenever Hitler was feeling especially vengeful, he would revert to the idioms of an Old Testament he otherwise wished to expunge, where he could revel in endless examples of inter-tribal enslavement and mass murder, although this is omitted from every contemporary study of the relationship between Nazism and what it derives from Christianity alone.[100]

Like many Catholic Nazis, lapsed or otherwise, Hitler loathed many features of his own Church, while being indulgent towards Protestantism, especially in its theologically liberal and socially conscious varieties. He claimed to have had no feeling for Protestantism, but, since he was talking to a Catholic archbishop at the time, we should probably not take that too seriously. There may have been something of the immigrant's over-compensation here, rather than the fancier explanations that have been going around. From Hitler's Austrian vantage point, Protestantism was one of the essential props of the Reich to the north, as well as being integral to Germany's providential story in which being German meant being Protestant. Although the young Hitler admired the populist talents of Vienna's Christian Social mayor, Karl Lueger, his emotional sympathies were with the 'Away-from-Rome' pan-German nationalist Schönerer, whose visceral Protestantism had limited his appeal in strongly Catholic Austria. This debility would not apply in Germany itself, where 'Protestantism will always stand up for the advancement of all Germanism as such, as long as matters of inner purity or national deepening as well as German freedom are involved, since all these things have a firm foundation in its own being.' A Party that would extinguish freedom lauded the freedom of the Reformation while deploring the slave mentality of Rome.[101]

Above all Hitler the politician had an astute appreciation of the limitations of religious sectarianism, whether Christian or neo-pagan. He was contemptuous of racialist sectaries in the broader right-wing movement in and around Munich, people who imagined they were living in AD 700 rather than the 1920s, brandishing their replica Teutonic 'tin swords', but liable to flee at the first sign of a Communist rubber-truncheon. One could not build a mighty political movement from a crowd of tweedy and weedy academic cranks, whose obsessions resembled the sandal-wearing crowd that Orwell thought discredited socialism. The left laughed at the nutty right; Hitler wanted his enemies to fear and hate him.[102] Similarly he had no time for the racist mysticism of Alfred Rosenberg, author of the impenetrable *Myth of the Twentieth Century*. Clerics were especially exercised by it, ignorant of Hitler's verdict that a book he probably never read consisted of 'stuff nobody can understand'. Goebbels was more succinct, calling the book an 'intellectual belch'. It followed from this that Hitler disapproved of anyone seeking to transform National Socialism into an arcane and mystical cult as opposed to a national–racial Church.[103] He learned this lesson early on.

Artur Dinter came from a Catholic Silesian family that had resettled in Alsace in the wake of the Franco-Prussian war. He renounced his faith and studied the sciences at university before becoming a teacher, although he harboured literary ambitions on the side. He managed to get a few execrable plays performed in Berlin before becoming co-founder of the association of German playwrights, whose remit was to protect dramatists' copyright. Immersion in the thought of Houston Stewart Chamberlain, which combined anti-Catholicism with antisemitism, persuaded him that Jesus was an 'Aryan' racial reformer and the Protestant Germans mankind's saviours. Shortly before the First World War, Dinter leaped up amid the large audience watching a play called *Miracle*, interrupting the performance with antisemitic imprecations against its director Max Reinhardt. After a brief period of undistinguished service in the army, Dinter returned to the cultural front-line in Thuringia, whence he directed vicious attacks against such cultural tyros as Reinhardt and the press baron Mosse. In 1918 he published his second novel. This was called *Sin against the Blood*. It sold around 230,000 copies, making it a Weimar bestseller. It was a peculiar sort of book. Although in every other respect like chalk and cheese, the blond German hero Hermann Kämpfer and his various dark Jewish antipodes were both attracted to blonde females. When Hermann, who is transparently modelled on the author himself, marries one of these blondes, he is horrified to discover that their child has dark eyes and black crinkly hair because his wife is half Jewish. Hermann then produces another child with the blonde woman, who by that time has become his second wife. Twice she produces dark children, because of an earlier liaison with a Jewish lover. Mother and the child die during the second delivery. Hermann did not mourn, but he did kill his wife's Jewish lover.

After the war, Dinter became a popular public speaker, calling for the extermination of the Jews and regretting that not all of his former Jewish comrades had been killed at the front. He advocated the withdrawal of citizenship from Jews and a prohibition on marriages between Germans and Jews. Jewish immigration was to cease and Jews were to be forbidden from owning landed property. Politically, Dinter travelled rightwards from the nationalist conservatives to the more overtly racist German Racial Defence and Offence League, which helped peddle more of his ravings, this time about Jews and ritual murder. He found himself frequently before the courts. By this time, he had shifted the emphases in his 'thought' from biology to religion. According to him: 'In the beginning God created a world of pure spirits.' Mankind was divided into higher and lower spirits, of which the highest was Christ. These categories were reflected in different races. He broadcast these new tidings in a novel called *Sin against the Spirit*, while his social philosophy appeared as *Sin against Love*, by which time wits called the author

'Sin-Dinter'. His more earthy comrades in the *völkisch* movement thought Dinter belonged in an asylum, since he was now ranging himself in the company of Jesus, Luther and Galileo.

By this time, Dinter had become a devotee of Paul de Lagarde, who, in contrast to the Teutomaniac tendency in the *völkisch* movement, thought to rescue Christianity from the falsifications of 'the Jew' St Paul, and from the divisions between Catholic and Protestant. Before long, Dinter – who produced an idiosyncratic edition of New Testament highlights – was proclaiming a Spiritual Christianity, with Christ as the 'Aryan–Germanic hero', which would complete the task of Luther and become the religion of a future *völkisch* state. Late in 1922 he co-founded the German Racial Freedom Party, becoming leader of its tiny caucus in the Thuringian parliament two years later. He quickly fell out with his colleagues. In autumn 1924 he joined an organisation that was designed to help the Nazis over Hitler's incarceration; he had already taken the precaution of sending the prisoner a selection of his literary endeavours. A year later he took the various *völkisch* cells in Thuringia into the NSDAP, receiving from a grateful Hitler the coveted number 5 in the membership lists of the refounded Nazi Party. Hitler also held the first Reich Party day in Weimar, the capital of Thuringia, paying the Dinter family a three-hour visit in their home. Dinter took the opportunity to read his 197 theses for the completion of the Reformation.

A year later he founded a Christian-Spiritual Religious Association in Nuremberg, together with a journal called *Spiritual Christianity*. Dinter was replaced by Fritz Sauckel as Thuringian Gauleiter, for the 'religious struggle' rather than administration was his forte. He devoted himself full time to propagating the idea that a Third Reich would result from a religious movement based on his own crackpot notions. In 1928 he was subjected to internal Nazi Party disciplinary proceedings after he had cast aspersions on the integrity of Graf Reventlow, who had also moved from the German Racial Freedom Party to the NSDAP. Dinter was told to withdraw his aspersions and to tone down his attacks on the Churches. A testy Hitler reminded him that he was 'bold enough' to claim the same infallibility in political matters that Dinter claimed as a religious reformer. Dinter responded by castigating Hitler for his complete lack of understanding of the 'immense' political significance of Dinter's *völkisch*-religious movement. As for Reventlow, Dinter was not going to apologise to a Party member whose number was higher than 50,000 rather than the talismanic 5. Dinter also had the temerity to suggest that the Nazis needed a senate consisting of older members who would advise the Party leader. Hitler responded by denouncing divisive discussions of religious questions; Dinter replied with a tract entitled *Religion and National Socialism* in which he called for outright war against the 'Jewish-Roman Pope's Church' and deprecated the spiritual tone of the entire Nazi leadership.

In October 1928 Dinter was expelled from the Party, not for being religious, but for being too religious in a divisive sectarian way.[104] He was not a quitter. He published explicit attacks on Hitler, whom he accused of being under Rome's spell and a 'coolly calculating demagogue who practises his speaking gestures in front of a mirror'. The Nazi Party was like the Catholic Church – a means of keeping the masses stupid so as to pursue political ends. Once Hitler came to power, Dinter used the offices of Winifred Wagner in an attempt to work his way back into the Führer's good books. He failed and in 1937 his Party card was inscribed 'never receive him back'. By then his paper had been closed down and his books removed from public libraries. He lost his house, and his wife had a nervous breakdown. In October 1938 he committed a minor traffic violation in Weimar, informing the policeman, 'If you knew at all who I was, you wouldn't mess me around like this. I am the founder and the former Gauleiter of the Party's Thuringian region.' When he received, through Himmler's malign intervention, a swingeing fine of one hundred Marks, he protested, 'In Nazi Germany there can't be two types of law!' – unaware that indeed there were. Himmler eventually let him off the fine, on compassionate grounds, but warned him not to boast about his Nazi days. In 1940 he and his wife were arrested for membership of a proscribed organization – his German People's Church. The Special Court found extenuating circumstances in his 'struggle against Jewry and miscegenation' since 1914 and waived his three-month sentence. He died in 1948.[105]

As the absurd story of Artur Dinter also suggests, Hitler was keen to avoid reigniting the dying embers of the Kulturkampf, preferring to blame the Jews for tensions between Catholics and Protestants. Speaking in Passau in 1928 he said: 'We are people of different faiths, but we are one. Which faith conquers the other is not the question; rather, the question is whether Christianity stands or falls ... We tolerate no one in our ranks who attacks the idea of Christianity ... in fact our movement is Christian. We are filled with a desire for Catholics and Protestants to discover one another in the deep distress of our own people.'[106] This is why Hitler the politician regarded Nazi Protestant sectarians as expendable, since too close identification of Nazism with Protestantism would alienate Catholics.[107]

So far we have said little about Hitler's own God, or his credulity towards a very reductionist form of 'science', two subjects whose tensions clearly taxed his limited intellectual capabilities. He subscribed to the view that science had largely supplanted Christianity, without rationalism eradicating the need for belief, or undermining the existence of a creator God in whom he continued to believe. Christianity had been progressively subverted by science, which he understood as a series of heroic discoveries by titanic figures rolling back the frontiers of ignorance. Science was akin to a ladder of enlightenment, from whose ascending rungs one could perceive a wider world, in which God was revealed in and through the 'laws of nature', the chief of which was the God-decreed verities of race against which mankind had sinned. Science merely narrowed down the infinity of ultimate questions. Hitler was palpably irritated at drawing this rather commonplace blank: 'What comes naturally to mankind is the sense of eternity and that sense is at the bottom of every man. The soul and the mind migrate, just as the body returns to nature. Thus life is eternally reborn from life. As for the "why?" of all that, I feel no need to rack my brains on the subject. The soul is unfathomable.'[108] His literal-minded solution to resolving these mysteries was to equip every village with a telescope, with a giant observatory reserved for his (and the astronomer Kepler's) hometown of Linz. The pediment would bear the inscription, 'The heavens proclaim the glory of the everlasting,' which did not say much either.[109]

But we have not quite done with Hitler's God. Hitler himself believed in a God, despite having parted from the rote Catholicism of his Austrian childhood in his early teens, allegedly after innumerable rows with his priest–teacher.[110] He certainly referred to God often enough, whether using God reflexively as in 'by the Grace of God' and 'God knows', or in resorting to such sayings as 'God helps those who help themselves.' He had a growing sense that his own destiny was providentially guided, that he was 'doing the Lord's work'. Speaking before Christmas 1925 in Dingolfing in Bavaria he compared events in the political present with Christ's birth into 'a materialistic world polluted by Jewry'. For Hitler believed that Jesus had been a blond, blue-eyed Aryan rather than a Jew, on the rather shaky ground that he had cleared the Temple of 'Jewish' money-lenders:

Then too, victory did not come by virtue of the power of the State, but through a redemptive doctrine, whose herald was born under the most wretched circumstances. Despite this, people, of Aryan blood, still celebrate this birth. Christ had Aryan blood. Today, we have also given birth to a poisonous period with the State being totally incapable of mastering the situation . . . We National Socialists see in the work of Christ the possibility of achieving the unimaginable through fanatical belief. Christ arose in a rotten world, preached the faith, was scorned at first, but out of this faith a great world movement arose. We want to bring about the same thing in the political sphere.[111]

His sallies into theological matters were unimpressive, the musings of a saloon-bar bore. Hitler's God was not the Christian God, as conventionally understood: 'What is this God who takes pleasure only in seeing men grovel before Him? Try to picture to yourselves the meaning of the following, quite simple story. God creates the conditions for sin. Later on He succeeds, with the help of the Devil, in causing man to sin. Then He employs a virgin to bring into the world a son who, by His death, will redeem humanity!'

Since Hitler thought that heaven housed life's failures or 'women of indifferent appearance and faded intellect', it was probably not the right place for the German Führer.[112] He saw himself on Olympus, surrounded by historical figures of equivalent stature; hell, of which his understanding was primitive, held no terrors for him.[113] While, ironically enough, he respected the Ten Commandments, his attitude to Catholic dogma was that 'A negro with his tabus is crushingly superior to the human being who seriously believes in Transubstantiation.'[114] Nor did his attitude towards Christianity consist of a lapsarian view that it had fallen from the purity of the catacombs into its corrupt present state: 'Pure Christianity – the Christianity of the catacombs – is concerned with translating the Christian doctrine into facts. It leads quite simply to the annihilation of mankind. It is merely whole-hearted Bolshevism, under a tinsel of metaphysics.'[115]

Hitler was rabidly anticlerical, rarely missing an opportunity to make snide and vulgar comments, in private, about the pope, priests and pastors: 'The biretta! The mere sight of one of these abortions in cassocks makes me wild!' The clergy were 'black bugs'.[116] He might, just, have done business with the Borgias, but he thought his Italian co-dictator had been wrong in not throwing the present popes out of the Vatican.[117] He regarded the clergy of both major Christian denominations as devious, effeminate, hypocritical and venal. Their public subsidy should be drastically curtailed, the residue unevenly distributed so as to promote further clerical backbiting and political tractability:[118] 'We can make this clerical gang go the way we want, quite easily – and at far less cost than at present.'[119] During the war, his feelings towards clerics became as murderous as his feelings towards just about everyone else:

> I'll make these damn parsons feel the power of the State in a way they would never have dreamed possible! For the moment I am just keeping my eye on them; if I ever have the slightest suspicion that they are getting dangerous, I will shoot the lot of them. This filthy reptile raises its head wherever there is a sign of weakness in the State, and therefore it must be stamped on whenever it does so. The fate of a few filthy, lousy Jews and epileptics is not worth bothering about. The foulest of the carrion are those who come clothed in the cloak of humility, and the foulest of these is Count Preysing! What a beast! The Popish inquisitor is a humane being in comparison ... The Catholic Church has but one desire, and that is to see us destroyed.[120]

Hitler believed that he had a special relationship with God and Providence, or the belief that all things are ordered and regulated by God towards His purpose. God's will had guided Hitler's personal odyssey from Austrian obscurity to being the German Führer. Speaking in Würzburg on 27 June 1937, Hitler shed light on how this Providence functioned:

> As weak as the individual may ultimately be in his character and actions as a whole, when compared to Almighty Providence and its will, he becomes just as infinitely strong the instant he acts in accordance with this Providence. Then there will rain down upon him the power that has distinguished all great phenomena in this world. And when I look back on the five years behind us, I cannot help but say: this has not been the work of man alone. Had Providence not guided us, I surely would often have been unable to follow these dizzying paths. That is something our critics should above all know. At the bottom of our hearts, we

National Socialists are devout! We have no choice: no one can make national or world history if his deeds and abilities are not blessed by Providence.[121]

Scenes from what was retrospectively dubbed 'the time of struggle' were celebrated in Nazi art. Hermann Otto Hoyer's *In the Beginning was the Word* (1937) depicted a besuited Hitler speaking on a raised dais in a modest dark room, one hand on hip, the other raised to define an idea, while 'the word' illuminated the faces of the men and women seated nearest to him.[122] There was nothing uniquely National Socialist in the transposition of terms such as 'belief', 'creed', 'confession', 'faith', 'resurrection', 'sacrifice' and 'witness' from the religious to the political domain. Any politician worth his or her salt in any Western liberal democracy does the same nowadays. Hitler's heavily stylised biography, often interpolated in his speeches, exploited a narrative that intersected at various points with the life of the Messiah in a way that was blasphemous. Like Christ in provincial Galilee, Hitler came from a humble backwater on the peripheries of an empire. The Great War was the authentic experience that emotionally reconnected the listless drifter with millions of ordinary Germans who, like him, had also returned to the chaos and political strife of the Weimar Republic.[123] It was a two-way process, like people trying to touch each other in a dark room. Hitler's early supporters had 'found their way' to him, their faith giving their lives 'new meaning and a new goal', or something akin to the transforming experience of a religious conversion.

This mass appetency was culturally and historically determined. For just as some people voted for the Nazis for reasons of socio-economic self-interest, so others attached more importance to cultural, moral or religious factors. This is hardly surprising. There was a long-standing desire on the German populist right for an authoritarian and charismatic leader, albeit one better suited to an age of mass politicisation than Bismarck or the Kaiser, in that this leader had to be both representatively demotic and 'extraordinary' in his personal powers. That is what one early Nazi meant when he said: 'I did not come to Hitler by accident. I was searching for him. My ideal was a movement which would forge national unity from all working people of the great German fatherland ... The realization of my ideal could happen through only one man, Adolf Hitler. The rebirth of Germany can be done only by a man born not in palaces, but in a cottage.'[124] Hitler could articulate what many people were feeling in terms that resonated. In a letter written to the Führer in 1926, Goebbels declared: 'You gave a name to the suffering of an entire generation who were yearning for real men, for meaningful tasks . . . What you uttered is the catechism of a new political credo amid the desperation of a collapsing, godless world. You did not fall silent. A god gave you the strength to voice our suffering. You formulated our torment in redemptive words, formed statements of confidence in the coming miracle.'[125]

The language and imagery of the Bible were essential to this process of search and discovery in a culture that had yet to become illiterate in its own Christian heritage. Hitler had found the German people, and they had found him, in their mutual hour of need, a miraculous encounter that he spoke of at a Nuremberg rally on 13 September 1936: 'That is the miracle of our age – that you have found me [lengthy applause], that you have found me among so many millions! And that I found you, that is Germany's good fortune.'[126] As J. P. Stern showed in one of the few studies concerned with Hitler's use of language, entire passages from the Lutheran Bible were incorporated into Hitler's speeches, the original allusions being reinserted here as references in parentheses:

> How could we help but feel once more in this hour the miracle that brought us together! Once you heard the voice of a man, and that voice knocked at your hearts, it wakened you, and you followed that voice. For years you pursued it, without ever having seen the owner of that voice; you simply heard a voice and followed it. [Luke 3:4 and John 20: 19–31]
>
> When we meet here today, we are all filled with the miraculousness of this gathering. Not every one of you can see me, and I cannot see every one of you. Yet I feel you, and you feel me! [John 16: 16–17] It is the faith in our Volk that has made us little people great, that has made us poor people rich, that has made us wavering, discouraged, fearful people brave and courageous; that has made us, the wayward, see, and has joined us together! [Luke 7: 22]
>
> Thus you come from your little villages, from your small market towns, from your cities, from the mines and factories, leaving the plough; one day you come into this city. You come from the limited environment of your daily life-struggle and your struggle for Germany and our Volk, to have for once the

feeling; now we are together, we are with him, and he is with us, and we are now Germany. [Matthew 2: 6 and John 14: 3][127]

As this suggests, the fundamental structure of the Nazi creed was soteriological, a redemptive story of suffering and deliverance, a senti-mental journey from misery to glory, from division to mystic unity based on the blood bond that linked souls. Again, the first part of this is unsurprising, since any politician in most Western liberal democracies has a vested interest in painting the recent past as black as the future will be bright, although the blood business is quite barmy. The message of sentimental belonging, by virtue of race and nation, resonated deeply in a society that had been politically unified only in the late nineteenth century, whose monarchical institutions had been swept away after the Great War, and which was riven with confessional, cultural and political conflicts. So did the belief that God had chosen not the German people, but the 'Aryan-Germanic race' for His divine purposes, something that chimed with a long-standing Protestant German belief in the nation's divine chosenness in good times and bad.[128] It also chimed with a liberal Protestant theology whose God-decreed 'orders of creation' had been extended from the family, nation and state to include 'race'. Where this led is clear from a passage by Paul Althaus, one of Germany's most distinguished Lutheran theologians:

> As a creation of God, the Volk is a law of our life ... We are responsible for the inheritance, the blood inheritance and the spiritual inheritance, for Bios and Nomos, that it be preserved in its distinctive style and authenticity. We are unconditionally bound to faithfulness, to responsibility, so that the life of the Volk as it has come down to us not be contaminated or weak-ened through our fault. We are bound to stand up for the life of our Volk, even to the point of risking our own life.[129]

Much the same could probably have been heard from the Evangelical or Reformed theologies of slavery, segregation and apartheid in respectively the US South or South Africa. Those Lutherans who opposed Nazism, and played a part in the Confessing Church, equated Nazism with Jewish nationalism or Zionism, as 'a regression from universal humaneness', a trope that has subsequently proved highly popular among left-wing antisemites.[130]

It followed that many Protestant Christians had few theoretical difficulties with pseudo-scientific doctrines and policies that the Catholic Church explicitly condemned on the ground that the immortal individual soul always took precedence over 'supra-individual forms of life'. A theology of 'orders of creation' accommodated eugenics well before the Nazis issued sterilisation laws. The sanctification of earthly collectivities inevitably led to the desanctification of individual life. 'The mistake', the influential Protestant scientist Bernhard Bavinck explained, 'of many Christians is that they do not or cannot see that populations and races have the same standing in God's creation as individuals and therefore have the same claim to existence and protection from extermination ... God's creation obliges us with all our might to protect the well being of that whole to which we as individuals are subordinate: our Volk.' Indeed, such collectivities as race or nation had a higher claim than the individual 'under certain circumstances' and in the light of a new 'organic ethic'. The object of the latter was to treat Christian injunctions to brotherly love as a form of false consciousness. Others argued, as early as 1924, that eugenic intervention meant interfering not with God's handiwork, but with the consequences of the sins, such as alcoholism or sexually transmitted diseases, that the individual had brought upon himself or herself.[131]

These were not abstract concerns, and many of the Evangelical Christians who most counted in welfare circles agreed with him. In 1930–1 the Inner Mission – the main Protestant welfare association with a ramified network of charitable institutions across Germany – decided that sterilisation was 'morally and religiously legitimate', indeed a moral duty vis-à-vis future generations. The Inner Mission supported the decriminalisation of voluntary eugenic sterilisation in the November 1932 Prussian draft hereditary health law, although the consent of a person's legal guardian would suffice, suggesting too that in some circumstances the use of castration or X-ray sterilisation should replace the surgeon's knife. These measures should be restricted to people whose behaviour indicated that their children would be 'anti-social'. Because of this background, it is unsurprising that these Protestant circles raised few, if any, objections to the Nazis' introduction of compulsory sterilisation in order to fortify the race, rather than as an act of Christian concern.[132]

Even where Nazism appeared most indebted to modern science, namely in claiming that its racism was 'scientific', this discourse was as much cultural and religious, as anyone who can be bothered with, say, Houston Stewart Chamberlain's *Foundations of the Nineteenth Century* can readily establish. Take the notion of an 'Aryan'. There was nothing 'scientific' about this idea, which derived – oddly enough – from the age-old quest to discover what language was spoken in the Garden of Eden. The nineteenth-century discipline of philology secularised that quest, by transforming it into the search for the *Ursprache*, the ultimate ancestor of our modern European languages. Affinities with Sanskrit led to the hypothesis that these languages were 'Indo-Germanic' or 'Indo-European' in origin, fundamentally different in structure from what came to be known as Semitic languages, which included Arabic as well as Hebrew. The parallel discipline of speculative racial anthropology ensured that these Indo-Europeans or Aryans next acquired a face, a genealogy and characteristics, a process that largely involved transposing the most creative, dynamic and noble aspects of the ancient Greeks on to these mysterious peoples, while a lot of negative and static features were heaped on the Jews (the Arabs were quietly omitted) as the Aryan's 'spiritual' antipode.[133]

It was not a large step to invest these hypothetical Aryans with divine characteristics. Hitler's own description of the Aryan was as 'the highest image of the Lord'.[134] By Aryan, he meant 'the founder of all higher humanity . . . the prototype of all that we understand by the word "man". He is the Prometheus of mankind from whose bright forehead the divine spark of genius has sprung at all times, forever kindling anew that fire of knowledge which illumined the night of silent mysteries and thus caused man to climb the path to mastery over the other beings of this earth.'[135] The Aryan was creative and had the innate capacities for self-sacrifice and social cohesion, which explained all his conquests and creative endeavours throughout time. In a word, Aryans were synonymous with idealism.[136] Aryans were the eternal core of the Germanic race, whom God had chosen to carry out a redemptive mission on earth. Hitler derived his power in the symbolic sense from being both the prophet who identified that destiny and the leader who would fulfil it. Failure meant a planet on which all higher human life would have perished, left to orbit without purpose in a dark void. As the executive arm of the race-nation, the state's duty was to bring these residual Aryan elements together, husbanding the stock of their blood (a substance Hitler spoke of with mystical fervour) through eugenic regulation of marriage and procreation. Aryans faced perpetual dilution through mixing with lesser races, a process Hitler identified as the greatest sin: 'In the end, however, the [Aryan] conquerors *transgress* against the principle of blood purity, to which they had first adhered; they begin to mix with the subjugated inhabitants and thus end their own existence; *for the fall of man in paradise has always been followed by expulsion.*'[137]

The Aryan's maleficent counterpart was 'the Jew', for invariably Hitler used the singular whenever he spoke of Jews. This is not the place to rehearse the history of Nazi antisemitism, so only a few of its relevant characteristics need be mentioned here. 'The Jew' was the negation of the Aryan's God-given properties. As Goebbels uncharmingly put it: 'The Jew is indeed the Antichrist of world history.' Speaking in 1921, Hitler had already transformed Christ into an Aryan and 'the Jew' into the Devil: 'I can imagine Christ as nothing other than blond and with blue eyes, the devil however only with a Jewish grimace.' 'The Jew' was allegedly a materialist rather than an idealist, lacking culture-creating capacities – an anarchic, egoistic and individualistic 'destroyer of culture'. Hitler quoted Goethe's *Faust* to suggest the satanic: 'his intellect will never have a constructive effect, but will be destructive, and in very rare cases perhaps it will at most be stimulating . . . the prototype of "the force which always wants evil and nevertheless creates good"'. Christ himself was invoked to darken this picture further: 'Of course the latter made no secret of his attitude towards the Jewish people, and when necessary he even took to the whip to drive from the temple of the Lord *this adversary of all humanity,* who then as always saw in religion nothing but an instrument for his business existence.'[138] This suggested that National Socialism had a deeper understanding of religion than the Churches, and that its socio-economic doctrines could be presented as an attempt to realise 'true' Christianity, something many Christian Germans were only too eager to believe.

One final aspect of Nazi antisemitism deserves comment, especially since it derives from an insight of Sigmund Freud's. He had not deigned to write about this unsavoury phenomenon until the Nazis attacked his own 'Jewish science' of psychoanalysis and his own family came under direct threat. In his *Moses and Monotheism,* published in exile in 1939, and partly inspired by Michelangelo's great sculpture of Moses in St Peter's, Freud tried to distinguish between the superficial 'causes' of antisemitism – such as xenophobia – and what he thought were the deeper reasons. Unsurprisingly he highlighted the place of the Jews in 'the unconscious of the peoples', arguing that it was the Jewish claim to chosenness and moral superiority – symbolised by Moses – which caused others to resent them to the point of hatred. Pagan barbarians who had bitterly resented the coercive imposition of Christianity upon them projected this on to the Jews, a move made easier by the Gospels, which largely described Christian Jews.[139]

Of course, the subject of Nazism is not exhausted by reference to antisemitism. Just as National Socialism sought to transcend the confessional divide, so it looked for a Third Way between the Scylla of liberal capitalism and the Charybdis of Marxism, both regarded as twin offspring of the will to power of 'international Jewry'. German Protestantism provided at least one forerunner for such a project; the Christian Social Workers' movement of court preacher Adolf Stoecker before the First World War. Like all attempts to capture Protestant Germans within a political party, this one failed, although the electoral profile of the NSDAP would finally reach this holy grail, for in many respects the Nazis did for Protestants what the democratic Centre Party had done for Roman Catholics. While leading Nazis sometimes spoke of their socialism of the deed – to distinguish it from the ineffectually theoretical variety – they were more likely to claim that it was an attempt to implement a pure form of Christianity. In the depths of the Depression Hitler said: 'As Christ proclaimed "love one another", so our call – "people's community", "public need before private greed", "communally minded social consciousness" – rings out through the German fatherland! This call will echo throughout the world!' Ethics would prevail over economics in the sense that voluntarism would cure most social-economic ills through a combination of sentiment and will. Charity had other virtues. It enabled the Nazis to divert welfare resources elsewhere, and demonstrated the newly found spirit of community in action through the mobilisation of positive sentiments, even if individual charities were subsumed into a bureaucracy every bit as coldly impersonal, and a great deal crueller, than the welfare state apparatus that was abolished.[140]

This ostentatious subscription to Christian charity resonated among Protestants who identified the Weimar Republic not only with social atomisation and self-seeking, but with a more thoroughgoing moral breakdown. This was as true of Protestant women as of men, since organisations like the Protestant Mothers' Association or the Protestant Ladies' Auxiliary were staunch in Hitler's support. Noted Protestant theologians claimed that they had been fighting the good fight for many years. As Paul Althaus wrote:

Theology has waged a determined struggle against the individualistic and collectivistic attack on single marriage, against
irresponsibility, contraception and abortion, against the liberal-
capitalist and Marxist spirit in the economy and society, against
deflation of the state, against pacifist effeminacy of political
ethos, against the destruction of penal law and the surrender
of the death penalty – in general, for the order of God as the
standard for the shaping of common life.[141]

When, on coming to power, Hitler claimed that his revolution was a
moral restoration, there were many Protestant Germans all too eager
to believe him: 'The national government will regard its first and
foremost duty to restore the unity of spirit and purpose of our *Volk*. It
will preserve and defend the foundations upon which the power of
our nation rests. It will take Christianity, as the basis of our collective
morality, and the family as the nucleus of our *Volk* and State, under its
firm protection.' As if to confirm that the Nazi revolution coincided
with a religious revival, the worrying numbers of people formally leaving
the Protestant Churches under Weimar was replaced by large numbers
of people joining them. Protestant clerics and associations compared
the 'national uprising' with the Reformation.[142]

There was a final way in which Nazism deserves the epithet political
religion – its liturgy or rituals and use of sacred spaces. If churches
are built to encourage individual contemplation, these Nazi ceremonies
were intended to induce paroxysms of mass emotion that are hard to
recover today. They were also designed to contrast with the mood before
the Nazis came to power, reducing the brief democratic experience
to something resembling clinical depression in which colour leaches out
into an all-pervasive grey.

In August 1924 a listless Goebbels wrote in his diary:

A grey day has arisen. The rain falls and trickles in long streams
down the window. Autumn has descended upon Germany. Grey
autumn. Strength freezes in the veins, and life no longer pulses
so strongly in the heart. Faith has become poor and hope has
dried up. We no longer see the stars. Darkness. Evil has entered
into his realm. The bright light has vanished. We must rest and
find new courage. Dark day. The dawn breaks in grey. Will there
ever be light again?

Since Orwell's *Nineteen Eighty-Four,* totalitarian regimes have been synonymous with Soviet drabness, with a world of grimy overalls and oily gin. Monotone was hardly characteristic of Nazi Germany before the Second World War. The Nazis launched a multi-hued bombardment of the senses, disciplining masses of people into choreographed formations, and concussing their eardrums with rousing marches and choruses. It is quite difficult for an age that has outgrown participation to understand this. Hans Kohn, a great historian of nationalism, wrote in 1938: 'Fascism is a continuation of "the stupid nineteenth century", of its sense for mass movements and their dynamic quality, its love of quantity, noise, and acceleration, its desire for gigantic size and stupendous manifestations of power.' He might have added its sinister glamour, the very quality that fascinates television commissioning editors and producers, if no one else. Of course, there was more to Fascist or Nazi politics than seduction, for the seduced were hardly innocents abroad, least of all in relation to the real and symbolic successes of these regimes in the economy or foreign policy, but one cannot leave out of an account of political religions the rites and rituals through which they built their version of community. These provided a rhythm and tone, although the range was limited to the aggressively military or the appallingly plangent.[143]

Hitler and his propagandists created a 'Führer-cult', often relying on venerable tropes of the ruler–ruled relationship, which became the focal point of a regime of commemorations and celebrations that blurred Party and state, and subtly incorporated such rivals as the Christian calendar or the international labour movement's May Day. The propaganda chief Goebbels and others, such as the courtier architect Speer, worked out the details, but Hitler took a personal interest in National Socialist festivities, as in most aspects of design, above all in the tilting swastika symbol – to suggest movement – and the appropriation of the socialists' red as the most stirring colour.

The inner-worldly Nazi Church had 'Blood' as the centre of its creed. Then came the carriers of the Blood – the *Volk* – followed by the Soil that sustained them – *Blut und Boden* being the favoured slogan – and then the Reich which gave the People political form and the Führer who embodied and represented them. The holiest symbol was the swastika flag, or rather the blood-stained swastika carried on 9 November 1923, which Hitler used to consecrate lesser flags by rubbing them together. That was the essence of the faith which rituals were designed to communicate.

Festivals relied upon either existing or purpose-built sacred spaces, the former being made over with the appropriate blend of archaising symbols such as swastika banners, pylons and urns. The Nazi festive year commenced on 30 January, with celebrations marking the 'seizure of power', the high point being a reprise of the torch-lit parades that had dramatised the *Machtergreifung*. Commemoration of the promulgation of the Party's programme on 24 February did not catch on. The Weimar Republic had instituted a 'Day of National Mourning', or Remembrance Sunday, in 1925 to commemorate the dead in the Great War. Since the Nazis regarded this as too 'negative', they replaced it with a more upbeat Heroes' Memorial Day on 16 March, appropriating the war dead as harbingers of Germany's resurrection under Hitler's dictatorship. The last Sunday in March was reserved for induction of children into the Hitler Youth or League of German Maidens. Political paladins of the NSDAP celebrated Hitler's birthday on 20 April by renewing their personal loyalty oaths.

The socialist May Day was rejigged as the 'Day of National Labour' before becoming a general national holiday celebrating the transition from spring to summer. Mothers' Day on the second Sunday of May celebrated these Stakhanovites of the maternity wards. Neo-pagan enthusiasts, notably Himmler and Rosenberg, were allowed their own special day, on 21 June, when they used the element of fire to celebrate the summer solstice.

The Nazi Party celebrated itself for an entire week in early September, at the annual rallies held in Nuremberg, which was designated 'capital city of the Movement'. It reminded outsiders of Mecca.[144] Although Nuremberg had traditionally had a Social Democratic Party majority, from 1927 onwards it was decided to hold the annual rallies there, probably because the local police chief was notoriously indulgent towards the Nazis. There were also historical continuities with the medieval imperial Diets, and Wagner's *Meistersinger*, that could be exploited, and in any case the city had good communications with the rest of Germany. The prospect of regular business meant that the city administration smiled benignly on Speer's architectural projects, which turned extramural meadows into vast cultic sites upon which the faithful converged.

Each annual rally had a theme – 1935 Day of Freedom, 1936 Day of Honour and so forth – with the content of speeches squeezed into this straitjacket. For example, in 1936 Hitler closed a 'dishonourable' chapter in German history with the restoration of the nation's sovereign 'honour' through the remilitarisation of the Rhineland. Each day of the rally highlighted a particular Nazi formation, sometimes sending out a message about where it stood in the pecking order, like the 'cleansed' paramilitary SA after the summer 1934 purge. Leni Riefenstahl's chilling film of 1934, *Triumph of the Will*, shows the developing format, beginning with Hitler's aircraft delivering him to the expectant multitudes as the sun pierced lowering skies. After an afternoon of official receptions, Hitler spent the second day greeting the Hitler Youth and opening the Party Congress, with solemnities for the Party's martyrs. On the third day Hitler was saluted (with raised spades) on the Zeppelinfeld by fifty thousand members of the Reich Labour Service, who began by crying out the region they hailed from, in symbolic exemplification of the notion of 'national community'. The ceremony culminated in a series of injunctions and responses. 'No one is too good,' droned the loudspeaker; 'to work for Germany!' came the massed response. The fourth day was devoted to sports and gymnastic displays on the Zeppelinfeld, followed by a torch-lit procession to Hitler's Deutsches Hof hotel.

After the Party's capillary formations had separately convened on the fifth day, the night was given over to a mass rally, with Speer's 'cathedral' of blue-tinged electric light vaulting like streams of ice into the night. This ring of light protected the participants against the dark Walpurgis in which hovered Bolsheviks and Jews. The night finished with a procession of Party standards, some of which were equipped with their own up-lighters, the effect being a richly intense experience of densely saturated colour reminiscent of a 'flowing stream of glowing lava' in which the individual was lost in the strength-giving mass. Hitler mounted the altar cum podium to honour the Party's dead once again and to deliver a brief speech. The sixth day was devoted to the massed ranks of the SA on the Luitpoldhain. Hitler walked through these formations, to have solitary communion with the Party's most sacred relic, the 'blood banner of the Movement'. He then consecrated the new standards of the SA and SS by rubbing them against the 'blood banner', a magical gesture accompanied by manly handshakes and unwavering dictatorial eye. Just about everything worth knowing about Nazism is contained in that moment.

The philologist Viktor Klemperer glumly watched this moment in a cinema, remarking: 'This whole National Socialist business is lifted from the political realm to that of religion by the use of a single word. And the spectacle and the word undoubtedly work, people sit there piously rapt – no one sneezes or coughs, there is no rustling of sandwich paper, no sound of anyone sucking a sweet. The rally is a ritualistic action, National Socialism is a religion – and I would have myself believe that its roots are shallow and weak?'[145] About 120,000 men then marched past Hitler in Nuremberg's main square, including the white-gloved giants of his own SS bodyguard smashing their jackboots down on the cobblestones – stirring scenes guaranteed to erase memories of pot-bellied functionaries or the occasional formation that made a hash of marching past the Führer. These SS 'zigzag men' were partly like ancient dancers, partly something jagged taken over from Expressionism, like the high-voltage warning signs cum runic symbols that decorated their collars and helmets. The final day was given over to the armed forces, who demonstrated their military prowess on the Zeppelinfeld. The rally ended at lunchtime with a final speech by the Führer, the focus of every one of the week's successive celebrations.[146]

Using architecture, sound, light and quasi-liturgical responses, these rallies were the nadir of Nazi attempts to replace politics as rational conversation with affect and sensation. The choice of being actor or audience was nullified by making everyone a participant. Although every audible or optical effect was carefully managed, that was hidden from the ranks of participants, who found themselves in a world of aesthetic and emotional intoxication or *Rausch*, qualitatively distinct from the state of the sots in the upper echelons of the ruling Party. The Christian Harvest Festival in early October was replaced by syncretic celebration of fertility – animal, human and vegetable – notably at Bückeberg near Hameln.

Martyrs were an essential element of all three totalitarian political religions. Düsseldorf tried to get in on the act by creating a cult of relics connected with Albert Leo Schlageter, who had been shot by the French in the occupied Ruhr. His bed was reconstructed, and Hitler received a silver reliquary, allegedly containing the bullet with which he had been killed. This cult never took hold.[147] The most solemn Nazi festival of martyrs was 'Memorial Day for the Fallen of 9 November', whereby the Nazi Party commemorated the sixteen men killed in the abortive 9 November 1923 putsch. This was a very subtle blending of wartime remembrance days with Corpus Christi processions, whose purpose was to transform a squalid fiasco into one of the most significant events in German history. The defeat of the putsch became a victory because the dead men's 'sacrifice' heralded the Nazi 'seizure of power' a decade later. The shots fired by Munich policemen had only succeeded, as Hitler unfortunately put it, in 'stirring the river of blood that has flowed ever since'. Their blood, he explained in 1934, was 'the baptismal water' of the new Reich. That year, he merely laid a wreath at the Feldherrnhalle. By 1935 altogether more elaborate arrangements had been made, which never changed thereafter, whenever Hitler had to commune with his sixteen 'Apostles' – for naturally he had to go four better than the original Messiah.

The religious parallels began on the evening of 8 November, when Hitler and his 'old guard' had a 'Last Supper' in the historic Bürgerbräukeller. The next day, a silent procession snaked through the streets of Munich, a procession literally signifying the Movement, with only drumbeats marking its progress. The procession passed 255 portentous-looking pylons or stelae supporting urns from which smoke rose, and on which the names of all the Party dead were inscribed. The lower floors and shop fronts were covered by red cloth to mask distractions, while banners hung from the upper floors and criss-crossed the streets. After pausing to honour the dead at the first cult site, the Feldherrnhalle, the procession turned into a triumphal march to the Königsplatz, the march symbolising the Nazi 'seizure of power' in 1933. Paul Ludwig Troost had constructed two mausoleums, each with a sunken chamber containing eight of the iron sarcophagi in which the sixteen martyrs were buried. These were exposed to the elements, so that both God and 'the Reich' could see them. Dedicating these temples in 1935, Hitler plumbed uncharted depths of bathos:

> Because they were no longer allowed to personally witness and
> see this Reich, we will make certain that this Reich sees them.
> And that is the reason why I have neither laid them in a vault nor
> banned them to some tomb. No, just as we marched back then
> with our chest free so shall they now lie in wind and weather, in
> rain and snow, under God's open skies, as a reminder to the
> German nation. Yet for us they are not dead. These pantheons
> are not vaults but an eternal guardhouse. Here they stand guard
> for Germany and watch over our Volk. Here they lie as true
> witnesses of our Movement.[148]

A roll-call of the martyrs' names was taken, with the Hitler Youth responding 'Present!' Hitler walked up the steps of the mausoleums to commune silently with the not-really-dead, who became figuratively present in the SS guards who took up stations after Hitler had left.[149]

The SS were the avantgardistas in seeking to synthesise hyper-bureaucratic rationality with an almost postmodern mix of beliefs ahistorically derived from pagan, Christian and non-European cultures. The beliefs were like one of those silly 1980s buildings that merge snatches of the Egyptian or Greek with tubular steel, glass and concrete. The 'order' itself was based on such exemplars as the Teutonic Knights, the Jesuits and the Japanese samurai, with a grudging nod towards the Bolshevik NKVD. Himmler's own cranky interests, which extended to mad theories about Aryans emerging from beneath a global ice shield, accounted for the extreme unrelatedness of SS cultic sites. Some rocks near Detmold called the Externsteine were supposed to have been an ancient Germanic pagan sanctuary, but there was also Quedlinburg cathedral and the tomb of Henry the Fowler, with whom the Reichsführer-SS was in mystical communion. Himmler alighted upon Wewelsburg castle near Paderborn in January 1933 while resting during an election campaign. He planned to restore it as an SS version of the Vatican, a spiritual redoubt for the forthcoming war with 'Asia'. The content was derived from his conversations with an elderly SS officer called Karl Maria Wiligut, who had spent forty years in the Habsburg army before retiring to produce an antisemitic paper called the *Iron Broom*. The SS suppressed the intervening four years in a Salzburg asylum as a certified paranoid schizophrenic. Wiligut, or Weisthor as he preferred, claimed to be in 'ancestral-clairvoyance' with the original Germans in 228,000 BC. These 'live reports', so to speak, tantalised Himmler, who gave the old lunatic the honour of designing the SS 'Death's Head' ring and promoted him to brigadier.[150] Some of this non-sense remained arcane and restricted to the SS chief's court circle, while other parts streamed into the broader culture.

The Nazi year ended with Party formations celebrating the winter solstice on 21 December, an occasion called the Yule Festival within the SS, where senior SS officers could look forward to the gift of a Yule Light from their own leader, an object more fitting for Halloween. The Nazis intended to strip Christmas of its Christian associations, turning it into a general celebration of goodwill and the advent of the New Year, a goal pursued nowadays in Britain mainly by local government. Worryingly, within a relatively brief space of time, these festivals – which confused the Party with traditions stemming from Christianity, the German state and the labour movement – showed signs of being accepted and established. Doubtless the pattern would have become entrenched had the war turned out differently, with the possibility of a real war on God becoming a reality, as it did where the Nazis could operate without restraint.

Speaking of which, in the last decade several younger German academics have devoted almost obsessive attention to studying people much like themselves in terms of age, education and social mobility – that is, the leadership group of the SS Reich Security Main Office and of such organizations as the SS Security Service or SD, from which many organisers of the 'Final Solution' were drawn. Many of the conclusions of such studies are unexceptional, concerning as they do the ways in which Nazism licensed any number of expert professionals and technocrats to implement on a colossal scale their fantasies of power and control. Doctors who felt that prescribing sedatives and suppositories did not conform with their inflated sense of professional self became biological sentinels of the nation, watching over the flow of the gene pool.

These men were socially chippy, highly ambitious and morally autistic, and above all 'unbounded' in what they might do to others to get ahead. A career in the SS administration meant that one did not have to wait for some ineffectual greying professor, to whom one hitherto had to crawl and slime, to keel over with a heart attack. And this was the 'real' world too. The SS found new bureaucratic models to get over the rule-bounded nature of the state bureaucracy, enabling these thrusting young men to exercise initiative and implement the ideas of the centre in any given local context. They could crop up anywhere, gingering things up with their unique brand of amoral fanaticism. These men then found themselves in the occupied and war-torn East where civil norms no longer applied and where they proved themselves insanely fertile in destructiveness, for their subscription to the codes of their own bureaucracy was never incompatible with the most irrational, pathological fantasies. Behind depredations so carefully recorded in graphs, flow charts and statistical tables lay the holy mysteries of blood with which we started. But we should resist further discussion of a part in favour of the whole.[151]

These reflections upon the totalitarian political religions have occasionally alluded to thinkers who saw clearly that this was indeed what these movements and regimes were. Although many of them were political scientists or philosophers rather than historians, they were really trying to identify psychological commonalities to understand the violent passions unleashed. That is why people read them with profit decades later. Let's conclude by discussing some of these remarkable people in greater detail.

Waldemar Gurian was born in St Petersburg in 1902 into an assimilated Jewish family. After moving to Germany in 1912, his mother converted to Catholicism, into which faith Gurian followed two years later. He worked as a freelance journalist and writer in Bonn. In 1931 he published *Bolshevism. Theory and Practice*, which proclaimed:

> Bolshevik atheism is the expression of a new religious faith, the faith in an earthly absolute, which, its adherents believe, renders a God, Creator and Lord of the World and the Final Cause to which everything earthly, indeed the entire universe, is ordained, superfluous, an empty hallucination ... The new 'God' is the Socialist society, the first principle of Communism ... Faith in this new 'God' is the power which determines the entire edifice of Bolshevism ... It enables them to pass over failures, and admits of no compromise of principles, but only breathing spaces in the battle.[152]

Gurian had to flee Germany in 1934. He went to Switzerland where he published a number of important books and pamphlets. With a fellow exile, Otto Knab, he published a series of 'German Letters' about conditions in their homeland. Totalling two thousand pages of print, these provide one of the most important analyses of Nazi Germany from a Catholic point of view. This rather saintly, bumbling fat man, who spoke every language with a Russian accent, would find a sort of peace at the university of Notre Dame.[153]

At roughly the same time, the self-styled 'pre-Reformation Christian' Eric Voegelin published a short but Olympian essay entitled *The Political Religions*. Nothing could be further removed from the shelves filled with swastika-adorned 'mob-literature' on the Nazis that people consume nowadays along with endless trashy television programmes devoted to that phenomenon made by people who are unaware that they are debasing our culture by recycling the Nazis' own propaganda, intercut with less than illuminating reflections from sundry geriatric parties too young at the time to have exerted real power or influence. Indeed, in some provocative lectures he gave in Munich in the 1960s, in which he called the entire pre-war German elite a 'rabble', Voegelin said that many historians of Nazism were the problem, rather than the solution, in the sense that, blind to the possibility of its recurrence, they focused on trivia or unwittingly reproduced its own self-dramatising teleology for modern audiences, almost reinfecting future generations with the virus. Although his thought is immensely complicated, there was one powerful moral consideration that drove it:

> A further reason for my hatred of National Socialism [other than its fraudulence] and other ideologies is quite a primitive one. I have an aversion to killing people for the fun of it. What the fun is, I did not quite understand at the time, but in the intervening years the ample exploration of revolutionary consciousness has cast some light on this matter. The fun consists in gaining a pseudo-identity through asserting one's power, optimally by killing somebody – a pseudo-identity that serves as a substitute for the human self that has been lost.[154]

A man of formidable erudition, who would learn Russian just to read Dostoevsky, Voegelin had already fallen foul of the Nazi regime by writing in support of the Austrian *Ständestaat* of Dollfuss, on the ground that its authoritarianism was a defensive reaction against totalitarian ideologies that might have evolved in a more democratic direction. More particularly he published a devastating critique of Nazi racial 'science' as being no science at all but something he called 'scientism'. Influenced by both the satirist Karl Kraus and Max Weber, Voegelin thought that there were fundamental commonalities between human beings across vast reaches of time. From Kraus he learned to be alert to the debasement of language that long preceded, and made possible, something so ignorant and vulgar as Nazism. While he respected Weber's preoccupation with establishing the truth, he also thought that the demoralisation of social science meant that scholars were emasculating themselves regarding evil, immoral and unethical political ideologies. For him, Evil was a palpable actor in the world. Periods in the US, where he studied the pragmatic philosophers, only increased his impatience with what he saw as the profound provinciality of German academic culture – although, as friends of mine who were his pupils aver, even Mrs Voegelin always remained the Frau Professor.

In his *Political Religions*, which was published in 1938 and promptly confiscated, Voegelin argued that totalitarian ideologists were in the same tradition as the political religions of ancient Egypt, when Akhenaton had briefly transformed himself into a god, and the medieval and early modern millenarian perversions of Christianity. These were secular, temporal attempts to recreate a religious community to assuage mankind's spiritual needs. Even as they denied divine reality, they sought to impose a perverted temporal reality on humankind. The ideology and the Party–Church that incarnated it provided a surrogate affective community based on the terrible pathos and plangency of class, race or nation, in which the lonely individual could re-experience the warm fraternal flow of the world. The positive symbolism of the political Church community was accompanied by the 'anti-idea', or the Satanic foe who opposed the ideology embodying the Good.

In exile in the US, Voegelin spent the war largely focused on Nazi Germany. He came to the conclusion that Nazism was a form of emotional tribalism: 'Tribalism is the answer to immaturity because it permits man to remain immature with the sanction of the group.' He revised his earlier analysis of political religions so as to accommodate Marxism. Specifically, he turned to ancient and medieval Gnosticism – the belief in hidden certainties vouchsafed to the few – to explain the awful sureness of modern ideologists similarly seeking salvation from life's existential uncertainty. Ideological explanations of reality were in fact deformations of that reality since they limited the 'explanation' solely to the temporal world. Without a moral code derived from a transcendental God there was nothing to inhibit them. Any means were justified, from lying propaganda to physical mass murder, to bring about the desired realm of Good on earth, that being the key to the moral insanity that Communism and Nazism unleashed on the world, for massive violence was rendered unreal within the ideological dream world their devotees inhabited.[155]

It is impossible to impose a simple left–right framework on those who argued that totalitarian regimes were political religions. Although Voegelin was an instinctual conservative who fled the East Coast Ivy League for Baton Rouge in Louisiana on the ground that the former harboured too many spiritual totalitarians, many of his ideas were also shared by the Austrian heterodox leftist Franz Borkenau, most famous for one of the best books ever written about the Civil War in Spain. To the left's horror, this was a man who changed his mind. A renegade or 'loose cannon'.

Borkenau was the archetypal political renegade, with all the fervour of the adult convert. He was actually called Franz Pollak, but having converted to Christianity his judge father thought that the surname Borkenau might help the son in a military career. Borkenau gained a doctorate in history, joining the German Communist Party in 1921, and played a leading role in the KPD's adoption of the nationalist Leo Schlageter, who had been executed by the occupying French, as a martyr figure. Largely driven by his hatred of his own solidly bourgeois background, as he freely admitted, Borkenau worked for the research department of the Comintern organisation of world Communists within the Soviet's Berlin embassy. After the Nazis came to power, he embarked on a life of exile that took him to Vienna, London, Paris, Panama and Spain. There was even an interlude in Australia after the British threatened him with internment as an enemy alien. Eventually he would work as part of the British and US propaganda campaign against Nazi Germany. Returned after the war to journalism in Frankfurt, he was an early recruit to the Congress for Cultural Freedom, responsible for such marvels as the journal *Encounter*.

In 1940 Borkenau published *The Totalitarian Enemy*. This rejected every marxisant explanation of 'Fascism' while linking the two extreme ideologies together in a way that the left found heretical: 'The essence of these revolutionary creeds is the belief that the final day of salvation has come, that the millennium on this earth is near; that God's chosen instruments must make an end of all the hierarchies and the refinements of civilization in order to bring it about; and that complete virtue, simplicity, and happiness can be brought about by violence.' Of the Nazis, Borkenau wrote that they were 'negative Christians', in a state of 'ferocious revolt against the tenets of Christianity and therefore worshippers of all that in the Christian tradition is regarded as Satanic'. Rather daringly he asked: 'What else is the belief in the special divine election of the German people, but the Jewish idea of the Chosen People, transferred to Germany? And what else is Hitlerism unless it is these two credos: first, that the Germans are God's Chosen People, by nature superior to all other people, predestined to rule the world and to bring salvation to it; and secondly, that Hitler is the chosen prophet of the chosen people?'[156]

Finally, the subject of political religions occupied one of the finest minds in twentieth-century France, the liberal conservative sociologist and journalist Raymond Aron. If the French mandarin elite have any excuse to exist, Aron is probably it, since his entire output of articles and books is characterised by an impassioned but limpidly expressed lucidity. During the months of phoney war Sergeant Aron, as he became, was in charge of a meteorological unit on the Belgian frontier. He used the lazy days to think about Pareto, the sociologist who had briefly touched on socialism as a species of religion. He joined the general rout when the Germans invaded, washing up in Bordeaux. As a Jew, and a known opponent of Nazism, the thirty-five-year-old Aron fled to England, where he joined the Free French. A fluent German-speaker, he knew little or no English, although even on the boat out he developed a respect for the calmly confident behaviour of a people whose language he could not understand. Throughout the war he worked for the journal *La France Libre* in South Kensington, a journal which would have nearly eighty thousand subscribers, not counting the reduced-format edition which was air-dropped by the RAF into occupied Europe.

In July and August 1944 Aron published a two-part analysis of 'the secular religions' in *La France Libre*. By this he meant 'doctrines that, in the souls of our contemporaries, take the place of the faith that is no more, placing the salvation of mankind in this world, in the more or less distant future, and in the form of a social order yet to be invented'. He was interested in the psychological and moral effects of political enthusiasm: 'Partisans of such religions will without qualms of conscience make use of any means, however horrible, because nothing can prevent the means from being sanctified by the end. In other words, if the job of religion is to set out the lofty values that give human existence its direction, how can we deny that the political doctrines of our own day are essentially religious in character?' Aron realised that, despite its scientising pretensions, Marxism confused facts with desires, even as its claims to objectivity concealed a highly moralising view of the world and what amounted to a form of prophecy. An anti-socialist religion also emerged, based on a sort of revolutionary-salvation 'lite', which borrowed some of socialism's cast of villains, but dispensed with the apocalyptic revolution. That was National Socialism, although it would in fact be responsible for its own apocalypse.[157]

While the far-sighted began to map out the ways in which totalitarian regimes mimicked the soteriology and rituals of the Churches, the Churches – which adopted much of this analysis – faced the problem of how to respond to these novel challenges. To that complicated subject we turn next.

CHAPTER 3

The Churches in the Age of Dictators

I THE WIDER WORLD

To understand why the Catholic Church responded as it did to Europe's dictators requires a wide-angled view of a global institution. In the spring of 1938, the Catholic novelist Graham Greene visited revolutionary Mexico. Greene's first religious thriller, *Brighton Rock*, had stalled at the last five thousand words, while a film review he had published, suggesting that Shirley Temple was an adult midget with an odd appeal to middle-aged men, had resulted in the launch of a libel action by Twentieth Century-Fox. A magazine Greene relied on had folded. It was a gloomy time.

Greene also went to Mexico out of 'a desire to be a spectator of history', especially history that appeared to revolve around acute religious tensions. He had missed the chance to observe the 'religious war' in Spain, but Mexico promised a similar experience. In fact, he missed the boat again, as far as religious persecution was concerned, arriving in Mexico at a time when the worst anticlerical violence had abated. He loathed the country and the people, a revulsion extending to the biceps-clutching whenever friends greeted each other, gestures he attributed to the need to establish that one's 'friend' could not easily draw a weapon. He took to reading Trollope, to remind himself of a softer English civilisation, while columns of ants marched off dead beetles from his hotel-room floor. He claimed that his depression deepened when he lost his only spectacles, a state of mind consisting largely of 'the almost pathological hatred I began to feel for Mexico'.

The Lawless Roads is a typical 'Greeneland' of bed bugs, beetles and a cast of washed-up characters, a moveable feast of warm decay that he dyspeptically transferred to any number of exotic settings. From this grew *The Power and the Glory*, his first fully achieved religious novel, which closely corresponds to what he observed on his Mexican journey.[1]

The murderous conflict between Church and state in Mexico derived from the 1917 Queretaro Constitution, which was modelled on the 1905 French Separation of Church and state. However, it took a decade for this to explode into open warfare since the anticlerical laws were patchily enforced. Article 3 secularised education; article 5 banned religious orders; article 24 confined worship to the churches; and article 27 restricted the Church in its ownership of property. But it was article 130 which caused most ill-will. This forbade the wearing of clerical garb, and banned the clergy from voting, criticising government officials or commenting upon political affairs in Catholic publications.

While Mexico's president Alvaro Obregón was no friend of the clergy, he was astute enough to enforce these measures only in areas where the influence of the Church was weak, restraining himself wherever he anticipated opposition. His successor, Plutarco Elías Calles, was of Lebanese extraction, and hence known as 'the Turk', a dark and morose man who had alighted upon various careers before discovering his undoubted aptitude for politics as Obregón's protégé in the frontier state of Sonora. Illegitimacy is held to account for his visceral anticlericalism, which was tarted up with all the usual nostrums of scientism acquired when Calles trained as a teacher. His drunkard father's rakish antics presumably accounted for Calles's hatred of the bottle. Calles collected damaging information on his foes, real or imagined. He had a collection of clerical love letters, including some from a bishop to a Sonoran lady, which he hoarded for future anticlerical misuse. Although he was in other respects a modernising reformer, his face would redden and his fists pound the table whenever the clergy were mentioned in his presence. He associated the Church with everything negative and oppressive in Mexico's history, and was determined that the state would win any showdown with these forces of darkness. Within weeks of his coming to power, 73 convents, 92 churches and 129 religious colleges had been closed down. Instead, gimcrack museums of atheism proliferated, like the one Greene visited in Chiapas, a simple booth depicting eager monks flogging rubicund naked women; Trotsky in plus-fours; and the rough and smooth waxen hands of a worker and a priest juxtaposed. A mock-crib showed a dying woman, her baby and husband, with their empty food bowl being blessed by a priest, with the legend 'Their capital

50 cents and they must pay one and a half pesos for a Mass'.[2] Even place names were not spared, with Vera Cruz (True Cross) being contracted into Veracruz, a small but significant act of anti-religious malice.

Calles's regional sidekicks included governor Tomás Garrido Canabal in Tabasco, whose calling cards described him as 'the personal enemy of God'. Canabal's hatred of the Church was inherited. His father had begun burning images of saints when the intercession of a priest failed to assist a son who had broken his neck falling from a horse. The father had also started to drink – with Canabal junior joining Calles as a militant teetotaller. In his Tabasco, freed 'from clerical opium, ignorance and vice', it was forbidden to wear a crucifix or to use the traditional 'adios' because God figured in it. Crosses were removed from graves and a thousand women were encouraged to make a bonfire of statues of saints under the watchful eye of Canabal's youthful Red Shirts. The ferocity of anticlericalism in some of the federal states reminded Pius XI, in his second Mexican encyclical *Acerba Animi* (1932), of the persecution 'raging within the unhappy borders of Russia' – although that had not prevented the Vatican from trying to negotiate a concordat with the Soviets in the 1920s.[3]

In 1926 Calles ratcheted up tensions on a federal scale with the Law for Reforming the Penal Code, which fined priests five hundred pesos for wearing clerical garb and sentenced them to five years in jail for criticising the government. The 1917 Constitution began to be enforced, even in areas where this was likely to offend Catholic sensibilities. Although the Constitution proclaimed liberty of thought and conscience, it permitted the individual States effectively to establish the numbers of clergy by a registration system. This resulted in one state having a priest to every thirty-three thousand faithful, but others, such as Chiapas or Vera Cruz, having one priest to minister to sixty or a hundred thousand people. In July 1926, the Mexican bishops responded to these measures by suspending all public worship and with a boycott of entertainment and the public transport system. Catholics boycotted goods on which indirect taxes were levied. As a result in Guadalajara sales of clothing fell by 80 per cent, while the motorcar trade in Mexico City fell by 50 per cent. Although this boycott collapsed, proliferating clashes between federal forces and outraged ranchers resulted in priests being shot by the police, and then in outright Catholic rebellion, especially in the west-central Mexican states of Jalisco, Colima, Zacatecas, Guanajuato and Michoacán.

The rebel battle cry 'Long live Christ the King' led their opponents to dub them the Cristeros. Uniquely in modern Mexican history these rebels were not named after an insurgent caudillo, although they did have talented commanders, like the murderous father José Reyes Vega or Victoriano 'The Fourteen' Ramírez, who owed his sobriquet to the fourteen posse members he killed after a successful jailbreak. Ironically, general Enrique Gorostieta, who became the main Cristeros commander, was a freemason and liberal agnostic who liked to stretch out on a pew for a smoke after liberating churches.

These rebels fought a three-year guerrilla campaign against the federal army that Calles had recently modernised, penning government troops into urban strongholds, and sometimes defeating even the best of them in open combat. About seventy thousand people were killed, including ninety priests who were executed by virtue of their office. Graham Greene chronicled the fate of one twenty-five-year-old Jesuit, father Miguel Pro, who was picked up and shot in 1926, his short life reduced to seven photographs, four – including the melodramatic *El Tiro de Gracia* – chronicling his martyrdom. Neither the (exiled) Mexican hierarchy nor the Vatican were enthusiastic about the rebellion. Despite pleas from the Vatican, only Brazil, Chile and Peru openly criticised Calles, for Europe's powers did not want to jeopardise their investments. The cool response of the British Foreign Office may have been connected with the care Calles took not to include Protestant denominations in his anticlerical rampage.[4] Eventually Vatican lobbying in the US resulted in the appointment of a special agent to liaise with the US ambassador Dwight Morrow, who played a key role in trying to lower the temperature in Mexico. Morrow's nickname was 'Ham'n' Eggs' from his deft use of the working breakfast. The Church pinned its hopes of resolving the conflict on the return to the presidency of Obregón that under Mexico's peculiar alternating system of brokering power was scheduled for 1928. Unfortunately, the newly re-elected Obregón was shot by a Catholic artist–assassin, who had been sketching his likeness at a celebratory banquet in La Bombilla restaurant. Morrow patiently arbitrated between the conflicting parties, who on 21 June 1929 reached a series of 'arrangements'. Under these terms, religious worship resumed, while the government conceded the right to receive religious instruction, not in schools but in the churches, and permitted the clergy, as reinstated citizens, to petition for the reform or derogation of any law. When church bells were heard for the first time in nearly three years, ambassador Morrow turned to his wife and said: 'Do you hear that, Betty. I have reopened the churches in Mexico.' And what of former president Calles? He became addicted to golf and travelled in Europe, mutating from revolutionary into an admirer of Hitler. In April 1936, his successor Lazaro Cárdenas had him arrested (Calles was reading the Spanish version of *Mein Kampf* at the time) and deposited over the US border. After five years' exile in San Diego, Calles was allowed home. The former arch-rationalist spent the last years of his life playing golf and attending the weekly seances of the Mexican Circle of Metaphysic Investigations to commune with the dead about his political legacy.[5]

Graham Greene narrowly missed the opportunity to visit Spain, the scene of the most shocking anticlerical violence outside Bolshevik Russia and revolutionary Mexico in the 1930s. His leftist Catholic sympathies inclined him towards the Basques. When the Nationalist forces of General Mola surrounded Bilbao, Greene tried to fly into the besieged city from southern France, to report its dying days for the BBC. However, the pilot decided at the last minute that Nationalist anti-aircraft fire had become too deadly and refused to fly.[6]

Spain was ruled between 1923 and 1930 by an improbable military dictator, Miguel Primo de Rivera, who once sagely remarked: 'Had I known in my youth that I would one day have to govern this country, I would have spent more time studying, and less fornicating.' Primo was followed into exile by king Alfonso XIII, except that the latter chose Rome over Paris as his temporary domicile. The Second Spanish Republic was proclaimed, based on a provisional coalition cabinet of Republicans, reformist Socialists and the conservative Catholics, Maura and Zamora, whose token presence was supposed to reassure the upper classes. This government embarked on a programme of agrarian and military reforms during a global Depression, which alienated the intransigent right, without satisfying the raised hopes of its lowliest supporters, who in their disillusionment turned to anarchist or revolutionary Socialist alternatives. Both the Vatican and the papal nuncio Federico Tedeschini greeted the advent of the Republic with near equanimity, since Tedeschini had brokered contacts with the Republicans before they came to power. Indeed, when the Spanish primate cardinal Segura delivered a provocatively pro-monarchist sermon and was declared persona non grata by the government, the Vatican acquiesced in his expulsion and found him an alternative career as a Curial cardinal in Rome.

Anarchist and left-wing anticlericalism enabled the highly fissiparous right to regroup, while also reconnecting with millions of ordinary Catholics whose sensibilities had been offended by mobs which, in May 1931, sacked and burned about one hundred Church properties in Madrid and other cities, allegedly in response to earlier monarchist provocation. Opprobrium spread to a government, including its Catholic Republican members, which not only refused to stop the incendiaries, on the ground that 'all the convents in Madrid are not worth the life of one republican', but needlessly ordered the removal of all religious symbols from school-rooms.[7] Although the exiled monarchists tried to recruit Pius XI to their cause, the pope remained steadfastly neutral. The marriage of Alphonso's daughter to an Italian prince was celebrated in the Jesuit Church in Rome by cardinal Segura. The Spanish government protested. When the assembled throng of monarchists sought a papal audience, the pope kept them hanging around in the January chill for the briefest of blessings from an upstairs window.[8]

The newly elected left–Republican and Socialist coalition in June 1931 further provoked the religious with controversial articles in the new Constitution, Spain's first experiment in democracy. This went much further than a legal separation of Church and state. It extruded the Church from education, restricted its property rights and investments, and dissolved the Jesuits, who played a role in liberal and leftist mythology equivalent to that of freemasons, Jews and Marxists in the demonology of their opponents. This last measure was a bitter pill to swallow in the homeland of St Ignatius Loyola. Civil marriage and divorce were legalised, while the agreement of the authorities was henceforth necessary for any public celebration of religion – another indigestible measure in a society where religious processions were a highly developed art form. A supplementary law in 1933 nationalised all Church property, including secularising the cemeteries by putting them under local authority control and dismantling the walls which separated the dead religious from their non-believing fellows. Having nationalised Church property, thereby ignoring the wishes of those who had donated it, the government then taxed the clergy who used it. Measures against Church charities simply hurt poor people. The government also closed all religious schools, which since they educated 20 per cent of Spanish children, and were not replaced by secular alternatives, sat oddly with the Republic's expansion of education.

Although these measures were implemented with varying local intensities, there can be no doubt that preventing the ringing of church bells, removing religious symbols from classrooms, and bureaucratising the procedures for those wanting religious funerals grievously irked many Catholics. Officious insistence that dying people fill out forms to get the send-off they wanted failed to charm their friends and relatives.[9] These measures were condemned by Pius XI in the forceful 3 June 1933 encyclical *Dilectissima nobis*, which, while carefully professing indifference to forms of government, stressed the hypocrisy of these measures in terms of 'those declared principles of civil liberty on which the new Spanish regime declares it bases itself'. These laws were the product of 'a hatred against the Lord and His Christ nourished by groups subversive to any religious and social order, as alas we have seen in Mexico and Russia'. Republican Spain had become part of a 'terribile triangolo' whose object was the eradication of religion.[10] Anticlericals in the Cortes responded in kind, with snide remarks about the 'Mercantile Society of Jesus', while the Socialist leader Azaña crowed that with these 1931–3 measures Spain had ceased to be Catholic.

Of course, things had been tending that way far longer than the wave of measures introduced in 1931–3 may suggest. In 1881 the Churches had lost control of the universities. In 1901 religion had become optional within the curriculum leading to the school leavers' certificate. In 1913 non-Catholic parents could exempt their children from religious instruction. With a few exceptions, the arts and intelligentsia were dominated by secular-minded people. The Catholic presence among the urban working class and the southern rural poor was also exiguous. In 1935 a Jesuit calculated that, taking the eighty thousand parishioners of a Madrid working-class suburb, 7 per cent attended mass on Sundays; 90 per cent died without the benefit of the sacraments; 25 per cent of children were unbaptised; and of couples marrying, 40 per cent could not recite the Lord's Prayer. Similar levels of indifference and ignorance were revealed in studies of Bilbao and Barcelona. The Church was also like an alien presence in the villages of Andalucía, with anarchist and Socialist activists converting peasant indifference or quasi-pagan superstition into outright hostility. Churches were falling into disrepair, when they even existed, and priests were poorly paid with government stipends equivalent to the lowest grade of janitors. The priesthood was not an attractive career option, with recruitment for seminaries falling by 40 per cent between 1931 and 1934.[11]

Although there were a handful of Catholic Christian Democrats, and journals such as *Cruz y Raya* dedicated to reforms of the most egregious socio-economic inequalities, Catholic opinion was overwhelmingly ranged on the side of conservatism, however that may be understood in this context. The advent of the democratic Second Republic made the creation of a mass conservative party imperative on the part of those rightists who subscribed to an 'accidentalist' view of affairs – that forms of government were evanescent – and that power should be pursued through legal channels. Eventually, in February 1933 Gil Robles succeeded in bringing some forty rightist groups into an umbrella organisation called the Spanish Confederation of Autonomous Right-Wing Groups (or CEDA).[12] This ran the gamut of conservative opinion from a sprinkling of Christian Democrats to others who were indistinguishable from the 'catastrophist' right of Carlists, Alfonsists and Falangists seeking the violent overthrow of the Republic, and who were linked to foreign dictators and to Spain's own home-grown aspirants in the army. The CEDA availed itself of the Republic's recent enfranchisement of women – who became the most effective footsoldiers of the Party machine – while the examples of Mussolini and Hitler inspired Robles to call himself 'Jefe' and to hold what amounted to Spanish versions of the Nuremberg rallies. These had deeply impressed him as an observer. The Party machine used highly modern forms of propaganda, with films, posters and tons of printed leaflets. Disgracefully, some of its clerical supporters helped disseminate the view that the Republic was the result of a Judaeo-masonic–Marxist conspiracy, a view they could hold while simultaneously rejecting Hitler's division of mankind into higher Nordic Aryans and lower Slavs and Latins.[13]

Only in the Basque country and Catalonia were there minor political parties that combined deep-seated Catholicism with republicanism, chiefly because the conservative *cedistas* were implacably opposed to their goal of regional autonomy. Even here in the north-east corner there were splits aplenty. Straddling the Franco-Spanish border, the Navarese stuck with the reactionary Carlists after a brief flirtation with Basque autonomists based on their common Catholic identity.[14]

While the CEDA was prepared to win elections with the help of both the 'catastrophist' right and the Radicals who had been alienated from the Republicans and Socialists, the left ostentatiously decided to go it alone in an electoral system that gave an enormous number of parliamentary seats to whomever secured the slimmest majority. Slight electoral pluralities that turned into huge numbers of seats in the Cortes were then misinterpreted as a mandate for the most radical changes, a sure way of appalling middle-class opinion that turned further to the right. In the November 1933 elections, the CEDA became the largest party, imposing their conservative social programme on a government of Radical Republicans, the second largest parliamentary grouping. Of them the Spanish Socialist leader quipped: 'if they had not been in jail, [they] deserved to be'.[15] When in October 1934 three *cedistas* finally joined the government, the left reacted as if there had been a 'Fascist' coup, for in their minds the CEDA represented 'clerico-fascism', just as conservatives regarded all liberals and leftists as blood-crazed Bolsheviks. Robles certainly used a rhetoric hard to distinguish from that abroad in Italy or Germany:

> We must reconquer Spain . . . We must give Spain a true unity, a new spirit, a totalitarian polity . . . It is necessary now to defeat socialism inexorably. We must found a new state, purge the fatherland of judaising freemasons . . . We must proceed to a new state and this imposes duties and sacrifices. What does it matter if we have to shed blood! . . . We need full power and this is what we demand . . . To realize this ideal we are not going to waste time with archaic forms. Democracy is not an end but a means to the conquest of the new state. When the time comes, either parliament submits or we will eliminate it.[16]

There were nationalist and working-class risings in Catalonia and Asturias, the former being quelled without bloodshed, the latter – in which Asturian miners killed thirty-four clergy – repressed with great brutality by General Franco and the Spanish Foreign Legion. With politics polarised, stalemated and increasingly violent, the president called elections for February 1936. Although the Popular Front won a plurality of 1.5 per cent of the vote, the electoral system gave them two-thirds of the parliamentary seats. The Socialists refused to participate in another 'bourgeois' government, leaving a minority left–Republican regime to reach the limits of its imaginative capabilities, against a backdrop of labour militancy and political violence. While the Socialists imagined that power would fall to them by default, the right abandoned the legal path to power in favour of ever less fanciful military conspiracies.

Often contemptuous or ignorant of political ideas, the generals sought to restore backbone to a polity that one distinguished Spanish philosopher had once dubbed 'invertebrate'. News of the July 1936 military uprising, which stalled into a bloody and protracted civil war when the government armed the working classes, resulted in the largest example of anticlerical violence in modern history. Of course, nobody denies that the Nationalist rebels shot teachers and union leaders in their desire to make a clean sweep of all traces of 'anti-Spain'. If there was a difference between their respective atrocities, it was that Republican outrages were committed by anarchists and criminals whom the Republic had amnestied, and who ran amok wherever public order broke down, while killings in the Nationalist-dominated areas were premeditated and carried out by the responsible authorities. Although it did not take these anticlerical outrages to make many Catholics rabidly hostile to the Republic – the cynosure of developments they had long hated – they undoubtedly prompted a simple prudential calculation in such circles that their survival depended on the success of the military uprising.

In the Republican-held areas, nearly seven thousand clerics were murdered, the majority between July and December 1936, in anticlerical atrocities that eclipsed those of the Jacobins. Over four thousand of the victims were diocesan priests, as well as thirteen bishops, but they also included 2,365 male regulars and 283 nuns.[17] Contrary to mythology, these nuns were not sexually assaulted or raped, but that they were shot suggests a remarkable depth of feeling. There was no evidence that the clergy had aided the military uprising, nor that houses of God were misused as rebel arms dumps. While a few priests made public broadcasts supporting the rising, there is also scant support for the idea that sharp-shooting priests took potshots from their belfries. Churches were routinely the tallest structures in towns and villages, so all combatants automatically made a beeline for them to get the most advantageous firing positions. Clergy who found themselves caught up in the fighting simply by dint of being trapped in a church were routinely executed in a war that developed into one with few prisoners.

Anticlerical violence struck at buildings and images, as well as afflicting both the quick and the dead. One of the most heavily publicised photographs was of Republican militiamen lined up to shoot at the statue of the Sacred Heart of Jesus atop the Cerro de los Angeles outside Madrid. The statue was subsequently blown up because, having been dedicated by Alfonso XIII, it vividly symbolised the hated union of throne and altar. According to Franz Borkenau, in Sitges and other coastal places pious people were forced to bring objects of worship to beachside bonfires where children amused themselves by defacing the statues before burning them. The sight disgusted him.[18] The pro-Republican *Daily Telegraph* correspondent, Cedric Salter, described terrible scenes in Barcelona:

On my way down [into town] I passed a burning church. The flames had only caught at one end of the building and I pushed my way into the entrance. Flames were licking up round the altar, on which stood two beautiful wrought silver candlesticks gleaming through the clouds of black smoke. From the high carved stone pulpit an elderly priest swung very slowly to and fro by his sickeningly elongated neck. He had offered resistance, a guardia told me, when they had seized the Sacred Wafer and hurled it into the flames, and had died cursing them. Around the walls the pale painted faces of the Saints slowly distorted into nightmare grimaces as the heat melted the wax of which they were made.

Lower down, just above the British consulate, a crowd had formed outside the entrance to a convent. I went in with them, and found a long wall lined with coffins from which the lids had been stripped. The poor, century-old bodies of the nuns were exposed, and what flesh still clung to the bones was slowly blackening in the hot sun. Fresh coffins were being excavated from the convent burial ground and a peseta was being charged for the hire of a long stick with which to strike or insult with unnameable obscenities these sightless, shrunken relics. A charnel-house stench and my own sick horror drove me back into the street.[19]

Barcelona and the Catalan provinces of Lérida and Tortosa were where the largest proportion of clergy were massacred, nearly 88 per cent of diocesan clergy in the former and 66 per cent in the latter. Barcelona had fifteen schools run by around 150 Marist brothers. While some of these were arrested, others hid with friends and family. By early October thirty-six of them had been killed. The order's superiors then struck a deal with the local anarchist committee, who agreed to allow the Marists to go to France, in return for a payment of two hundred thousand francs. After half of this had been received, 117 Marist novices went to the frontier, where all those under twenty were allowed to exit Spain. The older minority joined the rest of their brethren in Barcelona, having been assured that as soon as the second payment was made they could depart by sea. One hundred and seven Marists eventually assembled on the quayside for embarkation on a steamer. When the boat failed to sail, the Marists were disembarked and taken to a convent that was being used as a prison. The following night, forty-five of them were machine-gunned in a cemetery. The sixty-two survivors were transferred to a Barcelona prison.[20]

Throughout Republican-controlled territory priests and religious were subjected to brutalities, some characterised by a bestial drunken savagery induced by over-indulging in communion wine. According to Hugh Thomas, when the parish priest of Torrijos explained to his tormentors that he wished to suffer like Jesus, they beat him mercilessly, tied a beam to his back, poured vinegar down his throat, and placed a crown of thorns on his head. The militia told him, 'Blaspheme and we will forgive you.' When he forgave them without blaspheming, the militiamen contemplated crucifying him, but then shot him instead. Other clergy had their ears stuffed with rosary beads until their eardrums burst, or had their ears cut off by their tormentors before being murdered. A woman who had two cleric sons died by having a crucifix rammed down her gullet. In Bellmut del Priorat the priest and his house-keeper were forced to undergo a mock marriage, after which they were both murdered.[21]

This level of violence requires explanation. Anticlerical violence certainly had a tradition in Spain, with a total of 235 clerics being killed in 1822–3, 1834–5, 1868, 1873, 1909, 1931 and 1934, not to speak of five hundred or so churches being burned down over the same period.[22] Whereas the capitalists and landowners were often absent or remote figures, clergy were highly visible, although their image had changed since the time of Goya. Mainly from the Spanish lower-middle class, they were regarded as toadies of the powerful, with little in common by way of education or social origins with their more lowly parishioners. As the oligarchic right wrapped its naked self-interest in the cloak of religious values, so inevitably violence began to home in on the most visible exponents of militant Catholicism. Myths abounded about the wealth of the Church – much of which went to charity or education at a time when state provision of both was negligible – while what went on behind cloistered walls was fantasised in a manner that Diderot or Voltaire would have been proud of. As some commentators have claimed, at some inchoate level this explosion of anticlerical violence may have reflected a perverted religious instinct, an expression of outrage against people whose (greatly exaggerated) corporate wealth and hypocritical personal morality offended a very literal and primitive understanding of Christianity. Arson, iconoclasm and the desecration of the dead were simultaneously a purging of excrescences resembling Reformation Europe several hundred years earlier, and a modernising eradication of what seemed like superstitious mumbo-jumbo. That was how one Republican writer regarded it: 'Those buildings had lasted long enough; their mission was completed; now they were anachronisms, weighty and obstructing, casting a jailhouse stench over the city. The times condemned them to death and the people carried out the execution of justice. These burnings were the autos da fé necessary for the progress of civilisation.'

The identity of the rebels and Catholicism was by no means total, given that only four of the ten members of their ruling Junta were identifiably Catholic, while the presiding Nationalist general Miguel Cabanellas was a moderate liberal freemason. The youthful Franco showed few signs of religious fervour, preferring the example of Beau Geste to Jesus. This changed with marriage in 1923 to the devout Carmen Polo, although as his eldest niece remarked with considerable understatement: 'his . . . way of understanding the gospels might leave a good deal to be desired'.[23] The initial absence of references to religion in the rebel platform was born of a desire not to antagonise large numbers of moderate middle-class anticlericals on the part of career soldiers who were much clearer about what they opposed than what they stood for.

The Civil War also revealed striking anomalies. Glaringly, the ultra-Catholic Basques were allied with the Republican camp, while the rebels had imported Muslim mercenaries from north Africa, who rather touchingly hedged every bet by resorting to patches of cloth portraying the Sacred Heart of Jesus in the hope that these would deflect a bullet. In the Basque country, moreover, in late 1936, the insurgents committed their own anticlerical atrocity by shooting fourteen priests who had sided with the Basque autonomist allies of the Republicans. But this identification between the rebels and Catholicism quickened because of the anticlerical atrocities and the rebels' need for a noble cause that would fire the imaginations of potential middle-class supporters. The rebel leader general Mola began to intrude references to religion into his broadcasts and speeches, while several senior clergy made statements supportive of the Nationalist enterprise. The Spanish primate, cardinal Isidro Gomá, archbishop of Toledo, was especially prominent in reducing the complexities of the Civil War to a clash of antagonistic 'spirits': 'This most cruel war is at bottom a war of principles, of doctrines, of one concept of life and social reality against another, of one civilization against another. It is a war waged by the Christian and Spanish spirit against another spirit.'

Gomá knew perfectly well from his recent experience of the Basque region, where he had gone to take the waters for a kidney ailment, that matters were far more complicated than that. Given that Catholic Basque autonomists were now fighting ultra-Catholic Carlist Navarrese, he tried to get the two local bishops to pressure the Basques into switching their allegiances to the Nationalists. While the bishop of Pamplona capitulated, his colleague in Vitoria, bishop Mateo Múgica, was torn between his Basque identity and his support for the Nationalists. One of his brothers had also been killed in Madrid's spasm of anticlerical fury. Múgica was worried that Gomá's letter might prompt anticlerical violence on the part of the autonomists and hence refused to allow his clergy to publicise it. By this time, the Nationalists were plotting to kidnap and murder the bishop, whom Gomá had transferred to Rome, partly for his own protection, but also to remove a man the Nationalists regarded as a thorn in their flesh, standing in the way of reintegrating the Basques into a consolidated Catholic camp. With Múgica ensconced in the Vatican, and with Nationalist forces bearing down on the autonomous Basque government, both Rome and cardinal Gomá tried to negotiate a non-violent resolution of the Basque–Nationalist conflict. This was obviously deeply damaging to the Nationalist cause abroad, since the rhetoric of a 'crusade' against the massed godless sat oddly with using Muslim troops to suppress the impeccably Catholic Basques, a paradox that was causing some of the world's leading Catholic intellectuals and writers to desert the Nationalist cause. Franco was also lobbying the Vatican to secure its condemnation of the Basques, a condemnation that would undermine foreign aid for the Republic.

Like all leaders, Pius XI and secretary of state Pacelli were bombarded by conflicting advice and assessments from interested parties, with the complication in their case that relief for war-ravaged Spain depended upon diplomatic neutrality. Different religious orders often backed different political horses. The Polish head of the Jesuits was pro-Nationalist while his Dominican equivalent was more equivocal. The Spanish hierarchy and the exiled Alfonso XIII may have been anti-Republican, but then the pope was also being lobbied by the exiled Basque and Catalan primates Múgica and Vidal. Both men powerfully argued that, if the Church lined up with the Nationalists, then Spanish priests would pay the consequences. Pius' first utterances on the conflict were at an audience for five hundred Spanish clerics and laymen at Castel Gandolfo, his summer palace high above one of Lazio's lakes, in September 1936. He deplored the anticlerical atrocities, linking them with an axis of evil stretching from Mexico to Russia. While he blessed 'the defenders of God and religion', he also warned them that 'it is only too easy for the very ardour and difficulty of defence to go to an excess . . . Intentions less pure, selfish interests, and mere party feeling may easily enter into, cloud, and change the morality and responsibility for what is being done.' His Nationalist auditors threw their copies of the speech to the floor as they departed, while only a very attenuated version of the address was reported in Nationalist-occupied zones. The Vatican continued to recognise the Republic, although neither side had representatives, while denying accreditation to the Nationalists' envoy, a stance that led Franco to drag his feet in rescinding the Republic's anticlerical legislation. However, relations between the Vatican and the Republic were never normalised. Despite the efforts of the one remaining Catholic minister, Manuel de Irujo, to defuse the religious issue by restoring freedom of worship, the Vatican was not convinced that the Republicans had sufficient control of extremist elements to warrant the risks of open worship. Increasingly desperate attempts by Catholic Republicans to persuade the Vatican that religious persecution was over were met with hesitation and scepticism.

Meanwhile, although Pacelli rebuffed Franco's request for explicit support, he floated the idea of a collective letter by the Spanish hierarchy explaining the incompatibility of Catholic Basques fighting alongside Communists, while simultaneously seeking to mediate a separate end to the Nationalist–Basque conflict. The collective episcopal letter was dated 1 July 1937. It was nominally a lengthy response to the concerns expressed by foreign clergy regarding the anticlerical outrages. Gomá systematically refuted the notion that the Spanish Church had brought this catastrophe on its own head, for even then it was fashionable to blame the victims, at the same time dilating upon the failings of the Second Republic and the existence of a Comintern conspiracy to make Spain Communist. The cardinal argued that the war was between two antagonistic 'spirits' and rolled out Thomas Aquinas to argue for the theological legitimacy of the rebellion. Catholics should support what he described as the 'civic–military movement'. Conceding that this movement was sometimes responsible for excesses, Gomá made it clear that these paled into insignificance beside the lurid atrocities of the other side. There was one qualification to his support for the Nationalists. So long as their intentions were restorative and traditionalist, they could bank on the Church's support, or rather, the Church would accept their offers of protection. But there was a warning should the content shift in favour of one of the foreign ideologies on offer to European rightists:

> With respect to the future, we cannot predict what will take place at the end of this struggle. We do affirm that the war has not been undertaken to raise an autocratic state over a humiliated nation but in order that the national spirit regenerate itself with the vigor and Christian freedom of olden times. We trust in the prudence of the men of government, who will not wish to accept foreign models for the configuration of the future Spanish state but will keep in mind the intimate requirements of national life and the path marked by past centuries.[24]

All of the Spanish hierarchs signed the letter, with the exception of two exiles, Múgica and his Catalan colleague cardinal Francesc Vidal i Barraquer, who was exiled from revolutionary Barcelona to Lucca in Italy. Both men were sympathetic to regional automism, and aware of atrocities rather than 'excesses' on the nationalist side. They felt that Gomá was dragging the Vatican into a situation where neutrality was the lesser evil, and were alive to the anticlerical repercussions that were likely to flow from such obvious sympathy with the Nationalist cause.

As the fortunes of war tilted in the Nationalists' favour, so the Vatican began to adjust its line on recognition of the Nationalist government in Burgos. The Vatican initially dispatched a chargé d'affaires whose role was to aid Basque prisoners and to repatriate twenty thousand Basque children who had been evacuated abroad. Full diplomatic relations were restored in May 1938 after the Nationalists revoked the Republic's anticlerical legislation. Religious symbols returned to classrooms, where religious instruction was now compulsory, the Jesuits came out of hiding, and laws on marriage, divorce and abortion were reversed. The clergy were omnipresent at Nationalist celebrations and to the fore in propagandising the war as a religious 'crusade' or 'holy war', a theme that particularly excited Franco's limited military imagination since he saw himself as a latterday El Cid, the epic hero of medieval Spain's struggle with Moorish invaders.[25] Foreign Catholic Churches, like that of Ireland, pumped money into the Nationalist cause in the knowledge that it was being used to purchase munitions rather than bandages. Catholic military chaplains sometimes evinced an unChristian zeal to shoot people, while pulpits resounded to blood-curdling exhortations to eradicate the 'satanic' enemy. Some priests, disgracefully, participated in post-war purge committees, set up to exterminate Republican sympathisers in implacable detail, while others passively dispensed the last rites at what amounted to massacres. Others held people's livelihoods, or lives, in the balance, by issuing or refusing 'certificates of catholicity' that became obligatory in some areas.

The Church's ideological role did not stop with singing Te Deums at Carlist or Falangist meetings, or in apotheosising Franco as the spiritual embodiment of the most Catholic monarchs Ferdinand and Isabella. Nostalgia for a vanished golden age in which hard-faced Catholic monarchs had expelled Moors and Jews from their rural idyll that had paradoxically created a world empire lay at the heart of the Francoist vision of what Spain should be. Catholic intellectuals struggled to distinguish the Spanish brands of Fascism and totalitarianism from the this-worldly varieties on offer in Italy and Germany. An influential book laboured to explain with sophomoric pretentiousness:

> The New State must be founded on all the principles of traditionalism in order to be genuinely national and Spanish. Thus in Spain the Falange must become the technique of traditionalism. Our fascism, our Hegelian juridical absolutism, must necessarily be grounded in its form on a historical–Catholic– traditional basis. Spanish fascism thus becomes the religion of religion. The Italian and German fascisms have invented nothing new for us. Spain was already fascist four full centuries before them. When it was united, great and free, Spain was truly so; in the sixteenth century, when state and nation were identified with the eternal Catholic ideal, Spain was the model nation, the alma mater of western civilisation.[26]

In the Falange Franco found a prefabricated political party, with the added bonus that its playboy leader José Antonio Primo de Rivera had been arrested before the uprising and then tried and shot in November 1936. There was no love lost between the two men. Franco regarded Primo as a rich dilettante, while Primo thought Franco was a plodding soldier. Although as the aristocratic heir of Spain's former dictator Primo had always been deferential to Spain's oligarchs, the Falange – which itself was a fusion of Fascist grouplets – had radical residues that disappeared as the ideology was diluted with traditional integralist Catholicism. It became Franco's main arm of repression and his personal claque. Sometimes a change of names speaks volumes. The name change undergone by what became the sole Francoist state party – Traditional Spanish Phalanx – indicated that the more radical aspects of Spanish Falangism had been ditched in favour of a demobilised 'Fascism' heavily permeated by authoritarian and traditionalist Catholicism. The diehard Falangist nucleus became the only residual focus for anticlericalism in Nationalist-controlled areas, with its youthful activists waylaying the occasional passing religious procession. In the long run more useful dead than alive, Primo was mythologised as a martyr – literally 'the absent one' (*ausente*) and as the herald of Spain's stocky and long-lived Caudillo.[27]

In contrast to Spain, where the presence of Catholics on both sides of a vicious civil war dictated a cautious response by the Vatican, there were two countries where Pius XI's vision of a 'golden mean' between invasive totalitarianism and weak democracy was apparently being realised by authoritarian governments.[28] The first was Portugal, where in 1911 the new Republican regime had introduced some of the most anti-clerical legislation in Europe. The Church was an easier target for urban radicals, many of whom were freemasons, than either the army or the large landowners of the south. Church property was nationalised, the university of Coimbra's famous theology faculty was abolished, and feast days were restored to the world of work. Foreign priests and Jesuits were expelled, and both civil marriage and divorce were introduced. Religious teaching was prohibited in all schools. Both women and the 65 per cent of the population who were illiterates were disfranchised to destroy any potential Catholic voting base. Virtually the entire hierarchy were either exiled or expelled and in 1913 the Republic broke off diplomatic relations with Rome.

The fight-back against the efforts of a radicalised minority to impose French-style laicisation began among students in the devoutly Catholic north, and particularly among students at Coimbra. Two leaders emerged, Manuel Gonçalves Cerejeira and António de Oliveira Salazar, of the Academic Centre for Christian Democracy, out of which developed a political party called the Portuguese Catholic Centre Party. Circumstances enabled this moderate party to make its influence felt. During the First World War, the Republic needed the Church to provide chaplains to its army of Catholic soldiers, while missionaries became crucial to the retention of a vast overseas empire at a time when the military was over-stretched. By 1919 the Republic and the Vatican had restored diplomatic relations.

Initially, Salazar, an academic economist, subscribed to democracy as 'an irreversible phenomenon', but by the 1920s he and many others had become disenchanted with what was a highly corrupt local version of it. When he was elected to parliament in 1921, his revulsion for the opening session was so great that he walked out and returned to university teaching the same day.[29] Between 1911 and 1926 Portugal had eight presidents, forty-four governments and twenty attempts at coups or revolution.

In 1926 the parliamentary republic was finally overthrown by general Manuel de Oliveira Gomes da Costa. Within two months, he was in turn deposed by general Carmona. Since Carmona had few ideas of his own, he depended upon conservative lay Catholics, including Salazar, who was twice brought in to right Portugal's parlous finances. Salazar was careful to separate his political ambitions from his Catholicism, even if this meant tensions with his old student friend Cerejeira, who had become archbishop of Lisbon. When Salazar told the latter that he represented 'Caesar, just Caesar, and that he was independent and sovereign', Cerejeira shot back that he represented 'God . . . who was independent and sovereign and, what's more, above Caesar'. Salazar's dictatorship retained the Republic's separation of Church and state.

Restoring the Portuguese economy at a time when the world was sliding into depression lent Salazar a wizardly mystique, which he used to civilianise the military dictatorship from within. In 1930 he proclaimed a new National Union, an authoritarian non-party whose primary purpose was to demobilise opinion. One of its first casualties was the Catholic Centre Party, which, Salazar argued, would impede the march to dictatorship. Thereafter Catholic Action became the main vehicle for the Church's plans to reconquer Portuguese society for Catholicism.

President Carmona appointed Salazar premier in July 1932. He proclaimed the New State a year later. Catholic corporatist teachings, however misunderstood, were combined with a form of integral nationalism derived from Charles Maurras.[30] The quiet professorial dictator, who avoided public speaking and staffed his regime with numbers of fellow academics, faced one remaining challenge. Portuguese disillusioned with the low-key tone of Salazar, and suffering under his austere economic policies, turned to the National Syndicalist Blue Shirts, who modelled themselves on the Fascists and Nazis. Salazar dealt with this radical Fascist threat deftly. He co-opted its more opportunist members into the regime, and then in July 1934 dissolved the remainder. This hostility to the National Syndicalists was similar to that of the Portuguese Catholic hierarchy. Referring to their desire for a 'totalitarian state' he asked: 'Might it not bring about an absolutism worse than that which preceded the liberal regimes? . . . Such a state would be essentially pagan, incompatible by its nature with the character of our Christian civilization and leading sooner or later to revolution.'

Salazar saw little difference between the Communists, Fascists and Nazis, all of whom were wedded to a totalitarian ideal 'to whose ends all the activities of the citizen are subject and men exist only for its greatness and glory'.[31] Portugal had no imperial ambitions – its empire was already the world's fourth largest – and the regime dissociated itself from Nazi antisemitism, welcoming Jewish refugees fleeing their oppressors. The regime's object was to entrench and intensify conservative Catholic values rather than to experiment with a 'new man' or woman. That lack of ambition, which extended to an aversion to modernising the nation's economy, may partly explain why Salazar remained in power in this backwater until 1968.

Another European state to receive the Vatican's blessing was the 'State of Estates' – or 'Ständestaat' in German – created by Engelbert Dollfuss in the ruins of the first Austrian Republic. Since the turn of the century, Austrian politics had been dominated by a clash between 'Red Vienna', where the atheist and militant Social Democratic Party held sway, and the provinces, where the parties that made up successive governing coalitions – that is, the Christian Socials, the Pan-Germans, and the Agrarian League – had their greatest support. In this respect, Austrian politics resembled other countries with a 'Red' metropolis hated by many provincials, notably Berlin and Madrid in the same period, although it is important to note that since the days of Mayor Karl Lueger the Christian Socials had support among Vienna's petit-bourgeoisie who were drawn to his demagogic antisemitism, anti-liberalism and deference towards the Catholic Church. The intellectual and political leadership of the Party was also based in the capital.

Although in their 1926 Linz programme the Social Democrats distanced themselves from the freethinking that had supplanted nineteenth-century Liberal anticlericalism, their success in persuading significant numbers of people to leave the Church meant that the clergy and the Christian Social Party regarded them with deep suspicion. The

Christian Social leader was a cleric, Ignaz Seipel, who between 1922 and 1924, and then again from 1926 to 1929, was chancellor of the Republic. By the late 1920s the Christian Socials' use of the term 'true democracy' indicated their coolness towards the failing parliamentary regime. In 1932 Seipel claimed that political parties were 'inorganic' temporary expedients in the absence of such 'organic' mediating bodies as socio-economic corporations which would repair the damage done by atomistic liberal individualism.

Both the Christian Socials and the Social Democrats had large paramilitary armies, which were soon augmented by the strong-arm groups of the Austrian National Socialists. The Christian Socials (and in some places the Pan-Germans) were close to many of the regionally based 'home defence groups', or Heimwehren, originally established after the war to protect villages from looters and deserters. These had evolved into a strike-breaking force financed by the employers and armed by the Italians and Hungarians. In the Korneuburg Oath, which they swore in May 1930, the Heimwehr leaders resolved to replace democratic government with an authoritarian corporative system modelled on the ideas of the political economist Othmar Spann. In 1923 the Social Democrats formed their own Schutzbund, after the Heimwehr had crushed a strike in Styria. The nature of the problem faced by the state becomes clear from the fact that its army of thirty thousand men faced sixty thousand members of the Heimwehr and ninety thousand equally well-armed members of the Schutzbund. In 1927, following the acquittal of Heimwehr men accused of murdering socialists, the latter stormed and set fire to the Courts of Justice during three days of rioting. The Heimwehr threatened a Fascist-style March on Vienna. Austria's domestic disturbances were intensified by the obtuseness of France and the Little Entente in blocking a customs union with Germany.

In May 1932 Engelbert Dollfuss, an able peasant boy and war hero who had risen to be agriculture and justice minister, was appointed chancellor. At thirty-nine he was Europe's youngest head of government; at four feet eleven inches he was also the slightest in stature. Dollfuss immediately negotiated a foreign loan of 300 million Schillings, only to find that the Pan-Germans voted against it, on the ground that renunciation of union with Germany was among the loan's conditions, while the Social Democrats also refused to support the government out of doctrinaire bloody-mindedness. He achieved a narrow majority only by bringing Heimwehr leaders into his cabinet.

In early 1933 the government clashed with militant railway workers whose union was a mainstay of the Social Democratic Party. The railway-men had discovered a mysterious arms shipment disguised as routine freight – which they thought was being sent by Mussolini via Austria to aid the Hungarians – going on strike when the employers penalised them on behalf of the government. The government's attempts to outlaw further rail strikes led to a parliamentary crisis, in which successive Speakers resigned without being replaced, and to the prorogation of parliament. That was the pretext for the creeping authoritarian reconstruction of the Republic.

Like chancellors Brüning and Schleicher, Dollfuss used emergency legislation to marginalise the defunct parliament. He resorted to a 1917 law that had originally been used by wartime governments to requisition food. The opposition press was silenced and demonstrations and meet-ings prohibited. In May 1933 Dollfuss appointed Emil Fey, a Heimwehr leader, secretary of state for security with responsibility for all police forces. Their pay was increased and police ranks were augmented with auxiliaries for the battle ahead, for Dollfuss was explicit in his desire to take on and defeat the left. He also struck at the Nazis, who had been emboldened by Hitler's coming to power in Germany. On 19 June 1933, Dollfuss banned the Nazi Party, and dissolved its various paramilitary formations. He closed various higher-education facilities to deny the Nazis one of their main sources of support among students. An intern-ment camp was opened at Woellersdorf in October 1933, where Marxist and Nazi militants were quarantined together in conditions that were not especially oppressive. Hitler responded to this challenge by raising the tourist visa fee to a thousand Reichsmarks, severely damaging the Austrian hiking- and skiing-based tourist industries. The Austrian Nazis launched a campaign of terror inside the country.

Dollfuss turned to Italy and the Vatican for external support against Hitler. He hastened to Rome to revive negotiations for a concordat that had been ongoing since 1931. It took about six weeks to finalise terms. The concordat, signed on 5 June, was incorporated into the new Constitution, and ratified on 1 May 1934 when the Constitution was promulgated. It reversed the entire Josephinist tradition, restoring religion to state schools and terminating government interference in the appointment of bishops. The state would henceforth recognise canonical marriages. Having seen what the Nazis had done in Germany, the semi-official Jesuit journal, *Civiltà Cattolica*, praised those who wished to preserve an independent Austria under the Cross of God rather than a pagan symbol.

Rather than relying for mass support on the Christian Socials, on 20 May 1933 Dollfuss established a new Fatherland Front, which was supposed to absorb all existing right-wing potential into one governing party, along the lines already essayed by Primo de Rivera in Spain and Piłsudski in Poland in the 1920s and by Salazar in the 1930s.[32] Its nominal membership eventually reached three million. The Front adopted a syncretic political symbolism, with a straightened-up version of the swastika called the 'Kruckenkreuz' and a Fascist-style authoritarian administrative structure. Dollfuss employed Catholic corporatist rhetoric and enjoyed the confidence of Pius XI, who spoke of him as 'a Christian, giant-hearted man . . . who rules Austria so well'. This was largely because Dollfuss claimed to have implemented the Social Catholic corporative alternative to Darwinian capitalism and Marxist socialism that the pope had outlined in the 1931 encyclical *Quadragesimo anno*, issued to mark the forty years that had elapsed since Leo XIII's *Rerum novarum*. In reality, the pope had not said a word about political, as distinct from economic and social, organisation, and the manner in which the new arrangements were imposed blatantly contradicted the principles of subsidiarity enshrined in the encyclical. The pope wished to diminish the powers of the state by restoring grassroots human fellowship; not to increase it through the establishment of a dictatorship.[33]

The regime faced two challenges: one from the left, which it won, and another from the Nazi 'brown Bolsheviks', which it eventually lost. In February 1934, the Heimwehr arrested Schutzbund leaders and expelled representatives of democratic parties from provincial diets. In Linz, the Social Democrats decided to fight back, and met police incursions into their headquarters with machine-gun fire. In Vienna, the socialist leadership dithered so that the general strike they declared was imperfectly implemented against a regime that was well prepared for just this eventuality. Martial law was proclaimed while Heimwehr troops surrounded working-class suburbs. A full-scale shooting war ensued, with artillery and tanks firing into housing projects with such resonant names as 'Bebelhof', 'Liebknechthof' and 'Karl-Marx-Hof'. One hundred and ninety-six workers were killed and 319 wounded, with 118 dead and 486 wounded on the government side. The government banned the Social Democrat Party and neutralised the trades unions by subsuming them into its own corporatist entities. Socialists were expelled from the national and provincial civil service. Courts martial were used to sentence twenty-one people to death – one of the nine eventually executed being taken to the gallows on a stretcher. Even Hitler managed briefly to occupy the moral high ground when he condemned 'the criminal stupidity of letting people shoot down socialist workers, women and children'. The Vatican secretary of state, Pacelli, intervened in vain on behalf of those sentenced to death.

There were fitful attempts to promote a culture reflecting the ideology of the 'State of Estates', in which a sense of common vocation would overcome class conflict. In the new Constitution promulgated on 1 May 1934, four advisory councils, whose members were chosen rather than elected, selected a federal diet which could approve rather than initiate legislation. Government was freed from any form of parliamentary criticism or scrutiny. All mayors and regional administrators were government appointees. The new Constitution began with the words: 'In the name of God the Almighty from whom all justice flows, the Austrian people accept this Constitution for their Christian, German, Federal State on the basis of a State of Estates.'

Entirely without imperial or military ambitions, Austria was to be the 'natural mediator' of German civilisation further east, and the best example of the happiness a state based on Christian principles could bring: 'We intend that this German land of the Alps and the Danube shall once again be a country which will prove to mankind that under a new form of government and with a social order inspired by the Christian ideal a people can be happy and contented.'[34]

Since the elite were suspicious of the Social Democrat masses, they were not especially adept at choreographing public events, which tended to be dominated by secular or ecclesiastical notables. Books, films and plays denigrated, or ignored, the mess of modern industrial urban civilisation, in favour of a picture-postcard, tourist-office idyll which depicted placid folk in traditional garb toiling away in a verdant alpine setting.[35] Beyond this was a lightly oppressive and omnipresent clericalism, symbolised by the joint press conference with the Catholic bishops which Dollfuss and the justice minister Kurt von Schuschnigg held in March 1934. State employees, and especially teachers, were obliged to take part in religious services. Clouds of incense marked whenever a building or hall was dedicated. In Salzburg about a dozen people who ostentatiously deserted the faith (presumably they were Nazis) were jailed for six weeks. Children who skipped confession received poor grades at school. Whereas twenty-nine thousand people had formally left the Church in 1927, in 1934 some thirty-three thousand were eager to join it.

The Austrian Nazis continued their terror campaign, which tragically reached the chancellor himself. On 25 July 1934 some 150 men assembled in a gymnasium on the Siebensterngasse in Vienna. They arrived in small groups and wore civilian clothes. Each carried a packet under their arm containing a change of clothes. In the gymnasium they donned military kit and armed themselves with guns which had been brought on a truck. They belonged to the 89th Standarte of the SS. They left in trucks, into the midday heat, heading for the government chancellery. Since information about this operation had ceased to be a secret, most of the cabinet had gone home early, although Dollfuss himself was informed of the plot only an hour before it burst in upon him.

The trucks drew into the chancellery just before 1 p.m., opportunely just as the guards had changed and the courtyard was deserted. Simultaneously, other SS men seized the Austrian Radio transmitter and broadcast that 'The Dollfuss government has resigned. Dr Rintelen has taken over the affairs of government.' Policemen shot these intruders about ten minutes later. At the chancellery Dollfuss tried to make his escape. Aided by a servant, he ran the wrong way down the corridors, and was shot by the putsch leader Otto Planetta. Although Dollfuss repeatedly called for a priest, he was left bleeding on the floor from the wound in his neck, until the assassins put him on a sofa where he died shortly before 4 p.m. As the day wore on, Schuschnigg, the justice minister, rallied the troops, who surrounded the chancellery.

After negotiations, the putschists were prevailed upon to surrender, with some of them being tried by a military court and executed over the next few days. It took until the end of July to suppress the simultaneous Nazi uprisings in several of Austria's outlying provinces. Although the wires of conspiracy reached back to Berlin's Foreign Ministry and the chancellery, Hitler immediately fired the Nazi leader Habicht and dissolved the Austrian Nazi Party, while his news agency denied any German involvement. Mussolini warned Hitler off any precipitate steps against his former client by moving a few divisions to the Brenner. The Church set about transforming the dead Austrian chancellor into a national martyr, as side altars filled with kitsch commemorating the 'minimetternich'. In Portugal, Salazar quietly crushed the National Syndicalists in retaliation for what Austrian (and German) radical

Fascists had done to Dollfuss. The pope was outraged by Dollfuss's murder. The Vatican daily said that National Socialism could be better described as national terrorism, and praised Mussolini for sending a deterrent force to the Brenner.[36]

On 29 July the Austrian president Miklas appointed Schuschnigg chancellor, with the Heimwehr leader Starhemberg as his deputy and leader of the Fatherland Front. Schuschnigg was a law professor from a distinguished family. He was a coldly intellectual Tyrolese, hiding himself from humanity, as Otto von Habsburg had it, behind 'the glass wall of his spectacles'. He dreamed of a federal central Europe in which Austria would be the cultural magnet for its neighbours. The authoritarian system acquired some of the trappings of Fascism, without entering into its spirit. Since a Jewish lawyer, Robert Hecht, was one of the main architects of the regime, and since its supporters included Sigmund Freud, it cannot be said to have reflected the antisemitism that was otherwise pervasive in Austria as a whole. The Fatherland Front acquired its own paramilitary force, which in turn spawned an elite troop, in dark-blue uniforms and with the motto 'Our will becomes law' that reminded many observers of the SS. There was also a politicised youth movement. Reactions to this indicate that relations between his government and the bishops were far from smooth. They protested against the militarisation of children under fourteen, and in the autumn of 1935 warned the government 'that fascism as a foreign import does not fit our circumstances and must be decisively rejected in its concept of the absolutist, totalitarian state'.[37]

Schuschnigg's dreams fell foul of a tectonic shift in Europe's diplomatic alignments in the 1930s as Mussolini moved closer to Hitler, who had torn up the military restrictions imposed by the Versailles Treaty and remilitarised the Rhineland. Mussolini exerted mounting pressure on Schuschnigg to cut a deal with Hitler as the only way of guaranteeing Austrian independence. In the July Agreement of 1936, Germany recognised Austrian sovereignty, while Austria agreed to conduct itself as 'a German state'. The evolutionary strategy favoured by Franz von Papen, Germany's ambassador to Vienna, and the 'moderate' Austrian Nazi leader Seyss-Inquart, effectively sanctioned a gradual Nazi coup, with Hitler browbeating Schuschnigg whenever necessary. One group Papen tried in vain to win over were the Austrian bishops. To his disappointment, in November 1937 they issued a public declaration of sympathy for the plight of their German colleagues, adding, 'we know that many are endeavouring to replicate here the conditions that have developed in your country in order to bring about a victory for godlessness'. When, very late in the day, Schuschnigg tried to call Hitler's bluff with a hastily improvised and far from democratically unimpeachable plebiscite designed to affirm Austrian independence, Hitler established Mussolini's benevolent neutrality, and ordered his forces over the frontier. They met policemen and soldiers already wearing Nazi insignia and in control of much of the country.

One of the first to welcome Hitler's homecoming was the leading spokesman of Austria's Protestant minority, who on 13 March pronounced 'in the name of Austria's more than 333,000 Protestant Germans': 'After a period of repression that brought back to life the most terrible times of the Counter-Reformation, after five years of the deepest suffering, you have come as the deliverer of all Germans here, without regard to the different beliefs they espouse. God bless your progress through this German land, your Heimat!' The Social Democrat leader and former chancellor Karl Renner was equally effusive as he urged his fellow countrymen to vote yes in the plebiscite that retroactively sanctioned the Anschluss:

> I would have to deny my entire past as a theoretical advocate of the right of nations to self-determination and as an Austrian statesman, if I didn't welcome with joyful heart the great historical deed whereby the German nation has been brought together ... I would vote 'Yes' as a Social Democrat, and therefore a champion of national self-determination, as the first Chancellor of the Austrian German Republic, and as the former president of your delegation to the peace [conference of] St Germain.

Signalling that they were in charge, SA squads put the archbishop of Salzburg under house arrest. Stones crashed through the windows of his palace. Vienna's cardinal Theodor Innitzer was summoned to meet Hitler. The latter expressed the hope that, after the failure of the German Catholic Church to prostrate itself, the Austrian Catholic Church might demonstrate greater (uncritical) loyalty. Innitzer told Hitler that Austrian Catholics would be loyal to the new state, but hoped the terms of the 1934 Concordat would be respected. Innitzer then presented his colleagues with a thoroughly unnecessary appeal to Christians in general to support 'the greater German state and its Führer' in his 'world-historical struggle against the criminal madness of Bolshevism' by voting for the Anschluss in a new plebiscite. He was summoned to Rome where Pacelli insisted that this declaration be redrafted. The Austrian cardinal was coolly received by Pius XI. Innitzer was obliged to publish in *Osservatore Romano* a denial that his statement contained an approval of anything incompatible with the laws of God and the rights of the Church.[38] His disclaimer added: 'that statement cannot be interpreted by the State and the Party as a duty of conscience of the faithful nor must not be used for propaganda purposes'. The reason Innitzer was forced to eat humble pie was that Western governments had misinterpreted his public statements as Vatican approval of the Anschluss. New sources from the Kennedy Library in Boston shed light on Vatican thinking. In mid-April Pacelli had a private interview with Joseph Kennedy, the US ambassador to London, who was on a private visit to Rome. He handed Kennedy a memorandum, indicating that it should be given to 'your friend'. Kennedy sent it to James Roosevelt with instructions to show it to the President. The memorandum categorically disowned the statement by the Austrian Catholic hierarchy, pointing out that it was probably drafted 'by a government Press bureau' and then signed under duress. Pacelli deplored the absence of references to the Kulturkampf in Germany, and the prospect that such a conflict would erupt in Austria after the Anschluss. Nothing Pacelli had experienced with the Nazis indicated that they dealt 'in good faith', which 'so far has been completely lacking'. He reflected that the 'Supreme Moral Powers of the World' feel 'powerless and isolated in their daily struggle against all sorts of political excesses from the Bolsheviks and the new pagans arising among the young 'Arians' [sic] generations'.[39] When Hitler visited Rome in May 1938 and wished to see the Vatican Museum and St Peter's, the pope ostentatiously repaired to Castel Gandolfo, distressed that another cross appeared to be adorning the city's streets.

One country that celebrated what Dollfuss had tried to achieve was the Irish Free State. As Vice-President Sean O'Kelly explained to a conference in Geneva in October 1933: 'The government [of the Free State] is now engaged in endeavouring to do for its people what Chancellor Dollfuss announced his government is trying to do for Austria. In the development of its programme of economic and political reform its work is founded on the same Catholic principles.'[40] Ireland was an authoritarian state in the sense that Sinn Fein, the main opposition party, had opted out of the Dáil by refusing to swear an Oath of Fidelity, thus leaving the government to do what it wanted. The first two years of the Free State's existence were marked by civil war between the Cumann na nGaedheal government of William T. Cosgrave, supported by the Catholic hierarchy, and the sizeable republican Sinn Fein remnant opposed to the Treaty that gave birth to a partitioned Ireland. The Church's condemnations of IRA violence were rewarded when its role was enshrined in the Free State's (and Eire's) constitutions. In 1926 Sinn Fein split. The majority adhered to Sinn Fein's former president Eamon de Valera's new Fianna Fáil (Warriors of Destiny) party. The remainder were addicted to pursuing romantic revolutionism outside constitutional politics, while flirting in respectively the long and short term with Marxism and Nazism. A compromise formula was found to enable Fianna Fáil representatives to take an oath that they regarded as meaningless.[41]

It would be easy to ridicule aspects of the Free State, such as its pervasive and puritanical clericalism, or the attempts to 'gaelicise' a culture where less than 20 per cent of the population had any grasp of native Irish. In fact, together with an ostentatiously neutralist international stance, these were essential to the cohesion of what one authority has described as both a post-revolutionary and post-colonial society, with a modest economy that could not sustain the generous levels of social welfare inherited from the British, and which was under constant internal threat from purist republican militants. The emphasis upon Irishness and Catholicism not only helped create a society that was ostentatiously unlike Britain, but also undermined those republicans who regarded the Free State as a sell-out.[42]

In 1932 de Valera's party came to power by offering a more positive vision of Ireland's future than the men who had secured independence. After an embarrassing interlude under the 'green Duce', former police commissioner Eion O'Duffy, who subsequently took his Blue Shirts off to Spain, Cumann na nGaedheal metamorphosed into Fine Gael, a party nostalgic for the Free State as a Dominion. The austerely devout de Valera – who had once toyed with a clerical vocation – presided over Ireland for several decades, first as prime minister and later as president, by celebrating its Catholicism and the virtues of small-scale family farming. Like Salazar, he tried to keep the modern world at one remove. However, attempts to impose a pseudo-medieval corporatist order on Ireland, as advocated in Father Edward Cahill's strange *The Framework of a Christian State* (1932), floundered in the face of opposition from both the civil service and the Catholic hierarchy. Why tinker with the social and political system when most of the government had the outlook of bishops and were closely connected with them in what was a tiny elite within a small society?[43]

From independence onwards, the Church's influence was strongly represented both through such as events as the 1932 Eucharistic Congress, which was attended by more than a third of the population, and through official and unofficial crusades aimed at moral regeneration which sometimes bordered on vigilantism. Such foreign pollutants as the English *News of the World* were dumped in the harbours as soon as ships unloaded. Neither increased duties on newspapers nor one bishop's advocacy of 'imprisonment or the lash' for erring news-sellers affected a circulation of nearly two hundred thousand, eager for tales of English high-society adulterers. Bishops constantly dilated upon the evils of rural dance halls – known in such circles as 'synagogues of Satan' – while a few priests in Kerry took more direct action by burning down, or reversing their cars into, the wooden dance platforms set up at crossroads.[44] Two Intoxicating Liquor Acts in 1924 and 1927 reduced opening hours and the number of bars in a country known for its love of a drink. The 1929 Censorship of Publications Act handed censorship over to local committees of the Catholic Church, whose enthusiasm for their task may have owed something to the fact that many prominent Irish writers, including W. B. Yeats, were Irish Protestants. In the early 1930s, the primate urged a general boycott of the cinema, one of the main sources of information about the world beyond, as well as of glimpses of calf and cleavage. In 1935 it became illegal to import or sell contraceptives, while the Public Dance Halls Act of that year introduced licensing for such premises. The new 1937 Constitution, which ripped up the 1922 Treaty in order to achieve a purely 'external association' with Great Britain, cemented Catholic influence, its preamble leaving little room for doubt: 'In the Name of the Most Holy Trinity, from Whom is all authority and to Whom, as our final end, all actions both of men and states must be referred'. Various articles of the Constitution were devoted to the protection of marriage and the family, by encouraging women to stay at home and prohibiting laws licensing divorce without a constitutional amendment and plebiscite. The Church was recognised as having 'a special position ... as the guardian of the Faith professed by the great majority of its citizens', although other 'Churches' were also recognised – including Jewish synagogues, notwithstanding the antisemitism that was rampant in Ireland at the time.[45]

Beyond the totalitarians and what might be called its favoured sons in the Iberian peninsula, Austria and on Europe's Atlantic periphery, the Church dealt with a further range of countries that defy classification. The Czechoslovak government's French-style anticlericalism, and nationalist enthusiasm for the proto-reformer or heretic Hus as a symbol of new nationhood, led to a temporary breakdown in relations, but by the late 1920s these had substantially improved with the signing of a 1928 modus vivendi that fell short of a concordat. Whereas Catholics had been part of the ruling majority in the Habsburg Empire, after 1918 they were a large minority in a Serb-Orthodox-dominated Yugoslav federation. Religious affairs were so complicated in the Kingdom that it took from 1922 to 1935 to negotiate a concordat with the Vatican, which in this instance ignored the local Catholic bishops. One of the sticking points was that, whereas the Catholic clergy were willing to abstain from political involvement, their Orthodox counterparts were not prepared to reciprocate. Throughout the 1930s, archbishop Alojzije Stepinac of Zagreb, the Church's youngest archbishop in what was Europe's largest archdiocese, managed to keep the clergy clear of politics, although he could do little to check the flow of radicalised young Catholic Croats into the Fascist Ustashe movement.[46]

Although there were Christian democrats, and a sprinkling of Christian Marxists, in inter-war Europe, most Catholic politics was conservative, and subject to a gravitational pull towards the authoritarian and anti-parliamentary right. This did not mean that Catholics were sympathetic towards either Fascism or German National Socialism. In Belgium, one of Europe's most staunchly Catholic countries, the vote of the Catholic Party held up well, despite the latent possibility of Flemish or francophone Catholics splitting off into rival nationalist groupings. These groups were highly coloured by Catholicism, with the Flemish National Union (VNV) slogan being 'Alles voor Vlaanderen, Vlaanderen voor Kristus' (All for Flanders, Flanders for Christ). Even the excitable crypto-Fascist students of Louvain university used the name of a Catholic publishing house – Christus Rex – as the name for their political party, the Rexists which were led by Léon Degrelle. Although this managed to scoop part of the Catholic Party's votes in 1936, denunciation of Rexism by Belgium's primate cardinal Van Roey the following year, after Degrelle had falsely claimed to have the cardinal's support, led to the precipitous decline of its support to less than 5 per cent in the last pre-war elections.[47]

The majority of French Catholics were deeply conservative, in politics, way of life and values. However, during the 1930s, the sense of being embattled gradually abated, to the extent that the political expressions of Catholicism were no longer exclusively identified with the lay lobby of the Fédération Nationale Catholique with its support for the parliamentary right. From 1924 onwards there was the Parti Démocrate Populaire, a centre-right gathering of Social Catholics and Christian Democrats inspired by the 'personalist' ideas of Paul Archambault and the 'popularism' of the Italian Luigi Sturzo. Although on the right, the PDP was implacably opposed to the Action Française, not to mention the various Fascist grouplets and it had no time for Nazism or appeasement. However, before considering these dramas, it is important to see how the tribal affinity of Catholicism and the right started to break down in the 1930s, largely over events in Spain.

Inter-war French Catholic intellectual life was not only vibrant but also bewilderingly diverse. It revolved around journals, newspapers, discussion groups and networks, some of which encountered problems at the Vatican concerning their orthodoxy, trouble which conservative French Catholics played a part in fomenting. There were two fine Dominican journals, La Vie Intellectuelle and Sept, which had a much larger circulation. Sept supported the League of Nations stance on Italian adventurism in Abyssinia, and refused to regard the Nationalist cause in the Spanish Civil War as a 'crusade'. In 1937, the year when Sept was closed down, it intimated that Catholics should support the Popular Front government of Léon Blum. Christian democracy was represented by Francisque Gay, who founded both La Vie Catholique (1924–38) and L'Aube (1932–), a daily. L'Aube severed the reflexive connection between Catholicism and the right, especially through its opposition to Mussolini's invasion of Abyssinia, its refusal to support the Nationalists in Spain, and its forthright opposition to Anglo-French appeasement of Hitler. Attempts to reconcile Christianity and Communism were the concern of Maurice Laudrain, who from 1935 ran a journal called Terre Nouvelle, whose cover depicted a white hammer and sickle superimposed on a red cross, which many regarded as needlessly provocative.

A more eclectic and maverick enterprise, claiming to be of the left but open to the thinking right, was represented by the philosopher Emmanuel Mounier, the guiding spirit behind Esprit, founded in 1932. His group is vaguely reminiscent of the sort of extreme Marxist sectarians, such as the contemporary British journal Living Marxism, whose detestation of most of the left has led them to being successful corporate consultants.

Mounier was a bright, shy and hulking boy from the Dauphine, who made it to the elite Ecole Normale Supérieure, where to the surprise of those who mocked his philosophical and religious interests he passed out second only to the towering figure of Raymond Aron. Mounier was mightily impressed by Jacques Maritain's 'primacy of the spiritual' and by the mystic Péguy's romantic admiration for the collective craftsmanship that built medieval cathedrals. Repelled by the arid rationalism of the universities, in whose faculty many Catholics felt that Protestants and Jews were over-represented, Mounier set about creating his own network of collaborators and sympathisers, first in France and then throughout Europe and North America. These included such figures as the exiled philosopher Berdyaev and Maritain, the painters Chagall and Roualt, a motley array of Catholics, Protestants, Russian Orthodox, Jews and non-believers, who subscribed to a journal that was bankrolled at its zenith by a sympathetic Jewish wallpaper manufacturer called Georges Zerapha.

Although he was not especially interested in Marxism, which he regarded as too materialistic, Mounier claimed to be a 'revolutionary' seeking to detach Catholicism from its intimate connections with a political right that was interested only in defending privilege. He hated money and the worship of it. The opening issue of *Esprit* declared: 'We are . . . revolutionaries, but in the name of the spirit. It is not force which makes revolutions, it is light.' He was dismissive of Christian democrats like Archambault for seeking to work within the parliamentary system rather than fundamentally transforming it. Deeply hostile to the slack, routinised religion of the Catholic bourgeoisie, he saw in both Communism and Nazism how people ceased to be mere individuals, becoming a new collectively aware 'person'. In other words there was something worth while in both movements, which appeared to attract large numbers of decent and idealistic people. Mounier connected these observations with the way individuals who joined religious communities took on the collective spiritual 'personhood' of their fraternity, institution or order. Here one can already detect how someone a little over-impressed with ideas could drift across the ideological frontiers.

For Mounier was certainly no proto-Christian democrat. He had connections to what he perceived to be the anti-Hitler wing of the Nazi movement, whether to the Strasser brothers, the Hitler Youth or to Otto

Abetz – Hitler's man in Paris – who in the 1930s was a Nazi cultural ambassador. He was involved in encouraging the many variegated little shoots from which a 'New European Order' based on Fascism might grow. With extraordinary ignorance and naivety, he thought that a form of national socialism might emerge, shorn of its extreme, Hitlerian, racist features. Apparently the significant number of Jews involved with *Esprit* agreed. Though things did not work out with the Strasser brothers (one of whom was murdered on Hitler's orders in 1934) Mounier set great store on developments in Belgium. After an initial enthusiasm for the Catholic Fascist Léon Degrelle, his interests drifted to the authoritarian socialists Henri de Man and Paul-Henri Spaak, who promised to fuse nationalism and socialism in a new synthesis. He welcomed the defeat of France – for someone who had been excoriating its decadence since 1932 he could hardly do otherwise – turning his mind to how its 'rape' by the Germans might give birth to an altogether healthier child once the war was over.[48]

The religious philosopher and Catholic convert Jacques Maritain's involvement with Mounier may be likened to the brake pedal in a car – every time Mounier waxed a little too enthusiastically about Communism or Nazism as metahistorical 'spiritual' events, Maritain would restrain him. Beyond the heady heights of *Esprit*, French Catholics obviously had opinions about the great developments of the day. Like conservatives elsewhere, they often regarded the advent of Mussolini's regime in a positive light, especially since the Duce appeared to respect religion, and had banned dual membership of the Fascist Party and the lodges of the dread freemasons. This enthusiasm turned to caution after the violence, domestic and external, of the regime became apparent, with the invasion of Abyssinia and intervention in Spain being landmarks in Catholic alienation. French Catholics across the political spectrum were less than enthusiastic about German National Socialism. The ultra-patriotic French right were often militantly germanophobic, managing to elide Germans and Jews, while many Catholics believed that Nazism was a form of 'neo-paganism'. Even when the right-wing Catholic press could countenance antisemitic policies designed to 'reduce' what they believed was an over-proportional representation of Jews in the economy or society of places like Austria, they were opposed to the violence that went with Nazi antisemitism in Germany.

Many French Catholics regarded Communism and Nazism as twin totalitarian evils, and were not overly impressed by the anomie engendered by modern liberalism and its political system of party-based democracy either. The advent of the Popular Front government in France in 1936, despite the best attempts to stop it by mobilising Catholic voters, was a particular challenge, since its leader was Léon Blum, a French Jew, and it rested upon the tacit support of Thorez's Communists. Some right-wing Catholics, like the FNC deputy from the Ardèche, Xavier Vallat, marvelled aloud when Blum announced his government on 5 June 1936: 'Who would have believed that this old Gallo-Roman country would be governed by a Jew!'[49] In fact, the Popular Front's bark was a lot worse than its toothless bite, and even Thorez 'extended his hand' to the Catholics, who refused to take it.

The Spanish Civil War sharply divided European opinion, with Communist depredations stirring the consciences of such left-wing renegades as Arthur Koestler and George Orwell, while conservatives had to avoid the trap of supporting the Nationalists without approving of either their atrocities or the Nationalists' Fascist and Nazi bedfellows. British Catholics generally supported Franco. They included cardinal Hinsley, who kept a photo of the Caudillo on his desk at Westminster cathedral, the influential *Tablet* journal, and the writers G. K. Chesterton, Hilaire Belloc – who called Franco 'the man who has saved us all' – and Evelyn Waugh. Waugh characteristically argued that if he were Spanish, he would have supported Franco, but as an Englishman he declined to choose between the twin evils of Communism and Fascism.[50] The Irish Catholic hierarchy positively enthused over the Nationalist 'crusade', with archbishop MacRory of Armagh claiming that the war was 'a question of whether Spain will remain, as she has for so long, a Christian and Catholic country or a Bolshevist and anti-God one'. The Irish Catholic press bought into the Nationalists' mythological mix of history and piety: 'It must be joyous to live in liberated Spain today, feeling that the spirit of the Cid is exultant in Burgos, that the sons of Santiago are freemen in Galicia again, and that the daughters of Aragon may give thanks for victory before the Virgin del Pilar in Saragosa.' The former Fine Gael leader and organiser of the Irish Eucharistic Congress, Eion O'Duffy, who had been ousted in 1934, led his Blue Shirts to the Spanish battlefields, although they were regarded as a joke by their Falangist comrades, who managed to kill four of them in 'collateral' incidents.[51]

Events in Spain had particular premonitory urgency in France, for they seemed to prefigure what might have become a French civil war.

The Catholic right insisted that the Nationalist side were waging a legitimate crusade against the forces of anticlerical Bolshevik darkness, the line propagated by the Spanish hierarchy, albeit without Vatican endorsement. Cardinal Baudrillart saw the Spanish Republicans as the lineal successors of the Jacobin anticlericals of 1792, while the FNC leader Castelnau spoke of the Spanish 'Frente Crapular'. Paul Claudel published a poem in June 1937 entitled 'To the Spanish Martyrs' which took some liberties with the number of clergy who fell victim to anti-clerical violence, but whose Nationalist sympathies were unmistakable: 'Sixteen thousand priests! The battalion formed in a single moment, and behold, heaven is colonized in a single burst of flame.' Bizarrely, Franco did not lack supporters among conservative French Protestants either. The monarchist Protestant organisation Sully published a bulletin that praised Franco for erecting a 'bulwark of Christendom' against the godless.[52]

Typically perhaps, Mounier and *Esprit* declined to support the Nationalists, but then alighted on the anarchists as Spain's salvation. However, this enthusiasm for the irresponsible and puerile was accompanied by a much more interesting shift in sympathies on the part of conservative Catholic writers and intellectuals, who began to have grave doubts about where their tribal loyalties as Catholics were leading them in Spain. Maritain may have been an erstwhile supporter of the ultra-right Action Française, but he registered its condemnation by Pius XI in 1926, and was appalled by the extra-parliamentary violence that rocked France in 1934, much of it attributable to crypto-Fascist Leagues. Although many French clerics did not like it, Pius XI was determined to eradicate Action Française. He made the Jesuit cardinal Billot resign after the latter sent the movement a sympathy note. On learning that the rector of the French Seminary in Rome, a member of the Holy Ghost Fathers, was also a sympathiser, Pius summoned the ancient head of his order and told him to sack the rector. 'I'll see what I can do,' came the vague reply. Pius grabbed the old man's beard and shouted: 'I didn't say, see what you can do, I said fire him!'[53]

Maritain was shocked by attempts to construe the Nationalist cause in Spain as a 'holy war' or 'crusade'. He wrote:

It is a horrible sacrilege to massacre priests, even if they are 'fascist' (they are ministers of Christ), out of hatred for religion; and it is another sacrilege, just as horrible, to massacre the poor, even though they are 'Marxists' (they are the people of Christ) in the name of religion. It is an evident sacrilege to burn churches and the images of the saints, sometimes in blind fury, sometimes, as in Barcelona, with cold anarchic method and in the spirit of systematic madness; and it is also a sacrilege (of a religious nature) to decorate Muslim soldiers with badges of the Sacred Heart so that they might kill in a saintly manner the sons of Christians, and to claim that God shares their own passionate hatred which considers the adversary unworthy of any respect or pity whatsoever.

Having attacked the claim that the Nationalist cause was inherently holy or sacred, Maritain concluded:

Let people invoke, if they wish, the justice of a war they are waging if they believe it just; let them not invoke its sanctity! Let them kill, if they think they have a duty to kill, in the name of the social order or of the nation; that is already horrible enough; let them not kill in the name of Christ the King, who is not a military leader, but a king of grace and charity, who died for all men, whose kingdom is not of this world.

Maritain was supported in this stance by the Catholic novelists Georges Bernanos and François Mauriac. A supporter of the Nationalist rebellion, the conservative monarchist Bernanos – whose sixteen-year-old son Yves volunteered to fight for Franco – was appalled by the Falangist purges he witnessed on Majorca, especially when they were conducted with the clergy's blessing.[54] Mauriac utterly condemned such horrors as the massacre in the Badajoz bullring on the feast of the Assumption or the bombing of Guernica by the German Luftwaffe. While he expected the godless left to commit atrocities, he expected better of the Christian right, and loathed Franco until the day he died.[55] He knew that his support for the Catholic Basques could be misused by the Communists. The left-wing writer Julien Benda took a more morally absolutist stance, refusing to condemn Republican massacres of 'Fascists' lest this indirectly aid Franco. Rather grimly he announced, 'I am for the extermination of a principle which is incarnated in some human lives. I am not a humanitarian, I am a metaphysician, which is just the opposite.' Mauriac disagreed, arguing that one should condemn all manifestations of barbarity, regardless of whom such a condemnation might benefit: 'I have suffered to have seemed to carry water, or rather blood, to the Communist mill . . . but a Christian people is lying in the ditch, covered with wounds. In the face of their misery it is not playing into the Marxist hand to manifest to all the world the profound unity of all Catholics. This is the vine and these are the branches. One of the branches is threatened with destruction and the whole vine is suffering.' Inspired by his friend Alfred Mendizabal, whose book on Spain is still useful, in 1937 Maritain established a Committee for Civil and Religious Peace in Spain, which Bernanos and Mauriac joined. This Committee was instrumental in getting Vatican relief assistance for Basque and Catalan children whose lives had been affected by war. As a result of their stand, all three writers were systematically slandered by supporters of Franco, who, like the interior minister Serrano Súñer, did not fail to stoop to playing upon the fact that Maritain's wife Raissa was a converted Jew.[56]

II THE VATICAN, COMMUNISM AND FASCISM

As the Austrian, Irish and Portuguese examples suggest, the anticlerical fury in Russia, Mexico and Spain did not mean that the Catholic Church – a worldwide religious communion – turned to Fascism or Nazism as the lesser evils, or as its putative saviours from godless Bolshevism. This would not be true of, for example, most Anglo-Saxon Catholics, and it was not true, as we have seen, among prominent French Catholic intellectuals. It was also not true of the papacy. At no point did the Vatican ever entertain the idea of entering into a 'pact' with Nazi Germany to combat the greater evil of Bolshevism, for the elementary reason that the Vatican regarded both regimes as alien totalitarian ideologies.[57] The only pact worth talking about is the one in August 1939 between the Nazis and their Soviet friends that precipitated the Second World War. The Church had spiritual goals which took precedence over evanescent temporal governments, regarding whose precise forms the Church professed a lofty indifference. That was especially true of both Achille Ratti, who was elected as Pius XI in 1922, and his secretary of state and successor, Eugenio Pacelli, who became Pius XII in 1939.

Pius XI signalled his desire to see the restoration of Christ's Kingdom with the encyclical *Ubi arcano Dei*, his answer to rampant materialism, secularism and nationalism, being the re-Christianisation of society through such non-political vehicles as Catholic Action, the introduction of new feast days, and the canonisation of exemplary figures. Inter-state concordats, of which the Vatican concluded forty in the 1920s, were to provide the legal framework for this ambitious apostolic mission.[58]

It has become commonplace among historians of Communism, Fascism and National Socialism to emphasise that these regimes were not monoliths, but consisted of unstable and warring factions that were susceptible to pressures from below, even though such revisionism hardly detracts from the totalitarian aspirations these regimes harboured, or from the psychopathic violence that a dynamic combination of ideology and bureaucratic rivalries unleashed. Curiously enough, many critics of the Catholic Church imagine that it functioned in the way a scholar in the 1950s might have imagined a totalitarian state, and as if the popes were in the same position as the Duce or Führer. In fact, the Vatican itself was a Babel of conflicting views, not to speak of the religious orders also represented there, or the hierarchies in each country, who were in turn susceptible to shifts in clerical and lay opinion. On a number of occasions, Vatican initiatives were retracted at the urging of the national episcopacies concerned. Matters were further complicated in bi-confessional or predominantly Protestant countries, notably Germany, where the Church was constantly wary of a Protestant backlash. These reminders of the historical reality caution against any loose generali-sations about the 'Catholic Church', about which any number of crude and stereotypical prejudices seem to be acceptable among people who spend most of their time denouncing prejudice.[59]

Both future popes were Vatican diplomats, involved in negotiating the concordats that the Holy See insisted upon, after three former European empires (four including Turkey) were abruptly replaced by eleven successor states and the Soviet Union, whose official creed was atheism. The advent of these new states, which often included substantial ethnic or religious minorities, not only played havoc with historic diocesan boundaries, but involved new constitutions in which relations between Church and state would have to be negotiated anew. It is worth emphasising that these concordats were not signs of special papal favour, but a means of defining relationships with what might be called problem (or rogue) states through solemn legal documents.[60]

Achille Ratti served as apostolic visitor and then papal nuncio to

Warsaw for three years between 1918 and 1921. His younger colleague Pacelli was papal nuncio to Munich until 1925, when he moved to Berlin as nuncio to the German Reich, a post he occupied until 1929. Ratti's task was to restore the structures of the Catholic Church in newly sovereign Poland and the three Baltic States, while as a papal monitor of the plebiscites in Upper Silesia and East Prussia he had to prevent the Polish and German Catholic clergy from involving themselves in politics.[61] Ratti has not been spared the antisemitic slurs that have been directed at Pacelli, for no sooner is one criticism of Pius XII confounded than the battleline is shifted elsewhere by critics who have a fundamental animus against the Catholic Church. Inevitably, the former reported on tensions between Polish Christians and Polish Jews – it being in the nature of what diplomats do to report on local opinion – although his personal dealings with Jews involved 'being as friendly with Polish Jews as he was with the Christians. On no occasion would he allow anybody to recognize a difference.'[62] Another contemporary observer, Lord Clonmore, confirmed this when he recalled:

> He [Ratti] did not confine himself to Catholics, but met large numbers of Jews as well; as everybody knows, the Jewish problem in Central and Eastern Europe tends to become acute, and one knows that the reactions to it are sometimes barbarous and cruel, as in Hitler's Germany; Ratti *made it quite clear that any anti-Semitic outbursts would be severely condemned by the Holy See*, though from what one hears of Poland during the last few years, his wishes have not been respected as they should be ... All through his visit he was on the best of terms with the Jews, and on one occasion a chief rabbi specially asked for his prayers on behalf of himself and his people.[63]

Pacelli's nunciatures to Bavaria and then Berlin were designed to negotiate concordats, guaranteeing the rights of the Church with the various German federal states and with the Reich as a whole. Separate concordats were concluded with Bavaria in 1925 and with Prussia in 1929, with negotiations under way with Baden that were only successfully concluded in 1933, all preparatory for a future concordat with the Reich government. Pacelli also kept an eagle eye on the German bishops, as well as on what was being published by Catholic scholars. Obviously, he was also concerned with such 'cultural' developments in Weimar as public displays of nudity and an artistic culture based on provocation and sensation which unsurprisingly appeals to our own time.

The missions of both clerics to Poland and Germany at this time also meant that they were confronted by the economic distress of war and its aftermath, the threat of Bolshevism, and – partly related to this, for some Jews sought salvation from persecution in revolutionary politics – the antisemitism that was rife in both places. We have already seen that Ratti seems to have been actively condemnatory of this contagion, but what about Pacelli?

Pacelli had first-hand experience of a rogue Bolshevik regime: the short-lived Munich Soviet, one of those perennial objects of academic left-liberal nostalgia. In reality, an elegant southern German city briefly slid into the hands of fanatics and maniacs. On one occasion the nunciature was sprayed with machine-gun fire; on another a group of Bolsheviks broke in and threatened Pacelli at gunpoint. These political thugs also attempted to expropriate his official car, but his chauffeur disabled the transmission. They returned to tow the vehicle away. In response to this blatant disregard for the extraterritoriality of embassies and missions, Pacelli's assistant, Luigi Schioppa, went with the Prussian ambassador to Bavaria, to meet Eugen Leviné, the head of the local Soviet Republic. The meeting was ugly from the start, since monsignor Schioppa was evidently inexperienced in dealing with radicalised young women, who constituted the leading revolutionary's political 'groupies'. Pacelli signed off on Schioppa's report, which contained derogatory remarks about his rough and rude interlocutors, some of whom, including Leviné, were Jewish, although the Italian original of this document is less sensational than it has been made to seem in some English translations. 'Gruppo femminile', for example, as even non-Italian readers may sense, is perhaps not best rendered as 'female rabble'.[64]

It requires a major stretch of the imagination to regard this single document from 1919 as evidence of Pacelli's alleged antisemitism, rather than of his assistant's distasteful experience at the hands of Bolsheviks, many of whom, in Munich and elsewhere, were indeed Jewish radicals, or to connect it with his responses to the Holocaust, which began, by most respectable accounts, in 1941 – that is some twenty-two years later and two years after Pacelli had become pope. A mass of evidence from those intervening decades undermines whatever this letter is supposed to insinuate rather than prove.

The report does not even tell us much about the Vatican's responses to Bolshevism, which can hardly be described as motivated by purblind anti-Communism. Both nuncios, Ratti in Warsaw and the younger Pacelli in Munich (until 1925, when he moved to Berlin as nuncio to the German Reich), were closely involved in Rome's diplomatic initiatives with the Soviets. The Vatican initially welcomed the fall of the Romanovs, believing that this would herald a new era of freedom and opportunity for the Roman Catholic Church in the debris of the Tsarist Empire. Benedict XV employed Ratti to contact Lenin on behalf of persecuted Catholic and Orthodox clergy. In late 1921, the Vatican offered the Soviet Union humanitarian assistance, hurriedly incorporating a broader secret agreement which, capitalising on the disarray of the Orthodox Church, would – they imagined – have enhanced Roman Catholic activities in Russia. The aid was provided, but the wider agreement remained a dead letter. Assisted by the German government, which saw relations with Russia as a means of terminating Germany's pariah status, the archbishop of Genoa held talks with the Soviet foreign affairs commissar Chicherin on board an Italian cruiser with a view to negotiating a concordat. A further series of meetings took place at Rapallo, based on Vatican calls for freedom of conscience and Soviet demands for diplomatic recognition. Effortlessly overcoming the extreme distaste for German (Jewish) Bolsheviks that he is alleged to have expressed in 1919, Pacelli secretly met Maxim Litvinov, the Soviet Union's (Jewish) foreign minister, at the Berlin villa of the brother of the German ambassador to Moscow.

When Mussolini recognised the Soviet Union on 8 February 1924, and was quickly followed by, among others, Britain, Norway, Austria, Greece and Sweden, the Soviets ceased to regard negotiations with the Vatican as important except for the question of aid. Pacelli continued to negotiate with the Soviets in Berlin until mid-August 1925 when the execution in Leningrad of a Polish Catholic priest complicated matters. However, he met Chicherin twice in 1925 and 1927, discovering that his Soviet interlocutors were prepared to concede less and less, and such talks abruptly stalled under Stalin, to whom the Vatican was an irrelevance.[65]

While the historic Church has often been hostile or lukewarm in its attitudes towards individual liberty, democracy and popular sovereignty, which it associated with Jacobin mobs, it has also zealously patrolled the respective patrimonies of God and Caesar. Pius XI distinguished between what he dubbed 'objective' and 'subjective' totalitarianism. A state could, if it so desired, insist that 'the totality of the citizens shall be obedient to and dependent on the State for all things which are within the competence of the State'. However, it could not make 'objective' and total claims upon the citizen's whole life, whether domestic or spiritual. According to Pius such claims would be 'a manifest absurdity in the theoretical order, and would be a monstrosity were its realization to be attempted in practice'. These were precisely the sort of claims made by Mussolini and the Fascists. In a speech to the Party's Quinquennial Assembly, the Duce said: 'The State, as conceived and realized by Fascism, is a spiritual and ethical unit for the organisation of the nation, an organisation which in its origin and growth is a manifestation of the spirit . . . Transcending the individual's brief spell of life, the State stands for the immanent conscience of the nation.'[66]

The aspirations of the Fascist state went far beyond the bothersome meddling of traditional erastianism, or the studied indifference of classical liberalism, in that it sought to determine life's ultimate goals and to reorder fundamental moral meanings. Such pretensions were unacceptable to the Catholic Church, for they intruded into precisely those areas where the Church itself claimed primacy. There were further radical incompatibilities. If it was opportune for Mussolini to claim that Fascism itself was 'Catholic' in a society where 99.5 per cent of the population described themselves as such, Fascist ideology included several components that were hard to reconcile with the doctrines of the Church. Fascist enthusiasm for the ancient Roman Empire grated with a Church that liked to stamp such sites of pagan barbarism as the Colosseum with proclamations of its gospel of universal love. The Church frowned too on Fascist usurpation of religious forms, notably the 1925 catechism of the Balilla youth movement, which parodied the Christian original in blasphemous fashion:

> I believe in Rome the Eternal, the mother of my country, and in Italy her eldest Daughter, who was born in her virginal bosom by the grace of God; who suffered through the barbarian invasions, was crucified and buried; who descended to the grave and was raised from the dead in the nineteenth century; who ascended into Heaven in her glory in 1918 and 1922; who is seated on the right hand of her mother Rome; and who for this reason shall come to judge the living and the dead. I believe in the genius of Mussolini, in our Holy Father Fascism, in the communion of its martyrs, in the conversion of Italians, and in the resurrection of the Empire.

The Church also took a dim view of Fascist glorification of war and violence as aids to Fascist character formation, dubious enthusiasms which the education system and youth movement sought to inculcate even in the very young. Furthermore, a politics based on national egoism and hatred of others was surely hard to reconcile with the Christian precepts of charity and brotherly love. Even in those few areas where the concepts used by the Church and the Fascist state bore some superficial terminological similarity, closer inspection of the content reveals acute differences. For example, both Church and Party sought to restructure industrial relations through corporate structures that would transcend class conflicts between employer and employee. Corporatism would help chart a steady course through the Scylla of laissez-faire economic liberalism and the Charybdis of socialist collectivism and massification. The Church and Fascism may have resorted to the same term, but they invested it with radically different content. Where the Church sought a flowering of voluntary organisations, the Fascists identified an opportunity for enhanced state control over the economy. Hence when in 1931 Pius XI issued the encyclical *Quadragesimo anno*, in celebration and elaboration of Leo XIII's *Rerum novarum*, he criticised what had become an excuse for a further inflated Fascist bureaucracy:

> We feel bound to say that to Our knowledge there are some who fear that the State is substituting itself in the place of private initiative, instead of limiting itself to necessary and adequate assistance. It is feared that the new syndical and corporative organisation tends to have an excessively bureaucratic and political character, and that, notwithstanding the general advantages referred to above, it ends in serving particular political aims rather than in contributing to the initiation and promotion of a better social order.[67]

These fundamental differences – which of course were accompanied by others in outlook and temperament between politicians and the majority of clerics – were reflected in a series of clashes between the Church and Fascist regime that deserve notice. Since the Popolare Party had been disbanded along with the other democratic parties in 1926, shortly followed by the Catholic Trade Union Confederation, tensions between the regime and the Church revolved around the lay Catholic Action organisation, whose autonomy had been ostensibly guaranteed under article 43 of the 1929 Concordat. Since many former Popolari activists and politicians had regrouped within Catholic Action (next to a much smaller group of regime-supporting clerico-Fascists), any growth in that organisation was seen as doubly threatening, by a regime that had otherwise greatly diminished the size of Catholic financial and media interests by the late 1920s. In Fascist eyes, the increasing membership of Catholic youth organisations from 394,251 in 1928 to 713,623 in 1930–1 was worrying, especially since many of the recruits were in the eighteen-to-twenty-one age range, for whom there was no Fascist analogue to bridge the gap between graduation from the Balilla and entrance into the Fascist Party upon reaching the age of majority. Both the Church and Balilla aggressively competed for young members, with the Church often winning because of the superiority of its recreational facilities.[68] Tensions came to a head as the regime endeavoured to use an oath of loyalty to bind schoolteachers to the state, while selecting school heads and university presidents only from those who had been long-term members of the Party. Symbolically, the leadership of the Ballila was incorporated into the Ministry of National Education. These moves took place against a background of enhanced surveillance and harassment of Catholic organisations. The Pope's response took the form of the January 1930 encyclical *Rappresentanti in terra* which asserted the primacy of the family and Church in the education of youth, while insisting that Sundays should be spared the paramilitary exercises to which the Fascist Balilla exposed young people of both sexes. If on this occasion tensions did not reach boiling point, because the honeymoon after the Concordat still lingered, this was not true a year later.

Conflict was made more likely by changes in key personnel. The appointment as Party secretary of Giovanni Giurati and his deputy Carlo Scorza was paralleled by the advent of Eugenio Pacelli as Vatican secretary of state. Pacelli's lack of recent experience of the Italian scene – he had been in Germany – ensured that the voices of both the pontiff and of prominent Popolari in the Secretariat were unmodulated by such a practised diplomat as the retiring secretary of state Gasparri. The imminent celebration of the fortieth anniversary of *Rerum novarum* served also to galvanise Catholic efforts to reconstitute its presence in the workplace, by effectively infiltrating residual professional associations. Such crypto-unionisation irked Fascist labour organisations, whose paper *Il Lavoro Fascista* said: 'a particular object of these activities was the conversion of workers from a Fascist to a Catholic allegiance' inspired by 'Popolare elements'. The more upbeat the Church became in the run-up to the anniversary of *Rerum novarum* the more Fascist moods darkened as the Italian economy stalled under the impact of global depression. Both Catholic Action and the Vatican itself seemed to be operating as safe havens for former Popolari and outright opponents of the regime who the political police thought were engaged in conspiracy to collapse the government at a very vulnerable moment. In 1931, Mussolini bowed to pressure from Giurati and Scorza to abolish the Catholic Action youth movement, as the most subversive challenge to Fascist totalitarianism.

Having been hit at a particularly neuralgic spot, Pius XI responded with his 29 June 1931 encyclical *Non abbiamo bisogno*, one of only two encyclicals written in Italian. This systematically refuted the notion that Catholic Action had engaged in party politics, while lambasting such features of life under the Duce as the controlled press: 'the only press which is free to say and to dare to say anything and is often ordered or almost ordered what it must say'. It also sharply rejected the totalitarian aspirations harboured by the Fascist youth movement:

> The resolve (already in great measure actually put into effect) to completely monopolize the young, from their tenderest years up to manhood and womanhood, for the exclusive advantage of a party and a regime based on an ideology which clearly resolves itself into a true, a real pagan worship of the State – the 'Statolatry' which is no less in contrast with the natural rights of the family than in contradiction with the supernatural rights of the Church.[69]

Ironically, the pope's firm stance even won him the solidarity of foreign freemasons and Protestants, no mean feat since he had earlier been urging the Fascist government to tighten restrictions on both groups within Italy. Having gone to the brink of war, both parties pulled back and in the September Accords found a revised basis for the continued existence of Catholic Action.

III THE CATHOLIC CHURCH AND GERMAN NATIONAL SOCIALISM

As nuncio to Bavaria, Eugenio Pacelli sent regular reports on the activities of the National Socialists to Rome. These have received less publicity than what he is alleged to have said about Bolsheviks in 1919. In October 1921 he linked the former with extreme nationalists in remarks cited in the *Bayerische-Courier*: 'The Bavarian people are peace-loving. But, just as they were seduced during the revolution by alien elements – above all, Russians – into the extreme of Bolshevism, so now other non-Bavarian elements of entirely opposite persuasion have likewise thought to make Bavaria their base of operation.' Five days after the Munich putsch by the Austrian Hitler and the Prussian Ludendorff, Pacelli reported on it to secretary of state Gasparri. The supporters of the putsch had turned their rage against Munich's cardinal Michael Faulhaber: 'The attacks were especially focused on this learned and zealous Cardinal Archbishop, who, in a sermon he gave in the cathedral on the 4th of this month and in a letter of his to the Chancellor of the Reich published by the Wolff Agency on the 7th, had denounced the persecutions against the Jews.'

Faulhaber had indeed issued such denunciations, in letters to Gustav Stresemann and Bavaria's Heinrich Held, as well as in public sermons. The Nazis managed to blame Faulhaber's machinations for the failure of their putsch.[70] One might add that the head of the Franciscan seminary in Munich, Erhard Schlund, had written a critique of the Nazi Party programme, in which he specifically denounced its unChristian antisemitism. There were demonstrations against Faulhaber for an entire weekend, as well as at the university, where students voiced denunciations 'of the Pope, of the archbishop, of the Catholic Church, of the clergy, of von Kahr, who, even though he is a Protestant, was characterized by one of the orators as an honorary member of the Society of Jesus'.[71]

In a further report, Pacelli informed secretary of state Gasparri about the 'vulgar and brutal campaign' that Hitler's supporters were waging in the press against Catholics and Jews, whom he linked as victims of Nazi persecution. He followed the trial of Ludendorff closely, regarding the general as the epitome of 'the blind fanaticism of intolerant Protestantism'. Nationalism, he wrote, 'is perhaps the most dangerous heresy of our times'.

In 1928, that is six years after Ratti had become Pius XI, the Holy Office issued a binding condemnation of 'that hate which is now generally called anti-Semitism'.[72] During one of his audiences with Sir Ivone Kirkpatrick, Britain's representative to the Holy See before 1933, Pius displayed his temper – the warning signal being that he pulled his cap over his ear, telling Kirkpatrick 'in pungent terms what he thought of Hitler's persecution of the Jews'.[73]

Both the Austrian and German Catholic bishops were more condemnatory of Nazism than may be popularly realised. In 1929, bishop Johannes Gföllner of Linz warned the faithful against the 'false prophets' of Nazism: 'Close your ears and do not join their associations, close your doors and do not let their newspapers into your homes, close your hands and do not support their endeavours in elections' being as unequivocal as one could reasonably expect, although it was not incompatible with his advocacy of 'ethical antisemitism'. The Austrian Catholic newspaper *Volkswohl* even parodied life in a future Nazi state in a manner that seems extraordinarily prescient. Every newborn baby's hereditary history would be checked by a Racial-Hygienic Institute; the unfit or sickly would be sterilised or killed; dedicated 'Aryan' Catholics would be persecuted: 'The demonic cries out from this movement; masses of the tempted go to their doom under Satan's sun. If we Catholics want to save ourselves, then it can never be in a pact with these forces.'[74]

The German bishops were similarly condemnatory of National Socialism when in 1930 the Nazis broke through the ceiling that separated a marginal sect with less than 3 per cent of the vote from a mass political party. Adolf Bertram of Breslau warned Catholics in 1930 against the Nazis' radicalism, 'racist madness' and their schemes for a single supra-confessional 'national Church'. The archbishop of Mainz went further, by declaring that Nazism and Catholicism were simply irreconcilable:

> The Christian moral law is founded on love of our neighbour. National Socialist writers do not accept this commandment in the sense taught by Christ; they preach too much respect for the Germanic race and too little respect for foreign races. For many of them what begins as mere lack of respect, ends up as full-blown hatred of foreign races, which is unChristian and unCatholic. Moreover the Christian moral law is universal and valid for all times and races; so there is a gross error in requiring that the Christian faith be suited to the moral sentiments of the Germanic race.

The provinces of Cologne, Upper Rhine and Paderborn warned clergy to have nothing to do with the Nazis, and threatened the leaders of parties that were hostile to Christianity with denial of the sacraments. The Bavarian bishops banned Nazi formations from attending funerals or services with banners and in uniform, while condemning both Nazi racism and their eugenic contempt for unborn life.[75]

The statements of these bishops so shocked the Nazis that Göring was despatched to Rome to smooth things over. Since Pius XI instructed Pacelli not to meet him, Göring had to vent his grievances against the Catholic Church on Pacelli's under-secretary. His approach was to combine defence with attack, the latter diplomatically couched as 'regrets', such as the claim that many of the priests who belonged to the Centre Party were attacking Nazism in private. At the same time he disowned the writings of Rosenberg. Interestingly, as a prominent and sincere Protestant, who had married his wife Emmy in a Lutheran ceremony and whose daughter Eda underwent a Lutheran baptism, Göring tried to justify Nazi racism with reference to the theology of orders of creation, 'for races had been willed by God'. He contrasted the silence of the Lutheran Churches with the 'attacks' the Party had received from the Catholic clergy, warning that the Nazis would defend themselves.

While Protestants voted for the Nazis in greater numbers than Catholics, the latter were not immune to a general disillusionment with the Weimar Republic, which manifested itself in the fashion for authoritarian solutions, involving the state acting decisively to restore community, in the heady realm of political thought.[76] Like the conservative and liberal parties, the Catholic Centre Party responded to this shift in opinion by moving to the right, a shift symbolised by its appointment as leader of the conservative priest Ludwig Kaas in 1928. Two Catholic politicians, Heinrich Brüning and Franz von Papen, were also instrumental in the creeping demolition of Weimar democracy, and in bringing Hitler within the purlieus of political respectability, although they were hardly unique in desiring either, or in the delusion that this political wild man could be tamed. After Hitler's appointment to the chancellorship in January 1933, the votes of the Centre Party in the Reichstag were crucial to his passage of the Enabling Law on 23 March, allowing him to govern without recourse to parliament for four years. Fearful of what the Nazis might do, and lured by their vague assurances regarding religion, Kaas ensured that seventy-two deputies of his party voted for the Law. In a major speech, Hitler made great play with 'the political and moral cleansing of our national life', and with his government's goal of 'creating and securing the conditions for a really deep and inner religious life'. Christianity, he said, was 'the unshakeable foundation of the moral and ethical life of our people'. Within a few days, in late March, the Catholic bishops rescinded their earlier 'no' to National Socialism with what might be called a 'yes, but': 'Without revoking the judgements of our previous statements against certain religious and ethical errors, the episcopate nevertheless believes it can now cherish the hope that the previous general warnings and prohibitions need no longer be considered necessary.'

It is tempting to see betrayals and conspiracies at every stage of this story. Many (mainly Protestant) historians have claimed that the Vatican did a squalid deal with the Nazis, first persuading the Centre Party to vote for the Enabling Law, in return for the prospect of a concordat, only to throw the Centre Party to the wolves once the Concordat had guaranteed the Church's own institutional interests. Thanks to the scholarship of Rudolf Morsey and Konrad Repgen, we know that none of these speculations is true. The prospect of a concordat played no part in the negotiations between the Centre Party leadership and Hitler that led the former to support the Enabling Bill. Nor when the Vatican responded in early April 1933 to vice-chancellor Papen's offer of negotiations for a concordat was the intention either to abandon the Centre Party or to go along with the Nazis' wish to stop all clerical participation in politics. The Vatican also took the opportunity of condemning the persecution of the Jews.[77]

Papen certainly sought such a total ban, which was why he took a copy of article 43 of the 1929 Lateran Treaties with him to Rome, but his interlocutors then spent three months trying to find ways for bishops to permit some clerics to engage in politics. That position was abandoned only when the Centre Party, and its Bavarian analogue the BVP, had abolished themselves on 4 and 5 July. As Pacelli subsequently wrote, the first he knew of this development, of which he disapproved, was when he read about it in the newspapers.[78] It seems inherently improbable that an experienced diplomat such as he would urge the abolition of a party whose continued existence was a vital bargaining chip in his negotiations with the German government.[79] Quite independently of Vatican diplomacy, the auguries were not good for any political party, every one of which, it should be emphasised, whether left, centre or right, had made a contribution to the demise of democracy in Germany. The Communists, who had sometimes aided and abetted the Nazis in rendering parliament unworkable, had been forced underground in February 1933. The Social Democrats were proscribed on 22 June and the conservative-nationalist DNVP followed within days. The Bavarian Catholic Party and the right-liberal DVP were gone in the first week of July. In these circumstances, the dissolution of the Centre Party was a foregone conclusion.[80]

A more important question is perhaps what both negotiating parties hoped the Concordat might achieve. The government side harboured various agendas. As a Catholic, Papen may have wished to remedy his own anomalous membership of a government whose creed the bishops had condemned. Hitler's concerns were narrower and he could always disown the Concordat as Papen's initiative should the negotiations collapse. The Führer hoped that a concordat would fatally weaken the Centre Party by removing clerics from its leadership, an objective already pulled off by his Italian counterpart Mussolini. The Centre Party's auto-destruction meant that Hitler had achieved his primary goal before the Concordat was ratified, which was why in early July he began applying the brakes to the negotiations. Papen, in Rome for the final talks, had to call Hitler to convince him of the advantages stemming from 'recognition of the youthful Reich by the supranational power of the two-millennia-old Church'.[81]

Its utility to the Nazis was, firstly, to undermine what had proved to be relative Catholic immunity to the 'movement', at least before they achieved power, and secondly, to build on the international recognition they were receiving from other European governments. There had been trade agreements with the USSR in May 1933, and Britain, France and Italy were on the verge of concluding the Four Power Pact. The key consideration for the Vatican was that, since the emergency legislation enabled Hitler to suspend the Weimar Constitution, its provisions protecting freedom of religion were effectively null and void.[82]

Thanks to Konrad Repgen's researches, we know that opinion in the Vatican was divided on how to respond to the German initiative. An influential Jesuit, Robert Leiber, outlined the different positions in a report to the Austrian ambassador to the Vatican, whose government was worried about the implications of a concordat with a regime that was actively undermining it. One group was against any negotiations that might lend prestige to a regime that was so inimical to Catholicism, negotiations that would inevitably demoralise German Catholics. A second group argued that, while Nazism was based on principles utterly alien to the Vatican, the latter was duty bound to protect the existence of

Catholicism with whatever legal guarantees it could secure. The search for such agreements would be a source of moral strength in the persecutions that were likely to come. A third group thought that, while the Church should in no way endorse Nazism, it could use negotiations to squeeze out as many benefits for Catholicism as possible. If, as the Vatican fully expected, the Nazis infringed the terms of such a treaty, then the Vatican could deal the regime's prestige a major blow by formally abrogating it. Recent developments in Germany had also effectively rendered the three existing concordats null and void, so there was a practical need for new arrangements with the Reich. The constant reports of Nazi infringements of Catholic rights meant that many Vatican officials were in two minds whether to ratify the Concordat at all. Then there were the gloomy reports from Orsenigo, the nuncio in Berlin, who claimed that large numbers of young Catholics were no longer immune to the 'fascination' to which many Protestants had already succumbed, separating what they liked about Nazism's political programme from the 'ideological–political' issues that had bothered their bishops. Secondly, Orsenigo was worried about the prospect of a Protestant Reich Church under a single Reich bishop, which would consolidate 'into one giant mass' some forty million Protestants, with the Catholics relegated to being a powerless minority.[83]

The German Catholic bishops were similarly divided, especially when Hitler launched a 'charm offensive' by granting personal interviews to Berning of Osnabrück and Bertram of Breslau in late April 1933. He told the former:

> I am personally convinced of the great power and deep significance of Christianity, and I won't allow any other religion to be promoted. That is why I have turned away from Ludendorff and that is why I reject that book by Rosenberg. It was written by a Protestant. It is not a Party book. It was not written by a Party man. The Protestants can be left to argue with him ... As a Catholic I never feel comfortable in the Evangelical Church or its structures ... you can be sure: I will protect the rights and freedoms of the churches and not let them be touched, so that you need have no fears about the future of the Church.[84]

Some thought that a concordat was the 'last hope' for protecting Catholic interests in a climate that threatened to develop into a Nazi version of the nineteenth-century Kulturkampf, naturally enough the most vivid example of persecution that these elderly sons of the Kaiserreich could imagine. Their folk memory of those times included bishoprics left vacant, bishops on the run or in prison, and both clergy and laity subject to petty bureaucratic chicanery. The Church's top men in Germany were under no illusions about deals with this government. Cardinal Faulhaber remarked that 'With the Concordat we are hanged, without the Concordat we are hanged, drawn and quartered.' Arch-bishop Gröber warned Pacelli that, if the Vatican walked out of these negotiations, 'everything we have will soon be smashed. Catholics would say: the Holy See could have helped us and did not. The government would publish the text of the Concordat and blame the Holy See for blocking the accomplishment of such a good work.' Others hoped that a concordat would bolster moderate Nazis against their more radical comrades, a common delusion at the time. A third, rejectionist group thought the Concordat would not only reduce the Church to being the 'trainbearer' of National Socialism, but make it co-responsible for whatever evils the Nazis might commit in future.[85]

The chief accomplishment of the Concordat for the Church was that Hitler's government formally acknowledged a sphere notionally beyond the totalitarian aspirations of the Nazi state. That concession was unique, even if the Nazis proceeded to breach the agreement at every turn. Both signatories naturally regarded the Concordat in a very different light. Notwithstanding his views of clerics, Hitler was confident that the Concordat had secured 'the unreserved recognition of the present regime'. This view was transformed into fulsome Catholic recognition of National Socialism in the regime's press. Pacelli immediately issued a disclaimer in the official *Osservatore Romano*, stating that the Concordat did not signify 'approval or recognition of a particular political direction or teaching', an important rider that the German press was told not to report. Privately, Pacelli told the British representative to the Holy See that 'a pistol had been pointed to his head and he had no alternative. The German government had offered him concessions, concessions, it must be admitted, wider than any previous German government would agree to, and he had had to choose between an agreement on their lines and the virtual elimination of the Catholic Church in the Reich.' According to Ivone Kirkpatrick, Pacelli responded to his stated hope that Hitler might calm down in power: 'I am afraid not, we shall see that with every year power will make him more extreme and difficult to deal with.'[86] In November, nuncio Orsenigo submitted to Pacelli a draft of the New Year address he was to deliver to Hindenburg and Hitler as doyen of the diplomatic corps. Pacelli deleted the entire paragraph on Hitler and recommended that 'the praises contained in the address should undoubtedly be moderate, in consideration of the grave difficulties to which the Church is now exposed in Germany'.[87] There, although archbishop Gröber welcomed the Concordat with jubilation, cardinal Faulhaber coupled praise of Hitler's 'statesmanly breadth of vision' with the expectation that the terms of the agreement would be respected by the Führer's minions, and a plea for an amnesty for concentration-camp inmates – a plea unnoticed by Faulhaber's contemporary critics.

It is worth emphasising that there was no Catholic equivalent of the Nazi–Protestant German Christian movement, which had nearly six hundred thousand members, and in whose creed antisemitism played a crucial part. The Catholic intelligentsia also emerge favourably from any comparison with their Protestant counterparts. While a book called *Catholic-Theologians in Nazi Germany* contains numerous examples of opponents of the regime, and rather few examples of supporters, an equivalent study of Lutheran Germany's finest consists of supporters of Nazism.[88] Some of the most eminent Catholic theologians, such as Engelbert Krebs, Wilhelm Neuss, Karl Rahner and Romano Guardini, lost their university teaching posts under the Nazis. Krebs not only published articles reflecting his positive view of Judaism, but was denounced in August 1934 for saying at a private gathering in his brother's house, 'We are being governed by robbers, murderers and criminals,' a remark that resulted in several years of harassment, the loss of his job, a trial and imprisonment.[89]

The ecclesiastical historian Joseph Lortz was one of the very few Catholic intellectuals who took things further than was seemly. Author of an influential history of the Church, Lortz thought that Western Christian civilisation had been on the skids since the twelfth century, when the progenitors of modern subjectivist liberalism had shattered the harmony of Christendom. He thought things had picked up again in the twentieth century. There were signs that the nineteenth-century age of 'doubt, hypercriticism, or subjectivism' was passing, while 'ethically' 'the trend is from unrestrained freedom to authority, from the egoism of individualism to communal thinking'. Finally, in the political sphere, professor Lortz of remote Braunsberg thought that parliamentary democracy was being edged aside by 'the principle of leadership, in the form of dictatorship, or government without parliamentary majorities, or nonparty government (Fascism, Nazism)'. He claimed that Nazism promised to restore a lost harmony. In a lecture entitled 'The Catholic Approach to National Socialism' which he delivered at Königsberg in 1933, he recommended that Catholics recognise the positive sides of Nazism as Germany's deliverance from Communism and parliamentary chaos. He enumerated what the Church and Nazism allegedly had in common, namely an antipathy to individualism, hedonism, liberalism, Marxism and 'the cesspool of the capital' Berlin. Above all, he found the Nazis' use of religious rhetoric persuasive:

> It is now extremely valuable for the Catholic understanding of religion as revelation that through National Socialism the formal attitude of a creedal standard which had almost completely dropped out of circulation, is once again in place in the widest ranks of society as a valuable attitude. In fact, this is true with an unexpected intensity. 'Faith' no longer appears as something of lesser worth or weak, but rather as momentous and heroic, through which humanity realizes the best dimension of itself.[90]

In contrast to professor Lortz, many leading lay Catholics were redoubtable opponents of Nazism. Someone clearly thought they were worth killing. Erich Klausener, the general secretary of Catholic Action, was shot dead by the head of Hitler's bodyguard on the express orders of SD chief Heydrich, and Adalbert Probst, the director of the Catholic Youth Sports Association, was also killed in the course of the 'Night of the Long Knives'. The bishop of Berlin protested against Nazi claims that Klausener had committed suicide and interred his ashes (his cremation being an offence in itself to Catholic sensibilities) with a solemn requiem, obliging every church in the diocese to read out Klausener's obituary. The bishop wrote three letters to Hitler defending Klausener's honour; all but the first received no answer. Another Catholic victim deserves more extended consideration.[91]

Fritz Gerlich was originally from a solidly Calvinist family in Stettin. Settling in Munich, he first worked for the Bavarian State Archives, before becoming editor of the *Münchner Neusten Nachrichten* in 1920. In that year, he published a path-breaking account of Russian Communism as a form of political messianism with roots in the chiliastic tradition of the Middle Ages.[92]

Gerlich was an overworked and hardbitten journalist who naively tried to camouflage his alcoholism by drinking wine from stone beer mugs in the editorial office. In 1927 this religious sceptic went to report on a village stigmatic called Therese Neumann, who had allegedly cured local people of many illnesses. She also made cryptic utterances in Latin, Greek and what appeared to be biblical Aramaic, while her diet seemed to consist solely of communion wafers. This crucial encounter led Gerlich to write a two-volume refutation of Neumann's medical and scientific detractors. In 1931 he converted to Catholicism, consulting Neumann about every important decision he made thereafter. Cardinal Michael Faulhaber, who became one of his strongest supporters, received him into the Church.

In the meantime, Gerlich had been sacked for holding too many one-man carnivals in the office. Fortunately, the young Furst Erich von Waldburg zu Zeil agreed to finance a new Catholic paper as part of Pius XI's Catholic Action movement. Gerlich and his patron purchased an existing Sunday illustrated, which by coincidence shared its presses with the Nazi *Völkische Beobachter*. Week after week Gerlich published attacks on the Nazis, including a spoof interview in which 'Swiss' friends asked him to compare Hitler with Kaiser Wilhelm. This went very near the wire when the interviewer asked: 'Doesn't the penetration of homosexuals into leading positions in the [Nazi] Movement and in the intimate circles of the coming Caesar provide a further shocking parallel to the Eulenburg era of Wilhelm II [Eulenburg being the Kaiser's homosexual favourite]?' The reactions of Hitler to the headlines 'Has Hitler got Mongol blood?' or 'Lock up the Führer' are unrecorded. In 1931–2 Gerlich changed the paper's name to *Der Gerade Weg*, an allusion to Christ's words to Ananias. Its circulation rose from about forty thousand to ninety thousand in 1932. The Nazi response escalated from character assassination, via libel writs, to death threats. Gerlich acquired an arms permit. In March 1933 fifty SA men burst into and demolished the editorial offices, and Gerlich was very badly assaulted, by among others Max Amann, the Nazis' publisher in chief. The raiding party was looking for information Gerlich had on Hitler's niece Geli, and on Ernst Röhm's dealings with Anglo-Dutch oil. Gerlich was dragged off to Munich's Stadelheim prison, where some time on the Night of the Long Knives he was transferred to Dachau and immediately murdered.[93]

The Catholic Church was at once confronted after January 1933 by Nazi antisemitic and eugenic policies. The German bishops equivocated on how they should react to the 1 April 1933 boycott of Jewish businesses, on the grounds that these were political questions and that, since the boycott ended after three days, the Jews seemed well able to look after themselves. This seems a pharisaic response, not least because some of those affected were converts to Catholicism. A week later the Nazis introduced the Law for the Restoration of the Professional Civil Service, designed to remove political opponents, including 'political Catholics' as well as 'non-Aryans' from public service. Here, the secretary of state in the Vatican took an initiative. On 4 April Pacelli instructed nuncio Orsenigo, following approaches to the pope from various Jewish dignitaries, that he was to 'see if and how it is possible to take the matter up in the desired way'. He reminded Orsenigo that 'it is the tradition of the Holy See to exercise its universal mission of love and peace among all peoples, regardless of their social circumstances or their religion, and where necessary for its charitable establishments to intervene'. Orsenigo duly reported the limp response of the German hierarchy, although three bishops had issued a statement deploring the suffering of 'many loyal citizens' (meaning Jews) and 'conscientious civil servants'. On 12 April the archabbot of Beuron wrote to Pacelli enclosing an impassioned letter from the Catholic convert Edith Stein regarding the Church's 'silence' about the persecution of the Jews, which she wanted put before the pope. The archabbot endorsed her views on the gravity of the situation in Germany. In his reply on 20 April, Pacelli informed the archabbot that Stein's plea had been seen by the pope, with whom Pacelli had prayed for God's protection for the Church as 'the precondition for a final victory'. In October Stein became a Carmelite nun. She was murdered at Auschwitz and in 1998 was declared a saint. Nothing indicates that Pacelli did not take her concerns seriously and responded accordingly.[94]

During the same cabinet session that approved the Concordat, Hitler's government implemented eugenic sterilisation policies that ran contrary to fundamental Catholic teaching as most recently stated in the 1930 encyclical *Castii connubi*, which had been issued because of the alarming number of eugenic sterilisations in many US states. On 14 July, the Nazis introduced the Law for the Prevention of Hereditarily Diseased Progeny. As a result of this, some 350,000 people would be eugenically sterilised before the Second World War, because of conditions whose hereditary nature was sometimes a matter of scientific faith. *Osservatore Romano* warned in November 1933 that governments should not degenerate into cattle-breeding laboratories.[95] When two Catholic academics at Braunsberg's Staatliche Akademie, the theologian Karl Eschweiler and canon lawyer Hans Barion, publicly supported the sterilisation law, cardinal

Pacelli instituted canonical proceedings against them and they were suspended from teaching seminarians.[96]

Compulsory sterilisation directly affected both Churches because the Catholic Caritas Association and the Protestant Inner Mission were responsible for extensive health and welfare networks ranging from special homes for disadvantaged or sick children to institutions for alcoholics, epileptics, geriatrics and psychiatric patients. Organisational circumstances partly explain the differing response of these Catholic and Protestant networks to eugenic policies. A highly decentralised Protestant Church seems to have lost interest in controlling its own charitable apparatus, to the extent that pro-eugenic professional enthusiasts within its ruling councils went increasingly unchallenged as the Depression made their arguments for sterilisation seem economically compelling. Protestantism was generally more prone to worrying about seeming out of step with scientising modernism – and other secular trends – than a Catholicism steeped in Natural Law doctrines, and in which the autonomy and integrity of the family was so central. The Roman Catholic response was made relatively easier by virtue of their Church's authoritarian and international structure – in other words, those characteristics that many liberals most hold against it.

The Vatican hierarchy mostly came from Latin countries that found these policies instinctively reprehensible, even when they were ruled by authoritarian or Fascist governments themselves. *Casti connubii* was unambiguous in its assertion that the right of families to children overrode the state's desire for eugenically unimpeachable citizens. Even before the Nazis' promulgated their Law, *Osservatore Romano* ran articles with such headlines as 'Dangerous Eugenic Plans'.

The representatives of the Catholic bishops, Berning and Gröber, secured from the interior minister the right to continue making the Church's views on sterilisation known, as well as an exemption stipulating that Catholic doctors would not be legally compelled to initiate a person's sterilisation. Nurses from religious orders and Catholic clinical sisters were not allowed by the Church to take part in sterilisations in any shape or form. Much to the annoyance of the regime, throughout the 1930s Catholic diocesan papers routinely published the judgements of the Hereditary Health Courts, thereby drawing attention to them. This was seen as a subversive act. So too was simply arguing the Catholic case. The following story reminds us that Nazi Germany was not a free society where one could talk about what one liked. In 1935 Ludwig Wolker of the

German Catholic Young Men's Association shared a train compartment with Marta Hess of the Racial Political Office of the NSDAP. They talked about the sterilisation law, which Wolker disagreed with. Three days later Wolker received a letter from Hess inviting him to lay his concerns before Walter Gross, the leader of the Racial Political Office. Wolker declined on the grounds that it was not his remit and that he had only anecdotal evidence of malpractices. Gross took over the correspondence, demanding the written proof that would support Wolker's criticisms, and threatening to hand the matter over to 'other offices' if Wolker did not co-operate. Wolker was worried now and replied that he had only had a conversation on a train. He had reason to be worried since Gross had already invited the SS Reich Security Main Office to see whether his loose talk was actionable. The SD then handed the materials on Wolker to the Gestapo, who arrested him in November 1935 and kept him in custody until May the following year.[97]

On the whole, the Catholic hierarchy seems to have had a tighter grip on the Caritas welfare network than was the case with its Protestant counterparts. There were, however, ways of weakening it. The state used various forms of chicanery to close Catholic homes and institutions. These ranged from adversely changing their charitable tax status to using the Gestapo to suborn children to make accusations of sexual abuse against those in charge of them.

Between September 1933 and March 1937 secretary of state Pacelli signed over seventy notes and memoranda protesting against Nazi violations of the Concordat.[98] The Nazis almost immediately began chipping away at the autonomy of Catholic lay organisations which had apparently been secured by the Concordat. They suggested that Catholic youth organisations fuse with the Hitler Youth, which would then allow some unspecified degree of religious instruction. While Nazi formations proselytised among the nation's youth, often with such inducements as a new inkpot, Catholic youth organisations were banned from seeking recruits. Catholic organisations were either pressured into allowing themselves to be absorbed by monopolistic Nazi formations or had to adapt their internal structures to chime with the all-pervasive intro-duction of ideology and hierarchy into areas that had done very well without them. Members of Catholic youth groups were set upon by Nazis, who sought to intimidate the former into no longer displaying their emblems and pennants. In February 1936 the leadership of the Catholic Young Men's Association were charged with treasonable involvements with the proscribed Communists. One after another, Catholic newspapers and journals were closed or transformed beyond recognition, while the size of diocesan newspapers was curtailed on the pretext of paper shortages. By 1935 none of the four hundred Catholic daily newspapers existed. Between January 1934 and October 1939 the number of Catholic weeklies and periodicals fell from 435 to 124 by being either suppressed or starved of newsprint.[99] By contrast there was paper in abundance for officially produced anticlerical smut, with the famous 'Black Madonna' of Częstochowa dismissed as 'a middling thing between a negress and a Mongol woman'. Although it is often unremarked, Nazi publications frequently played upon the identity of Jesuits and Jews, or claimed that secretary of state Pacelli was in league with the Bolsheviks. When in 1935 bishop Bornewasser of Trier visited Kreuznach, his entourage were assailed by Hitler Youths shouting 'Jew Bishop! Bolshevik Pimp!' The SS organ frequently insulted the pope, as in this 'poem' entitled 'The Chief Rabbi of all Christians':

> Go bury the delusive hope
> About his Holiness the Pope.
> For all he knows concerning Race
> Would get a schoolboy in disgrace.
> Old, muddle-headed, doddering, ill,
> His knowledge is precisely nil.
> And, gone in years, he can but keep
> His motley flock of piebald sheep;
> Since he regards both Blacks and Whites
> As children all with equal rights
> As Christians all (whate'er their hues)
> They're 'spiritually' naught but Jews.
> . . .
> A pretty picture all men know –
> The form of 'Juda-Rome and Co.'
> An 'Old Man' e'er can tell the tale
> And, sure, his pity will not fail.
> The banner is at last unfurled:
> 'The Chief Rabbi of the Christian world'.[100]

Notwithstanding the guarantees of freedom of religious education in the Concordat, the Nazis attacked confessional schools as deleterious to the unity of the 'national community'. Parents were pressured by the prospect of losing their jobs in the state and municipal sectors, or by propaganda campaigns, into voting at registration meetings for the transformation of their schools into non-denominational 'community schools'. Home visits from Nazi block wardens helped them make up their minds. Since it is otherwise inexplicable how a solidly Catholic region like Bavaria could end up having no Catholic schools by 1939, one must assume these votes were rigged. Thus a vote held on 25 October 1937 in an Upper Palatinate village produced a 47:9 majority for community schooling. In fact, of the sixty-five qualified voter parents, twenty were absent, eleven left the meeting because discussion was prohibited, and, of those who remained, sixteen voted for community education and nine against. Sixteen became forty-seven by counting all these absentees as present. Another way of achieving the laicisation of education was simply to dismiss nuns from the profession, as the Ministry of Education in Bavaria did in May 1936. This caused sporadic protests. In the village of Glonn, for example, straw-filled images of the two Nazi replacement teachers (wearing red shirts and with Soviet hammer-and-sickle cap badges) were attached to the school's lightning conductor by way of popular protest.

Having frequently promised to maintain religious instruction in the non-denominational sector, the regime contrived the laicisation of religious instruction and the curtailment of the time devoted to this subject. The flow of qualified instructors was interrupted by closing all religious teacher-training colleges at the same time as the number of students reading theology at university fell from 6,388 in 1933 to 1,335 five years later. Since only four of Germany's theology faculties were regarded as being ideologically sound, considerable effort went into reallocating posts in this discipline to such burgeoning fields as racial science. In 1939 Romano Guardini, one of Germany's most distinguished Catholic philosopher–theologians, was dismissed from his Berlin professorship in the Christian worldview on the ground that Nazism had its own worldview which brooked no rivals.

Within schools the time devoted to religious instruction was cut, while that given over to gymnastics or sport rose. Nazi irreligion and racial dogmas could be insinuated into children's minds through a variety of subjects, such as biology, history and mathematics, where children were encouraged innocently to compute how many houses could be built for 'healthy national comrades' if one eliminated the 'unhealthy' population of 'luxury' lunatic asylums. If all else failed, religious instruction could be damaged, simply by scheduling rival Hitler Youth events simultaneously. State subsidies to private schools were abolished, and, when this sector was 'co-ordinated' in 1939, Catholic institutions were forbidden to belong to the association that controlled them. The public presence of the Catholic Church, which was highly visible at certain times of the year in western and southern Germany, was reduced. Impossibly high visa charges – designed to wreck the Austrian tourist economy – made it impossible for German Catholics to attend the major Catholic rally which in 1933 was scheduled to be held in Vienna. Insofar as Catholic religious processions could not be banned, the presence of Nazi photographers who would then display their handiwork in a person's workplace was an attempt to deter participants. Uniformed hooligans disrupted respectable religious gatherings. There was no point in appealing to policemen among the onlookers, given that, as a banner on the police headquarters of Essen proclaimed, 'The police stand by the Hitler Youth.'[101] Religious processions were rerouted or passed along streets devoid of decorations. In some regions, wayside shrines and crucifixes were vandalised, while in entire areas the authorities sometimes sought to remove religious images from school classrooms, replacing them with portraits of the Führer. Such a bold step was essayed by the government in north-west German Oldenburg in November 1936. Civil disobedience spread very rapidly and led to ugly scenes at the public meetings organised to persuade people of the desirability of removing religious images. Gauleiter Röver received such a rough ride at one such gathering in Cloppenburg that he immediately ordered the restoration of these images to the Catholic Münsterland's classrooms. That the protesters were Lower Saxon farmers of racially unimpeachable stock, and included at least one holder of the NSDAP's gold insignia, only added to the regime's desire to backtrack as quickly as possible.

In the mid-1930s these various measures were given a more vicious accent by government-sponsored campaigns against the Catholic Church involving those old standbys of money and sex. Much of the responsibility for these campaigns can be traced to Himmler's SS, specifically SD Office II 113, which was responsible for religious affairs, in tandem with the executive arm of the security services, in this case the Gestapo's parallel office called II 1 B 1. The SD religious affairs office included two former priests and four former monks. Its first leader was an elderly former priest, who was a cousin of Himmler's, and then the renegade priest Albert Hartl, who in 1937 married one of Heydrich's cast-off girlfriends. Hartl had been ordained in 1929 by Munich's cardinal Michael Faulhaber and went to work in the seminary at Freising. Hartl opportunistically joined the Nazi Party in the spring of 1933 having become disillusioned with the Church. Late that year, he denounced the director of the seminary for frustrating his transfer to a teacher-training college. At a private supper the director Josef Rossberger had said that Nazi Party members were 'the scum of humanity'. Rossberger was jailed for eight months on Hartl's testimony.

In 1934 Hartl joined the SS and became head of the section of the SD that dealt with religious affairs, including sects as well as Churches. As Faulhaber accurately surmised, Hartl's SS colleagues neither respected nor trusted him, partly because he seemed so short, bespectacled and bloated in the company of the sleek. He spent the later 1930s writing attacks on the Christian Churches and pamphlets outlining the religious–political ideology of an SS whose spiritual avatars eclectically included the Jesuits, the Samurai, the Teutonic Knights and Bolshevik commissars. In early 1941 he became the nominal chief of section IV B of the Reich Main Security Office, which dealt with ecclesiastical affairs, freemasonry and Jewish issues, although the officer responsible for the latter – Adolf Eichmann – could bypass him whenever he chose. That summer Hartl was accused of improper contact with a female staff member. In early 1942 he was assigned to the staff of Einsatzkommando C in Kiev, although his first brief was to report on 'the nature of the Russian soul'. After being wounded by a landmine, he was sent home. For the last two years of the war, he worked for the foreign intelligence service of the SD, which regarded his reports as unreliable. Old habits resurfaced in US captivity. After 1945 Hartl emerged as a key prosecution witness in countless trials of his SS colleagues, his preferred pose being that he himself had been too weak to carry out the atrocities in the East that they were convicted of. He was also the source of the unfounded rumours that the Catholic bishops had given the green light to the Nazi 'euthanasia' programme, claims that gullible commentators have taken at face value.

As Hartl's unit included seven theology graduates, it was jokingly said that they had transferred 'from the heavenly to Himmler's host', a play on *Himmel*, the German word for heaven, and the name of the SS leader. Hartl's team relied on the usual secret-police techniques of black-mail, disguise, eavesdropping and infiltration, even despatching suitably attired agents to the 1937 Oxford World Ecumenical Conference. They usually had the texts agreed at the Fulda Bishops' Conference in their hands by the same evening, although they eventually managed to plant bugging devices in the room where the Conference was convened. These techniques were brought to bear on the Catholic Church, whether against its clergy or lay affiliates, since 'political Catholicism' was their chief obsession. [102]

In 1935 the SD intercepted a letter permitting the steering body for Catholic overseas missions to send foreign currency to Rome. The SD instructed the Gestapo to block such permissions in future, and to investigate past instances where currency had gone abroad.[103] This resulted in several senior members of religious orders – which of course were international – being convicted of illegally transferring monies abroad in breach of Germany's strict foreign currency laws. The SD regarded this as a deliberate attempt to 'force the Third Reich to its knees'. The guilty received long jail sentences and hefty fines running into hundreds of thousands of Reichsmarks. These convictions fuelled an avalanche of snide commentary which claimed that the Catholic Church was a gigantic money-making operation. The trials gave rise to so-called 'currency ditties': like this 'Song of Religious Life': 'Oh the cloistered life is jolly! Nowadays, instead of prayer, Smuggling money is the business; Forth on this sly sport they fare. / Swift they say a Paternoster, Priest and monk and pious nun. Swifter then with zealous purpose, Smuggling currency they run.' One obvious objective of these trials was to discourage people from donating money to Catholic charities – because it would end up being embezzled or improperly sent abroad – while encouraging them to give to Nazi charitable and welfare organisations. The next step was a major investigation of the entire finances of the Catholic Church in Germany, the explicit intention being to destroy its economic viability through a sort of death by a thousand cuts.[104]

Well-publicised investigations into these currency violations in turn triggered denunciations of the Catholic clergy for mostly homosexual, but also paedophile offences. Between May 1936 and July 1937 there were 270 prosecutions of such men, of whom 170 monks and 64 priests were convicted. A major trial was held in Koblenz in May 1936 which resulted in the conviction of past and present members of a lay nursing order, most of the evidence coming from a former member of the order who had joined the SD. The intervening Olympic Games led Hitler to drop further trials, which were resumed with a vengeance after Pius XI's encyclical *Mit brennender Sorge* was released in early 1937. Hitler immediately instructed the minister of justice to give priority to these 'morality trials'. The Ministry of Propaganda urged the press to treat these trials as evidence of pervasive perversity within the Catholic Church. The press, and caricaturists in particular, had a field day with illicit intimacies in the confessionals or tubby monks whose capacious cassocks concealed several pairs of dainty female feet. That summer Nazi publications also attacked secretary of state Pacelli, accusing him of using a visit to Lisieux in France to organise the 'moral encirclement' of Germany with the aid of 'friends' in the French Communist Party, who were shown holding his cloak. Pacelli responded to this when he had a three-hour meeting in Berlin with the former US consul Alfred W. Klieforth, whose papers are at Harvard. Klieforth reported: 'He [Pacelli] opposed unilaterally every compromise with National Socialism. He regarded Hitler not only as an untrustworthy scoundrel but as a fundamentally wicked person. He did not believe Hitler capable of moderation, in spite of appearances, and he fully supported the German bishops in their anti-Nazi stand.'[105]

Tendentious reporting of a small number of sex crimes (involving mainly lay staff) in Catholic boarding schools or religious houses enabled members of the government to claim that the Catholic Church was awash with sex fiends. There were few holds barred in gathering the evidence, which involved the SD and Gestapo interviewing disgruntled religious drop-outs, ex-pupils and orphans, with offers of sweets alternating with a head bashed into a wall or the threat of concentration camp to secure the appropriate testimony. On this basis, minister for the Churches Kerrl could claim that seven thousand clergy had been convicted of sex crimes between 1933 and 1937, whereas the true figure seems to have been 170, of whom many had left the religious life prior to their offences. This deliberate inflation of statistics was a favoured Nazi device for ramping up hysteria, as they would do in 1939 when they turned five thousand ethnic German victims of the Poles whose country the Nazis had invaded into '50,000'. There was no reporting of similar sexual transgressions involving members of Nazi formations.

Article 31 of the Concordat had said: 'Those Catholic organisations and societies which pursue exclusively charitable, cultural, or religious ends, and as such are placed under the ecclesiastical authorities, will be protected in their institutions and activities.' However, the same article left it to future talks to decide which organisations were to receive this 'protection'. These negotiations dragged on into 1935 and were characterised by mounting mistrust between the interlocutors. The German government used the threat of tearing up the Concordat to force the Vatican to devolve negotiations to the German bishops, who promptly divided on how to respond to the Nazis' combination of concessions and menaces. Bishop Berning was the most accommodating, offering to dissolve labour and sports organisations and drastically to curtail the activities of Catholic youth. Bishop Galen was the most hardline, recommending that if Hitler did not tone down Nazi attacks on Catholic organisations the Church should reinstate its earlier prohibitions on Catholic involvement with Nazi organisations.[106]

Continued Nazi infractions of the Concordat required a concerted response from the Church. In the autumn of 1936 the Fulda Bishops' Conference sought a collective audience with Hitler to discuss matters of common concern. After a month's silence, they were told their request was out of the question, and that the letter had been referred to the minister of Church affairs. The Ministry of Church Affairs suppressed a joint pastoral letter that sought to establish common ground on the issue of anti-Bolshevism while criticising Nazism itself. The German bishops also thought that it was time for Rome to issue an authoritative statement about events in Germany. On 4 November 1936 Hitler nonetheless decided to grant cardinal Faulhaber a lengthy private audience on the Obersalzberg to see whether a compromise could be reached. Hitler did most of the talking, using a global tour d'horizon to avoid any discussion of details concerning Christians in Germany. He warned Faulhaber that, in the light of a possible Communist victory in Spain, the fortunes of the Church and National Socialism were bound together. He was especially disappointed in the Church's responses to Nazi racial policies. Whenever Faulhaber tried to steer the conversation on to the present, Hitler raised the subject of the Church's hostility to Nazism before 1933, while dismissing attacks on the Church as 'small and risible bagatelles'. The conversation petered out in mutual incomprehension – 'I won't conclude any cattle deals. You know I am the enemy of compromises, but there will be a last attempt,' being the Führer's parting shot.

The German bishops discussed this dismal encounter at a plenary Fulda conference on 12–13 January 1937, once again issuing a statement condemning Nazi infractions of the Concordat. Immediately afterwards a deputation of the more resolute German Catholic prelates left for Rome, where they had meetings with the secretary of state. Pacelli discussed the problems of the German Church with these men, who reported that 'at that moment it was a matter of life or death for the Church; they want our destruction'. They argued that a letter to Hitler from the pope would be less effective than an encyclical. When the five German prelates and Pacelli met the pope on 17 January, it was Pacelli who recommended an encyclical dealing with German affairs. The fact that other encyclicals were being drawn up dealing with Mexico and Russia meant that the papacy would be seen to be politically neutral. It is also likely that while in Rome Faulhaber was informed of the contents of a 'syllabus of contemporary errors' which the Holy Office of the Inquisition had been developing between 1934 and mid-1937. The extraordinary length of time it took senior theologians to elaborate such a document was indicative of the way in which the wheels of Vatican bureaucracy turned very slowly. Ironically, by the time they had finished, the Congregation of the Holy Office, mindful of the disastrous precedent from 1864, decided not to issue the syllabus at all, but rather insisted that the pope himself speak out.[107]

The release of the preparatory documents enables us to see how the Holy Office extrapolated a series of propositions from Nazi and Communist ideology, many of them based on passages in Hitler's *Mein Kampf*, which were then systematically condemned as 'social heresies'.[108] The propositions condemned included:

> The races of mankind are so different from one another through their innate and unchangeable character that the lowest of them is more different from the supreme race of men than from the highest species of animal.
>
> The 'battle of selection' and the 'stronger force', if successful, by that fact give the victor the right to dominate.
>
> A religious cult, in the strict sense of the term, is due to the nation.
>
> The state has the absolute, direct, and immediate rights over everyone and everything that has to do with civil society in any way.[109]

Although this document was never issued as a formal decree, it did work its way into the major papal encyclical which Pacelli had in hand. First, he instructed Faulhaber to convert some notes he had made on developments in Germany into a draft document that could become the basis of a German-language encyclical. The Holy Office had already had extreme difficulty in translating terms like 'race' or 'totalitarianism' into Latin, and it was felt that use of the language of the offenders would have greater resonance. Faulhaber spent three nights working on a draft which condemned the Nazis' making a fetish of race and state, as well as their aspiration to extend Caesar's province into an empire. Pacelli then rewrote drafts of Faulhaber's text, probably in conjunction with the pope, and certainly incorporating themes identified by the Holy Office – of which he was an ex-officio member. He transformed Faulhaber's draft into the extremely trenchant but subtle condemnation of National Socialism, which the encyclical contrived never to name, but which was omnipresent in spirit. Pacelli was responsible for converting Faulhaber's 'with great concern' into the encyclical's more memorable opening 'with burning concern'.

Mit brennender Sorge complemented the encyclicals on Mexico and Russia. The latter was a forceful rejection of Communism, which was described as 'bad in its innermost core'. What may have seemed to some to be 'good' ideas merely masked Communism's evil intentions, as when leading Communists espoused peace 'while at the same time instigating a class struggle in which streams of blood have been spilled'. As in his encyclical to the German Catholics, Pius XI emphasised the primacy of Natural Law, the importance of the individual human personality, and of the sacrosanctity of the family when it came to such matters as a child's education. Communism reduced the complexity of each individual to a mere 'cog in a machine' while propagating the doctrine of class struggle as a form of 'crusade' whose reality was 'hatred and a madness to destroy'. In sum, Communism, according to Pius XI, was 'a new Gospel, that offers Bolshevik and atheistic Communism as a message of salvation and deliverance for humanity', though in fact it represented 'the taking away of rights, the debasement and the enslavement of the human personality'. As the Mexican encyclical also makes clear, the pope believed passionately in socio-economic justice, arguing that the social encyclicals of his predecessor Leo XIII were the best guide for how to 'rescue today's world from the sad collapse resulting from an unbridled liberalism' while avoiding the twin pitfalls of class conflict based on Marxist–Leninist terror or the arrogant misuse of the state's power – a veiled reference to Fascist 'statolatry'.

The German encyclical is an immensely astute critique of everything that Nazism stood for. It anticipates virtually all of the themes that contemporary scholars of Nazism, especially in continental Europe, are currently pursuing to comprehend this phenomenon. Consider these passages:

> Immortality in the Christian sense is the survival of man after temporal death as a personal individual for eternal reward or punishment. Whoever uses the word immortality to mean only collective survival in the continuity of one's own people for an undetermined length of time in the future perverts and falsifies one of the fundamental verities of the Christian faith and shakes the foundations of every religious outlook which demands a moral ordering of the universe. Whoever does not wish to be a Christian ought at least to renounce the desire to enrich the vocabulary of his unbelief with the heritage of Christian ideas.

That dismissed both the Nazis' notion of collective racial immortality (and by implication liberal Protestant notions of orders of creation) and their attempts to dress up their ghastly doctrines in the language of religious belief. Pacelli returned to this theme in a discussion of grace:

> Grace in a wide sense can be said to be everything which comes to the creature from the Creator ... The repudiation of this supernatural elevation to grace because of the alleged particular nature of the German character is an error, an open declaration of war on a fundamental truth of Christianity. To put supernatural grace on a level with the gifts of nature is to do violence to the language created and sanctified by religion. The pastors and guardians of the people of God will do well to oppose this spoliation of sacred things and this work of leading minds astray.

Worse, from the Nazi point of view, Pacelli pinpointed the tendency of the Führer-cult to elevate a man into a god:

> Since Christ, the Messiah, fulfilled the work of redemption, broke the dominion of sin, and merited for us the grace to become the sons of God, 'there is no other name under heaven given to men, whereby we must be saved' but the name of Jesus. Thus though a man should embody in himself all wisdom, all might, all the material power in the world, he can lay no other foundation than that which is already laid in Christ. He who sacrilegiously misunderstands the abyss between God and

creation, between the God-man and the children of men, and dares to place beside Christ, or worse still, above Him and against Him, any mortal, even the greatest of all times, must endure to be told that he is a false prophet to whom the words of Scripture find a terrible application: 'He that dwelleth in heaven shall laugh at them.'

The Nazis' contempt for Christianity's emphasis upon human suffering was robustly rebuffed: 'By foolishly representing Christian humility as a self-degradation and an unheroic attitude, the repulsive pride of these innovators only makes itself an object of ridicule.' Pacelli also found time to condemn the Nazis' obsessions with greatness, heroism, strength and so forth, not to speak of their athletic cult of the body, often cultivated at the expense not only of the mind but of those unfortunates the Nazis were compulsorily sterilising. He found a moment for a shaft of sarcasm: 'The Church of Christ, which in all ages up to those which are nearest to us counts more heroic confessors and martyrs than any other moral society, certainly does not need to receive instruction from such quarters about heroic sentiment and action.' Pacelli used Natural Law doctrine to confound the Nazi philosophy of 'Right is what is advantageous to the people.' The encyclical stated that 'the believer has an inalienable right to profess his faith and to practise it in the manner suited to him. Laws which suppress or render difficult the profession and practice of this faith are contrary to natural law.' Nazi attempts to monopolise the education of children at the expense of their parents or the Churches were attacked too: 'Laws or other regulations concerning schools, which take no account of the rights of the parents given them by natural law, or which by threats or violence nullify them, contradict the natural law and are essentially immoral.'[110]

Meanwhile, the Holy Office issued a general appeal to nuncios and bishops, calling for conferences and courses at which Nazi doctrines would be confounded. This was withdrawn once cardinal Faulhaber had warned of the potential consequences for German Catholics.

Copies of the encyclical were smuggled into Germany by couriers, enabling local printers to run off as many as three hundred thousand pamphlet copies before it was read from the pulpits. This thwarted the efforts of the Gestapo to stop dissemination of the pope's message. Hitler vented his fury at the incompetence of the security services which had let such a subversive document slip into the country. When the Gestapo struck at the printing firms – thirteen of which were closed down and nationalised – the intrepid resorted to hectographing the encyclical, or typing out multiple copies. Anyone caught disseminating or reading the encyclical beyond a certain date was liable to arrest. A chaplain in Berlin received a sentence of one hundred days' solitary confinement for distributing a thousand copies. A secretary employed by the German Labour Front who typed out eight copies in her lunchbreak was denounced to the Gestapo. A Munich teacher was dismissed from his post simply for reading out the encyclical to a class, entirely ignorant of the ban on extending the encyclical's impact after its initial reading.[111] On 12 April the German ambassador to the Vatican delivered a protest, which blamed the Church for infringements of the Concordat and accused the pope of 'attempting to summon the world against the new Germany'. In his response secretary of state Pacelli remarked that the German protest addressed none of the themes raised by the encyclical.[112]

In Germany the encyclical provided an opportunity for heightened harassment of Catholic clergy and laity as they went about their lawful business. Clerical residences were covered with graffiti along the lines of 'Hang the Jews, put the Blacks against the wall', 'Blacks' being the pejorative term for both clergy and the Catholic Centre Party. Services were interrupted and banners were snatched from those carrying them during processions. On 28 May 1937 Goebbels weighed in with a characteristically snide commentary on the 'morality trials' in a speech he delivered in Berlin's Deutschlandhalle. He seized the moral high ground: 'Today I speak as the father of a family whose four children are the most precious wealth I possess – as a father who therefore fully understands how parents are shocked in their love for the bodies and souls of their children, of parents who see their most precious treasure delivered to the bestiality of the polluters of youth. I speak in the name of millions of German fathers.' Seven years later he and his wife Magda poisoned all four of their 'most precious treasures'.[113]

Such vilification led bold Catholic activists to issue pseudonymous letters which dwelled upon the moral corruption within the NSDAP. Individual Catholic priests who spoke out against Nazi iniquities felt the force of the regime's terror. The priest in Lower Franconian Mömbris protested in 1936 against anticlerical and antisemitic slanders in the copies of Julius Streicher's notorious *Der Stürmer* that were displayed in public showcases. The holder of an Iron Cross, the priest refused to play the church organ, celebrate the mass or ring the church bells until these cases were emptied of this offensive material. He encouraged his parishioners to protest to the local mayor. They organised a petition which four hundred people presented to the mayor. The latter – himself the local Nazi chief – was 'insulted and threatened'. When forty SA men ventured a counter-demonstration, replete with a new *Stürmer* showcase, their marching songs were drowned out by catcalls and whistles. This sort of incident could not pass unnoticed. On 28 December the priest was arrested along with various of his parishioners. While awaiting his trial he was fêted as a local hero, a view confirmed when his bishop visited Mömbris later in the year.[114]

The Catholic Church is a worldwide institution. Nazi harassment of German clergy resulted in a forthright denunciation of the regime by cardinal Mundelein in the US at a well-attended diocesan conference:

> Perhaps you will ask how it is that a nation of 60,000,000 people, intelligent people, will submit in fear and servitude to an alien, an Austrian paperhanger, and a poor one at that, I am told, and a few associates like Goebbels and Goering, who dictate every move of the people's lives, and who can, in this age of rising prices and necessary high cost of living, say to an entire nation: 'Wages cannot be raised'.
>
> Perhaps because it is a country where every second person is a government spy, where armed forces come and seize private books and papers without court procedures, where the father can no longer discipline his son for fear that the latter will inform on him and land him in prison.[115]

Comments like that deserve to be included in any account of how the Catholic Church responded to Nazism. It is worth noting that they had no effect on those responsible for the persecution.

Catholic opposition to the regime in Bavaria crystallised around the figure of Rupert Mayer, a Munich-based Jesuit priest who had been badly wounded in the Great War, and who had been decorated with the Iron Cross. A renowned spiritual adviser to men, Mayer had already come to the attention of the Gestapo, on account of the oppositional content of his sermons. In April 1937 the Berlin Gestapo decided to ban him from preaching anywhere in the Reich, a ban which the Munich Gestapo interpreted as a prohibition on him preaching outside his own church. Mayer carried on preaching regardless, until in late May the Reichsführer-SS and the minister of justice specified a ban on him preaching 'in ecclesiastical or profane areas'. Mayer had to put his signature to a document acknowledging this ban. Oral permission was given to continue preaching in the Munich Jesuit church of St Michael's. Mayer's Jesuit superior Augustinus Rösch immediately recognised that this ban contravened the Concordat, and authorised Mayer to continue preaching. So did cardinal Faulhaber, who protested to the Ministry of Church Affairs. When Mayer let it be known that he intended to continue preaching, the Gestapo arrested him, a development that led to disturbances among the congregation at St Michael's and four hundred demonstrators outside the Munich police headquarters. Although these people were dispersed, 150 of them descended upon the Munich Gestapo headquarters where there were fights with local Nazi supporters. In the evening 250 more people mobbed the Gestapo headquarters. Cardinal Faulhaber tried to negotiate a tricky course between supporting Mayer and defusing a popular mood that was turning ugly. Mayer refused to compromise with the Gestapo and declared that if he were released he would preach throughout Bavaria. As his trial commenced, his superiors persuaded him to sign a document to the effect that he would obey the law, although he orally reserved the right to act in accordance with his conscience. The court sentenced him to six months' imprisonment, but this was suspended once the court had seen his declaration that he would obey the law and the Jesuit provincial had agreed to post Mayer elsewhere. In fact, Mayer immediately began preaching again, being arrested once more in early 1938. Upon his release in May, he ostentatiously left his Iron Cross First Class upon the table in his cell. The Gestapo rearrested him in November, after he refused to tell them the names of visitors he had received whom they suspected of treason. Mayer was eventually released from Dachau two years later, a sick man, and retired to the monastery at Ettal.

Under a regime which tried either to tempt the masses into its own liturgical spaces or to leave them alone and frightened if they would not play ball, any large-scale public event that was not instigated by the Nazis was an implicit assertion of spiritual autonomy. In countless villages the *Primizfeier*, when a newly ordained priest celebrated his first mass in his home parish, became an occasion for demonstrations of Catholic solidarity. Supra-regional events, such as the St Viktorstracht in Xanten held from 18 August to 18 September 1936, took on the character of mass demonstrations. What happened in Xanten was a very public affair, consisting of two huge processions, meetings of various Catholic associations, and four Sunday pilgrimages. The SD and Gestapo, having failed to have the celebrations banned, monitored every event, on the look-out for illegal political activity. Seventy thousand people were bussed in or caught special trains to attend the septennial celebrations of the consecration of Bamberg cathedral in May 1937, and an amazing 750,000 to 800,000 Catholics attended the Heiligtumsfahrt in Aachen in July of the same year. One hundred thousand people watched the final procession consisting of twenty to twenty-five thousand men.

The SD was also responsible for a large part of the work of 'public enlightenment' designed to undermine Catholicism through a slew of books and brochures whose content was then recycled in Nazi newspapers. No issue of the SS journal *Schwarze Korps* was complete without sensational revelations about the Catholic Church. Some of this was quite cleverly pitched. The mass processions in Xanten were used as evidence that the Church was not being persecuted, even though the SD had done its best to prohibit them. Articles on the Jesuits falsely attributed to them the maxim 'the end justifies the means' and praised the Order for turning an idea into organisation, before attacking 'Jesuit' casuistry and the Order's designs on 'conquering the world'.[116]

Under a police dictatorship, it was dangerous for people to express their opinions freely, but no such constraints affected exiles abroad. One of the most distinguished was Waldemar Gurian.[117]

A freelance writer and journalist, Gurian initially specialised in writing about Bolshevism, which he regarded as 'a new religious faith'.[118] Because in July 1933 a Nazi journal used him as an example of how 'German Catholicism has allowed itself to be heavily judaised', he decided to emigrate to Switzerland. Together with another exiled German journalist, he published the 'German Letters', chronicling conditions in Nazi Germany. These tracts, which total over two thousand pages, were then smuggled into the Reich. Some of them were remarkably prescient. For example in 1935, the year Gurian was stripped of German citizenship, he wrote: 'The Nuremberg Laws appear to be only a stage on the way to the full physical destruction of Jewry.'[119] In August 1934 he published a pamphlet which pointedly contrasted the reactions of the German bishops to Hitler's murder of the SA leaders and several prominent Catholics with St Ambrose's condemnation of the emperor Theodosius' slaughter of people who rioted at a Roman circus. In 1936 Gurian published *Hitler and the Christians*, which astutely saw that the Nazis' bleak racial doctrines would be camouflaged for popular consumption in the charitable guise of 'positive Christianity'. He also realised that the Churches' preoccupation with Rosenberg and the neo-pagans was misdirected, as it might propel them into the arms of a more reasonable-seeming Hitler, who was not slow, as we have seen, with public professions of goodwill towards the Churches. The neo-pagans were just reconnoitring 'robber bands' used to conceal the real deployments of the main Nazi army.[120]

Exiled Catholic clergy and laity were also responsible for the journal *Kulturkampf. Reports from the Reich*, which was disseminated from Paris, and then London and New York. Separate issues dealt with such themes as the 'idolisation of Hitler' in the Führer cult, shrewdly pointing out to foreign observers that beyond the scenes of mass orgiastic jubilation there were 'the tears of shame, the bitterness and the suffering, of those who stayed at home, hiding themselves behind their flag-bedecked windows'. *Kulturkampf* also devoted several pieces to whether or not Nazism was a religion; disputing the claims of more secular commentators that the ideology was a 'stage set' or smokescreen for more common concerns with raw power. Cogently, the journal argued that Nazism was not some heretical deviation from Christianity, nor merely a 'substitute for religion', but rather a 'substitute religion', an *Ersatzreligion* rather than a *Religionsersatz*. The Germans were not living in an atheistic state, but in one where a religion other than Christianity had burgeoned within the public domain. This religion of Nazism may have been incoherent and as flimsy as a 'house of cards', but it was the daily reality for those who lived within its shadow, as palpable as a cathedral dwarfing the neighbouring houses of its close. Despite all their rhetorical invocations of God, not to speak of neo-pagan sectarians, the Nazi 'God' was the power of nature, conceived of as the brutal rule of the strong, with the Führer as a tangible focus for a party that was like a religious order or Church. The Party's function was totally to conquer the state, converting this into another 'member' of the cult. When this Church had achieved its dystopian ambition of restoring natural inequalities and hierarchies, then God's kingdom would have been partially realised on earth: the kingdoms of nature, empire and heaven would be one.[121]

Pius XI was over eighty in the year he died. Having abolished the corps of papal physicians, he refused to acknowledge that he was ill. When a cardinal gently suggested that he might need to rest, Pius acidly replied: 'The Lord has endowed you with many good qualities, Salotti, but he denied you a clinical eye.'[122] Despite repeated cardiac arrests, he continued to condemn Nazi persecution of the Church and both Fascist and Nazi racism literally until the very end, since just before he died he compared Hitler to the emperor Julian the Apostate. Every writer who claims that this racism was simply an outgrowth of Christian anti-Judaism has to reckon with the fact that this argument was much favoured by the Fascists and Nazis themselves, and was robustly rebuffed by the Church. Roberto Farinacci's newspaper constantly unearthed embarrassing evidence from the early modern papacy, to demonstrate that 'the Church today finds itself in strident contradiction with its past'. That version of the Church's history was highly selective, omitting as it did those popes – Innocent IV, Gregory X, Martin V and Paul III – who had condemned such notions as the 'blood libel'. It also had to omit such modern popes as Pius X, Leo XIII and Benedict XV, who by any criteria could not be called antisemitic. That Farinacci's paper came to resemble *Der Stürmer* was no coincidence, since he and Julius Streicher were close friends. Quite why Mussolini suddenly decided to emulate the racism of his northern counterpart need not detain us, but it was in marked contrast to previous Fascist policy. After all, about ten thousand Italian Jews belonged to the Fascist Party, including the only Italian university professor to belong to the Party before 1922, who went on to be rector of Rome university, and Carlo Foa, the editor of the Fascist paper *Gerarchia* – not to speak of Guido Jung, Mussolini's minister of finance down to 1935, and one of Mussolini's many mistresses – a subject passed over in silence by the critics of the papacy. Although the Fascist media had begun to adopt a more favourable tone towards Nazi racism a year before, it seems to have been Hitler's visit to Rome in May 1938 that led the Italian Fascists to abandon any residual cultural snobberies regarding lessons on 'race' from people whose ancestors were living in forests when the glories of imperial Rome were built. Hitler arrived with an entourage of five thousand. He watched army, air force and navy displays of might, including a march past by troops doing the *passo romano* on the Via dell'Impero, with trips to the Bay of Naples and Florence.

Hitler ostentatiously avoided a courtesy call to the Vatican, which led the pope to bring forward his summer escape to Castel Gandolfo, leaving instructions not to admit the German dictator's entourage to the Vatican's treasures in his absence. His parting shot, published in the Vatican newspaper, was 'The air here makes me feel sick.' Pius condemned the public display of 'another Cross which is not the Cross of Christ', while the Vatican newspaper published extracts from German racist tracts in which the Latin races were treated disparagingly. The pope was consistently opposed to racism, under which rubric he subsumed antisemitism. Unless one assumes that the only racism abroad in the world in the 1930s was antisemitism, which would hardly encompass the gamut of racism around in, say, the European colonial empires or the US, then there is no reason why the pope should have registered disapproval solely of that phenomenon rather than 'racism' in general. In April 1938 Catholic universities and theological faculties were informed that the pope condemned certain propositions. These included the notion that 'purity of blood and race had to be maintained with every means; everything that serves that goal is justified and permitted', or the view that the aim of education was 'to develop racial quality and passionate love of one's own race as the highest good of mankind'.[123]

This became urgent when Mussolini introduced racial laws in 1938. These were a shock to the highly assimilated Italian Jewish minority, especially to the quarter of adult Italian Jews who were Fascists. Two hundred and thirty Jewish Fascists were proud of having participated in the March on Rome, while three Jews were counted as Ur-Fascist martyrs. Fascist Italy had been a haven for Jews fleeing totalitarian persecution. Refugees included the Russian ancestors of the historian Alexander Stille, and the German ancestors of the historian George Mosse.[124] This climate changed as the regime introduced racial laws – primarily, it seems, to provide a harsh definition of race relations in its instant colonies, although those Fascists who were antisemitically inclined soon ensured that the Jews were also encompassed.

The response of the Church was unequivocal. Pius condemned a report on 'Fascism and Racial Problems' produced in July 1938 by a number of Fascist academics, as being contrary to fundamental Catholic doctrine. Later that month he told chaplains of Catholic youth organisations: 'If there is anything worse than the various theories of racialism and nationalism, it is the spirit that dictates them. There is something peculiarly loathsome about this spirit of separatism and exaggerated nationalism which, precisely because it is un-Christian and irreligious, ends by being inhuman.'[125] Both the pope and secretary of state Pacelli materially assisted Jewish scholars who were affected by these laws. When the cartographer Roberto Almagia was dismissed from a post at Rome university which he had held since 1915, he was immediately appointed director of the cartographical section in the Vatican Library, in charge of reproducing a fine sixteenth-century map. Another appointee in the library was professor Giorgio del Vecchio, who, despite having been a Fascist since the Party's inception and a former rector of the university, had been dismissed from his chair in the law school. He was joined by the leading Arabist Giorgo Levi della Vida, who was given a post cataloguing the Vatican's Arabic manuscripts. When the Fascist Academy of Science refused membership to Tullio Levi-Civita, a Jewish physicist, Pius XI insisted he become a member of the Pontifical Academy of Sciences. Pacelli personally invited him to speak on Vatican Radio about the latest developments in his field.[126] Pacelli was also quietly active in helping Italian Jewish academics to emigrate. He helped the leading mathematician Vito Volterra, who worked in the Vatican Library, to flee to the US, and he persuaded a Latin American university to offer a job to his childhood friend Guido Mendes, in whose family home the young Pacelli had celebrated the Jewish sabbath. As pope he would help Mendes to get immigration certificates enabling him to go to Palestine, where he would develop a successful medical practice. This rather militates against the notion that either cleric was antisemitic.[127]

That summer Pius read a book called *Inter-Racial Justice* by an American Jesuit called John La Farge, about the dire state of race relations in the US. Pius commissioned La Farge and two other Jesuits, Gustave Desbuquois and Gustav Gundlach, to prepare drafts of an encyclical to be entitled *Humani generis unitas*. This occupied them from July to September 1938.

In mid-September, and despite mounting Fascist attacks on his stance on this issue, Pius told a group of Belgian pilgrims: 'The Promise made to Abraham and his descendants was realized through Christ, of Whose mystical Body we are the members. Through Christ and in Christ we are Abraham's descendants. No, it is not possible for Christians to take part in anti-Semitism. Spiritually we are Jews.' This message was taken up even by senior clerics who were regarded as sympathetic to the Fascist regime, such as the patriarch of Venice, and Milan's cardinal Schuster, who attacked 'the heresy born in Germany and now insinuating itself almost everywhere'. Like the pope, Schuster was especially exercised by the thought of Italians being slavishly imitative of mere Germans, echoing Pius in ascribing to the ancient Roman Empire a spirit of tolerance it certainly never possessed.

The pope kept up his attacks on Fascist and Nazi racism until he ceased to draw breath. While the Fascists were careful to shape their antisemitic enactments around the rock represented by the Church, their decree-law of November 1938 signified a collision since it banned 'intra-racial' marriages in flagrant violation of both the Concordat and canon law. The pope planned to make this a central feature of the speech he was to make in February 1939 on the tenth anniversary of the Lateran Treaties. However, during that winter of 1938–9, Pius XI's health rapidly declined. Despite heart attacks, diabetes and a persistent cold, he insisted on sitting up all night working on this address. This exertion resulted in his death on 10 February 1939. In a message of condolence to cardinal Hinsley, the British chief rabbi Hertz wrote: 'Jews throughout the world will revere the Pope's noble memory as a feared champion of righteousness against the powers of irreligion, racialism and in-humanity.' The London *Jewish Chronicle* mourned 'the loss of one of the stoutest defenders of racial tolerance in modern times'. Those sober contemporary verdicts seem to be lost on Pius XI's modern detractors – verdicts which, it should be noted, predate the founding of the state of Israel.[128]

III THE MIRAGE OF PROTESTANT UNITY.

In 1933, almost 67 per cent of the German people described themselves as Protestants. This designation concealed a bewildering array of pos-sibilities, including what is called cultural Protestantism, that is, a set of attitudes that were not necessarily accompanied by any religious practice. The Protestant Churches included major denominations, some of which brought Lutheranism and Calvinism together in a Union that existed only in the largest state of Prussia, or observed separation for reasons of doctrinal purity; others, and the list is long (it includes the Adventists, Jehovah's Witnesses and Methodists), hovered on the borderlines between Church and sect. These denominational niceties need not over-occupy an account of the politics of the main groupings.

Book after tendentious book traduces the (worldwide) Catholic Church for its alleged responses to Hitler. Since they were not a mono-lith, the German Protestant Churches are harder to group, beyond a few highly atypical figures who resisted. Of course these resistors sometimes had views that many might find questionable were they to know anything of them beyond a few selective quotations from Martin Niemöller which have entered into the contemporary sermoniacal repertory. This imbalance may strike many readers as curious, as will the near-total 'silence' that greeted Richard Steigman-Gall's study of how liberal Protestantism and Nazism interacted with one another. US Protestant Christians have so far not been asked to conduct a 'moral reckoning' for what their co-religionists did, or did not do, in Germany more than seventy years ago. Perhaps the fact that conservative Protestant Christians are stalwart supporters of Israel, while the Catholic Church has to weigh the interests of Arab Christians, may have much to do with this.

There is no evidence that the Nazis persecuted the Protestant Churches, as distinct from some of the fundamentalist sects, despite what happened to a few dissenting individuals. They did not object to 'political Protestantism' in the way they sought to destroy 'political Catholicism', meaning the dispersed wreckage of the Centre Party and Catholic Action, because most 'political Protestants' were either Nazis or conservatives – and many of the latter's views were hard to distinguish from those of the former. All the evidence speaks of a keen desire, on the part of senior Nazi leaders, to find some accommodation with fractious Protestant clerics, and points to a restraining of the security services, which, because of their obsession with Catholicism, had little knowledge of how Protestant Churches even operated.

At their own request, the Federal Organisation of Protestant State Churches had met with Franz Stöhr of the executive of the NSDAP in 1931 to ascertain the Party's position on religion. They were told that the Party was 'supported and led by Christian people who seriously intend to implement the ethical principles of Christianity in legislation, and to bring them to bear on the life of the people'. Entirely contradicting what Hitler would tell Catholic bishops a few years later, Stöhr explained that 'the party leadership was shaped by Protestantism'. Even the Catholics in that leadership, he alleged, inclined more to Protestantism. The Protestant League was the first Christian organization to give its support to the Nazis; there was no equivalent support from Catholic organisations.[129]

Conservative German Protestants overwhelmingly welcomed the 'government of national concentration' in January 1933, with some bishops issuing too fulsome statements. It appeared to be a regime of the conservative elite that had merely co-opted Hitler and the Nazi Party to lend itself the appearance of popularity. Even if both the chancellor and vice-chancellor were thought to be Roman Catholics, which Papen indubitably was, there was a solid Protestant as president, field marshal Paul von Hindenburg, to whom former corporal Hitler seemed to doff his hat. Moreover, at one time or the other, over the next few years, the cabinet included Göring, Frick, Blomberg, Dorpmüller, Rust, Seldte, Neurath and Schwerin-Krosigk, who were all Protestants, clearly still outnumbering such lapsed or nominal Catholics as Goebbels and minister of justice Gürtner.[130]

Although members of the elite, including Papen, imagined they could exploit and jettison Hitler, in fact he both altered the constitutional framework to ensure his dominance and skilfully broadened his support beyond the Nazi Party. Hitler constantly stressed an entirely spurious apostolic succession, from Frederick the Great, to Bismarck, to Hindenburg and on to himself, which reassured the historically minded that all was back in order after the chaotic interregnum of the preceding fourteen years. The Day of Potsdam symbolised those continuities. The religious, who thought their values had been mocked under the Weimar Republic – from which in reality they had not done badly at all – further liked the sound of the Decree for the Protection of the German People and State of February 1933. This criminalised assemblies or demonstrations that offended the religious.[131] In one of his many speeches that spring, Hitler's rhetoric became indistinguishable from a sermon:

> I cannot divest myself of my faith in my Volk, cannot dis-associate myself from the conviction that this nation will one day rise again, cannot divorce myself from my love for this, my Volk, and I cherish the firm conviction that the hour will come at last in which the millions who despise us today will stand by us and with us hail the new, hard-won and painfully acquired German Reich we have created together, the new German kingdom of greatness and power and glory and justice. Amen.

If Protestants appreciated Hitler's claim finally to have overcome the ideology of Marxism, they also liked the bluntness with which he told political Catholics 'what's what'. If the Catholic Centre Party was so concerned about threats to religion, he asked, what had they been doing in coalition governments with Marxists? Having had a prolonged period of ascendancy under Weimar, political Catholicism was abruptly returned to the cold. Protestants also welcomed the end of the

'Party-state', in which the Centre Party had played a leading role, and the onset of a moral revolution that would reverse the excessive individualism, whether in the arts, gender roles or sexual mores, that characterised a few urban enclaves during the Weimar Republic. Since much Weimar culture was deeply tedious in its puerile provocations, who can blame them?

German Protestantism was subjected to three pressures after 1933, which were designed to de-Judaise it, to heroise it and to unify it. These came from within, although beyond the Churches there were clusters of neo-pagans whose clamorous agitations encouraged Protestant Nazi sympathisers to 'Nazify' their own Churches before they were replaced by something wholly unrelated to Christianity.

The idea of fusing extreme racist nationalism with Christianity was not new; a League for a German Church had been founded in 1921 precisely for that purpose. Some 120 Protestant pastors belonged to the Party by 1930, eight having stood as candidates in elections. Wilhelm Kube, the gauleiter of Brandenburg, was both leader of the Nazi caucus in the Prussian parliament and an active member of the synod of the diocese of Berlin. In late 1931 he suggested the formation of 'Protestant National Socialists', a Church party not formally integrated with the NSDAP itself. Hitler thought that 'German Christians' would be less contentious. From their inception in 1932, the German Christians, a group of clergy and laity, sought to impose an ecclesiology defined by race rather than grace, blending 'traditional' anti-Judaism with new-fangled scientific racism to establish a new 'Church of blood'. They wished to revivify Protestantism by incorporating those things that had made Nazism itself such a potent force. Their banner consisted of a cross and the initials DC with a swastika in the centre. This was not the first or the last time that a Protestant Church inclined towards a secular creed in the expectation that its adoption might fill empty pews, a cycle those Churches have endlessly repeated with environmentalism, campaigns against the Bomb and soft Marxism ever after.

Since the German Christians seemed to give empty churches a new lease of life – albeit by introducing the lurid razzamatazz of Nazism into places of worship – they were welcomed by some senior Protestant clergy as a way of restoring the popularity of religion. Bishop Theophil Wurm of Württemberg was not alone in imagining that Nazism might represent a revival of the fusion of nationalism and religiosity that had last been seen in Germany during the Wars of Liberation.

The democratic electoral structures of many of the Protestant Churches served to give the most radical German Christian faction control, as a prelude to introducing the 'Führer-principle' to Church governance. Instead of sparking a spiritual awakening, the German Christians seemed bent on the total politicisation of religion. A sweaty, militarised and uniformed pastorate, in brown shirts with swastika armbands, would remasculinise Christianity, thereby counteracting the notion – often grounded in various European realities – that the faith had become feminised as men had turned to politics and the pub. Their preferred example of a Christian was Horst Wessel, the son of a Protestant pastor and war veteran, who despite being killed in a squalid brawl over a prostitute was immortalised by Goebbels as a Nazi martyr, his story being turned into the infamous 'Horst-Wessel-Song'. Finally, and this was the sticking point for their opponents, they sought to disbar 'non-Aryan' Christians from the ministry (this measure affected about thirty-seven individuals in a pastorate of eighteen thousand) and to downgrade or expunge anything that reminded Christians of the Jewish fundaments of their religion.

Since the German Christians promised not only to simplify and Nazify a bewilderingly complex landscape of Protestant Churches, but to create one supraconfessional Church including Catholics, they initially received the regime's backing. In May 1933 Hitler supported a moderate German Christian, the former naval chaplain Ludwig Müller, for the new post of Reich bishop. Wits shortened his title to 'the Reibi' a play on 'rabbi'. Müller was a protégé of gauleiter Koch of East Prussia, himself president of a regional Protestant synod. The appointment of a Reich bishop would be the prelude to uniting Germany's twenty-eight provincial Protestant Churches, some of which were ruled by bishops, others by more democratic consistories.[132] While Müller was backed by the Lutheran provincial bishops, a rival candidate, Friedrich von Bodelschwingh, of the renowned Inner Mission charitable institutions at Bethel, garnered the support of the Churches in Prussia. German Christian and Nazi Party protests led Bodelschwingh to abandon his candidacy. Müller returned to the fray in Church elections that summer. He managed to convince Hitler that those clergy who were hostile to his own candidacy were also opposed to the Führer's regime. This was untrue, since many who objected to the Nazification of the Protestant Churches, either through the German Christians or by 'co-ordination', often had no difficulties with other parts of the Nazi platform, whether this meant reducing the putative influence of Jews in German life or restoring the nation's position within the European system. The German Christians received two-thirds of the vote, especially since Nazis had been encouraged to reacquaint themselves with their Church's democratic procedures. Müller was elected Reich bishop at Reich Synod that September.

This coup d'église, which was followed by the introduction of an 'Aryan paragraph', triggered a response on the part of those classes customarily used to governing the Protestant Churches, namely senior clergy and civil servants, academics, doctors and lawyers, major landowners, bankers and businessmen.[133] A Young Reform Movement, which in some places had stood in the July elections as 'Gospel and Church', metamorphosed that autumn into the Pastors' Emergency League consisting of about sixty members. Its leading light was the former U-boat commander Martin Niemöller, pastor to the great and the good in Berlin's fur-coated suburb of Dahlem. It should be noted that these pastors objected not to antisemitism, but to the state's arrogation of the right to dismiss pastors on racial grounds. Otto Dibelius, a leading figure in the Confessing Church, had denounced those countries that had objected to the Nazis boycott of Jewish enterprises.[134] One final straw broke the camel's back. The German Christian leader in Berlin, Reinhold Krause, addressed a monster rally in the Sportspalast, at which he got carried away with his own rhetoric. 'Those people [Nazis] need to feel at home in the Church,' to which end he demanded 'liberation from everything unGerman in the worship service and the confessions – liberation from the Old Testament with its cheap Jewish morality of exchange and its stories of cattle-traders and pimps'. If Nazis refused, in good conscience, to buy a tie from a Jew, 'how much more should we be ashamed to accept from the Jew anything that speaks to our soul, to our most intimate religious essence'. The negative response to this forced Müller to abandon introduction of the 'Aryan paragraph' and to dismiss the German Christian leader.

In early January 1934, Hitler held a meeting with Müller, Niemöller and various other opponents of the Reich bishop. He took the rug from under Niemöller by asking Göring to read out the morning's intercepts of a telephone conversation in which Niemöeller had revealed that he hoped to play off Hitler against Hindenburg. Affirming that 'inwardly he stood closer to Protestantism', Hitler upbraided Niemöller, shouting at him: 'You leave concern for the Third Reich to me and look after the

Church!' Reich bishop Müller's star was back in the ascendant, which meant that the German Christians moved aggressively to capture Church governments or to dragoon provincial Churches into his Reich Church. He also reintroduced the 'Aryan paragraph', thereby alienating all those who thought that baptism conferred equal membership of the Church regardless of a person's ethnicity.

By 1934 the Emergency League had developed into the Confessing Church. This was a network of between five and seven thousand like-minded pastors, whose 138 delegates held their first synod at Barmen. The Swiss Reformed theologian Karl Barth, then teaching at Bonn, who regarded the authoritarian Niemöller as almost as bad as the German Christians, took the lead in drafting the Barmen Declaration which defined the respective spheres of Church and state, rejecting the claim that the state should be 'the single and totalitarian order of human life'. Article 5 read:

> We reject the false doctrine that the state, over and above its special commission, should and could become the single and totalitarian order of human life, thus fulfilling the Church's vocation as well. We reject the false doctrine that the Church, over and above its special commission, should and could appropriate the characteristics, the tasks, and the dignity of the state, thus itself becoming an organ of the state.

By asserting that the 'Church must remain the Church', the Confessing Church implicitly rejected the totalitarian claims of the Nazis, as well as the German Christian attempts to incorporate the Church within the Nazi state. This forthright stand was compromised by the Confessing Church's refusal to form a 'free Church', that is one financed by its own congregations rather than through Church taxes that the state collected. Their concern to maintain the Church undefiled by Nazi Christians sat uneasily with their continued espousal of teachings of which the Nazis would scarcely have disapproved. In 1935, for example, Niemöller told his Dahlem congregation: 'the Jews have caused the crucifixion of God's Christ . . . They bear the curse, and because they rejected the forgiveness, they drag with them as a fearsome burden the unforgiven blood-guilt of their fathers.'[135]

If the Confessing Church thwarted attempts to Nazify Protestantism, simply by taking German Christians to court to establish the illegality of their actions, so the opposition of some 'intact' south German Lutheran

Churches that had not been taken over by German Christians frustrated Müller's attempts to force the Protestant Churches into a single structure. This conflict became one between local people and outsiders. On 23 August 1934 the Bavarian Provincial Synod unanimously supported bishop Meiser in resisting amalgamation with the Reich Church. The Reich Church struck back by seizing bishop Wurm of Württemberg, who together with Meiser, apprehended a few days later, was placed under house arrest. Civil disobedience ensued, especially in Protestant Franconia, where Nazi support was stronger than in Bavaria as a whole, beginning with a demonstration by ten thousand people in Julius Streicher's Nuremberg. Farmers sent deputations; Party offices were inundated with letters and telegrams; and even holders of the Party's Golden Badge of Honour, as well as ordinary 'national comrades', handed in their Party membership cards. This was a catastrophe in the making, which Hitler resolved as quickly as possible by ordering the release of the two bishops.[136]

German Protestants were part of a wider ecumenical community that stirred at the thought of bishops under house arrest. Some US Lutherans, not to mention anti-Communist Canadian Mennonites and francophone Canadian Catholics, took an indulgent stance towards Nazism. But the vast majority of official Christian opinion in Canada and the USA was condemnatory. The influential *Christian Century* condemned Nazism for a 'Christian nationalism' worthy of 'ancient Israel' in its virulence; while in 1934 the World Baptist congress 'deplored and condemns as a violation of the law of God, the Heavenly Father, all racial animosity and every form of oppression or unfair discrimination toward the Jews, toward coloured people, or toward subject races in any part of the world'.[137] One of the most informed and intelligent critics of totalitarian political religions was the Swiss Calvinist theologian Adolf Keller, who took upon himself the task of enlightening Americans about events in Europe at the time. His published lectures at Princeton Theological Seminary, *Religion and the European Mind*, and a major book, *Church and State on the European Continent*, were extraordinary explorations of the political consequences of mass insecurity:

> The multitudes tremble in such a situation. They have fear in their hearts, and fear is hatred; fear is defiance; fear is superstition; fear is the ghastly flight of men running for their lives. They feel behind them the lash of an invisible whip. They feel

homeless. The soul of this generation is like Noah's raven, which went forth to and fro and found rest nowhere, because the earth was still covered with water, as in the beginning when creation began.[138]

Another influential voice was the acerbic Swiss theologian Karl Barth, who in 1935 fled his post at the university of Bonn for a life of exile in Basle. He was one of the main influences upon Frederick Voigt, the former *Manchester Guardian* foreign correspondent turned High Tory, whose 1938 book *Unto Caesar* was one of the most perceptive English-language commentaries on totalitarian political religions.[139] These perspectives found their way into political currency, as when addressing the Leeds Chamber of Commerce in 1937 Winston Churchill said:

> It is a strange thing that certain parts of the world should now be wishing to revive the old religious wars. There are those non-God religions Nazism and Communism ... I repudiate both and will have nothing to do with either ... They are as alike as two peas. Tweedledum and Tweedledee were violently contrasted compared with them. You leave out God and you substitute the devil.[140]

The Anglican clergy were deeply hostile towards totalitarianism, with the sole exception of bishop Headlam of Gloucester, the former professor of divinity at Oxford, who not only urged German Protestants to find a 'modus vivendi' with Hitler, but even in 1938 continued to believe that the latter was 'profoundly religious'. So he was, though not in terms comprehensible to an Oxford professor.[141] No Anglican leaders were sympathetic to Nazi views on race. In 1930, the Lambeth Conference officially welcomed the stance in J. H. Oldham's 1924 *Christianity and the Race Problem*, which took the view that all men are brothers under the skin. Leading laymen, such as the vice-chancellor of Birmingham university, Sir Charles Grant Robertson, denounced the totalitarian claims of the Fascist and Nazi states with as much passion as the exiled Luigi Sturzo.[142] In general, the English clergy were attracted to social radicalism, and repelled by Fascist and Nazi violence, especially when they witnessed the Mosleyites in action on their own turf. Explicitly Tory bishops, a minority in the Church of Lang or Temple, were in the vanguard of denouncing Nazi antisemitism.

The most outspoken Tory bishop was Herbert Hensley Henson of

Durham, who as early as May 1933 attended a meeting in Sunderland to protest the Nazi persecution of the Jews. In the mid-1930s Henson opposed the Italian invasion of Abyssinia – justified in some appeasing circles in Britain by Abyssinia's practice of slavery – taking advantage of the Vatican's silence on the issue to traduce the Roman Catholic Church too. In November 1935 he spoke passionately at the Church Assembly about events in Germany, recalling that as a boy he had lived two miles from Sir Moses Montefiore, a great Jewish philanthropist, whose largesse had benefited the people of East Kent. The news from Germany put Henson into a 'blind rage', making him wish to draw the sword to help the lowly against the mighty. He regarded the Nazis as neo-pagan 'pederasts' and the Fascists as 'bullies' on a par with British trade union-ists. When the English socialist publisher Victor Gollancz published a collection of documents on Nazi Jew-baiting, Henson provided the introduction, even though as a High Tory he did not care for the book's red cover. He wrote a blurb for the journalist Konrad Heiden's brilliantly deflationary *Der Führer*, which remains the most outstanding biography of Hitler. In letters to *The Times* Henson sought to have British uni-versities break all contacts with German institutions, including Durham university, of which he was official visitor. Remarkably, for a clergyman, Henson had few qualms about political assassination. In February 1936 he wrote: 'Who could deny the morality of a patriotic Italian who, for public reasons, killed Mussolini? Or who would not applaud the German who, in the interest of elementary morals, killed Hitler? I should give them Christian burial without hesitation.' In 1938 his rebuke to archbishop Lang during a debate in the Lords on appeasement was so intemperate that he was reminded of his august 'position' by foreign secretary Halifax.[143]

Bell and Henson were so forthright against the treatment of Niemöller that Hitler bearded the British ambassador about the two outspoken English bishops. The Anglican bishops were particularly exercised by the arrest of Meiser and Wurm. Bell and Cosmo Lang made forceful representations to the German embassy, threatening to break off con-tacts with the 'official' Protestant Church. Since English, French and Swedish protests against the imprisonment of the two bishops might have adversely affected the outcome of the plebiscite in the Saar, foreign minister Neurath – himself a prominent Protestant – prevailed on Hitler to restrain bishop Müller in his zeal to incorporate the two south German Lutheran Churches in the emergent Reich Church. Hitler ordered the release of Meiser and Wurm, granting them an audience to reassure them of his moderate intentions.

When he retired from Durham in 1938, Henson summed up:

> I shared in full measure the sentiments of disgust and detestation which the abominable persecution of the Jews in Germany stirred in generous minds throughout the English-speaking world; and I did not hesitate to give public expression to my feelings. Less barbarously cruel but, perhaps, even more luminously suggestive of the ethical quality of Hitler's regime, was the cunning and continuous oppression of the Christian Churches, both Roman Catholic and Protestant. I did my best to bring home to the English people the fact and the significance of a religious persecution within modern Christendom that reproduces the policies and procedures of ancient pagan violence. Indeed Hitler was showing himself to be the true successor of Decius, Diocletian, and Julian the Apostate, though wholly without their excuses.[144]

In 1935 Hitler tried one last time to reconcile the warring Protestant clerics, whose quarrels, he claimed, had spoiled too many of his breakfasts. On a tour of inspection, during which he talked about the 'Church Struggle', he took up the suggestion of Hanns Kerrl, the former Prussian minister of justice, that he be allowed to sort the Protestant Church out. Kerrl's appointment as minister of Church affairs was supposed to mean that clerical heads were going to be bashed together. Actually, Kerrl was too politically lightweight to achieve a goal which underestimated the fractiousness of the clergy.

He tried to establish a Reich Church Committee and committees for the provincial Churches, which sidelined the Reich bishop, who kept his title and salary but lost his office and official limousine. Kerrl sought to introduce proportional representation of the various factions in the governance of the Churches, while also rationally deciding which faction was most entitled to use Church buildings. He sought to find common ground between the factions while restraining the Gestapo from persecuting Confessing Church pastors. His moderation split the Confessing Church between moderates and 'Dahlemite' radicals, or between those like the Lutheran bishops of 'intact' Churches who would co-operate in Kerrl's committees and those who would not.

The more radical Dahlemites occasionally ventured criticism of the regime, albeit addressed to Hitler, whom they mistakenly regarded as a moderate man surrounded by maniacs. In May 1936, they sent Hitler a memorandum seeking clarification whether the 'de-Christianisation' of schools, the persecution of the Jews and the use of concentration camps for political opponents were official government policy. The memo even challenged the deification of the Führer: 'Only a few years ago the Führer himself disapproved of placing his pictures on Evangelical altars. Today his opinions are increasingly accepted as normative not only in political matters, but in matters of morality and law, and he is being surrounded with the religious dignity of a national priest and hailed as an intercessor between God and the *Volk*.' This memo, which went unanswered, was leaked to the foreign press, appearing in the *New York Herald Tribune* on 16 July, and then circulated in Germany. Three of those responsible for the memorandum and its distribution were sent to concentration camps, where one of their number – a lawyer – was murdered for being Jewish while the other two were released.[145]

Having failed to achieve Protestant unity through Reich bishop Müller, by 1937 Hitler was wearying of minister Kerrl too. In the spring he unleashed the power of the state upon the more radical pastors within Prussia, including Martin Niemöller, who, despite being acquitted at his trial, was incarcerated, none too onerously, in Sachsenhausen concentration camp. Theology courses and seminars where the Confessing Church line was dominant were closed down. Effectively, Hitler abandoned the quest to unite the Protestant Churches at this point. He would build his own rival religion instead.

Sneering at the ambivalences of authority has become habitual since the 1960s. There is almost a will to believe that something sinister is always afoot. In fact relationships between the Churches and the totalitarian political religions were infinitely complicated and require considerable effort to reconstruct. At the time some of the greatest intellects found themselves revising their own views, the capacity which made them great in the first place. On the eve of war, two men wrote about a Catholic Church that neither had greatly admired or liked. Indeed Sigmund Freud had written a powerful polemic against religion as such, even as he established a discipline that has become a modern cult. In February 1938, he wrote that it was the Catholic Church 'which puts up a powerful defence against the spread of this [totalitarian] danger to civilisation'. In a second letter to his son, he added the hope that 'the Catholic Church is very strong and will offer strong resistance', although a month later Austrian church bells would peal welcoming the return of the prodigal Führer.[146] Two years later the exiled physicist Albert Einstein would make a remarkable admission in *Time* magazine: 'Only the Church stood squarely across the path of Hitler's campaign for suppressing the truth. I had never any special interest in the Church before, but now I feel a great admiration because the Church alone has had the courage and persistence to stand for intellectual truth and moral freedom. I am forced thus to confess, that what I once despised, I now praise unreservedly.' The following chapter explores whether the Church deserved such unqualified praise.[147]

CHAPTER 4

Apocalypse 1939–1945

I BEGINNING AND ENDINGS

The Second World War presented a still nominally Christian Europe with unprecedented challenges. For the British, these were indirect and existential, for there was no significant domestic constituency of totalitarians. Under a great wartime leader, whose religion combined the Whig version of the island race's saga with a powerful faith in the convergence of his and the nation's divinely ordained destinies, in 1940 the British narrowly avoided invasion and occupation, and the destruction of their way of life.[1] Churchill was not a conventionally religious man, preferring, as he once said, to offer the Church the support of a flying buttress – that is, from without – and dismissive of the abilities of most of its leaders, the robust Tory Herbert Hensley Henson, whom he persuaded to move from Durham to a canonry at Westminster abbey, and the Catholic primate cardinal Arthur Hinsley being the notable exceptions. In oratory that sometimes seemed overblown in peacetime, Churchill captured the urgency of the times by speaking of ultimate things that the British usually preferred to leave unstated:

> What General Weygand called the Battle of France is over. I expect that the Battle of Britain is about to begin. Upon this battle depends the survival of Christian civilisation. Upon it depends our own British life, and the long continuity of our institutions and Empire. The whole fury and might of the enemy must very soon be turned on us. Hitler knows that he will have to break us in this island or lose the war. If we can stand up to him, all Europe may be free, and the life of the world may move

forward into broad, sunlit uplands. But if we fail, then the whole world, including the United States, will sink into the abyss of a new Dark Age, made more sinister, and perhaps more protracted, by the lights of perverted science. Let us therefore brace ourselves to our duties, and so bear ourselves that, if the British Empire and Commonwealth last for a thousand years, men will still say, 'This was their finest hour.'[2]

By 1940, when Britain faced its summer of peril, Henson was seventy-seven years old – as he put it, 'an unserviceable onlooker at this supreme crisis'. Although still a vigorous walker, he needed to supplement his spectacles with a powerful magnifying glass, especially in the uncertain light of Britain's wartime churches. His autobiographical diary is a vivid account of the war from the standpoint of this aged patriotic cleric, including such horrors as shortages of coal or tea and the trials of his regulation gasmask. On his visits to London, Henson's nights in the Athenaeum were routinely interrupted by air-raids, the novelty of watching dogfights from the club balcony overlooking St James's being superseded by sleepless nights as the crumps of high explosives and the sound of shattering glass came nearer; nor was sleep to be had if he stretched between chairs in the club basement, thanks to the 'persistent snoring of one of their number'.[3]

The moral issues the British faced were unambiguous, that is a fight between good and evil, a stance encouraged by the fact that all the Allies were victims of Axis aggression. An intelligent few, notably those who met from April 1938 onwards as the informal 'Moot' forum, reflected on the paradox of defending a Christian civilisation, a concept that bulked large in wartime rhetoric but which some British Christians felt no longer existed in really.[4] The Anglican poet and leading Moot member T. S. Eliot caught this very well when he wrote to his Jewish friend Karl Mannheim: 'We are involved in an enormous catastrophe which includes a war.'[5] Not for the first or last time, assault from without led to urgent reflection about core beliefs in a society not especially prone to such ruminations. A leader in *The Times* contrasted the investment of Hitler and Stalin in disseminating their respective 'faiths' among the young with the parlous state of religious education in Britain. The war witnessed an extraordinary efflorescence of ecumenical activity, with the foundation in 1942 of the British Council of Churches and two years earlier of the Sword of the Spirit movement through which Catholics sought to involve Anglicans in bringing a religious perspective to bear upon democratic society at a time when English Catholics were briefly suspected of Francoist or Vichyite sympathies. A courageous minority, whose representative figure was the Anglican bishop George Bell of Chichester, refused to view all Germans as Nazis, and worked tirelessly, in Bell's case through his ecumenical contacts in Sweden and Germany, to convince the British government that there was 'another Germany' ready and willing, if not able, to displace Hitler. Bell also took the lead in publicly denouncing the indiscriminate bombing of German cities as morally reprehensible, a stance that incurred the enmity of Churchill, and probably cost Bell the archbishopric of Canterbury when in 1944 William Temple died at the age of sixty-three.

In Nazi-occupied Europe, moral choices were much starker or more slippery, depending on whether one moved from east to north or west, for race ultimately dictated differential treatment of Europe's subject peoples. In Poland, which was crucified between two thieves, both the Communists and the Nazis sought to extirpate Christianity, although only the Nazis attempted to reduce the Poles to helotry in the remnants of their former state. White Europeans were treated 'like the blacks in the colonies', as the metropolitan of Lwów put it. Six million Poles were killed, half of them Christians, half of them Jews. At 220 wartime deaths per thousand, proportionally this was a far greater loss than any other nation in the Second World War. That huge death toll included a fifth of Poland's Catholic clergy.[6]

At the other extreme, in France, where thirteen people per thousand died, Christians had to deal with German military occupation or with the collaborating French regime at Vichy that adopted much of the rhetoric of conservative Catholicism. Elsewhere, Christianity – of various types – was integral to a rabid and religoid integral nationalism. The puppet regime in Slovakia was lead by a Catholic priest. In the Balkans, the Romanian Orthodox Church illustrated what happened when a Church threw itself wholeheartedly behind a war of extermination, while the Catholic Church in Croatia had intimate involvements with the murderous Ustashe. Only people with no understanding of how the Catholic Church operates can hold the Vatican responsible for fanatic elements of its own lower clergy, whether in Croatia or Ireland.[7]

In the western parts of the Soviet Empire, Christians of various persuasions were confronted by the invidious choice of whether to welcome the Nazis and their multinational confederates as 'liberators' or to adhere to an equally murderous Marxist–Leninist regime, which for tactical reasons belatedly recognised the mobilising power of an Orthodox Church it had almost annihilated in previous decades. Christians in Germany had to deal with the satanic reality of a political religion that had successfully confused itself with the nation's history, identity and destiny. German Protestantism had no external hierarchy (beyond the fraternal admonitions of British, Scandinavian or US Christians) or theological resources to enable it to withstand even the most outrageous aspects of Nazi policy. Christianity in Germany survived this ordeal, perhaps by never forcing people to choose between nation, race and faith, for the outcome might have been bitterly disappointing to all the Churches. The German Churches may have emerged dishonoured, but that they emerged at all was perhaps a slight achievement when measured against the crimes they had witnessed in silence. And when the fighting, rather than the trauma, was over?

The experience of near total Nazi and Fascist hegemony cured the overwhelming majority of continental European Christians of their instinctual predilection for a politics that was cool or hostile to liberal democracy. The process of distancing Churches from anti-Judaism (as distinct from a newfangled antisemitism to which few had subscribed), which had commenced in the inter-war period, became absolute after the Nazi charnel houses were fully exposed. Many European intellectuals, for example in Poland, dropped their reflexive anticlericalism. The experience of resisting Nazism, in which Christians everywhere had played a distinguished part, led to an appreciation of the virtues of liberalism and democratic socialism, and a willingness to work with such people in future within a democratic framework. That political eclecticism was true of some of the leading resistance groups in Germany itself. Except for a few peripheries, where religious sectarianism continued, the war also made a virtue of ecumenical contacts, which would assume political form as varieties of Christian Democracy in much of post-war Europe. The war also gave an enormous fillip to the expansion of the state, a process that the introduction of mass welfare entitlements perpetuated after the war. Although Christians played a key role in supporting such a development, they were paradoxically contributing to their own eradication from activities that the bureaucratic state now regarded as pre-eminently its own, with faith-based welfare henceforth having, largely unsuccessfully, to fight its way back in. Despite deploring war, the papacy emerged with greater influence and prestige after the

Second World War than it had ever enjoyed before in modern times. The world henceforth had many religions but only one paramount spiritual leader. The never ending 'Pius Wars', about the wartime conduct of Pius XII, do not seriously affect that conclusion, and it is to this controversial figure that we turn first.[8]

II THE DILEMMAS OF A DIPLOMAT

As an international institution, the Catholic Church had to negotiate every political context, protecting the rights of Catholics in all belligerent countries through the mechanism of concordats; rendering assistance to a much wider range of humanity; and balancing its diplomatic cum spiritual objectives with the role of moral prophecy. Perhaps no one could have performed the multiple roles of pope to universal satisfaction in such circumstances, and the legacy of Pius XII, who faced these challenges, is still disputed, as was that of Benedict XV during and after the First World War.

Nazi racial exterminism has become so dominant in the historiography of the last two decades that it has eclipsed every other aspect of the war, including attempts to prevent, contain or mitigate it. That downgrades most of the activities that were of paramount concern to all Europe's Churches in the two years before the 'Final Solution' started under cover of a war that had raged since September 1939. One of the chief activities of the papacy was to prevent war at all, an activity that sometimes had the support of Mussolini, as well as the European democracies and the US. This papal diplomatic activity is relatively straightforward to understand, while in its sheer unassuming scale the relief and rescue work is difficult to get a purchase on despite the abundance of documentation.

Unconscious of the ironies involved, countries based on a separation of Church and state, or whose Protestant historical identities were bound up with resistance to Rome, supported Pius XII in his quest to maintain peace.[9] Such solicitations, and his own undoubted skills as a negotiator, may have led Pius XII to place too much faith in Vatican diplomacy. It encouraged him in a diplomatic posture to which neutrality was appropriate, but, once it was clear that the time for talking had passed, he did so arguably at the expense of his obligations as universal witness. Pius

XII was eminently suited, by background and temperament, to the role of mediator. He had immense experience as a diplomat and had been Vatican secretary of state for nine years when war came. Pius XII was the first pope to have been to the US, and president Roosevelt addressed him in his letters as 'dear friend'. But even his admirers do not claim that this austere, scholarly figure was robust enough for the prophetic role, although one wonders whether statements in which outrage did not need to be coaxed out from the finely chosen phrases, but which might have made matters worse, would have made the slightest impression, at a time when so many people were infected by extreme hatreds and nationalist passions. Calling Hitler 'a motorised Attila', as one senior Vatican cleric did, sounds good, but most likely Hitler would have regarded it as a compliment.[10]

In May 1939 the pope sought to convene a conference with France, Germany, Britain, Italy and Poland to resolve disputes that divided those countries (but not Britain). The Italians were the most enthusiastic, while the Western Powers feared another Munich, and Hitler disavowed any aggressive intentions, thereby making such a meeting superfluous. He told the papal nuncio Cesare Orsenigo, who had flown down to Berchtesgaden, that other leaders might benefit from similar recuperation in the alpine air and pastures, and sent him away frustrated. The only basis for a deal would have been to persuade Poland to surrender Danzig and the Corridor, while extending German 'protection' over the aggrieved ethnic German minority in Poland, something that a newly independent nation could never have accepted.[11]

The August 1939 Molotov–Ribbentrop Pact increased Hitler's belief that he could attack Poland without a European war. Since it appeared to undermine the British and French guarantees to Poland, this seemed an opportune moment to persuade the Poles that concessions were better than defiance. Fully conscious that it might be accused of being in Mussolini's pocket, or for having sponsored a second Munich, the Vatican proposed that the Poles abandon Danzig, while the pope broadcast a final call for peace:

> We address Our most pressing appeal to governments and nations, imploring them to lay down their arms and forswear their threats, and try, instead, to hammer out a remedy for these conflicts in the only procedure that is left, negotiation. We appeal to them to explore with goodwill, calm and serenity, the

pacific methods that are still possible and to let the force of reason prevail over the violence of arms for the triumph of justice. Conquests not founded on justice cannot be blessed by God. Politics emancipated from morality betray those who desire it to do so. Danger is imminent but it is still not too late. Nothing is lost by peace. Everything can be lost by war.[12]

Diplomatic traffic was still passing through the autumnal darkness that Hitler's forces used to cover their attack on Poland. The British foreign secretary, Halifax, by then moodily contemplating Prussian jackboots resonating on the stones of village churches in Yorkshire, reassured Pius that he had done everything humanly possible to avert war.

The pope, informed of the invasion of Poland, retreated to his chapel to pray. The war immediately raised urgent humanitarian problems. On 30 September Pius addressed Polish pilgrims: 'Before our eyes pass as a vision frightened crowds and, in black desperation, a multitude of refugees and wanderers – all those who no longer have a country or a home. There rise towards Us the agonised sobs of mothers and wives.' He established the Pontifical Relief Commission, whose remit was to provide war refugees with food, clothing and shelter. To take one example, the US Catholic dioceses collected US$750,000 which the bishop of Detroit sent to the pope for distribution among Poles in Poland and scattered throughout Europe.[13] He also revived the Vatican Information Bureau, its aim being to reunite people separated by warfare, including prisoners of war – about whom the families everywhere were desperately anxious. The Bureau received a thousand items of correspondence per day, requiring a staff of six hundred to process it and conduct the ensuing inquiries. Its card index contains the names of over two million prisoners of war whom it helped locate and support.[14] Like the parallel work of the International Red Cross, such labour involved a certain suspension of open moral judgement if it was to be at all effective. Vatican Radio also broadcast nearly thirty thousand messages a month in the search for missing persons.

Vatican documents are quietly eloquent on the papacy's variegated interventions on behalf of so many victims of the Second World War, whether the despatch of food to Greeks starving because the Italians had made off with all the available food and the British were blocking ships bringing grain; exchanges of sick or wounded British prisoners in Italian captivity in North Africa; or, when the war had reached the Pacific theatre, having nuncio Morella in Tokyo organise medical supplies from Hong Kong for British prisoners of the Japanese. The Greek famine, in which one hundred thousand people starved to death, is instructive. The Germans handed over control of Greece to the Italians in the summer of 1941. Bulgaria had occupied some of the main grain-producing areas, while the Italians had commandeered much of the food stored. The 1941 harvest was poor. The British blockaded Greece, stopping grain shipments from Australia and preventing the arrival of 320,000 tons of grain that the Greeks had bought. Into this extremely complicated set of circumstances, where enemy nations were passing the buck on to their opponents while Greeks died, came monsignor Roncalli, the apostolic delegate to Greece and Turkey who was based in Istanbul. He visited senior German commanders, celebrating a mass for wounded German troops and visiting British POWs, so as to win the confidence of his interlocutors. Simultaneously he urged the Holy See to intervene with the US and British to bring about a temporary lift of the blockade. This persuaded the Germans to allow food to go to Greece via neutral Turkey; they also promised that any future food shipments would go exclusively to the civilian population. The British finally allowed a one-off shipment of eight thousand tons of wheat and flour. Meanwhile, in Athens, Roncalli organised soup kitchens that served twelve thousand meals a day, with supplies purchased by the Holy See in Hungary. Because of these measures fewer people died. It was complicated, undramatic work, in which each side blamed the other for the plight of the Greeks, and it resulted in an agreement between the belligerent powers to put in place mechanisms to ensure that the famine was not repeated.[15]

Even before the war started, efforts were made by the Holy See to help 'non-Aryan' Catholics to emigrate from Germany, a group that was especially isolated since Jewish relief organisations offered these 'renegades' no assistance while they were the group of refugees that Catholic states were most likely to favour. The Holy See encouraged the formation of national relief committees to assist baptised Jews who managed to get out. In Germany, the St Raphael Society, which had existed since 1871 to aid Catholic emigrants, assisted Catholic victims of racial persecution to leave. Although the Holy See had no success in urging the US to relax its stringent visa requirements, which in fact were tightened to the point of impenetrability during the war, it did manage to persuade the Brazilian government to issue visas for three thousand people. It was not the Vatican's fault that every government involved in a refugee's complex passage across Europe to South America seemed to put bureaucratic obstacles in his or her way: this passport had expired; that document was invalid; this piece of paper lacked the requisite stamp. At the same time the Vatican was inundated with more or less hair-brained schemes to resettle the Jews in Australia, Africa, the Caribbean, Latin America or Alaska. As a Vatican official acidly noted: 'The author of this scheme does not seem or want to know that we have been quite unable to obtain even a single visa for Australia.'[16] With some percipience, Vatican diplomats foresaw that a Jewish state in Palestine would lead to enduring international problems. After Italy went to war in June 1940, the pope personally sent money that autumn to bishop Giuseppe Maria Palatucci of Campagna to distribute 'preferably to those who suffer for reasons of race', which meant foreign Jews interned in the concentration camp at Ferramonte-Tarsia. On 14 April 1942, the Jewish internees at the camp profusely thanked Pius XII for the gift of clothing and bedding which he had sent for the five hundred Slovakian Jewish children who had been fished out of the sea off Rhodes when their ship went down, as well as for a previous gift of money and the help the Vatican Information Service had given to families ripped apart by war. That was within the bounds of what was possible.[17]

In November 1941 secretary of state Maglione outlined the principles that underpinned such relief efforts:

> The Holy See, remaining by its very nature outside and above the armed conflict, is, nevertheless, profoundly sensible to the great suffering which follows in the wake of war. Therefore, without entering the sphere of purely political or military affairs, the Holy See has constantly had as its supreme and animating principle that human and Christian charity which embraces all men as brothers: consequently it has not only sought, whenever the occasion presented itself, to turn men's minds and hearts toward those noble and salutary sentiments, but has also dedicated a great part of its activity to alleviating, insofar as possible, the widespread sufferings caused by war. In harmony with this fundamental programme, the Holy See has endeavoured, above all, to carry out its beneficent activity wherever there was need for it, for the relief of every form of misery and privation, without distinction as to race or nationality, on behalf of Catholics and non-Catholics, recognising in their

common suffering a special title to the benevolent interest of the Apostolic See.[18]

Pius strove to prevent the extension of the war while the major belligerents were not fully engaged during the period of phoney war before Hitler attacked westwards. Insofar as Italy declared its non-belligerency, this strategy, which the pope pursued in tandem with US president Roosevelt, who despatched the steel magnate Myron Taylor as his personal representative to the Vatican, seemed successful.

Pius XII's first encyclical, *Summi pontificatus*, issued in October 1939, was, as the *New York Times* reported, 'a powerful attack on totalitarianism and the evils which he considers it has brought upon the world'. If the *New York Times* is to be believed, it 'is Germany that stands condemned above any country or any movement in this encyclical – the Germany of Hitler and National Socialism'. In other words, while the pope sought to remain impartial, he was not morally indifferent. Who else was he thinking of when he said: 'To consider treaties on principle as ephemeral and tacitly to assume the authority of rescinding them unilaterally when they are no longer to one's advantage would be to abolish all mutual trust among states. In this way, natural order would be destroyed and there would be seen dug between different peoples and nations trenches of division impossible to refill.'[19] The encyclical explicitly sympathised with the plight of Catholic Poland, and referred to the fundamental unity of the human race, notably article 48 which cited Galatians 3: 28 – 'There is neither Jew nor Greek, there is neither bond nor free, there is neither male nor female: for ye are all one in Christ Jesus.' The head of the Gestapo commented: 'The Encyclical is directed exclusively against Germany, both in ideology and in regard to the German–Polish dispute. How dangerous it is for our foreign relations as well as our domestic affairs is beyond discussion.' People frequently criticise the elliptical language of the papacy, but time and again those who were the object of papal censure knew whom the pope had in mind.[20]

In his address at Christmas 1939, by which time the German security services had killed fifty thousand Poles (including seven thousand Jews), Pius denounced as crimes:

> a calculated act of aggression against a small, industrious and peaceful nation, on the pretext of a threat that was neither real nor intended, nor even possible; atrocities (by whichever side

committed) and the unlawful use of destructive weapons against non-combatants and refugees, against old men and women and children; a disregard for the dignity, liberty, and life of man, showing itself in actions which cry to heaven for vengeance.[21]

The Nazis declared that the pope had abandoned any pretence at neutrality.[22] Nazi atrocities in Poland brought a further complication for the Vatican. Throughout late 1939 and early 1940 Vatican Radio broadcast accounts of Nazi crimes in Poland, whose content was summarised in the London *Tablet*:

> The New Year brings us from Warsaw, Cracow, Pomerania, Poznan, and Silesia, an almost daily tale of destitution, destruction, and infamy of all kinds, which one is loath to credit until it is established by the unimpeachable testimony of eye-witnesses that the horror and inexcusable excesses committed upon a helpless and homeless people . . . are not confined to the districts of the country under Russian occupation, heartrending as the news from that quarter has been. Even more violent and persistent is the assault upon elementary justice and decency in the part of prostrate Poland which has fallen to German administration . . . A system of interior deportation and zoning was being organised in the depths of one of Europe's severest winters, on principles and by methods which can only be described as brutal. Stark hunger stared 70 per cent of Poland's population in the face, as its reserves of foodstuffs and implements were shipped to Germany to replenish the granaries there. Jews and Poles were being herded into separate ghettos, hermetically sealed and pitifully inadequate for the economic sustenance of the millions destined to live there.[23]

According to US diplomat Harold Tittmann, the Polish bishops alerted the Vatican to the fact that 'the various local populations suffered "terrible" reprisals'. On 14 January 1940 the bishop of Danzig (and administrator of Culm) wrote to the pope reporting Gestapo allegations that cardinal Hlond's broadcasts were encouraging Poles to resist the Germans. As a result, Catholic priests and teachers 'have been arrested, shot or tortured to death in the most terrible ways, or deported to the furthest East'.[24] Their Polish superior ordered the Jesuits operating Vatican Radio to refrain from broadcasting these revelations, explaining:

'How I hated to have to give the order to stop these broadcasts, especially since I am a Pole myself. But what else could one do?' This stance would exasperate the exiled Polish government in London and Casimir Papée, their doughty ambassador to the Vatican, who continually urged the pope to speak more forthrightly against escalating inhumanities, unaware that Church sources within Poland were giving contradictory advice.[25]

An extraordinary series of events enables us to glimpse the strategic thinking behind Vatican diplomacy more clearly. In late 1939, a Bavarian lawyer, Josef Müller, who worked on Italian affairs for German military intelligence, contacted the exiled Ludwig Kaas, who was the superintendent of St Peter's basilica in Rome. A devout Catholic, who knew Pacelli, Müller had been inducted into the German conservative resistance to Hitler; his frequent trips to Rome were a useful cover for contacting representatives of the Western Powers.

Kass reported his dealings with Müller to the pope's secretary, the Jesuit Robert Leiber. Leiber asked Pius to inform the British government of the strength of the military opposition, and to ascertain on their behalf whether the British would offer Germany honourable peace terms should there be a successful coup. Perhaps the pope would care to guarantee such peace terms in person, since the generals were afraid of a repeat of the false dawn symbolised by President Wilson's Fourteen Points? It took the otherwise extremely cautious pope a day to decide to carry out this clandestine 'mission', which, it should be emphasised, directly involved him in a well-advanced conspiracy to overthrow the head of the German government. Leiber met Müller in the grounds of the Gregorian university where he worked; the pope himself dealt with the British.[26]

On 12 January 1940, Pius XII informed the British ambassador to the Vatican, D'Arcy Osborne, that he had met a representative of various German generals – omitting to add that the leading conspirator, general Ludwig Beck, was a friend from his days as nuncio – who were prepared to overthrow Hitler, thus pre-empting the latter's plans for an offensive in the west that February, an offensive in which the pope averred the Germans were planning to use 'microbes'. Pius was acutely conscious that any indiscretions regarding these conversations would result in the deaths of the generals involved and extreme sanctions against Leiber's Jesuits. The conspirators sought guarantees that they would receive an honourable peace settlement, based on the restoration of Czechoslovakia and Poland, and the retention by Germany of Austria. At a further meeting with Osborne, Pius was well informed about how the German conspirators saw events unfolding. There might be a civil war, followed by a military dictatorship, which would gradually hand power over to a democratic, conservative, federal government. The British should respond generously to this new regime by recognising the status quo ante as established at Munich. Osborne communicated this intelligence to Halifax, who relayed it to prime minister Chamberlain and the king. Chamberlain's response was critical. It was characteristically cautious, insipid and unimaginative – the response of a glorified clerk to schemes of some boldness. The British government was not going to act without informing the French, and the seriousness of the conspiracy would have to be more clearly established before any response could be made. Halifax, by contrast, instructed Osborne to inform the pope that the British were ready to discuss what was proposed, provided the French were engaged. From that point onwards, British interest in what was afoot faltered. The British had suspicions about the integrity of Ludwig Kaas in a Vatican that provided perfect cover for foreign agents; and more seriously, the legacy of an earlier struggle against the 'Hun' meant that they were unable to distinguish between conservative 'Prussian' generals and a demagogic Austrian upstart whose mind was clouded with bloodthirsty fantasies about the Jews. The more the British pressed for details of the conspiracy, the more the plotters equivocated, until the opportunity passed.[27]

'Hitler's pope' did not confine himself to being a reluctant intermediary for dissident German generals seeking to contact the British. When in March 1940 Müller informed his Vatican contacts of the date of the May offensive in the west, Pius immediately passed that information in encrypted form to the nuncios in Brussels and the Hague who relayed it to London and Paris as well as to the governments directly threatened. The pope was therefore directly involved in betraying the military plans of a wartime power to two of its opponents, as well as conspiring with Hitler's domestic opponents.[28]

While the pope was engaged in conspiracy, Orsenigo in Berlin regularly and persistently protested against the Nazis' systematic attempts to destroy the Polish elites, including the Catholic clergy, while simultaneously implementing a devastating programme of what is now called 'ethnic cleansing'.[29] Orsenigo was in many respects not a big enough man for the post he occupied, but we should not underestimate the difficulties he encountered in dealing with a regime for which lying was routine. In a meeting with Ernst Wörmann, the Director of the Political Department of the Foreign Ministry, Orsenigo was remarkably persistent as Wörmann gave him the usual bureaucratic 'run around', at the same time denying that the atrocities the nuncio raised with him had even taken place:

> He knew, he said, that as Nuncio he was not entitled to bring up this matter, but he felt obliged as a human being to do so . . . Things had recently occurred there which Germany, in its own interest, should not permit. He did not want to investigate here, he said, whether shootings of landowners which had taken place were justified or not; he was speaking only for the ordinary people. Women, children, and old people were being dragged from their beds by night and expelled, without having any other living quarters allotted to them. The Nuncio asked me if I could not advise him whom to approach in this matter.
>
> I replied to the Nuncio that I could not recommend him to approach high-ranking German personalities because they would perhaps not listen to him as quietly as I had done and would object at once that, as Nuncio, he had no right to speak of these things. Moreover, I said, I firmly believed that he was the victim of false information. The Nuncio disputed the last point, stressing how cautious he was in evaluating reports. He asked me at least to have some discussion with the State Secretary as to whether something could not be done.[30]

The moral contours of the Nazi 'new order' began to emerge across Germany's enlarged sphere of influence. The Warthegau was a huge territory, named after a tributary of the River Oder, consisting of some forty-six thousand square kilometres, created amid the ruins of Polish statehood. It had nearly five million ethnically Polish subjects, together with 340,000 ethnic Germans, although a ruthless programme of 'Germanisation', involving expulsions and the repatriation of the German diaspora, would ensure that by 1944 the German element had trebled. In their imaginations, the Nazis regarded this as a tabula rasa, a vision they encouraged by not allowing the German state bureaucracy to get a footing even though the Polish administrative apparatus had been swept away. The Nazis approached this laboratory for their principles with the 'exhilaration' of missionaries entering a new territory, although here they also claimed that the Germans had been before. Part of the experiment involved eradicating the Churches. To this end, the papal nuncio to Germany was excluded by the simple device of restricting the terms of the 1933 Concordat to the 'Old Reich'. Orsenigo's efforts to introduce papal representatives to occupied Poland were rebuffed and his own competence vis-à-vis events in the Warthegau disputed. Protestant Churches in the Warthegau were also formally cut off from their equivalent communions in the 'Old Reich'. The Catholic Church was denied any legal recognition, despite it being the overwhelming religion of the Poles. The Churches were denied funding through Church taxes, and all convents, monasteries and seminaries were closed. Children were forbidden to belong to Churches, and religious instruction in schools was proscribed. Schoolteachers and Nazi officials were compelled not to belong to a Church. Worse, Germans and Poles were formally segregated for religious as well as other purposes, with German churches bearing signs saying 'Poles forbidden'. In October 1941 this led to the creation of separate cemeteries. The drastically reduced number of priests had to minister, under appalling circumstances, to enormous numbers of people. Before the war, the diocese of Posen had 441 churches. During the war, fifteen churches were available to Germans (who constituted 10 per cent of the population), while the Poles had to make do with thirty; 828 pre-war Catholic clergy were reduced to 34. Only in Bolshevik Russia were clergy exposed to similar tribulations. In October 1941, the Gestapo began rounding up those Polish clerics who had escaped mass shootings of the Polish elites. A total of 2,700 Polish priests were detained at Dachau, where nearly half of them perished. There was a special concentration camp for nuns.

De-Christianisation and massacres were followed by state-sponsored mass murder involving modern technological methods. This was the product of apocalyptic and scientising strains within National Socialism, a synthesis of certitudes devastating for victims stigmatised as demons or pathogens by killers who, not least in the case of Hitler himself, switched unselfconsciously between the roles of redemptive prophet and of Pasteur.[31]

Beginning in July 1940, Lutheran clergy in a few regions began to receive reports of a covert and systematic policy to eliminate 'life unworthy of life' – that is, people deemed to be eugenic and economic burdens on the wartime 'national community'.[32] Following the example of bishop Theophil Wurm of Württemberg, Catholic bishops wrote to those members of the government whom they took to be susceptible to such influences, protesting against policies that were illegal under German law and reprehensible in the eyes of wider Christian opinion. They were in a better position than the pope to do so, since as members of the residual establishment, they had high-level contacts with members of the government who were adjudged not to be ideological fanatics, and as German patriots could argue that these policies were damaging domestic morale or Germany's international standing. In a letter to Heinrich Lammers in the Reich Chancellery, Breslau's cardinal Bertram warned that 'if this principle [the inviolable and absolute support and protection of the life of the innocent individual person] is once set aside, even with limited exceptions, on the grounds of an occasional need, then, as experience teaches us, other exceptions will be made by individuals for their own purposes'. A decade earlier, the papacy had condemned eugenic sterilisation, with the US – rather than Germany – in mind. On 6 December 1940 the Congregation of the Holy Office in Rome categorically denounced euthanasia killings on the grounds that 'this is contrary to both the natural and the divine positive law'. Although nothing in occupied Europe had reportedly pained and shocked Pius as much as these policies, he also insisted that the Holy Office remove any 'polemical' expressions – such as 'inhumanum' or 'nefarium' from what was a Latin document.[33] In a letter to bishop Preysing of Berlin, Pius indicated that he was behind this more temperate condemnation, adding, 'We would not think We had done our duty, if We had kept silent about such deeds. It is now time for the German bishops to judge what the circumstances of the time and place permit to be done.' The bishops continued to register their informal protests.[34]

Another response was public protest, a method employed by a handful of both Protestant and Catholic clergy. The most celebrated instance was bishop August Clemens Graf von Galen of Münster. In July 1941 the Münster gauleiter Alfred Meyer seized Church properties in and around Münster which had been badly hit by RAF bombing. On 12 July the Gestapo attempted to seize Jesuit property in the city, only to run into the imposing and outraged figure of the city's bishop. After this confrontation, Galen retreated to his study where he tapped out a sermon which he delivered the following morning. The sermon was a bold defence of justice, at a time when 'none of us is safe . . . he cannot be sure that he will not some day be deported from his home, deprived of his freedom and locked up in the cellars and concentration camps of the Gestapo'. Nothing suggests that Galen was speaking exclusively about Christians, rather than humanity as a whole, it being malicious to infer that he was somehow deliberately excluding Jews from his demand for justice, since Jews were not the principal victims of the assault against mental incompetents. He expressly said that 'it is not a specifically Catholic issue that I discuss before you today, but rather a Christian, yes, a general humanitarian and national religious issue'.

Galen's first sermon seems to have incited further seizures of Church property, including a convent where the Gestapo temporarily imprisoned his sister, the nun countess Helene. In his second sermon, on 20 July 1941, he trod a careful path between supporting the German war effort and condemning the Nazi enemy within, against whose hammer blows he advocated the resilience of the anvil. Although he had known about euthanasia killings since July 1940, he was now told of the imminent removal of patients from local asylums. There was much local anguish and anger at the prospect. In his third sermon, on 3 August 1941, Galen said that unlawful killing was still punishable with the death penalty. In order to avoid this outcome, an apparatus had been created to spirit people away, leaving no trace of the victims to be followed up by the police. A ghastly materialism informed the entire operation, as if people were like obsolete machines destined for the scrap heap. Such a mentality threatened endless swathes of people, including the elderly or wounded soldiers. It also menaced the entire moral order on which society rested.[35]

The pope wrote warmly to bishop Preysing of Berlin regarding his cousin Galen's protests. They demonstrated 'how much could still be achieved within the Reich through an open and manly public stance', which, however, was not open to the head of the Church, who had to be more restrained in what he said, because of the 'difficult and contradictory' general situation he had to deal with.[36] This declaration of his own position did not prevent Pius encouraging archbishop Gröber of Freiburg to protest against the Nazis' sinister 1941 feature film *Ich klage an*, which explicitly advocated and sought to legalise compulsory 'euthanasia'.[37] However laudable Galen's intervention, which Hitler and other Nazis wished to punish with his execution, it did not 'stop' euthanasia killings. Those responsible for these murders had already slightly exceeded the target figure they had set themselves before mass gassings started. With their surplus killing capacity, they were searching for other people to destroy, a search that brought them first into the orbit of the SS concentration camps, and then into evolving plans to murder Europe's Jews on an industrial scale. Murders in Germany's asylums continued on a decentralised basis, through starvation, neglect and lethal medication. No protest, no matter how forceful, and no matter how widely known, deflected the Nazis from their self-appointed mission to redeem 'Aryan' mankind through the elimination of racial pathogens. They showed a steely persistence in pursuing those goals even when their own world was collapsing around them.

Revelations of atrocities did not interrupt efforts to contain a widening war. Having overrated Mussolini's capacity to restrain Hitler, the Vatican concentrated on exploiting divisions among the Italian ruling elites to keep Italy out of the war. Both Roosevelt and Pius XII were at one in thinking that it was essential to maintain Italian neutrality. Pius wrote to Mussolini urging him to spare the Italian people the calamity of entering the war and congratulated him when this seemed to be the case. For a while their joint strategy succeeded. Although the prospect 'distressed' him, on 11 March Pius XII met Ribbentrop, who delivered a long oration about German strength and the inevitability of victory over Britain and France. Pius responded by chronicling 'with cold severity' the precise facts 'regarding the tortures which the invader had already begun to inflict upon the Polish people'.[38] According to the *New York Times*, the pope also took the opportunity to speak out in defence of the rights of the Jews.[39] Nothing suggests that he discussed either proposals to restore peace on the basis of Germany's existing conquests or anything so outlandish as the 'liberation' of a Soviet Union that the Nazis were still in alliance with. The meeting was an attempt by Ribbentrop to influence domestic Catholic opinion in Germany and Italy through the symbolism of what was a dialogue of the deaf.[40]

As Hitler triumphed in the west, the prospects of Mussolini staying aloof from the conflict diminished by the day. On 10 May 1940, as the German armies entered Holland and Belgium, Pius sent telegrams to their rulers, calling the invasions 'against all justice', while in its commentary the Vatican newspaper said: 'the total war launched by Germany has clearly revealed itself as a pitiless war of extermination conducted in defiance of the laws of war'.[41] The French ambassador thought the condemnation too tepid when he had an interview with the deputy secretary of state, Tardini:

I [Tardini] pointed out to His Excellency that the Holy Father had already expressed his feelings with great clarity, nobility and with great sympathy towards the stricken countries. I do not see what His Holiness could do with more potency, efficacy and compassion. The Ambassador admits that those telegrams are very good and have made a good impression on all: but – he says – sympathy towards the suffering is one thing; the condemnation of the crime perpetrated is another. When I showed surprise and remarked that who can read will find in those telegrams what the Ambassador was asking for, His Excellency, somewhat embarrassed, continued by saying that he was not speaking in order to obtain help in favour of France but, as at present the Holy See was enjoying such a high prestige, this condemnation was almost an obligation deriving from this prestige . . .'[42]

By contrast, Farinacci railed against the *Osservatore Romano* as 'the faithful interpreter of Masonic Jewish democratic thought'. On 13 May Pius granted an audience to the Italian ambassador, Dino Alfieri, who protested the pope's three telegrams to the Benelux rulers which had irritated Mussolini. He warned Pius that the Fascist bands were restive. Pius was uncharacteristically voluble in reply. He said he had been held at gunpoint once before (in Red Munich); he had no fear of concentration camps; he was not going to be intimidated by the Italian government. He added: 'The Italians know well enough what horrible things happen in Poland. We ought to speak words of fire against things like that. The only reason we don't speak is the knowledge that it would make the lot of the Polish people still harder.'[43]

Fascist bands roughed up sellers of the Vatican newspaper and cut the Vatican's mail. A few days later, Pius' car was stuck in Roman traffic, and he was mobbed by Fascist youths screaming 'Death to the Pope' into his impassive face. The telegrams, and Italian intelligence that the Vatican had tipped off Hitler's latest victims about the timing of the German attack, deafened Mussolini's ears to further papal appeals for peace. As the French ambassador to Italy reported: 'Pius XII did not conceal from me that he had used up all his credit; the Duce refused to listen to him and no longer reads his letters.'[44] The time for diplomacy was over. Nonetheless, by continued and punctilious adherence to the outward neutrality that underpinned it, Pius XII would lessen his capacity to play a prophetic role in the war, although by their words and deeds his diplomatic representatives across Europe undoubtedly reflected his thoughts and feelings.

On 10 June 1940 Italy declared war on the Allies, almost managing to lose a two-week battle with a French army whose stuffing had been knocked out by the Germans. Hitler politely declined the offer of Italian troops for his invasion of Britain. On 26 July the pope sought to establish whether the British, German and Italian governments would welcome his mediation to restore peace. Both the apostolic delegate to London and cardinal Hinsley of Westminster declined to pass on to the British government what they thought might be interpreted as an 'invitation to surrender'.[45] In October, Mussolini despatched an ill-prepared army into Greece, which after four months' fierce resistance by the Greeks almost managed to lose the captured Albanian territory from which it had started. In January 1941 Hitler and Mussolini met to discuss future strategy. The Germans were not impressed by general Guzzoni, for he had a paunch, a Jewish mistress and a dyed wig, but Hitler's liking for the Duce led him to overlook obvious Italian military shortcomings. In February, Rommel arrived in North Africa to help the struggling Italians. In April the Germans fell upon the Greeks and Yugoslavs from the north, enabling the Italians to salvage the semblance of victory.[46]

Following diversion into this Balkan sideshow, in the summer a mighty multinational force began rumbling through the cornfields of Russia, in a haze of heat, sweat and dust. Mendacious reports in the Spanish press claimed that the 'crusade' had the full blessing of the German bishops, which contributed to the recruitment of forty thousand men to fight in Russia as the 'Blue Division'. That autumn, France's cardinal Alfred Baudrillart, an octogenarian with vivid memories of the Commune, issued embarrassing calls for men to join the League of French Volunteers against Bolshevism: 'The Archangel Michael brandishes his avenging sword, brilliant and invincible, against the diabolic powers. With him march the old Christian and civilized peoples who defend their past and their future at the side of the German armies.'[47] This was met with an icy silence by his fellow French bishops, while Pius XII ostentatiously refused ever to declare the war in Russia a 'crusade', just as his predecessor had denied the same blessing to Spain's Nationalists in the mid-1930s.

Unlike Stalin, who suffered a mental collapse when the reality of Hitler's invasion of the Soviet Union penetrated his state of denial, on the very day of the attack metropolitan Sergei sent a message to every Orthodox parish. It reminded the Russian faithful of the heroic deeds of their ancestors, and of the saints Alexander Nevsky and Dimitri Donskoi, who had rescued Holy Russia in past crises: 'Our Orthodox Church has always shared the fate of the people. It has always borne their trials and cherished their successes. It will not desert the people now ... The Church of Christ blesses all the Orthodox defending the sacred frontiers of our Motherland. The Lord will grant us victory.'[48] There was even a coded barb: 'we, the residents of Russia, have been cherishing the hope that the blaze of war which has engulfed nearly the whole globe, would spare us'. On 26 June, again before Stalin bestirred himself, metropolitan Sergei addressed twelve thousand people in the cathedral of the Epiphany, condemning anyone who imagined that liberation by the Germans was an alternative to fighting for the Russian motherland. When Stalin did finally address the nation on 3 July, he spoke in the uncharacteristic tones of 'Brothers and sisters! My dear friends!' whose religious accents were unmistakable. He may have mentioned Lenin, but the radio address was much more like a simple priest sounding the village tocsin. In October, patriarch Sergei wrote a further address, as the Germans came within sixty miles of the capital. He condemned clergy who had defected to the enemy, notably metropolitan Voskresensky who had been despatched to the Baltic States before the war as part of a wider attempt to exploit Orthodoxy to integrate the newly acquired states into the Red Empire. On 11 November, Stalin harangued troops on Red Square as German troops battled their way towards suburban Moscow, invoking Nevsky, Donskoi, Suvarov and Kutusov, realising that common or garden patriotism and religion had greater mobilising potential than Marxist–Leninism. Typically, patriarch Sergei had been dragged from his sickbed a few days before and deported to Ulyanovsk.

Of the other two remaining Orthodox hierarchs, metropolitan Nikolai was brought back from the Ukraine to Moscow, where he became the regime's main clerical foreign policy propagandist, while metropolitan Alexei rallied the faithful during the terrible siege of Leningrad. The regime made a few cautious and parsimonious concessions to a Church that played a major role in maintaining wartime morale. It tolerated rather than encouraged religion. Overt anti-religious propaganda may have ceased for the duration, perhaps in rueful recognition of Pius XII's leading role in persuading sceptical US Catholic bishops of the legitimacy of their government's Lend–Lease aid to the Russian people despite his predecessor's comprehensive damnation of Communism, a stance that militates against the notion that anti-Communism was the overriding obsession of his pontificate.[49] Sunday was restored as a day of rest, and artists were allowed to repair damaged icons. In 1942 the presses of the almost defunct League of the Militant Godless were used to produce a tome called *The Truth about Religion in Russia*, in which the weary remnants of a Church the Soviets had tried to destroy were displayed for foreign consumption. Beyond this there were no concessions. At Easter 1942 churches in Moscow were allowed to hold candlelit processions as the curfew was raised for a night. This was a meagre gesture given the enormous role that the Churches had played in the war effort. Starting with Alexei in Leningrad, sermons became appeals to donate money to the war effort. By January 1943, over three million rubles had been raised in Leningrad alone. Another five hundred thousand rubles funded a tank column named after Dimitri Donskoi. By the end of the war, the Church had contributed 150 million rubles.

In November 1942 metropolitan Nikolai became the first cleric since 1917 to have an official function, when he joined a government commission to investigate Nazi war crimes on Soviet territory. That included putting his name to accusations that the Germans had carried out massacres at Katyn for which the NKVD had been responsible. In January 1943, patriarch Sergei sent a telegram to Stalin requesting permission to open a central bank account where the Church could deposit such monies. When Stalin assented, relaying the gratitude of the Red Army, the Church effectively received corporate legal recognition for the first time. It was a sign of the times that in the same month a senior Party official in distant Krasnoyarsk formally received a bishop, who was also a brilliant surgeon, the man still being a prisoner at the time.[50] In September, the exiled Sergei was surprised to find himself brought back to Moscow and installed in the former residence of the German ambassador. At 9 p.m. the following night, he and metropolitans Alexei and Nikolai, were driven to the Kremlin for a session with Molotov and Stalin. The former improbably asked what the Church might need. Recovering from the shock of this request, Sergei said the reopening of churches and seminaries, a Church council and the election of a patriarch. As if it had nothing to do with him, Stalin gently inquired: 'And why don't you have cadres? Where have they disappeared to?' Rather than pointing out that most of these 'cadres' had died in camps, Sergei quickly joked: 'One of the reasons is that we train a person for the priesthood, and he becomes the Marshal of the Soviet Union.' This set Stalin off on a monologue about his days as a seminarian which went on until 3 a.m. Stalin helped the elderly Sergei down the stairs, saying, 'Your Grace, this is all I can do for you at the present time,' although he also appointed Georgi Karpov as the regime's liaison with the Orthodox Church. Karpov was the NKVD official who had arrested and shot most of the clergy, though Stalin added, 'I know Karpov, he is an obliging subordinate.' At some point in the course of that night there was oral agreement regarding the future status of the Orthodox Church. Within four days nineteen bishops were found who elected Sergei patriarch, successor to patriarch Tikhon who had died in 1925. They issued a joint exhortation to Christians around the world to unite against Hitler.

During the following year, dioceses were re-established with the aid of bishops who emerged from exile or prison. Others were members of the schismatic Renovationist Church who had seen the error of their ways. Some forty-one bishops were available when, following Sergei's death on 15 May 1944, they gathered in early February 1945 to elect Alexei his successor. The first seminary opened at a monastery outside Moscow a month after Sergei's death. By September, Karpov was permitting parents to give their children religious instruction, or allowing them to visit the home of a priest for group instruction. The number of Orthodox churches climbed from four to sixteen thousand, with the number in Moscow increasing from twenty to about fifty. There was a revival too of religion in the vast areas that the Germans and their allies swept through. German intelligence and academic experts on the east had extensive contacts within the exiled Orthodox community. When the invasion commenced, the exiled metropolitan Seraphim of Berlin appealed to 'all the faithful sons of Russia' to join the crusade launched by 'the great Leader of the German people who has raised the sword against the foes of the Lord'.[51] In the Ukraine, as soon as the Red Army had retreated, priests made their presence known, having until then been working as artisans, masons and farm labourers. They held services in the chapels attached to cemeteries that had generally been spared the anti-religious attentions of the Soviets. In the major towns, such as Kiev or Poltava, they emerged to hold services in the few remaining churches. In purely statistical terms, the religious revival in the diocese of Kiev was most striking. Of the 1,710 churches which had existed before the October Revolution, there were only two left when the Germans arrived in September 1941; by 1943, roughly eight hundred churches were functioning, served by just under a thousand priests.[52] In Smolensk, where only 25,430 of the 150,000 inhabitants remained after ferocious fighting, all but 200 declared themselves to be Orthodox Christians in a census undertaken by the Germans. The local Wehrmacht commanders encouraged the festive reopening of the cathedral there and at Minsk.

However, while the army and military intelligence were alive to the possibility of using a revival of religion to win over the indigenous population, this was not how matters were regarded by the various power-brokers in Berlin with a finger in the eastern pie. The notional minister for the occupied eastern territories, Alfred Rosenberg, was notoriously anti-Christian, although he saw merit in encouraging auto-nomous and autocephalous Churches in the regions occupied by the Germans, which would bolster ethnic separatism and restrict Orthodoxy to the modestly proportioned ethnic Russian area that he envisaged for the post-war period. Such arrangements would extrude Catholicism to the west and Orthodoxy to the east, enabling the Germans to create huge satrapies running from the Baltic to the Black Sea. However, neither Rosenberg nor his subordinates counted in relation to the SS, which was implacably opposed to any revival of religion in territories they regarded as a tabula rasa on which they were going to impose the future. Hitler himself, the ultimate arbiter, opposed any large-scale activity by the Churches, lest it provide the organisational framework for opposition to the occupation. As for allowing the Catholic Church back in, as Papen rather than the Vatican was proposing, Hitler joked they should 'open the door to all Christian denominations; in all probability they would then proceed to bash each other's heads in with their crucifixes,' before reminding himself that nowadays it was 'the fanatical Communists rather than the clergy who were prepared to die for their convictions'.

In the Ukraine, the occupiers recognised two rival Churches, the autocephalous and autonomous, the former closely associated with national separatism and inflexibly hostile to the Moscow patriarch, the latter prepared to acknowledge his headship of a loose federation in which they could worship according to a modified Ukrainian rite. In Belorussia, which was predominantly Orthodox but with a strong Catholic presence in the west of the country, the occupiers tried to establish a new Belorussian Autocephalous Orthodox National Church, but, as the name suggests, such a confection was never going to be popular. Paradoxically, in the predominantly Lutheran Baltic States, the Orthodox Church was encouraged, especially since metropolitan

Voskresensky declared his willingness to call upon the Russians to fight Communism. German policy towards the Churches in the occupied east consisted of divide and rule, intensifying existing tensions and inciting more. However, despite the conflicts at the top, on a local level religious life flourished in ways it had not done since the Revolution. Like the Soviets, the Nazis were primarily interested in a revival of religion for its propaganda value; it had no intrinsic value in itself.

Nazi paranoia that the Vatican was bent on evangelising 'liberated' Russia was evident in Heydrich's claim to have unmasked the 'Tisserant plan', named after the French cardinal who headed the Congregation for the Eastern Church. According to Heydrich, the Vatican hoped to form a Catholic bloc in the east based on Croatia and Slovakia, which together with France, Italy and Portugal in the west would counterbalance German hegemony. Moreover, it sought to use military chaplains attached to the forces of the many Catholic countries serving on the Eastern Front to win the Russians for Catholicism. In fact, Heydrich's spies had conflated several distinct activities. Some of them predated the advent of Nazism; others sought to get a foot in the door before the Nazis' brand of state irreligion consolidated itself or the Soviets returned.

The Vatican had launched various small-scale missions to Russia in the 1920s, most of which ended with the priests being shot. In 1929 Pius XI created the Pontifical Russian College, or 'Russicum', and a Pontifical Ruthenian College for Ukrainians: along with a network of abbeys, these were designed to train priests to work as clandestine missionaries in the Soviet Union. Britain's future cardinal Heenan was one such volunteer, slipping into Russia in 1932 as a 'commercial traveller' with a fold-up crucifix hidden in a fountain pen. Heenan managed to bluff his way out when he was caught; most of these missionaries disappeared to Siberia. During the invasion, missionary priests managed to attach themselves to the Wehrmacht, for example as grooms for horses, and then slipped away to begin ministering to those natives who declared themselves Catholics.[53] Secondly, Catholic chaplains attached to the Italian army in Russia did indeed ignore Hitler's prohibition on contact with the locals, as did many of the Italian commanders on the Eastern Front who tolerated their activities. Finally, Tisserant was concerned about eastern churches which used a Slavic rather than Latin rite, but which were in communion with Rome. The Vatican army of Heydrich's imaginings consisted of eight Russian priests endeavouring to equip people with catechisms and liturgical works.[54]

On 10 July 1940 marshal Philippe Pétain was declared head of a French state whose watchwords were no longer 'liberty, equality and fraternity' but 'work, family and country'. The regime based at Vichy was recognised by the US, the USSR and the Vatican, with the papal nuncio Valerio Valeri remaining in situ for the duration.

Vichy used much of the moralising rhetoric that had been favoured by the French Catholic Church in the century since the Revolution. The regime denounced the 'esprit de jouissance' (pleasure-seeking) that was allegedly responsible for the defeat, promising a 'moral recovery'. This resonated with a Catholic tradition of moralising major events, as in 1789, 1870 and 1914. The Church welcomed the fall of a republic responsible for aggressively laicising legislation, with the archbishop of Chambéry asking, 'what did our country do in the past to merit the protection of heaven?' Many hierarchs, including the archbishops of Algiers, Carthage and Quebec, issued fulsome declarations of support. They celebrated Pétain as a man sent by Providence to preside over the nation's atonement. Since fifty-one of the French hierarchy were veterans of the Great War, including holders of major decorations for bravery, they viewed the victor of Verdun as an esteemed old comrade. Younger Catholics responded to the wider moral activism that the regime espoused, something they were familiar with from Catholic Action and youth movements in the 1930s. Not only were clergy ubiquitous at public occasions in the Vichy zone, but they played a major role in enveloping Pétain in an aura of pious kitsch. Photographs and other images of the marshal abounded, some equipped with such captions as 'Our pilot' or 'The burning light'.

The Catholic hierarchy converted a complex national disaster into a moralising myth, which suited what the Jesuit Henri de Lubac called the 'masochistic' spirit of those times. Victory, some senior ecclesiastics argued, would have led to yet further moral degradation; defeat afforded a 'heaven-sent' opportunity for regeneration. Victory in 1918 had proved a wasted opportunity; perhaps 1940 could be different? The Catholic writer Claudel regarded defeat as a form of deliverance, confiding in his diary: 'France has been delivered after sixty years from the yoke of the anti-Catholic Radical party (teachers, lawyers, Jews, Freemasons). The new government invokes God . . . There is hope of being delivered from universal suffrage and parliamentarism.'[55] Cardinal Gerlier of Lyons said of the regime's slogan, 'these are our words'; in November 1940 he welcomed Pétain to the city, saying: 'Pétain is France, and France, today, is Pétain.' This was some time before the marshal belatedly regularised his civil marriage to a divorcee in the eyes of the Church.[56]

The National Revolution was indebted to Salazar's Estado Novo as well as to the local adherents of respectively the Action Français leader Charles Maurras, Social Catholicism and syndicalism.[57] Regardless of its derivations, Vichy signified an attempt to restore both the family and 'organic' communities, in which duties would precede rights, while historic regions, religion and a sentimental 'rootedness' would prevail over secularism and cosmopolitan deracination. The peasantry were extolled as the repository of the nation's true values. As in Ireland or Portugal, the Church welcomed the Vichy regime's puritanism, while itself lobbying for bans on young women wearing shorts or ski-pants, curbs on alcohol consumption and dancing, and more stringent film censorship. Divorce was made harder, and virtually impossible in the first seven years of marriage when children might be expected. The Church also warmed to the 13 August 1940 prohibition of the free-masons, who in the clerical imagination had connived at the worst excesses of laicism. For the first time, a representative of the French bishops was attached to the government, and for a brief period a leading Catholic philosopher, Jacques Chevalier, was minister of education.[58]

The Catholic Church was not unique in welcoming the National Revolution, since in September 1940 the Council of French Rabbis drafted a statement of allegiance to Pétain while supporting the conservative moral revolution.[59] The return for the Church was modest since Vichy had more constituent strands than the 'clerical' regimes of Dollfuss, Salazar or Franco. Vichy regularised the status of unauthorised religious orders, and restored property that had been appropriated by communes and municipalities. It spent a bit of money on Church building, and provided scholarships for poor children to attend Church-run schools. Members of religious orders were allowed back into the teaching profession (and nursing), it being an article of faith in conservative circles that secular liberal schoolteachers had undermined the nation's traditional beliefs and will to resist. Under Chevalier there were attempts to reintroduce religion as a voluntary option in the school curriculum, a measure that was swiftly dropped when the classicist Jerôme Carcopino replaced Chevalier. That there was no grand renaissance of French monastic life, apart from the well-publicised return of aged Carthusians, was because many religious orders had relocated their headquarters to Rome and had ceased to be exclusively French in character, while the Vatican was not keen to see a revival of Gallican self-assertiveness. Given this new climate of appreciation for the clergy, it is unsurprising that the Assembly of Cardinals and Archbishops (ACA) made successive professions of loyalty, culminating in the formula adopted in July 1941 of 'loyalty without servitude to the established powers' in order to achieve a France that was 'strong, united and coherent'. Catholic laity were prominent in Vichy's efforts to create various cadres for young people. These included the Chantiers de la Jeunesse and the Compagnons de France, as well as the elite academy at Uriage near Grenoble. This little hothouse, albeit braced by the Alpine winds, was supposed to be the breeding ground for a functional elite, but its tone was that of the frothy moral, mystical and religious discourse of among others Mounier. A prominent Catholic deputy from the Auverne, Xavier Vallat, became the first head of the Légion Francaise des Combatants, which brought together veterans of the two recent conflicts.

France's three hundred thousand Jews, of whom approximately half were immigrants, constituted less than 1 per cent of the population. Their dispersal from the invading Germans, together with the difficulty the Germans had – even after November 1942 when they moved into the 'Free' zone – in bringing their will to bear evenly across such an immense and variegated country are among the reasons why 70 per cent of mainly indigenous French Jews survived the war, although eighty thousand did not, nearly half of that toll consisting of recent immigrants. Unlike Belgium or the Netherlands, whose Jewish dead were significantly higher at respectively 42 and 75 per cent, France enjoyed some geographical advantages. It bordered two neutral countries, Spain and Switzerland, while fifty thousand Jews were also protected from the Nazi mania to destroy by the Italian occupation of the south-east coast. France was a big country, with plenty of remote areas where people could be sheltered.

In August 1940 the Catholic minister of justice, Raphaël Alibert, revoked the 1939 Marchandeau Law, prohibiting incitement to racial hatred, thereby effectively licensing antisemitic propaganda, although such German products as the movie *Jud Süss* were not popular and in Lyons led to Catholic students shouting 'No Nazi films' when the projectors rolled. Mounier's journal *Esprit* also ran a harsh review of the film and was closed down two months later. France certainly had its share of ideological antisemites, some of whom were Catholics, but the presence of the Germans undoubtedly 'incentivised' the Vichy authorities to be more active in this area. German initiatives in the autumn included regulations that Jewish-owned businesses should display signs reading 'Enterprise Juive – Jüdisches Geschäft'. In October 1940, the Germans conducted a census of the Jewish population in the occupied zone. Although an individual's religion had not been recorded by the French state since 1872, virtually the entire Jewish population registered with the police authorities, their sense of duty – or pride – outweighing any reservations about how this information might be used to their detriment. Nine months later the Vichy authorities carried out a similar exercise, with much the same results. All of these people had their papers stamped with the word 'Juif'.[60]

Vichy's first Statute of the Jews, dated 3 October 1940, excluded Jews from the higher civil service, teaching, the media and the arts, where they were allegedly disproportionately represented, while quotas were set for Jews in the learned professions. Exceptional individuals could apply for exemptions, but only 10 out of 125 university professors who did so received them. There was no response from the Churches, nor from the Communists. Jews who had sought refuge in France from eastern Europe, or Germany after 1933, Austria after 1938, or Belgium and Holland after 1940, were especially vulnerable, given the xenophobia that was evident among the French, including highly assimilated French Jews. Forty thousand foreign Jews were sent to internment camps at Agde, Argelès, Gurs, Les Milles, Noë, Récébédou, Rivesaltes, Saint-Cyprien and Le Vernet, many opened in 1939 to contain Spanish Republican refugees fleeing from Franco. Very few French people knew of the existence of such camps – for communications in wartime France were massively disrupted and people were concerned with existing or with the fate of French prisoners of war – and even fewer made it their business to know. Conditions at Gurs, on a rain- and wind-swept plateau, were especially atrocious with people mired in deep mud whenever they left their bleak huts. Offers of assistance to the internees came from Jewish groups, US Quakers, the YMCA and the Swedish branch of the Red Cross. Both French Protestants and representatives of Lyons' cardinal Gerlier joined them in establishing the Committee of Nîmes, which organised supervised residences for people they managed to get released from the camp. Prompted by a Ukrainian Jewish immigrant who had become a Catholic priest, Gerlier also made a formal protest 'in the name of Christian charity and the prestige of France' about conditions at Gurs to the interior minister at Vichy. It was unavailing, and some three thousand souls perished there.[61]

In March 1941, Xavier Vallat, a devoutly Catholic former deputy from the Ardèche, became head of the General Commissariat for Jewish Affairs. A veteran of the Great War, who had lost an eye and a leg, Vallat had greeted the appointment of Léon Blum with the comment, 'Your arrival in power, Mr President of the Council, is incontestably an historic day. For the first time this ancient Gallo-Roman country will be governed by a Jew.' In that month, Marc Boegner, the head of the Reformed Church in France, wrote to admiral Darlan, the head of government, and to the grand rabbi of France, expressing to the former his misgivings about the law, and to the latter his solidarity. The misgivings were partly motivated by fear that Vichy's list of enemies could be extended from Jews and freemasons to Protestants themselves who had folk memories of persecution. It was also noteworthy that Boegner joined many Catholics in thinking there was a 'Jewish problem'. A second Statute on the Jews in June 1941 imposed further restrictions, and began the process of 'aryanisation' or the licensed theft of people's property on the basis of their identity. In October, cardinal Gerlier held an audience with Vallat to convey his misgivings about this second law. According to Vallat himself, Gerlier said, 'Your law is not unjust, but in its application it lacks justice and charity.' This conceded rather too much.[62] Speaking at a mass for lawyers killed in the 1940 campaign, Gerlier alluded to German shootings of hostages, many of whom were Jews, when he said: 'I know not whether they were of our religious faith, but I acknowledge in them my brothers in Christ, who died to expiate crimes of which they were innocent.'[63] In late November 1941 all Jewish organisations in both zones were dissolved; thenceforth the Union Générale des Juifs de France (UGIF) was the sole and compulsory corporate representative of Jewish people vis-à-vis the authorities. Several prominent Catholic intellectuals, from Claudel to Maritain and Mounier, expressed their distaste for and disapproval of these measures.

So did a group of Jesuit theologians in Lyons, who under the leadership of the Old Testament theologian abbé Chaine drafted a declaration by the Lyons Catholic Theology Faculty saying that the new racial laws were 'unjust' and 'offensive'. In Paris, the Jesuit Michel Riquet similarly presented the Assembly of Cardinals and Archbishops with a note on 11 July calling the Second Statute on the Jews 'a scandal to the Christian conscience as well as an insult to French intelligence'. Riquet had harsh words for the French hierarchy, who he thought were 'inspired far less by the Gospel than by the exigencies of a nationalism whose excesses the popes have long denounced and condemned'. The lack of protest from the episcopate was a 'scandal'.[64] This prompted the ACA to issue a vaguely worded defence of human dignity and freedom. Both cardinal Gerlier and Boegner, who liaised with one another, then intervened at the highest levels in Vichy. On 31 July rabbi Jakob Kaplan, assistant to France's chief rabbi, wrote to Vallat about the incongruity of a professedly Christian government discriminating against Jews. These representations may explain why Pétain took the otherwise strange step of asking the Vatican what it thought of legislation that had already been promulgated. In August 1941, Pétain asked Léon Bérard, Vichy's ambassador to the Vatican, to find out what the Holy See felt about the Vichy legislation on the Jews. Bérard dutifully asked around in Rome, and studied Catholic teachings on racism and antisemitism, reporting back on 2 September 1941. There did not seem much room for ambiguity in the finding that:

> There is a fundamental antithesis between Church doctrines and 'racist' theories . . . Every human being has an immortal soul which is upheld by the same grace and is summoned to the same salvation as all other souls . . . All these propositions are incompatible with an outlook which derives from the shape of the skull and the nature of the blood the aptitudes and vocations of peoples, their very religion itself, and finally sets up a hierarchy of races, at the apex of which appears a pure or royal race called 'Aryan'.

Nonetheless, the ambassador came to the conclusion that 'Nothing has been said at the Vatican that supposes either criticism or disapprobation on the part of the Holy See regarding the laws or regulations concerned.'

His unnamed interlocutors had allegedly reassured him that, since the Statutes did not make the Italian Fascist mistake of impinging on the sacrament of marriage, they had no objections in principle to the Vichy legislation, provided the measures were implemented 'according to the precepts of justice and charity'.[65] What Bérard claimed, was, of course, not necessarily what anyone had said, it being extremely suspicious that he did not seek to reassure Pétain by mentioning any big names in the Vatican, and that his account resembles an academic treatise, replete with discussions of Thomas Aquinas, rather than a diplomat's rehearsal of actual conversations. Perhaps he simply had a word with reactionary French clergy who in any event supported Vichy. In mid-September 1941, Pétain raised Bérard's findings with Valeri, the nuncio to France, at a diplomatic reception. The ambassadors of Brazil and Spain were within earshot, so Valeri was careful to ensure that Pétain did not ascribe views to the Church that it did not hold. In his report on these conversations to the Vatican, Valeri said, 'I reacted quite vigorously, especially because of those who were present. I stated that the Holy See had already expressed itself regarding racism, which is at the bottom of every measure taken against the Jews, and which, as a consequence, M. Bérard cannot explain in such simplistic fashion.' Pétain suggested that Valeri might be out of touch with how the Vatican regarded such questions and invited him to inspect Bérard's report. Writing to the secretary of state on 30 September, Valeri said, 'As you noted, the pro memoria is much more nuanced than the Marshal would have had me believe,' while in a note to Pétain, he observed: 'I call attention to the grave harm that, from a religious perspective, can result from the legislation now in force, a legislation which in other respects is rather confused.'[66]

The first mass round-ups by Vichy police and deportations commenced in early summer 1942 after the Germans had taken such steps as having Vallat replaced by the rabid antisemite Louis Darquier de Pellepoix, who liked rather than hated Germans, while René Bousquet was put in charge of a consolidated police force that could operate in both zones. The prescient among the Jews fled, or hurled themselves down stairwells and off the balconies of Parisian apartments. If they had no children these people were kept in a half-finished housing complex at Drancy in north-eastern Paris, while those with children were kept in an indoor stadium called the Vélodrome d'Hiver near the Eiffel Tower in the fifteenth arrondissement. If Drancy was a drear public-housing project, the interior of the 'Vel d'Hiv' was muted in a blue light – the glass cover had been camouflaged in that colour – and was unbearably close as the sun beat down on the roof. There was nowhere to wash and the lavatories were shut. From Drancy it was a short bus ride to the station at Le Bourget from where these people were transported to Auschwitz. Families with children under sixteen were removed from the Vel d'Hiv to Pithiviers and Beaune-la-Rolande, south-east of Paris. The parents and adolescent children were deported first, leaving 3,500 younger children virtually defenceless apart from the Red Cross.

Relocated to Drancy, one August dawn they were awoken and shipped to Auschwitz, where, after a three-day journey in sealed cattle cars, they were killed on arrival. In the unoccupied zone, Vichy officials selected foreign refugees from the internment camps like Gurs, and sent them to Drancy, whence they were returned to the hands of the Germans who had either deported them or from whom they had fled. These were killed in Auschwitz too. Renewed arrests of foreign Jews were designed to repopulate the internment camps for further deportations, the surrender of foreign Jews being the price Laval's government thought it was paying so that 'French Israelites' would not be affected.

The public nature of these arrests and deportations, which throughout involved French gendarmes in destroying families, carrying bewildered and fearful infants, or supervising elderly ladies as they dragged heavy suitcases a few yards at a time in the summer heat, provoked a reaction despite all the burdens that the occupation entailed. Almost from the start of the occupation there had been Christian underground news-papers, one of the first being *La Voix du Vatican* (Voice of the Vatican), which published what had been said on the French service of Vatican Radio by its lead broadcaster father Emmanuel Mistiaen. Evidently unaware of any official Vatican 'silence', Mistiaen frequently condemned any attempts to divide mankind into higher and lesser races.[67] In November 1941 a small group of Catholic theologians in Lyons produced a series of clandestine pamphlets called *Cahiers du Témoignage chrétien* (Christian Witness) with the unofficial understanding of cardinal Gerlier of Lyons. The moving spirit was father Pierre Chaillet who was simulta-neously involved in Amitié Chrétienne, an inter-faith group that forged tens of thousands of documents, hiding Jews and helping to smuggle them to Switzerland, activities in which the laymen Jean-Marie Soutou and Joseph Rovan took a distinguished part. A glimpse into the dangers of these activities can be had from the occasion when Chaillet faced interrogation by the Lyons Gestapo boss Klaus Barbie: standing with his face to a wall, Chaillet managed to chew and swallow highly incrimi-nating documents that he had hidden under his habit, which resulted in his being released after only a few random kicks and blows. He was subsequently placed under house arrest for three months in a mental hospital in the Ardèche.[68] The *Témoignages* were the necessary pendant to active rescue work since they combated antisemitism on a more intel-lectual and spiritual level. The first issue warned, 'France, take care not to lose your soul.' Issue after issue was devoted to denouncing antisemitism and racism in general, in line with the teachings of the Holy See. 'The Church cannot disinterest itself in the fate of man, wherever his inviolable rights are unjustly threatened. When one member [of the human race] suffers, the entire body suffers with him.'[69] The summer deportations coincided with the meeting in Paris of the archbishops and cardinals of the occupied zone. After the meeting, cardinal Suhard wrote to Pétain giving voice to 'an anguished cry for pity at this immense suffering; above all, for that which strikes so hard at mothers and children'. Following further deportations that August, which since they involved children, the old and infirm undermined the credibility of the official fiction that the Jews were going to work camps in Germany, both the Protestant Boegner and cardinal Gerlier registered their protests with Laval and Pétain. Gerlier had been informed by a representative of the chief rabbi that the Jews were not being sent to work camps in Poland, but were being killed in Germany, a misapprehension about the nature of the 'work camps' that indicates the difficulty of comprehending what was going on. Gradually members of the French hierarchy abandoned their reticence. Notably, the elderly and infirm Jules-Gérard Saliège, archbishop of Toulouse, issued a pastoral letter on 30 July which said:

'That children, women, fathers and mothers are treated like cattle, that members of one family are separated from each other and packed off to an unknown destination, it has been left until our time to witness such a sad spectacle. Why does the right of asylum no longer exist for our churches? Why are we defeated? ... The Jews and the foreigners are real men and women. Everything is not permitted against them, against these men and women, against these fathers and mothers. They are part of the human species. They are our brothers, like so many others ...

Saliège had an honourable record of denouncing Nazi racism, as when in 1933 he had categorically stated: 'Catholicism cannot agree that belonging to a specific race places men in a position of inferior rights.' In 1939 he had attacked 'the new heresy of Nazism, which shatters human unity and places a superhuman value in what it considers to be privileged blood'. Whatever the reasons for his intervention, which may have been prompted by a private warning from de Gaulle about the hierarchy's closeness to Vichy, or perhaps, a reflection of the fact that many intern- ment camps were in his archdiocese, a man who was so ill that he could neither speak nor hold a pencil caused shockwaves, not least because his words were broadcast twice on Vatican Radio and by the BBC. Laval was furious, coupling his suggestion that the archbishop be retired with the threat that deportations might encompass Jews sheltering in religious institutions. In September, eight Jesuits in Lyons were arrested for refusing to reveal the buildings where they were sheltering Jewish children. Pierre Marie Théas, bishop of Montauban, weighed in with an even more forthright condemnation, which was secretly hectographed and then distributed by clerics and a woman called Marie-Rose Gineste who bicycled throughout the diocese:

> I voice the indignant protest of the Christian conscience and I declare that all men, Aryan or non-Aryan, are brothers because created by the same God; that all men, regardless of race or religion, have the right to the respect of individuals and states. Now, the present antisemitic measures are contemptuous of human dignity, and a violation of the most sacred rights of the person and of the family.

In September 1942, the Vatican secretary of state Maglione summoned the Vichy ambassador to inform him that 'the conduct of the Vichy government towards Jews and foreign refugees was a gross infraction' of the Vichy government's own principles, and 'irreconcilable with the religious feelings which Marshal Pétain had so often invoked in his speeches'.[70] When Pétain at a lunch attended by nuncio Valeri tried to justify the round-ups by remarking 'the pope understands and approves my attitude', Valeri replied: 'The Holy Father neither understands nor approves.'

Rattled by these protests, which were broadcast by the French service of the BBC, Vichy adopted a twin-track response to the querulous prelates. Collaborationist newspapers launched crude personal attacks, while Pétain and Laval told the Church (and the pope) to keep their noses out of affairs of state. Simultaneously, the regime offered state subsidies to both Catholic and Protestant higher education and theology institutes. Despite repeated insinuations, there is no evidence that these blandishments were responsible for a cessation of episcopal protests against continued deportations; rather the protests focused on another area of Vichy policy, namely the introduction of compulsory labour service in Germany.

This did not mean a cessation of practical efforts to save Jewish people. Predominantly Protestant villages in and around Le Chambon-sur-

Lignon high on the Massif Central in south-central France managed to give sanctuary to five thousand Jews during the occupation. The inhabitants had collective memories of the persecution of the Huguenots, while the climate and topography (the area was cut off by snowdrifts in winter) facilitated such activities. Whenever the French police or Gestapo made an appearance, their vehicles could be seen from miles away, which enabled Jews to be dispersed into the deepest countryside. The police themselves were sometimes sympathetic to the victims, and informers were few and easily identifiable. There were similar instances of communal rescue in the Cévennes, although these involved both Catholics and Protestants. Representatives of a Jewish organisation that rescued children sought out archbishop Saliège of Toulouse, who provided them with a passe-partout which enabled them to hide Jewish children in Catholic institutions throughout the unoccupied zone. Similarly, the bishop of Nice, Paul Rémond, afforded every assistance to a Syrian Jew called Moussa Abadi, who ran a rescue service, including providing him with an office in his own residence where Abadi forged papers, and arranging dozens of hiding places in religious institutions along the Mediterranean coast. Catholic convents, monasteries and schools throughout France were deeply involved in hiding Jewish people throughout the occupation, including Lucie Dreyfus, the widow of Alfred Dreyfus, who survived the war as 'Madame Duteil' in a convent of the Sisters of the Good Shepherd in Valence. The people who undertook this work ran very grave risks. Lucien Bunel, or father Jacques, was director of a Carmelite boarding school near Fontainebleau. In 1943 he agreed to take five Jewish youths at the behest of mother Maria de Notre-Dame de Sion, whose own school could shelter girls but not boys. In January 1944, the Gestapo acted on information they had gained by torturing a former pupil of father Jacques, whom they had arrested for involvement in the Resistance. The Jewish boys were deported to Auschwitz, while father Jacques was sent to various camps, dying a month after his liberation from the ill-treatment he had received.[71]

Society in the Netherlands was organised into self-contained 'pillars' or *zuilen*, with their own political parties, press, unions, schools and universities. The Catholic pillar was the largest, including 30 per cent of the Dutch population, but it considered itself a minority next to three Protestant pillars and the two secular *zuilen* of liberals and Social Democrats. These centrifugal tendencies were partially countered by the widely admired monarchy, and by Holland's highly efficient civil service.

Ninety thousand of the 110,000 Dutch Jews lived in Amsterdam, where the majority eked out modest livelihoods. The arrival of over thirty thousand Jewish refugees from Hitler's Germany led to resentments, both among Dutch gentiles and among the indigenous Jewish community, which financed a government internment camp at Westerbork near the German border. It took five days for the Wehrmacht to overrun Holland. The government fled abroad, instructing senior civil servants to co-operate, within the bounds of the Constitution, with the German civil administration of Reich commissar Arthur Seyss-Inquart. The German objective was to win the co-operation of a people they regarded as racially cognate; that did not include the Jews, two thousand of whom were identified and dismissed from government employment. Various organisations took the hint, including Amsterdam's famous Concertgebouw orchestra, which moved Jewish musicians to the rear rows and ceased performing Mendelssohn.[72]

Jews were subjected to creeping restrictions designed to identify and isolate them from the rest of the population without causing upset. The smoothness of this process went awry when in February 1941 the outrages of Dutch Nazis encountered resistance from 'action groups' in the working-class Jewish quarter of Amsterdam. One of the Nazis was trampled to death. Shortly after, German uniformed police, under the SD chief in the city, surrounded the 'Koco' ice-cream parlour, which was owned by two German-Jewish refugees, who had been involved in the creation of the action groups. Mistaking them for Dutch Nazis, these two sprayed the Germans with ammonia gas, which resulted in the parlour being raked with gunfire and the arrest of the two Jewish men. This incident was an outrage that Himmler would not allow to pass. The following Sunday, six hundred German policemen raided the Jewish quarter, dragging out four hundred men, who after being badly beaten were sent to concentration camps where all of them perished. These raids were conducted in broad daylight, appalling many Dutch bystanders who shopped in the Jewish quarter. In response, the tiny Dutch Communist Party organised strikes, which were widely supported by non-Communists. These went on for two days, until the occupation of the streets by German police and SS prepared to shoot people brought this large-scale protest on behalf of the Jews to an end. Public protest achieved nothing, except to caution the Germans into acting more circumspectly.

All of the major Dutch Churches protested against the deportations of

Jews that commenced in July 1942. They were told that 'Christian Jews' baptised before 1 January 1941 would be exempt, although this proved not to be the case. In late July the Protestant and Catholic Churches resolved to read their protest letter from the pulpits. They were warned that this might have dire repercussions. While the Protestants tried to stop their pastors reading the message, the Catholic hierarchy positively encouraged priests to do so. The archbishop of Utrecht was among those who read out the following words:

> Ours is a time of great tribulations of which two are foremost: the sad destiny of the Jews and the plight of those deported for forced labour . . . all of us must be aware of the terrible sufferings which both of them have to undergo, due to no guilt of their own . . . we have learned with deep pain of the new dispositions which impose upon innocent Jewish men, women and children, the deportation into foreign lands . . . the incredible suffering which these measures cause to more than 10,000 people is in absolute opposition to the Divine Precepts of Justice and Charity . . . let us pray to God and for the intercession of Mary . . . that He may lend His strength to the people of Israel, so sorely tried in anguish and persecution.

As a result of this protest, on 1 and 2 August 1942, Catholic Jews were arrested and deported. This was said to be an act of revenge on the part of Seyss-Inquart against the Catholic bishops who had protested against the deportation not only of baptised Jews but of Jews in general. As his deputy Fritz Schmidt explained in a speech in August 1942:

> The representatives of Protestant and Roman Catholic Churches sent a protest requesting better treatment of the Jews. The Jews are Germany's most dangerous enemies. Dutchmen cannot defend themselves actively against them without considering the question through spectacles of silly humanitarian sentiment. Owing to the passive attitude of the Dutch we Germans have taken over the solution of the Jewish Question, and have begun sending Jews to the East . . . Everyone crossing the path which we consider right and necessary, or hindering us in the execution of our tasks, must, whatever his nationality, expect the same fate. In Catholic churches a document was read out criticising the anti-Jewish measures taken to safeguard our

struggle. It was apparently also read in Protestant churches in spite of the fact that the Protestant churches had announced that it was not intended to read it everywhere in public. Owing to these events, the Germans must consider the Roman Catholic Jews their worst enemies and arrange for their quickest possible transport to the East. This has already taken place.'[73]

Six hundred Jewish Catholics, including the convert nun and philosopher Edith Stein, were killed in Auschwitz within two weeks of the Catholic Church's intervention, a protest which had the prior sanction of the papacy. Because the main Hervormde Kerk withdrew its plans to read out the protest, it was able to secure exemptions for Jewish converts, moves to deport whom only occurred later in the war. Some of them survived; none of the Catholic Jews did.

Experiences such as this, and what had occurred when Vatican Radio broadcast reports of atrocities in Poland, were among the considerations that inhibited a forthright condemnation by Pius XII of Nazi persecution, not only of the Jews but also the Catholic Poles. As long as he did not know that the intention was to kill every Jewish man, woman and child in Europe – and that intention was not clear at the start – then the desire not to make matters worse may have been a crucial consideration. It is easy, with hindsight, to object that matters could not have been much worse, but this is an utterly unhistorical approach to events that for Pius were either in the present or in the future rather than sixty years in the past. The specific fate of the Jews in Nazi-dominated Europe emerged fitfully from a broader pattern of atrocities, especially the German shooting of hostages in reprisal for acts of resistance. It took time for facts to be filtered from improbable rumours; for bits and pieces of information to be verified, and construed as symptomatic of a pathology. The policy of deportations was carried out opportunistically as well as relentlessly, with complicated exemptions muting consciences. 'They deplore the fact that the Pope does not speak,' Pius told the Jesuit rector of the Gregorian university in December 1942. 'But the Pope cannot speak. If he spoke, things would be worse.' In June 1943 he gave the College of Cardinals a rare insight into his terrible dilemmas: 'Every single word in Our statements addressed to the competent authorities, and every one of Our public utterances, has had to be weighed and pondered by Us with deep gravity, in the very interest of those who are suffering, so as not to render their position even more difficult and unbearable than before, be it unwittingly and unintentionally.'

All Allied governments, faced with fighting a desperate war, and bombarded by every conceivable group arguing its unique victimhood, were frustrated, sometimes to the point of callousness, by what they took to be Jewish special pleading. Stalin solved the problem by entirely ignoring the Nazis' murderous assault on Jewry. The Soviet Union suppressed all reports of what had happened to the Jews in Nazi-occupied Poland, thereby leaving Soviet Jewry in total ignorance of the fate likely to befall them, and then deliberately downplayed the extent and specific nature of the Nazis' racial rampage once it had been extended to the Soviet Union's own territories. Stalin made one passing reference to the Jews in all the public speeches he delivered throughout the war.[74] In some accounts, US actions are made to seem far more resolute than they were, so as artificially to contrast them with the irresolution of the Vatican, as they similarly tried to extract facts from rumours about human beings converted into fertiliser or soap.[75] The US State Department's response to the Riegner Telegram in August 1942, relaying high-grade intelligence from a German industrialist of a conspiracy to murder Europe's Jews, was to prevaricate for four months, until independent confirmation of the initial intelligence made a joint Allied condemnation on 17 December of 'this bestial policy of cold-blooded extermination' unavoidable.[76]

The Vatican received similarly patchy information from its own diplomats in neutral and occupied Europe, as well as from Italians returning from the eastern theatre, but by the spring of 1942 the full scope of policies that had only been definitively determined by the Germans that January became apparent. On 9 March Burzio in Bratislava wrote that the projected Slovakian deportation of eighty thousand Jews into German custody in Poland meant 'certain death' for the majority of them. On 10 March, nuncio Bernardini in Berne wrote on behalf of the Orthodox Jewish Agudas Israel, which had received no practical assistance from Jews in the US or Britain, urging the Holy Father to intervene in Slovakia.[77] Two days later Burzio reported that, according to a Slovak military chaplain returned from Russia, the SS were taking Jews from their homes who were 'slain with bursts of machine-gun fire'.[78] On 13 March, nuncio Rotta in Budapest forwarded to the Vatican a plea from representatives of Slovak Jewry, who bereft of money and possessions were facing 'certain downfall and starvation'.[79] Following further contacts with Jewish organisations in Geneva, on 19 March Bernardini warned that these policies were affecting Jews across central and eastern Europe. The accompanying memorandum was a detailed tour d'horizon of the fate of Jews throughout Europe, although the reference to countries countenancing 'even the physical extermination of the Jews' suggests that the Jews themselves had not grasped the magnitude of what was under way.[80] On 9 April 1942, these same personalities (Riegner and Lichtheim) asked Bernardini to thank the Holy See for its efforts, unaware that the Vatican's interventions in Slovakia had had no effect.[81]

As the summer arrived, reports of pervasive Nazi atrocities quickened, and the grim reality slowly dawned that these reflected a coherent policy aimed at the entire Jewish population of Europe. It took time to bridge the chasm between knowing and believing, and to separate fact from modern warfare's profusion of atrocity tales. In late June, the *Daily Telegraph* was the first British newspaper to break the news that seven hundred thousand Polish Jews had been killed and to mention both a specific killing centre and the use of toxic gas. D'Arcy Osborne transcribed BBC news reports for the pope each night. On 27 June he recorded: 'It is announced that since October 1939 the Germans have killed 700,000 Jews in Poland as part of their deliberate extermination policy ... mass shooting, drowning, gas.' On 30 June: 'The Germans have killed over a million Jews in all, of whom 700,000 in Poland. Seven million more have been deported or confined to concentration camps,' this last figure being prospective at that juncture.[82] In July Orsenigo reported from Berlin that 'the situation of the Jews excludes charitable interventions' and he had been warned that 'the less he talked about the Jews, the better it would be'. The complete suppression of news regarding where deportees were going only added to the confusion:

> As can easily be imagined, this suppression of news leaves the door open to the most macabre suppositions about the fate of the non-Aryans. Unfortunately there are all sorts of rumours, which are difficult to verify, about disastrous journeys and even of mass killings of Jews. Every intervention, even in favour of Catholic non-Aryans, has thus far been rejected, with the usual argument that the water of baptism does not change Jewish blood.

Within the Vatican, foreign diplomats took up the suggestion of the Brazilian ambassador Accioly that they make a joint démarche to Maglione insisting that the pope speak out. In the autumn, the Germans demanded that the Italians hand over the Jews in their section of occupied Croatia, causing consternation in both Italian government and military circles that the Vatican could not fail to have picked up. While reading out this extraordinary request, a member of the German embassy in Rome quietly added: 'This would mean, in practice, their dispersion and complete elimination.' In his submission, Osborne wrote:

> It may be objected that His Holiness has already publicly denounced moral crimes arising out of the war. But such occasional declarations in general terms do not have the lasting force and validity that, in the timeless atmosphere of the Vatican, they might perhaps be expected to retain ... A policy of silence in regard to such offences against the conscience of the world must necessarily involve a renunciation of moral leadership and a consequent atrophy of the influence and authority of the Vatican; and it is upon the maintenance and assertion of such authority that must depend any prospect of a Papal contribution to the reestablishment of world peace.

On 29–31 August 1942 metropolitan Szeptycki of Lwów informed Pius that 'the number of Jews killed in our little country has certainly exceeded two hundred thousand' and that the numbers of victims increased as German forces had moved eastwards. Thirty thousand men, women and children had been killed in Kiev alone, while similar massacres had been occurring in the smaller towns of the Ukraine for over a year. Szeptycki added that, compared with the Soviet regime, the Nazis were 'almost diabolical', resembling furies or ravening wolves.[83] Further pressure came from Myron Taylor, who visited the Vatican for several days of talks in September 1942. His brief was to persuade the pope of US resolve to win the war and to counsel him against attempts to achieve a compromise peace. He also revealed the US's extensive plans for the reconstruction of post-war Europe, and endeavoured to persuade the Vatican that the Soviets should be admitted to the European family. Tardini acidly replied: 'Stalin would not be suitable as the member of any family.' Nazi atrocities were among six headings for discussion. In conversations with Maglione, Taylor suggested that the pope condemn them. While he was still in Rome, Taylor was sent eyewitness accounts of the liquidation of the Warsaw ghetto, which he reported to Maglione. Their references to 'butchery' were slightly undermined by such lurid imaginings as 'Their corpses are utilized for making fats and their bones for the manufacture of fertiliser. Corpses are even being exhumed for these purposes.' Tittmann summarised the arguments the Holy See used to avoid making too specific a statement. The pope was being constantly pressed by each side to condemn the other's atrocities. Before condemning anyone, the Vatican would have to investigate each claim, which would soon become a full-time occupation, given that claims of atrocities were coming fast and furious. Reports of atrocities were integral to each belligerent's propaganda campaigns. What if the pope spoke out against something that turned out to be untrue? He had already condemned wartime atrocities in general terms that were consistent with the very long view that the Church had of earthly affairs, it being obvious to whom these condemnations were addressed. Local hierarchies had also spoken out, with knowledge of local circumstances, and with the pope's authority. It was all very well for exiled governments to want the pope to speak out, but were they aware of the vengeance the Nazis would wreak on Catholics in the countries concerned?[84]

Still the pressure mounted. There was a major meeting in London to protest against the killing of the Jews, while in November 1942 the US Catholic bishops issued a statement that, along with Hinsley's contemporaneous protests, rather militates against the supposition of universal Catholic judaeophobia: 'We feel a deep sense of revulsion against the cruel indignities heaped upon Jews in conquered countries and upon defenceless peoples not of our faith . . . Deeply moved by the arrest and maltreatment of the Jews, we cannot stifle the cry of conscience. In the name of humanity and Christian principles, our voice is raised.'

On 8 December 1942 cardinal Hinsley said from his cathedral's pulpit: 'Poland has witnessed acts of such savage race hatred that it appears fiendishly planned to be turned into a vast cemetery of the Jewish population of Europe.'[85] Although there are reports that when Pius heard of the mass murders he 'cried like a child', when he spoke out the emotional content was muted.[86] In his Christmas address, he appealed to mankind to return to the rule of God, making a solemn vow to the dead on battlefields, mothers, widows and orphans, exiles, 'and the hundreds of thousands of innocent people put to death or doomed to slow extinction, sometimes merely because of their race or descent', and those civilians whose lives had been destroyed by bombing. Many, including Osborne and Tittmann, found this part of the address too periphrastic, although the Germans said it 'was one long attack on everything we stand for . . . God, he says, regards all peoples and races as worthy of the same consideration. Here he is clearly speaking on behalf of the Jews . . . He is virtually accusing the German people of injustice towards the Jews, and makes himself a mouthpiece of the Jewish war criminals.' The *New York Times* devoted a long editorial to the Christmas message of 'this lonely voice crying out of the silence of a continent'. The address was 'like a verdict in a high court of justice'.

During this same period it was not Pius XII's 'silence' about the Jews that occasioned adverse comment – in fact contemporary Jewish leaders were highly appreciative of his endeavours – but his apparent unwillingness to say anything about the plight of the Poles. Before the war his efforts to enjoin reasonableness on the Poles vis-à-vis what were then Hitler's limited territorial demands led to suspicions that Pius was biased in favour of the Germans. He had then halted Vatican Radio broadcasts regarding German atrocities in Poland, although as we have seen this was at the behest of the Polish bishops. In November 1941 archbishop Adam Sapieha of Cracow urged the pope to speak out so as to bolster Polish morale. He was told that Pius preferred to work through diplomatic channels; a few months later Sapieha himself suppressed a papal message to the Polish bishops because of the possibility of German reprisals against the clergy and laity. He handed a letter detailing the horrors of concentration camps to the abbé Scavizzi, an almoner on an Italian hospital train, to deliver to the Vatican. He immediately thought better of doing so, and contacted Scavizzi telling him to destroy the letter, 'lest it should fall into the hands of the Germans, who will then shoot all the bishops and perhaps many others'.[87] Exiled Polish religious and political leaders were the least understanding of Pius' position. On 14 September 1942 bishop Karol Radoński of Włocławek, exiled in London, complained that 'the Pope is silent, as if he cared nothing for his flock'. When Maglione criticised Radoński for presuming to tell the pope what to do, Radoński brought up nuncio Valeri's conversations with Pétain regarding the pope's responses to the deportation of the Jews, arguing 'Are we less deserving than the Jews?' The exiled Polish president was similarly dissatisfied with the pope's public response to the plight of the Poles. Writing to Pius on 2 January 1943, Władysław Raczkiewicz said that Poles 'implore that a voice be raised to show clearly and plainly where the evil lies and to condemn those in the service of evil . . . the Apostolic See must break silence so that those who die, without benefit of religion, in defence of their faith, and their traditions may receive the blessing of the successor of Christ'.

Across Europe the military victories of Hitler, and Mussolini, encouraged the creation of new states, and the coming to power of either reactionary elites, as at Vichy, or of those one can safely designate as Fascists, although there was considerable fluidity, as well as conflict, between the two groups.[88] Some of these new states were wartime versions of the Catholic polities created by Dollfuss or Salazar in the 1930s, albeit in countries where antisemitism and intra-ethnic and religious conflict were more evident than in Portugal. Elsewhere, authoritarian regimes prevailed, while in Orthodox Romania a uniquely sinister fusion of religious extremism and political fanaticism illustrated what would really happen if a Christian Church threw its weight behind a war of racial exter-mination, although the Muslim grand mufti of Jerusalem, Haji Amin al-Husseini, a relative and mentor to Yasser Arafat, provides a more spectacular example of what such a religious leader would be like.[89]

Historically, strong rulers, such as Elizabeth I of England and Napoleon, have simply sloughed off even the papacy's ultimate sanction, which is why excommunication has never been essayed in the modern era. While it was impossible for the Vatican to exert any influence wherever the Nazis held direct sway, in some of these avowedly con-fessional states its interventions did delay or postpone the inevitable, with hundreds of thousands of lives saved in the process. The Vatican sought to influence policy in these states, especially if it had capable and dogged nuncios in situ, although these representatives also discovered that where nationalist passions were aroused they could not be sure of even influencing their own clergy.[90]

An independent Slovakian state resulted from the conjunction of Slovak separatism with Nazi imperial ambitions. After incorporating much of western Czechoslovakia as the Protectorate of Bohemia and Moravia, Hitler encouraged the creation in the spring of 1939 of an 'independent' Slovak state under German protection. The Catholic nationalist Slovak People's Party was renamed the Party of National Unity. It had an uneasy relationship with the paramilitary Hlinka Guard, in which militant fascists were more generously represented. In October 1939 Dr Jozef Tiso, a Catholic priest and theologian, became president of the self-proclaimed 'Christian national community'. A month earlier

Pius XII had touched on this very problem in conversation with a group of German pilgrims: 'For a priest, it is now more than ever before imperative to be wholly above all political and national passion; to console, to comfort, to help, to call to prayer and to penance, and himself to pray and to do penance.'[91] When Tiso was appointed, Pius said that he regarded this as 'inexpedient'. The Holy See was not pleased at the direct involvement of a priest at this level of politics, although it seems to have done remarkably little about it, or regarding the sixteen further priests who served in the State Council at Bratislava. The new regime, which was ostentatiously Catholic, set out to reverse the laicising measures of the Czechoslovak government in the inter-war period, denying legal recognition to the Protestant minority and reinserting religion into the education system at all levels. After a meeting with Hitler at Salzburg on 28 July 1940, the Slovak leadership resolved to set up a Nazi-style regime in Slovakia, including state-supported 'Aryanisation'. Most embarrassingly for the Church, on 7 September 1941 Tiso used the consecration of a church to announce that there was nothing incompatible about the social doctrines of the Church and Nazism. The Holy See contemplated removing Tiso's clerical status, but then did not pursue this.[92] While Tiso's popularity ensured that Hitler left him in power, he also exerted pressure on him through the more radical rightists Vojtech Tuka and Šaňo Mach, who had become foreign and interior ministers. This set-up may explain Tiso's decision to send Slovak forces to fight in Russia and his involvement in the developing 'Final Solution'. It was the price for his ascendancy over Slovak Fascists who were constantly jockeying to oust the clerical–conservative leadership of the Party, and the surest way of consolidating Slovakian statehood in the light of the prevailing values of Hitler's 'new order'.

In September 1941 Tuka and Mach implemented a Jewish Codex, which in terms of comprehensiveness outdid the antisemitic legislation that Tiso himself had already introduced. One of its consequences was the Aryanisation of over ten thousand Jewish-owned businesses, and the expulsion of fifteen thousand Jews from the capital. Giuseppe Burzio, the papal chargé d'affaires at the nunciature in Bratislava, telegraphed news of this Codex to the Vatican. The secretary of state, Maglione, sent a protest note to Karol Sidor, the Slovak emissary at the Holy See, expressing his 'lively sorrow' regarding legislation 'containing various provisions directly opposed to Catholic principles'.[93] After six months, the Slovak regime responded that, since the Slovak Jews were about to be deported, the issues raised by the Vatican were no longer pertinent. In June 1940 it had promised to supply Germany with 120,000 workers. By October 1941 there was a shortfall of forty thousand. The Slovaks offered ten to twenty thousand Jews instead. In early 1942 an agreement was reached whereby the Germans would take the Jews in return for payment of five hundred Reichsmarks to cover the cost of each individual's deportation. The Germans promised not to claim the property they left in Slovakia. The deportation trains began to roll in late March, allegedly delivering the Jews to work camps in the Lublin area of Poland. The Jews were mustered in a disused factory in Bratislava, and then transported by Slovak Hlinka Guards to six different transit camps near the border. From there they were shipped to Auschwitz, Belzec, Majdanek or Sobibor, where the majority were killed on arrival. Of the fifty-eight thousand Jews deported, six to eight hundred survived the war.

On learning of these deportations, on 28 February Burzio sought out Tuka. The latter 'vehemently defended the legitimacy of the measure and dared to say (he who paraded himself as a Catholic) that there was nothing inhumane and anti-Christian with this. Deporting eighty thousand people to Poland where they will be at the mercy of the Germans is the equivalent of condemning most of them to a certain death.' Both the premier's defensive remarks and the scepticism of Burzio suggest that this exchange was not a happy one. On 14 March 1942 Maglione summoned Sidor, the Slovak representative, to protest that he 'cannot believe that a country intending to be inspired by Catholic principles will take such grave measures which will produce such harmful consequences for so many families'. On 24 March Sidor was summoned again, and told on the direct authority of Pius XII to take immediate action with his government to halt the deportations.[94] The deportations continued, so Maglione instructed Burzio to seek out Tuka again. He added the thought: 'I do not know whether these steps will succeed in stopping . . . the lunatics. There are two of them. The first is Tuka who does things, and then there is Tiso . . . the priest who allows them to happen.' In April, Maglione had an interview with Sidor, who tried to justify the deportations, claiming that the Jews were part of a 'labour conscription' scheme. Maglione angrily brushed this aside: 'I told him that such actions were a disgrace, especially for a Catholic country.' At the end of April 1942, the Slovak bishops circulated a pastoral letter that unequivocally stated, 'The Jews are also people and consequently should be treated in a humane fashion,' although they also accepted the legitimacy of measures designed to curb the Jews' 'nefarious influence'. In May the Slovak parliament passed a law that retroactively legitimised the deportations, there being no dissent from any of the clerical deputies in the ruling party. Word of what was happening to the Slovak Jews in the Lublin region filtered back, and caused unease among the Slovaks.

The Vatican's intervention through Burzio, and the commotions it caused, induced the Slovak and SS authorities to halt deportations between October 1942 and September 1944, an unparalleled occurrence in the history of the Holocaust.[95] It was only a temporary respite since on 7 February Interior Minister Šaňo Mach vowed: 'March will come, April will come, and the transports will roll again!' When the Catholic bishops got wind that the deportations were going to resume, they issued a pastoral letter, which even father Tiso was obliged to read out in the parish he continued to occupy. The bishops rejected the notion of collective guilt that the regime used to justify the deportations. They cited the Slovak Constitution's guarantees of liberty 'without regard to ethnic origin, nationality or religion'. They refused to accept a distinction between Jewish converts to Catholicism and Jews and retold the parable of the Good Samaritan. On 7 April monsignor Burzio had a fractious meeting with the 'cynical Pharisee' Tuka. Ignoring Tuka's 'offensive and vulgar' responses, Burzio told him: 'Your Excellency is certainly aware of the sad news that is being spread regarding the atrocious fate of the Jews who have been deported to Poland and the Ukraine. Everyone is talking about it.' In May, Maglione personally protested to the Slovak government: 'The Holy See would shirk its divine mandate if it did not deplore these arrangements and measures which gravely strike at people's natural rights from the simple fact that these people belong to a specific race.' As a result of these protests, the regime promised the Vatican to limit deportations to Jews who 'endangered the State', while confining the rest either to labour camps in Slovakia or allowing them to continue with their professions. They offered to send a mission to inspect the condition of 'ex-Slovak citizens now in Poland', although that would have been difficult since most of them were dead. Indignant at Burzio's importunities, Tuka insisted that he would visit the pope in the Vatican to put him straight about the 'Jewish Problem'. His ambassador at the Vatican thought this ill advised because at a recent audience with prime minister Kállay of Hungary the pope had 'condemned the system and methods of the Germans, which independently of the war were inhuman and brutal especially towards the Jews, but also towards their own race' and had thanked the Hungarian for 'keeping Hungary from such inhumanity'.[96]

Tragically, in the autumn of 1944 the respite such protests had brought the Jews was nullified when the Germans effectively took over Slovakia after an uprising, and began deporting even those Jews who had been baptised. Burzio was joined by Roncalli, the nuncio in Istanbul, in protesting these measures, while Maglione relayed to Tiso the holy father's 'profound grief', warning him that such injustice would damage the Church itself. On 8 November 'Dr Josef Tiso *sacerdos*' sent a handwritten letter in Latin to Pius XII, defending the five-year record of his government. It sought to justify the measures taken against Czechs and Jews, while he recognised no incongruity between his conduct and his sacerdotal status. The pope saw this missive; there was no reply.[97] The reasons why were already evident in a memorandum on Slovakia written by under-secretary Tardini in July 1942: 'It is a great misfortune that the President of Slovakia is a priest. Everyone knows that the Holy See cannot bring Hitler to heel. But who will understand that we cannot even control a priest?'[98]

The creation of a demonstratively Catholic country amid the debris of Royal Yugoslavia also challenged the Church, especially since fault lines between Catholicism, Islam and Orthodoxy ran through it, and religion was integral to national identities. Axis forces invaded Yugoslavia in April 1939, with Belgrade exposed to a ferocious bombardment. German agents encouraged a local uprising on behalf of the exiled Croatian Fascist leader, Ante Pavelić, a lawyer and former deputy to the Belgrade parliament, who in 1929 had created the Ustashe (the Insurgency-Croatian Revolutionary Organisation, to give its full title). He and his entourage of terrorists immediately returned from Italy. The Ustashe's founding principles consisted of such assertions as 'the Croatian nation belongs to Western culture and to Western civilisation'. That sat ill with a crude nativism claiming that anyone not descended from a 'peasant family is not a Croat at all, but a foreign immigrant'. Like the Poles, the Croats viewed themselves as Catholicism's outer rampart, conveniently forgetting that the (schismatic) Serbs had fought the Muslim Ottoman Turks for many years at the behest of the Catholic Habsburgs. They also claimed they were wandering Goths, which served to elevate them above the 'slave-Serbs'. In practice they were nationalist terrorists who in the late 1920s undertook a terror campaign in Yugoslavia, to which the royal Yugoslav government responded with an assassination campaign of its own.

Pavelić had fled abroad after receiving a death sentence in absentia for urging the overthrow of the Yugoslav government. In 1934 he employed Macedonian terrorists to assassinate king Alexander (and the French foreign minister Louis Barthou) in Marseilles, which led Mussolini to imprison him, while closing the training camps of his followers who were interned on the island of Lipari. The Ustashe, who in the mid-1930s became more and more identified with Italian Fascism and National Socialism, rejected all attempts by Belgrade to afford Croatia a measure of autonomy, distancing themselves from the moderate and democratic Croatian Peasant Party.

The 'poglavnik' or 'leader' was installed as the head of the Independent State of Croatia or NDH after Vladko Maček, the Peasant Party leader, had refused Hitler's offer of the role of leader of independent Croatia. The Ustashe simply absorbed the Peasant Party, massively inflating its own numbers in the process. Independent Croatia was subject to joint German and Italian occupation, save for the parts of coastal Dalmatia that Mussolini directly annexed. In return, the Croat state was ceded most of Bosnia–Herzegovina, where its hatred of the Serbs resonated with local Muslims.[99]

The contemporaneous celebration of the thirteen-hundredth anniversary of Croatia's first links with the papacy led the primate of Croatia, archbishop Alojzije Stepinac, to detect 'the hand of God at work' in his country's deliverance from what many Croats regarded as alien Serb Orthodox tutelage, and a period in which Catholic Croats had been disadvantaged in terms of access to government appointments. In a pastoral letter dated 28 April 1941 Stepinac enjoined priests 'to fill the Poglavnik of the State of Croatia with the spirit of wisdom, so that he may perform the elevated and so responsible service to the honour of God and to the salvation of the people in justice and truth; so that the Croatian nation becomes the Divine nation, loyal to Christ and his Church built from Peter's cave'. He would maintain that loyalty to an independent Croatian state, even if he was increasingly critical of the actions of what he chose to regard as its wilder supporters.

By the end of April 1941 the regime had issued decrees designed to 'protect Aryan blood and the honour of the Croat people'. One of its first acts was to abrogate all constitutional and legal provisions that granted religious equality and freedom of conscience. The Croat Church warmed to prohibitions on abortion, contraceptives, freemasonry, pornography and swearing, and the introduction of compulsory Sunday observance.[100] The Ustashe state also assisted confessional schools and seminaries, and gave financial help to Catholic charities. But this was hardly de Valera's Ireland, even if many Ustashe fugitives would seek refuge there after the war.[101] It also welcomed the suppression of Cyrillic, the closure of Orthodox primary schools, and the expropriation of the Orthodox patriarchate, all measures that struck deep at Serb identity.[102] The response of the Holy See was more cautious. In keeping with its policy of withholding recognition of new states until the end of a war, the Vatican sent the Benedictine Ramiro Marcone as its representative to the Croatian hierarchy rather than to the Croat state. In return, Croatia was allowed to maintain an unofficial representative at the Vatican, who in contrast to the ambassador of Yugoslavia was kept at arm's length by the secretary of state. These carefully calibrated arrangements soon fell apart.[103]

On 18 May 1941 Pavelić was granted a half-hour private audience with the pope, the substance of which is unknown. It was carefully choreographed so as to exclude Pavelić's entourage. Given Pavelić's record, and the violence that had already occurred in Croatia, this meeting was met with widespread incomprehension. The Yugoslav legation sent two protest notes, one chronicling the persecution of Serbs by the Croats. The British foreign secretary complained to the apostolic delegate to Britain: 'I am much disturbed by this reception, and cannot accept the Vatican's description of Mr Pavelitsch [sic] as a statesman. In my view, he is a regicide. It is incredible that His Holiness should receive such a man.' It seems difficult to dissent from that view, or to understand why the Vatican simply accepted Pavelić's denials of involvement in the Marseilles murders.[104]

The spiral of violence unleashed from April 1941 onwards complicates any discussion of events in Croatia. Communist partisans and Serb Chetniks fought the Axis or the forces of the Serb collaborator Milan Nedić, finding time, between killing one another, to battle the Croatian Ustashe, who found themselves fighting other Croats. After the war, the victorious Yugoslav Communists tried comprehensively to impugn the Catholic Church for its recent record of collaboration with Pavelić. These hatreds have survived the collapse of Yugoslavia, since charge and counter-charge are littered across the internet, including such curiosities as a loyalist pro-Serb and anti-Croat website in distant Belfast.

The Ustashe regime was immediately and comprehensively vicious, as one might expect since (Catholic) terrorists and murderers were generously represented among its leading lights. The new education and culture minister, Mile Budak, set the grim tone when he warned: 'For minorities such as the Serbs, Jews and Gypsies, we have three million bullets.' The intention was to force half the Serb population to convert to Catholicism, thereby losing what was most integral to their national identity, and to deport or kill the remainder. As for the Jews, they would be killed by the Ustashe or handed over to the Germans for liquidation. From late April 1941 onwards the NDH introduced Aryanisation measures that discriminated against the country's Jews, up to and including the compulsory display of the Yellow Star of David. Even critics of the Catholic Church concede that Stepinac was quick to protest these last measures. On 22 May 1941 he wrote to interior minister Artuković requesting that the regime drop its insistence that Jews wear this badge, and insisting that 'the principles of their human dignity be preserved'. He added that 'the Holy See does not regard these laws with favour', especially since Pius XII had recently granted Pavelić a private audience. This was coupled with a clear warning that the Holy See would withhold diplomatic recognition from the fledgling state. Stepinac subsequently requested that deportees receive medical care and be kept in touch with their families.

On 3 May 1941, Pavelić decreed that conversion from one religion to another would require the permission of the civil authorities, who charged exorbitant fees for the privilege. On 15 May the chancellery of the archdiocese of Zagreb issued detailed guidelines to the clergy, insisting that the motives of each individual convert to Catholicism should be carefully explored, at a time when entire villages were rushing to convert to avoid Ustashe murder squads. In July the Franciscan responsible for religious affairs in the Ustashe government forbade the bishops to convert members of the Serb intelligentsia or Orthodox priests and seminarists. Stepinac protested against these incursions by the civil power into a domain that he regarded as the Church's own.[105]

Ustashe murders had all the vicious trademarks of intra-communal violence, resembling what the Einsatzgruppen were inciting in the Baltic States. It was a matter of knives and hatchets, and all the usual cruelties of the peasant imagination. It appalled Italian and some German onlookers, like general Dankelmann, to whom the Serb Orthodox authorities in Belgrade protested, after refugees had told them of what was happening in Croatia. Some senior Catholic clergy were also shaken, as well they might be, since it is estimated that the Ustashe killed three hundred thousand Serbs. In August 1941, bishop Alojzije Mišić of Mostar wrote to Stepinac, informing him of the seizure and murder of newly converted Serbs, with men, women and children thrown alive off cliffs or shot dead on the edges of enormous pits. He forbade his clergy to give absolution to anyone involved in these murders. Worst of all, Stepinac received unambiguous evidence that clergy, including Franciscan friars, were participants in some of the worst atrocities, being thick on the ground at the notorious Jasenovac concentration camp. The clergy were demi-educated peasants, who shared many of the communal hatreds of their fellow citizens, while many of the Franciscans had undergone a novitiate in the vicinity of Pavelić's erstwhile headquarters in Siena and had gone native as military chaplains to the Ustashe. The worst offender, father Filipović, was expelled from the Franciscans in 1942.

From 17 to 20 November 1941 Stepinac convened a national synod to discuss the issue of forced conversions. The bishops condemned the Ustashe's arrogation of the right to convert people to Catholicism and urged that the rights of the Orthodox Church be respected. They registered their disapproval in a letter to Pavelić, albeit dissociating him from the actions of 'irresponsible' subordinates in the Ustashe, while highlighting past experiences of 'artificial' conversions in the Byzantine Empire or early modern Spain. In a further letter to Pavelić, Stepinac warned that forced conversions were so damaging the Catholic cause that the Serbs might convert en masse to Islam instead: 'Precisely for this reason I think it is necessary to choose with special care the missionaries who are to be sent among the Serbs and not to entrust this charge to priests or religious who are not prudent and in whose hands a revolver might better be placed than a crucifix.'[106]

A three-man committee of bishops was formed to monitor conversions, but it was completely impotent in the face of Ustashe violence, which simply continued as if the bishops had never deliberated. Stepinac intervened more forcefully in the case of the Jews. In a letter to Pius XII he said: 'The Bishops' Synod discussed the affairs of those who suffer today and sent a letter to the head of state, in which it demands that he treat the Jews in a humane manner as far as possible, considering the presence of the Germans.' In early 1942, Stepinac heard rumours that the Ustashe authorities were conspiring with the Germans to deport those Jews who had survived the initial onslaught. He wrote to Artuković saying:

> If indeed such a course is being planned I take the liberty to appeal to you to prevent by virtue of your authority an unlawful attack on citizens who are not personally guilty of anything. I believe it will damage our good name if it becomes known that we have solved the Jewish question radically, i.e. in an extremely rough manner. The solution of the problem must refer only to crimes committed by Jews – meaning that innocent people must not be persecuted.

Given that Stepinac was writing to a mass murderer, it should not entirely surprise that he sought to avoid causing the Ustashe to lose face, by conceding the existence of a Jewish question or that some Jews were 'guilty' of unspecified offences. The reference to 'our good name' is also significant, as it shows that Stepinac's approach was to identify himself with Croatian nationalism so as to make his protest more effective. He also instructed the clergy:

> When people of Jewish or Orthodox faith who are in danger of death and wish to convert to Catholicism present themselves to you, receive them in order to save their lives. Do not require any special religious knowledge for Orthodox are Christians like us and the Jewish faith is the one from which Christianity originated. The role and task of Christians is first of all to save people. When these sad and savage times have passed those who converted because of belief will remain in our church and the others will return to their own when the danger is over.[107]

By April 1942 Stepinac was in contact with a Special Operations Executive agent representing the exiled Yugoslav government. The archbishop explained that although he could have withdrawn to a monastery, receiving plaudits for his defiance after the war, he felt he could help people most by staying at his post where he could exert influence. At the end of that month he went to Rome and had an hour-long audience with the pope to discuss events in Croatia. There can be no question that the Vatican was not thoroughly informed about what was happening there, for large numbers of Italian military personnel (and clergy) went too and fro across unimpeded land and sea borders. The Croatian representative to the Holy See, Rusinović, also reported to Zagreb that the Vatican was in possession of eight thousand photographs of Croat atrocities against the Serbs and that cardinal Tisserant, responsible for the Eastern Churches, had denounced the killing of 'three hundred and fifty thousand Serbs'. Shortly after this, Stepinac publicly attacked Ustashe atrocities: 'All races and nations were created in the image of God . . . therefore the Church criticised in the past and does so in the present all deeds of injustice or violence, perpetrated in the name of class, race or nationality. It is forbidden to exterminate Gypsies and Jews because they are said to belong to an inferior race.' A week later he used a sermon to remind Croats that 'The continuation of the proper order demands the proper treatment of neighbours; i.e. to treat men as God's creatures just like ourselves, and not as wild beasts.'

The deportation of those Jews the Ustashe had failed to kill began in mid-August 1942. Stepinac advised the chief rabbi to write an urgent letter to the pope. Rabbi Freiberger spoke of 'my and my community's deep gratitude for the sympathetic attitude of the Holy See's representatives and the leaders of the Church towards my unfortunate brothers'. On the ground, Marcone intervened with the Croatian authorities to save 'all those of mixed families, Catholics and non-Catholics', evidently with some success.[108] Although the Croatian regime could always deflect protests by the Catholic Church by claiming that the deportations were imposed by the Nazis, it was unsettled by what it claimed was the 'disloyalty' of the senior clergy. In November 1942 Pavelić sent the 'loyal' military vicar-general to speak with Stepinac about his private and public criticism of the regime. Stepinac replied: 'the Church obeys the laws of God. Cecelja [the vicar-general] may tell his government that the Church would continue to criticise terror acts against the population. The Croatian government will have to bear full responsibility for the growth of the Communist partisan movement . . . because of severe and unlawful measures employed against Orthodox Serbs, Jews and Gypsies, in imitation of German methods.' This was both accurate and prophetic, not least for Stepinac himself, who would be jailed by the victorious Communists.[109]

In early 1943 the Germans refocused on Croatia's converted Jews, or those hitherto protected by mixed marriages. On 6 March Stepinac wrote to Pavelić without leaving any room for ambiguity or equivocation:

Poglavnik, a great deal of panic has broken out in Zagreb and in provincial towns following the ordinance to register every non-Aryan. People fear that even couples, married lawfully in church, will be separated . . . I hereby proclaim that such proceedings are stark violence of which no good will come . . . how can a rational man believe that thousands of [Catholics] belonging to mixed couples will remain silent, while their beloved are being violently exterminated and their children exposed to an unknown fate? . . . As representative of the Church I ask you once again to issue instructions – to preserve the civil rights of everyone who was converted. I also appeal to you in the name of humane feelings to prevent that harm be done to innocent citizens of our state . . . I am convinced that these deeds were carried out without your knowledge, by irresponsible people, motivated by passion and the lust for revenge. If indeed these measures were imposed by the intervention of a foreign power in internal affairs of our state, I will not hesitate and raise my voice even against this power. The Catholic Church is not afraid of any secular power, whatever it be, when it has to protect basic human values.[110]

This stopped the arrest and deportation of Jewish spouses and the children of mixed marriages. It also led the German authorities to regard Stepinac as an enemy – among the reasons they gave being that he 'intervened many times personally on behalf of persecuted Jews and Orthodox'. The Germans' killing of hostages in reprisal for partisan attacks prompted Stepinac, whose brother had died fighting as a partisan, to speak out again on the subject of racism. At the end of October 1943 he said in a sermon:

The Catholic Church knows nothing of races born to rule and races doomed to slavery. The Catholic Church knows races and nations as creatures of God . . . for it the Negro of Central Africa is as much of a man as a European. For it the king in a royal palace is, as a man, exactly the same as the poorest pauper or Gypsy in his tent . . . The system of shooting hundreds of hostages for a crime, when the person guilty of the crime cannot be found, is a pagan system which only results in evil.

The education minister issued a furious rebuttal in which he said: 'the Ninth Symphony is certainly nearer to God than the howling of a cannibal tribe in Australia'. In March 1943 Stepinac attacked Pavelić in person, pointing out that *his* marriage to a Jewish woman appeared to be exempt from attempts by the regime and its German ally to dissolve such unions. In May Maglione drew up a list of the thirty-four separate interventions Stepinac had made on behalf of either Jews or Serbs.[111]

Few European Fascist movements went so far as to proclaim that 'God is a Fascist!' or that 'the ultimate goal of the Nation must be resurrection in Christ!' Romania was the exception. Romanian Fascists wanted 'a Romania in delirium' and they largely got one. The Legion of the Archangel Michael was founded in 1927 in honour of the archangel, who had allegedly visited Corneliu Codreanu, its chief ideologist, while he was in prison. It was the only European Fascist movement with religion (in this case Romanian Orthodoxy) at its core. In 1930 the Legion was renamed the Iron Guard. While rivalling only the Nazis in the ferocity of their hatred of Jews, these Romanian Fascists were sui generis in their fusion of political militancy with Orthodox mysticism into a truly lethal whole. One of the Legion's intellectual luminaries, the world-renowned anthropologist Mircea Eliade, described the legionary ideal as 'a harsh Christian spirituality'. Its four commandments were 'belief in God; faith in our mission; love for one another; son'. The goal of a 'new moral man' may have been a totalitarian commonplace, but the 'resurrection of the [Romanian] people in front of God's throne' was not routine in such circles. But then few European Fascists were inducted into an elite called the Brotherhood of Christ by sipping from a communal cup of blood filled from slashes in their own arms, or went around with little bags of soil tied around their necks. Nor did they do frenzied dances after chopping opponents into hundreds of pieces. Not for nothing was the prison massacre of Iron Guard leaders – including the captain Codreanu himself – by supporters of king Carol II known to local wits as 'the Night of the Vampires'. Although the Romanian elites emasculated the Guard's leadership, much of their furious potential was at that elite's disposal.[112]

Hitler's conquests in western Europe in 1940 led Carol II to abandon his country's alignment with Britain and to seek a role for Romania within the all-conquering German 'new order'. That June, the Soviet Union took Bessarabia and Bukovina under the terms of the deal it had struck with Hitler. Three million Romanian Orthodox Christians languished under an alien and atheist regime, a state of affairs that outraged opinion in the Old Kingdom. In September 1940 Carol invited the military strongman, General Ion Antonescu, to form a government, which within a month deposed the king in favour of his son prince Michael, who is still the claimant to the throne of Romania. Because, like Franco, Antonescu lacked a political base, he revived the Legion so as to provide a basis for what became the 'National Legionnaire State'. The Iron Guard leader, Horia Sima, became vice-premier, and the Guard gained five ministerial portfolios. For the ensuing five months the Guard attempted a stealthy coup from within, even as their corruption and violence created chaos. Since sections of the Nazi leadership favoured the Guard, the wily Antonescu knew where to turn.[113]

In January 1941, Antonescu flew to Germany for a meeting with Hitler, whose troops were massing in Romania for the projected invasion of the Soviet Union. The strong personal rapport between these two implacable haters of the Jews enabled Antonescu to provoke and crush a revolt by the Guard after he returned home; nine thousand were detained and eighteen hundred sentenced to imprisonment. The Guard was proscribed and the Legionnaire State abandoned. Antonescu assumed the title of 'conducator' used by the murdered Codreanu, while his son Mihai became vice-premier of a government largely consisting of antisemites of the National Christian Party, for in this respect the old elites were no different from the Fascists. Acting reflexively in its search for someone to blame, the Guard carried out a pogrom in Bucharest, killing 630 Jews, some of whose corpses hung in the capital's slaughterhouse as 'kosher meat'.[114]

In June 1941 Antonescu's troops joined the multinational invasion of the Soviet Union. The role played by the Romanian Orthodox Church was significantly different from that of the Catholic Church, to which multiple evils are routinely imputed, up to and including co-responsibility for the Holocaust. Romania shows what would have happened had the Church supported Hitler. Romania's Orthodox hierarchy had no inhibitions in calling the invasion of Russia a crusade, something Pius XII conspicuously omitted to do. The Orthodox Church's rhetoric was quite unlike anything one would have heard from the Catholic Church in western Europe for a few hundred years. God, according to the Orthodox metropolitan of Moldavia,

> had mercy on them [the inhabitants of the Soviet-occupied provinces] and sent His archangels on earth: Hitler, Antonescu

and [Finland's] Mannerheim, and they headed their armies with the sign of the cross on their chests and in their hearts in a war against the Great Dragon, red as fire, and they defeated him, chased him in chains, and the synagogue of Satan was ruined and scattered in the four directions of the earth and in their place they erected a sacred altar to the God of peace.[115]

By contrast, the papal nuncio, archbishop Andrea Cassulo, who had already protested about restrictions placed on converts to Catholicism, went on to give the Romanian Jewish community what help he could. That September, secretary of state Maglione reassured US bishops regarding their government's Lend–Lease shipments of aid to the 'Great Dragon', arguing that it was perfectly in order to aid the Russian people. That enabled Roosevelt to get the Lend–Lease legislation through Congress, which provided eleven billion US dollars' support to the Soviets. The apostolic delegate to the US also briefed the much respected archbishop of Cincinnati to make a speech supporting aid to the Russians at the annual conference of US bishops. Evidently, the pope's anti-Communism did not prevent him from appreciating the value to the Western Allies of the military might of the Soviet Union.[116]

The Jewish population of Romania, including the territories lost to Bulgaria, Hungary and the Soviet Union in 1940, was the third largest in Europe, consisting of three-quarters of a million people. In the opening days of the campaign attempts were made to evacuate Jews from the frontiers into the interior, because of fears that they would aid and abet the Soviets. Three days after the war started, rumours spread that Soviet parachutists had been seen near Iaşi. Deserters fired on Romanian soldiers investigating the town's Jewish quarter. In the resulting pogrom, thirteen thousand Jews were either killed in the city of Iaşi or left to die on sealed trains which crawled towards detention camps, stopping only to throw out the dead. Many people in Romania blamed the Jews for the Soviet occupation of Bessarabia and Bukovina, routinely alleging that they were spies or Communists. That included the clergy. As patriarch Nicodim explained:

God has shown to the leader of our country the path toward a sacred and redeeming alliance with the German nation and sent the united armies to the Divine Crusade against destructive Bolshevism ... the Bolshevist Dragon ... has found here also villainous souls ready to serve him. Let us bless God that these companions of Satan have been found mostly among the sons of the aliens, among the nation that had brought damnation upon itself and its sons, since it had crucified the Son of God. If by their side there had also been some Romanian outcasts, then their blood was certainly not pure Romanian blood, yet mixed with damned blood. These servants of the Devil and Bolshevism, seeing that their master, the monster called Bolshevist Russia, will soon be destroyed, are now trying to help him ... they disseminate among our people all sorts of bad new words.

Of the three hundred thousand Jews who lived in these territories, one hundred thousand had the presence of mind to flee with the retreating Russians. For these territories were now to be 'cleansed' of Jews in the most brutal manner. Romanian army and police units began by trying to herd twenty-five thousand Jews over the River Dniester, whose opposite bank was controlled by the Germans. The Germans were appalled by this prospect. When they learned that Antonescu was planning to deport sixty thousand Jews from Old Romania into Bessarabia, thus raising the nightmare of hundreds of thousands of Romanian Jews milling around behind forces attempting to kill the Jews in the Ukraine, they forced him to abandon this measure. In August 1941 the German and Romanian armies agreed that Romania could push the Jews from Bessarabia and Bukovina over the Dniester but not across the Bug. This effectively turned the whole of what was known as Transnistria into a vast network of camps and ghettos for Jews expelled under unimaginable conditions from Bessarabia and Bukovina, an ordeal in which alone twenty-five thousand people perished, with the survivors subjected to forced labour while living on a pitiful diet. Brutality, disease and starvation meant that by May 1942 two-thirds of the Jews in Transnistria were dead.

Meanwhile, Romanian troops were also responsible for the largest single massacre of Jews during the war after the Romanian headquarters in Odessa was destroyed by a mine left by Soviet forces. On 23–24 October 1941 troops acting on Antonescu's direct orders slaughtered as many as sixty thousand people. In the same month, president Filderman of the Romanian Jewish community protested to Antonescu that his measures meant 'death, death, death without guilt, except the only guilt of being a Jew'. A stony Antonescu replied, and the very act of replying was significant, that the Jews had welcomed the Soviets with flowers into Bessarabia and Bukovina, and had denounced Romanians to the Soviet police apparatus. He invited Filderman to reflect on these horrors before he accused the Romanians. Similar appeals to the Orthodox hierarchs were met with hostility or indifference. Only the metropolitan of Bukovina, who had witnessed deportations, seems to have protested to the government. There was an extraordinary encounter between the chief rabbi Safran and the patriarch, 'a grim, ruthless old man, of the old antisemitic priesthood'. Nicodim listened with suppressed hostility to Safran's pleadings. Sensing that he was making no impression, Safran, according to his own account, burst out:

> 'Don't you realise that I am talking about the lives of tens of thousands of absolutely innocent people? You shall bear an overwhelming responsibility if you allow such a striking injustice to occur! . . .'
>
> Unable to control myself, I collapsed and fell to the floor. He saw me kneeling in front of him and was extremely impressed. The Patriarch descended his throne and, together with his secretary, helped me up. There was a dramatic change in the atmosphere. The Patriarch began muttering, as if talking to himself, but talking to me at the same time: 'What can I do?'

The Jews in the Old Kingdom were meanwhile subjected to onerous financial exactions while their land and businesses were 'romanianised', the only ray of light being that, as the most venal country in Europe, those with money could mitigate their plight through bribery. Although the Germans themselves ensured that the Holocaust was the greatest act of larceny in modern history, passing the results of theft on to the German people on a massive scale, they were censorious about the Romanians' corruption and erratic approach to killing people. In July 1942 the SS were confident that deportations of the three hundred thousand Jews in the Romanian Old Kingdom to the Lublin region of Poland would commence in September. They thought they had the Antonescus' word for it. Everything was worked out, including the Jews to be initially targeted, and the railway timetables were prepared. At that point, the Romanians changed their minds. Several factors persuaded the Romanian government suddenly to reverse its course on the Jews. In such a backward country, it was not possible to eliminate large swathes of the professional intelligentsia without dire general consequences. The government was piqued that it was being asked to surrender its Jews before neighbouring Hungary. The Germans treated Radu Lecca,

Bucharest's Jewish affairs expert, offhandedly once they had the agreement of Romania's dictator. With so many troops committed in Russia, the Romanians were well placed to appreciate that the outcome of the war was uncertain. Above all, they were subjected to high-level interventions. The US secretary of state Cordell Hull warned them that there would be judicial consequences after the war. At the prompting of rabbi Safran, nuncio Andrea Cassulo joined the Swiss emissary René de Veck, the Turkish ambassador and the Red Cross in putting pressure on the queen mother Elena, who made her disapproval known to Antonescu himself at a lunch at the royal palace.[117] When the president of the Jewish community in Switzerland asked the papal nuncio there to intervene on behalf of Romanian Jews, that is precisely what the nuncio did.[118] Cassulo went to Rome, at the behest of Safran, returning to Bucharest with a message that Antonescu could not ignore. Shortly afterwards, the Romanians told the Germans that the inclemency of the weather made the deportations impossible.

Safran had no doubt that the Jews of the Old Kingdom owed their lives to the nuncio, whom he visited twice to convey the thanks of the Jewish community to the Holy See.[119] Typically, the Germans argued that Safran had 'bought' the nuncio.[120] In early 1943 Cassulo visited Transnistria, after which he wrote to the Romanian foreign minister drawing attention to the plight of eight thousand Jewish orphans. No one reading the published Vatican documents can doubt that Cassulo acted in close concert with Safran; that he repeatedly intervened with the Romanian government; and that he did all of this with the encouragement of the Holy See.[121] Throughout 1943 Cassulo also acted on behalf of the Jews in Transnistria, liaising with Angelo Roncalli, the papal nuncio and future John XXIII, in organising shipping and safe-conducts that took refugees to Istanbul and on to Palestine, while he also tried to establish the fate of Romanian Jews who were at the mercy of the Hungarians in northern Transylvania. The nuncios were also instrumental in seeking German safe-conducts for the SS *Tari*, which the Turkish government offered to send to a Romanian port to ship Jews to Haifa. The Turks noted that despite having huge merchant fleets, neither Britain nor the US were prepared to make available their own ships, nor even to replace the SS *Tari* should it be lost, yet they were 'posing before the world as the saviours of the refugees'. In early 1944 rabbi Safran of Romania and Herzog, the grand rabbi of Jerusalem, wrote to nuncio Cassulo thanking him and the pope for everything they had done for the Jews. Safran recalled that Cassulo would often visit the Antonescus twice a day, returning with some piece of paper that meant life to someone he did not know, and with tears of gratitude in his eyes, 'because I had given him the opportunity of doing a good deed'. That belongs in any discussion of the Vatican and the Holocaust too.[122]

In some respects developments in Bulgaria resembled those in other Axis satellite states: the introduction of antisemitic legislation in January 1941, and the creation of a commissariat for 'Jewish Affairs' in June 1942, followed by German pressure to deport the Jewish population. There the similarities end. Bulgaria played a clever game during the war, exploiting Germany to secure territory it coveted, but declining to join the other clients and satellites by refusing even to declare war on the Soviet Union.[123] Apart from the opportunism of its political elite, there are more benign interpretations of why Bulgaria's Jewish population was larger after the war than before. It retained a functioning parliament – the Sobranje – throughout the war which insisted on its right to scrutinise government measures, including whenever the government exceeded the terms of its own antisemitic legislation. Bulgarians had long experience of being an ethnic minority under the Ottoman Empire, and were little inclined to blame an unremarkable minority for their own travails. The tiny Jewish population in Bulgaria failed to excite anti-semitic passions, except among the country's dedicated Fascists, while the urge to rescue seems to have been more widespread than either indifference or the will to destroy. The documents reveal a warm-hearted people manifestly moved by the plight of the Jews.

The first operation to deport Bulgaria's Jews commenced in March 1943. But, while the Germans and their local accomplices succeeded in deporting denationalised Jews from Macedonia and Thrace, strenuous interventions stymied their efforts to round up Jews from 'Old Bulgaria' itself. The Orthodox bishops of Plovdiv and Sofia played a notable and honourable part in persuading king Boris III to prevaricate in his co-operation with the Germans. Metropolitan Cyril of Plovdiv secured the liberation of over a thousand Jews who had already been corralled in a schoolhouse. He offered baptised Jews asylum in his own house, and warned the local police that 'I, who until now have always been loyal towards the government, now reserved the right to act with a free hand in this matter and heed only the dictates of my conscience.' Metropolitan Stefan of Sophia was similarly appalled when he happened upon a train transporting Jews from Thrace across Bulgaria to Poland. He intervened successfully on behalf of the Jews of Dupnitsa. He felt that 'If our Church does not intervene, we should expect even worse outrages and acts of cruelty, which our people, who are good and kind, will one day recall in shame, and perhaps other calamities.' Forty-three members of the governing faction in parliament, including the assembly's deputy speaker, signed a letter of protest against the deportations. Deputy speaker Dimitâr Peshev, responsible for the first official mutiny against the persecution of the Jews, regarded the deportations not only as illegal, but as deeply shaming to his developed sense of national honour. Roncalli, the nuncio to Istanbul, who had been delegate to Sofia between 1925 and 1934, also made repeated interventions on behalf of Bulgarian as well as Slovakian Jews, while assisting Jewish organisations to provide Jews with improvised papers to transit Turkey en route to Palestine.[124] Subject to several conflicting pressures, king Boris agreed to disperse Sofia's twenty-five thousand Jews in the provinces as an alternative to their deportation to German-occupied Poland, where their fate would not be in doubt. By this juncture the strategic outlook for his Axis allies was not auspicious, following Germany's defeats at El Alamein and Stalingrad. The dispersal of Sofia's Jewish population coincided with the celebration on 24 May of the national saints Cyril and Methodius on the city's Alexander Nevsky Square. This turned into a mass demonstration in which four hundred people were arrested. Acting on the pleas of Jewish religious leaders, metropolitan Stefan went to every conceivable length to register his outrage with the government. Speaking in the imposing cathedral square, he said:

> We call out to the state authorities from this square and beg them not to shackle the democratic and hospitable spirit characteristic of the Bulgarian people, a spirit forged by humanity and broth-erly love, and hostile to foreign influence, foreign control, and foreign demands. Under the great roof of the Church, there were no problems with national minorities, because the Bulgarians were tolerant and their fundamental law was respect for freedom, which overcame the peculiarities of the minorities and the domination of the majority . . . It was our duty to prove ourselves worthy of the commandment that, in creating us spiritually, had educated Bulgaria in the gospel's truth, so that she could be, then as now, a state enlightened by the shining beacons of culture and built on a democratic and disciplined spirit.[125]

Metropolitan Stefan followed up his public protest – which had few direct consequences – with protests to the king, who referred him to the prime minister, who got the attorney-general to threaten the bishops with prosecution. Stefan responded to these threats by telling every church in Bulgaria to open its doors to the Jews. From then until September 1944, there were no further deportations of Jews from Bulgaria, and in late August the antisemitic legislation was rescinded.

Italy's Jews passed in the autumn of 1943 from a period of socio-economic discrimination under the Fascists, via a brief interval when the Allied invasion of the South seemed to promise liberation, to the disappointment of seeing German forces invest northern Italy, with the grotesque spectacle of German paratroopers encircling the Vatican. Pius instructed the Italian bishops to open all convents and monasteries to Jewish refugees. A huge number of Church buildings were used to secrete people, including 150 buildings in Rome alone, with canonical rules relaxed to allow members of the opposite sex to be hidden within single-sex institutions. Five hundred Jews were hidden in the papal summer residence at Castel Gandolfo, where Pius' private apartment became an obstetric ward. Many of these fugitives were subsequently killed in February 1944 when the palace was hit by Allied bombs. In this way, one-third of the Jewish population of Rome were hidden in buildings owned by the Catholic Church, including a large number of the Vatican's extra-territorial properties. It is inconceivable that such a large-scale rescue effort, which included religious houses across Italy, could not have been at the behest of the bishop of Rome.[126]

On 26 September 1943, the SS commander in Rome ordered the city's Jews to hand over fifty kilograms of gold within thirty-six hours or two hundred hostages would be deported. Kappler accompanied this demand with reassurances that the Reich was interested only in money. Pius offered to lend the Jews the gold when it seemed that they could not produce it. His offer was counter-productive, because it also encouraged the Jews to believe that the Germans had been bought off, and to remain where they were in expectation that the pope would protect them. Without any of the preliminaries that had been tried and tested across Nazi-occupied Europe, in late 1943 the SS began rounding up the Jews in several Italian cities, including Rome, using the community records they found in Rome's main synagogue to identify the addresses of the individuals concerned. Between October 1943 and January 1944 over three thousand Italian Jews were deported to Auschwitz, of whom forty-six survived the war. About eleven hundred of these people had been snatched from the ghetto and Trastevere areas of Rome, about a mile from the Vatican City, rather than 'under the very windows' of the pope as a German official had it.

The Holy See responded on two levels. Firstly, upon learning of the SS raids, through an Italian princess who hastened to the Vatican, convents were instructed to receive fugitives, with a special sign affixed to their doors warning that the buildings were under the Vatican's protection.[127] Secondly, Pius instructed Maglione to protest immediately to Ernst von Weizsäcker, the German ambassador to the Holy See. Weizsäcker's paramount concern was to prevent an open breach between the Vatican and the German regime over the subject of the Jews. To this end, he warned Maglione that 'any protest by the pope would only result in the deportations being carried out more vigorously', while forwarding to Berlin a copy of a letter written that day by the rector of the German College in Rome to the commander of the city's SS, warning that the pope might be compelled to make a public protest if these deportations did not cease. These sleights of hand stymied both parties. There were no more deportations directly from Rome, and no protest from the pope either, since the threat of such a thing had achieved its intended effect.

Critics of Pius, evidently unaware of such a thing as tactical lies, routinely alight upon a telegram Weizsäcker sent to Berlin on 28 October 1943 saying that 'he [the pope] has not allowed himself to be carried away making any demonstrative statements against deportations of the Jews'. After indicating that the problem 'has been disposed of', Weizsäcker pooh-poohed a recent article in the *Osservatore Romano* about the Vatican's charitable activities to all mankind as typically 'involved and vague'. Although his dismissive words have done considerable harm to the reputation of Pius XII, at the time they served the more urgent end of diverting Berlin's malign attentions away from the thousands of Jews hidden in Catholic churches and private homes in Rome, by insinuating that Germany had missed a papal protest only by a hair's breadth, while concealing the Church's rescue efforts within the Vatican's elliptical verbiage. Such serpentine stratagems were as normal to those who had to negotiate these shoals at the time as they are alien to the academic moralists who deplore them with the luxury of hindsight.[128] Even historians otherwise critical of Pius concede that the Germans would have had no hesitation in responding to an overt protest by invading the hundreds of Church properties where, in Rome alone, five thousand

Jews were sheltered. A clumsily handled protest would also have upset all the delicate arrangements which had been negotiated, often through the mediation of clerics, whereby converts and partners in mixed marriages were exempted from the Nazis' destructive rampage. It would further have jeopardised much of the post-war non-Communist leadership of Italy who were similarly hidden within the churches.[129] Jewish soldiers attached to the United States armies which liberated Rome were unequivocal in their praise of Pius for his role in protecting Jews: 'If it had not been for the truly substantial assistance and the help given to Jews by the Vatican and by Rome's ecclesiastical authorities, hundreds of refugees and thousands of Jewish refugees would have undoubtedly perished before Rome was liberated.' In recognition of Pius' efforts, Rome's chief rabbi Israel Zolli took the baptismal name Eugenio when he formally converted to Christianity in February 1945.[130]

Hungary further exemplifies the point that the Churches were capable of intervening only where the presence of a more or less independent government meant that the Germans had to respect local sensibilities. Although they had been subjected to years of discrimination, until the Germans arrived on 23 March 1944 the three-quarters of a million Jews of Hungary had been spared the fate of Jews elsewhere. Almost immediately, on 28 March and 5 April, the Vatican telegraphed its nuncio to Budapest, archbishop Angelo Rotta, warning him of the peril facing the Jews and enjoining him to take action to protect them.[131] Before the deportations commenced, both Rotta and the Hungarian primate cardinal Serédi intervened with premier Döme Sztójay on behalf of baptised Jews. When the deportations began, Rotta protested about treatment of people by virtue of their race, warning the prime minister: 'I hope that in his position as supreme pastor of the Church, as the one who safeguards the rights of all his children, and as the defender of truth and justice, he [the pope] will not be obliged to speak out in protest.' The Vatican encouraged Rotta to keep up his protests. On 5 June the nuncio ridiculed the government's claims that Jews were being deported for forced labour:

> It has been asserted that it is not a question of deportation but of compulsory labour. Once can argue about the proper term for these things but the reality is the same. When seventy year old, even eighty year old men, and also women, children, invalids, have to be dragged off, one is bound to ask, what kind of work is

it that such helpless creatures as these can carry out? ... When one also remembers that those Hungarian workers who go off to Germany to work are not allowed to take their families with them, it seems amazing that this great favour is accorded only to the Jews.[132]

This intervention did nothing to stop the deportations, which by mid-June had swept away three hundred thousand people without any sign of abating. The Hungarian hierarchy was dilatory in making a joint protest, with some arguing that the threat of such a thing would be more efficacious than actually making it. Acting on instructions from Maglione in Rome, Rotta insisted that the Hungarian prelates intervene. At that point, prompted by among others Britain's cardinal Griffin (Hinsley having died in 1943), who was responding to the requests of the World Jewish Congress, Pius XII decided to lend a hand.[133] On 25 June he sent an open or plain telegram to the Hungarian head of state, the Calvinist admiral Horthy: 'We personally address Your Royal Highness, appealing to your noble sentiments and being fully confident that you wish to do all in your power in order that so many unfortunate people be spared further afflictions and sorrows.'

The deportations ceased on 5 July, Horthy, following the pope's intervention, having been further bombarded by intercessions from the king of Sweden and the president of the International Red Cross, and New York's cardinal Francis Spellman having broadcast a powerful appeal to Hungarian Catholics. The papal nuncio, Rotta – according to the German plenipotentiary in Hungary – 'was seeing the regent and Szotajay several times a day'.[134] Jewish agencies were profusely grateful for the pope's intercession.[135] But power was being drained away from the Hungarian regent. On 15 October, after having briefly managed to extricate himself from a government dominated by Arrow Cross Fascists, Horthy announced an armistice with the Russians. The Germans deposed him and put the Fascist leader Ferenc Szálasi in his place. The hunt for Jews was resumed in the city rather than in its suburbs. The one hope for the Jews was an unusually active diplomatic corps, of which archbishop Rotta and the Swedish diplomat Raoul Wallenberg were the leading lights, together with the Catholic Church, which was influential in Hungary. Archbishop Rotta resumed his protests, while issuing thousands of letters of safe conduct, and joining the ambassadors of the neutral countries and the Red Cross in providing safe havens for Jews.

He took the lead in establishing an 'International Ghetto', consisting of several blocks of flats, in which twenty-five thousand people were housed under the protection afforded by Vatican, Swiss and Swedish emblems. A large number of religious houses in the capital hid Jews on their premises, at considerable risk to the monks and nuns involved, if they were denounced or otherwise caught. So did priests like Ferenc Kálló who issued Jews with certificates of baptism – he was shot. Since Hungarian Nazis tended to smash up religious classes being held for the large number of Jews converting to Catholicism, clergy – led by the Dominicans – ventured into air-raid shelters to carry out perfunctory baptisms which brought the all-important certificate that would enable Jews to evade destruction. Nuncio Roncalli in Istanbul simply sent thousands of such certificates to Hungary. Archbishop Rotta himself managed to issue fifteen thousand protective passes to those Jews who requested notional conversion to Catholicism, and he simply allowed a Red Cross official to take piles of blank but signed Vatican letters of safe conduct, to use as he saw fit in protecting Jews. The nuncio equipped a small Red Cross team that tried to release Jews being marched to Germany on what was known as the Hegyeshalom death march, with one cleric waving his pectoral cross at the Arrow Cross men who tried to prevent his ministrations.[136] As a result of these rescue efforts, fifty-five thousand Jews survived alongside the sixty-nine thousand in Budapest's 'Big Ghetto'.[137]

There are many criticisms one might make of the Catholic Church, but responsibility for the Holocaust is not among them. That was the devil's work of the Nazi government of Germany, and those who took the opportunity its evanescent continental empire afforded. Nor is there the slightest evidence to support the idea that Pius XII was 'Hitler's pope', a title more befitting 'Hitler's mufti', the antisemitic Haj Muhammed Amin al-Husseini of Jerusalem, if one seeks a spiritual leader who endorsed Hitler's racist views. Pius was actually involved in a conspiracy against Hitler which the Allies failed to support. Making use of the Holocaust as the biggest moral club to use against the Church, simply because one does not like its policies on abortion, contraception, homosexual priests or the Middle East, is as obscene as any attempt to exploit the deaths of six million European Jews for political purposes.

Where the Church could intervene, as in the smaller satellite states of eastern Europe, it did so, to the gratitude of the Jews concerned. Everywhere, those clergy who risked their lives by helping Jews attributed this to instructions they had received from Pius XII. That is why some people now argue that Israel ought to recognise him as 'Righteous among the Nations'. It seems extraordinarily mean-minded to claim that either the clergy who helped Jews or the Jews who praised and thanked Pius were acting according to some ulterior agenda – in the last case, allegedly seeking Vatican support for the creation of Israel.

Certainly, Pius XII is not above or beyond criticism regarding what he could or could not have done during the war. One has no investment in 'defending' him, except in cases where the criticisms are blatantly unfair or unjust. His attempts to maintain peace were noble, but largely ineffectual. For reasons either of personal character or of professional training as a diplomat, his statements were exceedingly cautious and wrapped up in an involuted language that is difficult for many to understand, especially in this age of the resonant soundbite and ubiquitous rent-a-moralists. A more robust character, like Pius XI or John Paul II, not to speak of medieval popes who took on emperors, might have said more in fewer words. One doubts that it would have had any effect. Perhaps, as the wise Owen Chadwick once observed, Pius was too good to comprehend the evil around him, or at any rate unable to distinguish clearly amid the simultaneity of so many evils – a deficiency in which he was hardly unique among world leaders at the time. For until we restore a panoramic perspective on all the death and suffering caused by the Second World War, we will not fully appreciate the magnitude of the dilemmas facing this most enigmatic of modern pontiffs.[138]

CHAPTER 5

Resistance, Christian Democracy and the Cold War

I THE SPIRIT OF RESISTANCE

Although in most countries resistance movements were militarily insignificant, everywhere – including Germany and Italy – they helped restore a sense of national self-respect, and concentrated minds on the shape of the future that Allied military ascendancy opened up. In honouring that resistance, we should not exaggerate its extent, or imagine that its strategic contribution was more than exiguous. Even a generous estimate of the highest number of resisters in France sets it at 2 per cent of all adults. In Germany there were isolated resistance groups, of various political colourations, and degrees of ineffectuality, although in Italy by early 1945 some quarter of a million people were ultimately involved in an armed anti-Fascist partisan movement that acted as the long arm of the Allies in the liberation of the northern areas of the country. They tied down fourteen of the thirty-one German and Fascist divisions in Italy; thirty-five thousand of them were killed and a further twenty-one thousand seriously wounded.[1]

Like altruism in wartime, resistance was as much a matter of outlook, upbringing, and temperament as of ideological conviction, a dangerous form of doing the decent thing. Determining when resistance commenced is as imprecise as gauging which acts constituted it, but in virtually every European country Christians were closely involved, the general trend, for them as well as for everyone else, being from passive to active resistance as the depredations and exactions of the occupiers grew more desperate and German defeat seemed more probable.

Resistance by Christians was complicated by ethical concerns that did not trouble many Communists, for whom the ends justified any means, regardless of their impact on innocent bystanders. Apart from Romans 13's exhortation to obey Caesar, more recently popes Gregory XVI, Pius IX and Leo XIII had strongly condemned any rebellion against legitimate authority, including rebellion by Catholic Poles or Irish. The experience of the Cartel des Gauches in France during the mid-1920s, and even more so of revolutionary Mexico, led Pius XI to qualify that position, although the Spanish Civil War rapidly indicated that the right of revolt – by Nationalist soldiers – could occasion greater evils than the Republic's initial injustices, and led the pope to reconfirm the original position. Beyond the pope, Catholic resisters could fall back on classical notions of resistance to tyranny, albeit complicated by the fact that modern tyrants were not morally degenerate monarchs, but democratically elected politicians, whose anti-constitutional activities enjoyed the widespread consent of people whose views could be manipulated with methods unknown in the ancient world. That also complicated the response of Lutherans, who, furthermore, had to transcend such local values as obedience, conscientiousness, fortitude and service to the community, values which the Nazi regime had made its own.[2]

There were other complications. Opposing foreign occupation was one thing, but should resistance be extended to such collaborating regimes as Vichy, about which many Christians initially harboured illusions, and not simply because of its ostentatious subscription to some of the Christian virtues, for Vichy corporatism was a room with many mansions? Lower clergy and religious who resisted were doubly disobedient, both to the legally constituted government and to their immediate ecclesiastical superiors, who necessarily played a more cautious game in proportion with their greater public responsibilities. Vichy enjoyed the virtually unanimous support of the French ecclesiastical hierarchy, who – with some exceptions – denounced resisters as 'dissidents', 'bandits' or 'terrorists', and refused to contemplate clergy acting as chaplains to the Maquis until the pope authorised them through the nuncio. Some of the most trenchant criticism of the feeble hierarchy came from Catholic intellectuals such as the diplomat and writer Paul Petit, whom the Germans guillotined in Cologne in August 1944.[3]

It is also difficult to determine whether a person's Christianity, or for that matter Marxism, rather than patriotic abhorrence for the Boche invader, was the crucial motivating factor in their lonely choice to resist.

Clearly it was important to the splenetic writer Georges Bernanos and cooler philosopher – theologian Jacques Maritain, who elected to oppose Nazism and Vichy from respectively Brazil and the USA. Between December 1940 and November 1941 Bernanos wrote a series of long impassioned letters to the English, and later to the Americans, regarding the 'fairy tale' of the little island's resistance to Nazism, 'a child's dream made real by grown men'. The letters were all the more moving in that their author had never visited England at all, celebrating an idea of it untroubled by the realities.[4] Maritain had travelled from support for the Action Français – proscribed by Pius XI in 1926 – to being an opponent of Fascism and racism (his wife Raissa was of Jewish extraction) and an advocate of a federal Europe that would be a reconstitution of medieval Christendom on the basis of human values, the right to work, free association and assembly, and free speech. Exiled in New York, Maritain acted as the unofficial ambassador of Free France, since Vichy continued to maintain an ambassador to Washington. Similar powerful voices from abroad were the Christian convert Maurice Schumann in London, and André Colin, the former secretary-general of the Association Catholique de la Jeunesse Française in Beirut.

For many Christian intellectuals in France, resistance to the occupation built on moral and religious objections to Nazism (and governmental appeasement of it) that they had repeatedly registered in the inter-war years as part of what some call the 'pre-resistance'. It is commonplace to say that no one read *Mein Kampf*; many Catholic clerics and thinkers who would resist Nazism clearly had and warned about the totalitarian drive of the modern state and the worship of class or race. In France, some of the most trenchant criticism of Nazism in the 1930s had come from Catholic journals, such as *L'Aube* or *Sept*, which metamorphosed under the occupation into primitively produced clandestine papers. Since the subscriber base for such journals always resembled the membership of a political party, their distribution networks formed the initial framework for organised resistance movements, which were distinct from much tighter military networks connected to the intelligence operations of the Allies.[5] The case of Edmond Michelet at Brive, perhaps the first resister, who began by distributing a simple typed tract days after the invasion, shows how one act led to another, for he was soon involved in spiriting Austrian and German refugees over the Spanish border, going on to command a resistance group operating in nine departments until he was deported to Dachau in 1943.[6]

Christian charity formed another route into resistance activity, when it involved helping French or Allied soldiers, as well as German and Austrian refugees, to reach the relative safety of the 'free' zone at Vichy, the route to neutral Switzerland or Spain. Archbishop Saliège of Toulouse had already afforded such a service to Spanish Republicans fleeing Franco. Country clergy were rightly celebrated even during the war for granting sanctuary to Allied airmen, Jews and resisters on the run. It was a short but fateful step, from performing the duty of asylum towards strangers to allowing bell-towers and crypts to be used to house a clandestine arms cachet or press, as was the case at the churches of the Nativity in Saint-Etienne or at Montbéliard. Such actions could, and did, result in arrest, torture, deportation and death in German concentration camps. We have already mentioned the Lyons- and Paris-based group of Jesuit theologians that produced *Cahiers du Témoignage chrétien*. Unlike parish clergy or Catholic politicians, members of such orders could think and write without having to bow to a constituency. The group that wrote these impressive tracts was motivated by the conviction that 'It is necessary to choose Christ or Hitler', and the belief that every human being was deserving of respect, regardless of their ethnic origins or religion. Several of the tracts contained searing critiques of Nazi ideology, notably of Hitler's *Mein Kampf*, which was described as 'Germany's holy book', while the accounts of persecution of the Jews, towards whom they proposed an entirely renewed theology, were remarkable in their detail. Issues 6 and 7 in April–May 1942 were devoted to a lengthy attack on the theory and practice of antisemitism, while issues 13 and 14 in early 1943, which were devoted to Nazi policy in Poland, included the intelligence that '700,000 Jews had been brutally murdered on Polish territory and that there could be no doubt regarding Hitler's plan to exterminate European Jewry completely'.[7]

Every resistance group thought about how they planned to reform France in the future after what most of them saw as the moral and systemic failures of the 1930s, a decade famously described as 'the hollow years'. The results of these urgent reflections mirrored the dominant political colouration of the resistance movement.[8] In general, Christian resisters came from left-wing Christian Democrat circles, certain religious orders – notably the Jesuits and Dominicans – Catholic trades unionists and, last but not least, independent-minded aristocrats and the bourgeois officer class, which was overwhelmingly Catholic and conservative. There was no single Christian Democrat resistance movement, but most resistance groups contained Christians. In the occupied zone, where the enemy was clearly defined, there was considerable co-operation between Christians, of various political persuasions, and socialists; in the south, Christian opposition to Vichy tended to come from Christian Democrats repelled by the regime's clerically tinged conservatism. After June 1941, when its tortuous 'line' towards the Germans and the Allies was eventually clarified by force of circumstances, the French Communist Party became an important element in the resistance with which many Christians were disposed to co-operate, despite the visceral anti-Communism of most of the French hierarchy.[9] The concept of 'humanism' provided common ground between Communists willing to soft-pedal atheism and Christians who subscribed to various forms of 'personalism', best described as a Christianised and socially conscious form of individualism, or who entered factories and trades unions as worker priests during the occupation.[10]

The resistance proved that Catholics could co-operate with freemasons, socialists, Protestants and Jews, and provided what would become the Christian Democrat Mouvement Républicain Populaire with an impressive range of leaders. Indeed, in June 1943 the Christian Democrat Georges Bidault became president of the Conseil National de la Résistance after the betrayal and arrest of Jean Moulin. The main advocate of a new broad-based political 'movement', embodying the ideals that had guided the resistance, was a young Catholic student in Lyons called Gilbert Dru, who was a disciple of Mounier and Sangnier. Dru was hostile to the old political parties of the Third Republic. He wanted to purge the right of its involvement with the extremist Action Française and big business, but he also had little faith in the Radicals and Socialists, whose contribution to the resistance had been negligible. Acknowledging realities, he wished simultaneously to collaborate with, and counterbalance the Communists, provided they abandoned their primary allegiances to a foreign country. Although Dru himself never lived to see the fruits of his deliberations – he was arrested and executed by the Germans in July 1944 – his various memoranda influenced the formation of the Mouvement Républicain de Libération, which subsequently became the Mouvement Républicain Populaire. Both the words 'movement' and (especially) 'republican' were significant, the first suggesting something more compelling than a mere political party, the second a very public Catholic recognition that the legacy of the Revolution was there to stay. In that the Mouvement was

Catholic, democratic and progressive, it represented a significant break with the mindset of the traditional Catholic right.

The Mouvement's founding manifesto called for a revolution that would not only free France from Vichy, but complete the imperfect Revolution of 1789 without its appalling violence. Its theorists sought to fuse freedom and justice so as to permit man to develop his full material and spiritual potential. In their view, man came first, followed by society, and then the state. The important groupings in life were anterior to the state. They attached enormous weight to 'pre-political' entities, beginning with the all-important family, and moving upwards through associations, communes, regions and trades unions, and above all the Church, which severally guaranteed man's rights against the 'spontaneous totalitarianism of the State'.[11] Although the Mouvement was in favour of welfare, in the form of family allowances and housing benefits, and favoured more nationalisation of industry and central planning than its more market-orientated counterparts in Germany and Italy, it wished to limit the power of the state through an emphasis upon human rights, bicameralism, regional devolution and limits on presidential power. While it wanted worker participation in management, it opposed worker control; while it desired a social security system, and succeeded in creating one, it wanted that system's beneficiaries rather than bureaucrats to be in charge. In foreign policy, which was controlled successively by Bidault and Schuman, the Mouvement initially flirted with the notion of being a mediator between East and West, while pursuing a hardline course towards Germany, but events transformed it into an Atlanticist and European-minded party, with the threat from Stalin soothing raw animosities towards the Germans.

The Mouvement joined the Socialists and Communists as one of the three main parties of the immediate post-war years. Initially, it enjoyed every prospect of success, since its leaders emerged with impeccable wartime credentials as part of the resistance, and its policies seemed to transcend conventional ideologies. The former history master Georges Bidault had been leader of its principal steering body, his autobiography a study in the discretion of concierges and close brushes with Milice and Gestapo.[12] François de Menthon had escaped from a prisoner-of-war hospital, Pierre-Henri Teitgen from a train taking him to a concentration camp. Maurice Schumann, a Jewish convert to Christianity and the Mouvement's first president, had been the BBC's 'Voice of France' in London for four years. Another resister, Robert Schuman, provided a focus for the Mouvement's more conservative supporters. The ramified organisations of Catholic Action and the Catholic trade union the CFTC (Confédération Française des Travailleurs Chrétiens) provided the Mouvement's lower-echelon leaders, although the eager support of the Catholic hierarchy was a more mixed blessing in a country where anticlericalism was entrenched and widespread.

The Mouvement did strikingly well during elections in 1945 and 1946, taking nearly 24 and 28 per cent of the vote in these contests, and managing to push the Communists into second place in that last election. This success was largely because the Mouvement was the only alternative open to conservative voters, whose traditional political parties had been discredited by the Vichy interlude. Wits claimed that MRP stood for 'Machine pour Ramasser les Pétainistes'. It also did well among Catholic women, who for the first time had the right to vote, and made some inroads in traditionally working-class industrial regions such as Alsace–Lorraine and the Nord. It seemed that the Mouvement would succeed in reconciling the Republic and the Church and the latter with the working classes. In reality, the Mouvement suffered from several problems. It was not Catholic enough to attract the majority of Catholic voters, but it was too Catholic to get anything other than Catholic votes or to overcome anticlerical resentments.[13] While most of its leaders were left-leaning Christian Democrats, many of its supporters were temporarily homeless conservatives, who departed in droves once de Gaulle's Rassemblement du Peuple Français provided a major authoritarian and bonapartist alternative. In 1947 the Mouvement lost 75 per cent of the support it had gained in municipal elections in Paris the previous year, and its vote also slumped in the legislative elections in 1951.[14]

II NIGHT PASSES AND EVIL THINGS DEPART:
ITALY AND GERMANY

In late July 1943 the Italian Fascist Grand Council passed a motion critical of Mussolini, who on 25 July was summarily dismissed by king Victor Emmanuel. For forty-five days Italians held massive popular demonstrations celebrating the fall of Fascism, burning down Party headquarters, and tearing posters and symbols from walls. The new government of marshal Pietro Badoglio concluded a secret armistice with the Allies before the latter were in a military position to exploit it. German troops poured into northern Italy, restoring Mussolini as head of a puppet republic on Lake Garda, and interning the 'Badoglio swine', that is Italian soldiers who had remained in their barracks rather than dispersing homewards. They were miserably treated. In a move that would soon seal the fate of the monarchy, Victor Emmanuel and Badoglio fled the capital for Brindisi in the south; in contrast pope Pius XII remained at his post in German-occupied Rome, defying Allied aerial bombardment, and sheltering opposition political leaders, and Jews, in Church buildings, actions which immeasurably enhanced his stature in the post-war period, when the Church seemed to many a rock of stability. In September 1943 the Vatican refused to recognise Mussolini's new republic at Salò, thereby decisively distancing itself from the Fascist regime.

In the north an armed 'anti-Fascist' movement emerged, like a chill wind, with links to half-a-dozen political parties across the political spectrum. Their numbers rose from about nine thousand in September 1943 to twenty to thirty thousand fighters by early 1944, and to eighty-two thousand later that summer.[15] They included many self-proclaimed Christian Democrats, including a large unit called Green Flame, and priests who sometimes recruited and organised partisans as well as sheltering and succouring them.[16] Partisans fought a hit-and-run campaign against the Germans as exhausted British, American and Polish troops inched their way northwards towards the formidable 'Gothic Line' which field marshal Kesselring had thrown across Italy. As elsewhere in Europe, partisan warfare unleashed a horrific spiral of violence involving savage reprisals by the Germans whenever their troops were assassinated. At Marzabotto, near Bologna, an entire village of eighteen hundred souls was wiped out during reprisals. People connected with the resistance were arrested and tortured, including lifelong friends of such senior curial officials as Giovanni Battista Montini, the future pope Paul VI, a useful reminder that the Vatican was not somehow miraculously insulated from (or oblivious to) the horrors of war.[17] The various partisan groups acquired a political face in the local and regional Committees of National Liberation that were established in areas they conquered from the Germans and their Fascist confederates. By Christmas 1943 nearly two hundred opposition politicians met in secret in the Lateran Palace to discuss Italy's post-war future. In March 1944 Palmiro Togliatti, the Communist leader, announced the 'shift of Salerno' which enabled the

PCI to support the royalist government, a shift that reflected Stalin's desire not to jeopardise an Anglo-American landing in France through the extension westwards of the civil war already raging in Greece, although on the ground many Communist partisan groups were bent on far more radical objectives. In April 1945 the partisans instigated mass uprisings in the major cities of northern Italy, largely to afford their political leadership a say in the shaping of the country's future. After saluting them at victory parades, the Allies quietly disarmed the partisans, so as to stymie plans for a thoroughgoing social revolution. This did not prevent a bloodbath of former 'Fascists', known as the *epurazione*, that claimed the lives of between twelve and fifteen thousand people.[18]

The discrediting of the Fascist experiment in forging a new Italian national identity meant that three alternative models (or myths) competed for dominance in the post-war period. Two of these were foreign in inspiration. Communism enjoyed enormous prestige, as the fighting prowess of the Red Army, and of wartime partisans, amplified older myths of the USSR as a workers' paradise and of Stalin as everyone's favourite uncle. Of course, few Italians – except the quarter of a million who fought alongside the Wehrmacht – had ever been to Russia, whereas millions of Italians had relatives in the land of opportunity across the Atlantic. Anti-Communism provided common ground for two other powerful forces that sought to shape a new Italian identity, although in other respects their relationship was uneasy: the United States, which had achieved an unparalleled economic and military dominance, and the Catholic Church, which had emerged from the wreckage of Fascism, as it had survived the fall of the Roman Empire fifteen hundred years earlier. For a brief period the Church thought that it might reconquer Italy for Christian civilisation, a project that only partially overlapped, in the matter of anti-Communism, with the ascendancy of the US in the peninsula.[19]

These grandiose visions triumphed over the more mean-spirited policy of the British, who initially were given the leading role in the peninsula. The British sought a weak Italian monarchical regime, dependent on Britain's diminished power in the Mediterranean; the US was concerned to create a self-confident Italian democracy and to restore Italian sovereignty as rapidly as possible. Whereas the British adopted an unpleasantly punitive attitude towards the Italians, Roosevelt's awareness of the fickle electoral loyalties of six million Italian-Americans contributed to the US's more sympathetic treatment of the anomalous 'alleato nemico'.[20] A 'special relationship' between Washington and the Vatican, which had begun with Myron Taylor's appointment as Roosevelt's personal representative, grew deeper as the Americans gradually abandoned their wartime indulgence of the Soviets and their optimism regarding Stalin's willingness to co-operate in the reordering of the post-war world, and came round to Pius XII's damning verdict upon Communism as the greatest danger for the future of Italy and European Christian civilisation as a whole. In December 1943 under-secretary of state Tardini formally advised Taylor that the Vatican had decided to abandon its prudent agnosticism towards forms of government in favour of democracy:

> Only this kind of consent offers sufficient guarantees for the control of government by the people; it accustoms people to self-discipline; it makes it possible for everyone, from whatever class they come, to enter public life; it embraces all the vital forces of the country; it can gradually educate the Italian people towards the habit of moderation in political rivalries so that the general harmony of the country will not be impaired.[21]

For a crucial period, the Vatican effectively represented Italian interests in Washington, while the conference of US Catholic bishops, and New York's redoubtable archbishop Spellman, whom Pius wanted to appoint as his secretary of state, also ensured that the voice of the Church was heard.[22] The Communists retaliated with crude smears against both Pius and the Church in general, regarding the Catholic Church's role in the war years and their aftermath. The Communist paper *L'Unità* 'revealed' the financial and material interests that were alleged to drive Vatican policy: 'The robes of the papal nuncio in the USA are saturated not only with incense but with oil' is representative of its leaden materialist manner. The Soviets meanwhile hired a professional anti-religious slanderer, Mikhail Markovich Sheinmann, to smear the reputation of the pope, an approach subsequently elaborated by the left-wing German playwright Rolf Hochhuth in his 1963 play *The Deputy*, which still influences uninformed views of Pius XII.[23]

The political vehicle for the defence of Christian values was the newly founded Christian Democratic Party of Alcide de Gasperi, who after a spell as foreign minister became prime minister in December 1945 and would remain in that post until 1953. The leadership of the Christian Democrats largely came from that of the former PPI. Based on clusters of like-minded individuals in Milan and Florence, its party statutes were adopted at a congress in Naples in July 1944. Without a party apparatus of its own, the DC initially broke with Don Sturzo's earlier attempts to maintain a distance between Party and Church, although by the early 1950s it would ignore attempts by the pope to dictate policy. In addition to its network of twenty-five thousand parishes, the Church had quietly extended its influence during the Fascist period to parts of the urban working class, while carefully nurturing a new generation of political leaders through FUCI, the federation of university students under its president Giulio Andreotti, in which Giovanni Battista Montini had played a distinguished part. It took some time for the Vatican to acknowledge that the Christian Democrats were the ideal political vehicle for the defence of Catholic interests: it initially favoured a monarchy or an authoritarian regime along the lines of Salazar's Estado Novo.[24] Pius XII seems to have thought de Gasperi 'too feeble' in his toleration of the Communists and Socialists. Many in the Vatican wanted the Party to move to the right, whereas its leaders viewed it as a centre party moving to the left. But once the decision to support the Christian Democrats was reached, the Church made an enormous contribution to a party that came to be regarded as an ark of salvation for more than the Italian middle class, largely by providing a mass following for a parliamentary party of notables. Luigi Gedda's Catholic Action capillary organisations quietly encompassed artists, businessmen, doctors, farmers and teachers, and Catholic welfare associations reached out even to the very poor. A party whose support was historically strongest in the traditionally 'White' areas of the north attracted southern notables, with their clienteles, and the sinister might of the mafia and camorra, through colossal reconstruction projects that poured money into the Mezzogiorno.[25]

It is important not to project on to the early Christian Democrats what they undoubtedly became by the mid-1950s: an opportunist party whose sole raison d'être was to occupy and hang on to power at any price.[26] Although Christian Democrats came in various hues, roughly corresponding to the conservative right and a quasi-socialist left under Giuseppe Dossetti, they placed a keen emphasis on freedom under the rule of law, a principle that had tremendous resonance after the oppressions of Fascism and as the US-led free world squared up to the more resilient totalitarian menace. Apart from a Christian faith that brought him to mass every day, and his daughter to a convent, de

Gasperi combined underlying principles with a definition of politics as a form of mediation. As a young man he had denounced the 'religion of nationalism', while his rejection of Fascism was based on first principles. 'It is the concept of the Fascist State that I cannot accept,' he explained to a Fascist tribunal, 'for there are natural rights which the State cannot trample upon. I cannot accept the annihilation, the disciplining, as you say, of liberty.'[27] Never again would bullying Jacobin minorities, or 'conventicles' as he had it, be allowed to take away the freedom of majorities without protest. The self-consciously centrist Christian Democrats were almost as unsparing in their condemnation of bourgeois materialism as they were of Communism: 'the bourgeoisie has given us mechanical progress and not civilisation, because civilisation has above all a spiritual connotation'.[28] They opposed Communism, not simply because of its atheism – although Palmiro Togliatti, the Italian Communist leader, was profuse in paying his respects to religion – but because everything, from education to property, would ultimately end up in the hands of the state, while all arrangements with the Communists would be preliminary expedients to the rule of one party. The Christian Democrats supported the family and small property ownership against the state, at the same time recommending the legal expropriation of both big business and large-scale landownership, albeit with adequate compensation for the former owners. Although the Party's left-wing intelligentsia sought a 'social state', its middle-class farming and small-business base, and its supporters in industry, were hostile to anything that smacked of state interference.[29]

A narrowly won referendum sealed the fate of the monarchy in favour of a republic, but much of the Fascist state apparatus was left in place, contributing to the feeling on the left that there had been no radical break with the past. Career policemen and prefects replaced partisans who had briefly usurped their functions, while as minister of justice Togliatti himself introduced an amnesty for those affected by the process of *epurazione*, that is purges of former Fascists. They were so ubiquitous at every level of Italian administration that removing them, and replacing them with technically incompetent former partisans, would have spelled national disaster, but there was something morally distasteful about turning a blind eye to serial torturers.[30]

In these early years, de Gasperi's primary concerns were with establishing a political framework for democracy; halting rampant inflation; reviving the economy with generous US assistance, which in the two

years 1945–7 amounted to nearly US$2 billion; and securing the consti-
tutional position of the Catholic Church through confirmation of the
1929 Lateran Accords.[31] In its efforts to extend a hand to the country's
majority Catholic population, the Communist Party supported article 7,
which guaranteed the Catholic Church's position in the new republican
Constitution, a measure which passed by 350 votes to 149. The Church
regarded the existence of these rights as sufficient basis for its own
extensive interventions in the post-war political process. In November
1946 the Vatican warned de Gasperi, who had several leftists in his
coalition government, that 'any kind of collaboration with the anticlerical parties, not only in the municipality of Rome but in the government, is no longer admissible. If the Christian Democrats were to
continue with such collaboration, they would be considered a party
favouring the enemy. The Christian Democrats would no longer have
our support or our sympathy.' Having secured everything he could from
a coalition with the left, and with an eye to president Truman's squaring
up to the Soviets on a global scale, in May 1947 de Gasperi decided to
reform his government without Communist and Socialist participation,
and to postpone the elections due in October 1947 until the following
April. Reassured by this step, the Church swung its full might behind
the Party. It was a turbulent period. The Peace Treaty with the Allies
was deeply unpopular and peasant unrest turned violent in Sicily; six
hundred thousand landless labourers went on strike in the Po valley. The
left managed to weaken itself when a pro-Western Italian Social
Democratic Party split from the Socialists, whose Marxist remnant then
decided to join with the Communists in contesting the election as the
Democratic Popular Front. They chose Garibaldi as their electoral
symbol, although the Christian Democrats transformed him into a
Janus-faced figure with the moustached monster in the Kremlin lurking
around the back.

During the latter months of 1947 economic discontent assumed
violent forms in the northern industrial cities, the Po valley and parts of
the south, where the mafia was used to suppress peasant discontent.
Although Togliatti was concerned to prevent the sort of civil war that
had gone badly for the Communists in Greece when the Allies took the
royalist side, both former partisans and his own more radical supporters
were clearly bent on toppling the government of de Gasperi, towards
whom the US finally abandoned any residual reserve. At a meeting
in February 1948 with the Irish ambassador to the Holy See, Pius XII

seemed worn out and deeply pessimistic. 'If they [meaning the Communists] have a majority,' he said, 'what can I do to govern the Church as Christ wants Me to govern?' Ambassador Walshe offered the pope asylum in Dublin. Pius replied: 'My post is in Rome, and, if it be the will of the Divine Master, I am ready to be martyred for him in Rome.'[32]

The US began to prepare for the possibility that the Communist Party might react to defeats in the elections by seizing power in northern Italy, particularly if these defeats coincided with the planned withdrawal of Allied forces which would leave battle-hardened former partisans with nothing more fearsome to face than a demoralised Italian army and the paramilitary Carabinieri. The US National Security Council anxiously contemplated a nightmare scenario in which an exiled democratic Italian government would continue to challenge a Communist-dominated mainland from the Mediterranean 'Taiwans' of Sardinia and Sicily.[33] Truman's government began supplying the Italian army with small arms, while US warships anchored off all of Italy's main ports in the weeks before the election. The US gave the Vatican Bank the one hundred million lire proceeds of the sale of surplus military equipment to put at the disposal of Catholic Action and the Christian Democrats.

Events in eastern Europe, notoriously the murder of the Czech foreign minister Jan Masaryk, which the Italian left tried to deny in weasel fashion, increased the attractions of the American way over a totalitarianism that was losing its romantic wartime lustre. Regarding Italy as a 'test case' in its struggle with Communism, the Truman administration ploughed in US$176 million interim aid, with US ambassador James Dunn omnipresent as each shipment of supplies arrived in Italian harbours. American-Italian 'friendship trains' then distributed the goods throughout the country. The US offered a carrot, in the form of Anglo-French agreement to the return of Trieste and the Valle d'Aosta to the Italians, while secretary of state George Marshall waved a stick by threatening to cut all aid if the left won the Italian elections. The Italian-American community was mobilised to write to the voters of the old country, and cardinal Spellman warned: 'I cannot believe that the Italian people ... will choose Stalinism against God, Soviet Russia against America.' The bishops of Dublin and Kerry raised £50,000 for the Christian Democrats, and Irish Catholics followed the elections with keen interest.

In Italy itself, the Church mobilised on a massive scale, believing that the success of an alien Communism would be inexplicable other than as a triumph of superior organisation.[34] Milan's doughty cardinal Schuster instructed his flock: 'Catholics must cast their vote for candidates who they know will preserve the rights of the Church. It is impermissible for a member of the Church to cast his vote for a candidate he knows to be hostile to the Church, or hostile to the application in public life of Christian moral principles.' Ricardo Lombardi, the Jesuit editor of *Civiltà Cattolica*, was so omnipresent and voluble in his attempts to rally the troops for what he regarded as an apocalyptic fight with the powers of darkness that he was known as 'God's microphone'. As his journal explained: 'Bishop and priest dare not wait until they are in front of the firing squad, and civil and religious liberties have been extinguished. If the Communists win the election, the Church will have to be administered from behind the Iron Curtain.'[35] Although the Concordat and Electoral Law prohibited clergy from attempting to influence voters, parish priests effectively became propagandists for the Christian Democrats, while Luigi Gedda's gigantic lay network of Catholic Action assumed the task of getting out the Catholic vote through the agency of 'civic committees' (*comitati civici*) which made up for the Christian Democrats' lack of a local Party apparatus. A network of nuclei, whose leaders were responsible for mobilising the Catholic vote, encompassed virtually every house, farm and factory in Italy. When the Communists accused the Church of bringing even the dead or the insane to the polls, the Church responded: 'The indignation of the Communists is as reasonable as that of the famous sword-dueller, proverbial in Italy, who shouted at his opponent, "But if you don't stand still, how can I run you through?"' 'Labour chaplains' were sent into the factories to combat Marxism at its source, while students from seminaries and universities were seconded to help the bishops in the task of getting out the huge female Catholic vote. Christian Democrat propaganda, whose main symbol was a shield with a cross inscribed 'Libertas', painted the not inconsiderable Red spectre on the wall, and reminded Italians that it was the US rather than Russia that was providing such prodigious quantities of aid as well as the alluring luxury items to be had on the US-fed black market. Posters showed 'Mongol' Red Army soldiers battering the Christian Democrat shield with their hammers and sickles, or despondent Italian families pondering the only aid the Russians gave – dynamite and pistols. 'Save your children' said a poster showing a gay little girl about to be crushed by an enormous Soviet tank. Pius XII never failed to broach the dark prospect of Cossacks watering their horses in Rome's delightful fountains.

Although the Communists still retained a certain leather-jacketed and neckerchiefed glamour, and dominated high culture and the universities for decades ahead, America – rather unfairly – signified cheap watches, chewing gum, chocolate, nylons, boogie-woogie, DDT and a relaxation of relations between the sexes rather than university faculties that included Einstein. It was no competition, although ironically the Church may have had greater sympathy with the austerity and sexual puritanism of the Communists than with the hedonism and materialism of the Americans.[36] In the rougher game routinely played in the Mezzogiorno, bishops and priests refused the sacraments (or a Christian funeral) to Communist and Socialist leaders, or banned them from the prestigious committees that organised the feasts of local patron saints, while mafia gangsters shot left-wing agitators or lobbed the occasional grenade into meetings of aggrieved peasants.

Whether by fair means or foul, the results increased the Christian Democrat poll from 35 to 48.5 per cent of the electorate, giving them an absolute majority of half the seats in the chamber. Having begun life as a party opposed to the system of liberal Italy, political Catholicism became – until the early 1990s as it proved – the main party of government. *Civiltà Cattolica* welcomed the outcome of the elections: 'On 18 April the Italian people decided for Christ and his representatives. With their own bodies the Italian people have erected a bulwark around the cliffs of the Vatican, thereby maintaining the sacred character of Rome and its centuries-old role as a centre of Christian culture.' Pius XII declared that 'the skies of Italy are now lightened with a new hope of tranquillity and order which will make speedily possible the material and social reconstruction of the country ... this day had also revived the confidence of Europe and the whole world'. However, the outcome had not been a Christian 'reconquest' of Italian society. Rather, the Church found itself battling a more insidious enemy in the form of Hollywood and American consumerism, while the Christian Democrats ignored the Church's desire for Italy to practise an 'equidistance' between the superpowers, as they opted for membership of NATO.

Unlike Italy no significant armed resistance developed in Germany; resistance was confined to lone individuals and isolated groups to which over-large labels are routinely attached for political reasons. As the SS secret monitoring of popular opinion reveals, large numbers of Christians had remained immune to the political faith of Nazism, sensing that it could offer no spiritual consolation or that it was actively satanic. The progress of the western Allies through France and of the Russians into East Prussia, together with relentless aerial bombardment and the non-appearance of wonder weapons, brought widespread disillusionment. Public hangings and shootings of deserters and dissenters, added to an epidemic of suicides by the true faithful, meant that in a metaphysical sense Nazism had collapsed before the Allies arrived.

Christians of all denominations had been active in the Kreisau Circle around Graf Helmuth James von Moltke, who presided over highly illegal discussions that sought to establish a moral framework for a post-National Socialist Germany. Merely being connected to such activities proved to be a death sentence, as the Jesuit Delp and Moltke himself discovered, once these discussions were deemed to have been part of attempts to assassinate the German Führer. Even expressing an abstract affirmative in the confessional to the question of whether tyrants could be killed, was enough to condemn the Munich priest chaplain Wehrle, since he should have known the identity of the tyrant concerned. A more exclusive group of practising Christians were among the brave individuals who in July 1944 made the most serious attempt on Hitler's life, a doomed enterprise that they felt they must undertake in order to testify to the existence of 'another Germany' uncontaminated by Nazism.

At the end of the Second World War, Germany was not simply physically ruined – a chaos of severed bridges, shattered masonry and twisted railways. Its people were also bewildered, exhausted and traumatised, not to speak of those Germans, slave labourers and refugees who were victims of National Socialism. Both Nazism's more than rhetorical 'national community' and the dislocations of total war had had what sociologists call a modernising effect, levelling hierarchies and bridging denominational, regional and class divides. Germany's traditional elite groups had been either irreparably weakened or destroyed, ensuring that they would not exert the deleterious influence they had manifested during the Weimar Republic. But there was a noteworthy exception to the fate that otherwise befell the armed forces, heavy industry and major landowners, which is important for a spiritual, as opposed to sociological, audit of Germany at the end of the war. Although the Churches suffered their share of human and material loss, in the form of bombed-out buildings, pastors and priests who had been imprisoned or killed, the disruption of their capacity to reproduce through seminaries and theological faculties, and, in the case of the Catholic Church, the virtual eradication of its lay organisations, they were the one organisation to survive the war relatively intact. Observers compared the situation of the German Churches to that of early Christians in the era of the Roman catacombs, an analogy that had the requisite odour of martyrdom and the promise that things could only improve.

In 1945 the Allied occupying powers and the broad German public had a greater regard for the conduct of the Churches under National Socialism than would be the case by the 1960s, the beginning of decades of therapeutic inquisition that has since become tawdry. As if to symbolise this, in July 1945 the BBC broadcast a remarkable memorial service celebrating the life of Dietrich Bonhoeffer from Holy Trinity Church in London's Kingsway, Bonhoeffer having been murdered a few months before. Virtually all sections of German resistance to Nazism had had a Christian presence, with a third of Catholic clergy coming into some sort of conflict with the regime, in the form of warnings, threats, fines, arrest or imprisonment. At the end of the war, both major Churches proved adept at transforming individual clergy who had resisted Nazism, such as the Catholics Delp and Galen, or the Protestants Bonhoeffer and Niemöller, into representatives of institutions whose corporate conduct was less gloriously heroic than the 'Lion of Münster' who, consistent to the end, was soon roaring at the petty injustices of the uncomprehending British. The Allies subjected the Churches to few restrictions and refrained from interfering with their internal organisation. Clergy were among the few Germans allowed to travel freely. Even the Soviets, whose conduct in occupied Germany was otherwise disgraceful, respected the sites and symbols of Christian worship. This was at once attributable to the co-operation of Christians and Marxists on the National Committee for 'Free Germany', as well as the residual Orthodox religiosity of many Red Army soldiers. Although bishop Preysing of Berlin ostentatiously refused to have any dealings with either the Soviets or the German Communists, whom he regarded as the moral equals of the Nazis, a certain pragmatic continuity in dealing with totalitarian regimes was guaranteed by bishop Heinrich Wienken, who had performed a similar function under Hitler.

Because of the moral regard they enjoyed, churchmen played a considerable role in the selection of people who would help in the reconstruction of Germany, providing testimonials of a person's political probity which helped smooth their way through the more or less stringent 'de-Nazification' procedures which each occupier adopted. Some clergy took the opportunity to correct what they thought was the

Allies' (and especially the British) predisposition towards the political left, a conservative clerical bias that, among others, the Christian Democrat Leo Schwering, who had himself been imprisoned by the Gestapo, highlighted.[37] Many churchmen were unhappy with the entire process of 'de-Nazification', since its use of clumsy categories failed to distinguish the innocent and the hapless from the guilty. As Martin Niemöller said in 1948, 'de-Nazification' also opened the floodgates for personal hatreds masquerading as civic virtue. Church-led opposition to 'de-Nazification' raised wider questions of collective and individual guilt. Even the radical theologian Karl Barth wondered what was the point of the exercise, since enthusiasm for Nazism seemed to have evaporated long before the end of the war. While the Protestant bishop Theophil Wurm was prepared to see war criminals punished, he thought the Allies were in breach of the maxim *nulla poena sine lege* as they sought to criminalise actions or expressions of opinion that were technically legal under German law before 1945. Bishop Galen of Münster used a sermon to protest: 'if people suggest that the entire German people, and each of us individually, are guilty of crimes which happened in foreign lands and in Germany itself, and above all those committed in concentration camps, then that is an unjustified and untrue accusation'. Pius XII concurred. In February 1946 the pope remarked, while investing Galen, Frings and Preysing with the red hat, 'that it is wrong to treat someone as guilty, when personal guilt cannot be proved, only because he belonged to a certain community. Ascribing collective guilt to an entire people and treating it accordingly is to interfere in God's prerogative.'[38] Both Niemöller and cardinal Frings were prominent in persuading the Allies to abandon their blanket 'de-Nazification' procedures. Frings, who proved himself a real thorn in the Allied flesh at every opportunity, denounced this 'Nazi inquisition' so forcefully that the chairman of the German Review Board resigned, warning that no Catholic lawyer would ever serve in his place. For once, clergymen demonstrated cold-blooded realism while the Allies floundered around in a woolly-minded self-righteousness. In reality, there was a more urgent, pragmatic reason for not carrying out wholesale purges of former Nazis, namely that a blanket juridical purge would make reconstruction almost impossible, a lesson apparently not learned either in contemporary Iraq. Of the twenty-one skilled personnel in Cologne's waterworks, only three had not belonged to the NSDAP; of the 112 doctors in Bonn, 102 were Nazi Party members. Simply to dismiss people on the basis of their membership of proscribed organisations was to invite chaos. By September 1948 'de-Nazification' had effectively been abandoned.[39]

The moral authority of the Churches was further boosted in German eyes by the role they played in averting widespread disease and mass starvation in the winter of 1945. Between 1945 and 1949 fifty-five thousand Protestants were involved in distributing sixty-two million tons of food and clothing, as well as processing the details of some ten million people who had lost touch with their families. They also organised youth camps for the large number of young people who might otherwise have fallen into crime, vice and delinquency. Before and after the watershed of May 1945, Europe witnessed the largest migrations it had undergone since the end of the Roman Empire. Some ten million ethnic Germans fled or were forcibly expelled from East Pomerania, Danzig, Lower Silesia and East Brandenburg, to be followed by those driven out of Czechoslovakia, Hungary and Poland. Since international refugee organisations were confined to helping non-German 'displaced persons', the burden of dealing with these huge numbers of indigent people largely fell upon Christian charities. The Vatican managed to send 950 goods trains loaded with food, clothing and medical materials. The Catholic Caritas Association distributed about 150,000 tons of aid between 1945 and 1962, and successfully relocated four hundred hospitals and charitable institutions that would otherwise have been lost beyond the Oder–Neisse line. The newly minted cardinal Frings endeared himself to many Germans when in a radio broadcast he allowed that to steal food or fuel when in dire need was not a mortal sin, which led to the new verb *fringsen*, or to steal for worthy reasons.[40] Protestants tried to prick the consciences of their ecumenical contacts with harrowing photographs of starving people, while simultaneously enjoining their own fellowship to spare something for the dispossessed from their meagre rations. After an unwarrantable delay, both Churches established dedicated agencies to assist those Christians who had been persecuted for racial reasons.

The Churches were also at the forefront of providing explanations for horrors that leading historians, judging from the octogenarian Friedrich Meinecke's pitiful *The German Catastrophe*, with its advocacy of Goethe Societies as the solution to Germany's spiritual crisis, seemed unable satisfactorily to explain. It is misleading to imagine that there was no serious reflection on the evils of the immediate past, although modern left-liberal historians have almost succeeded in popularising the view that they discovered the evils of 'Fascism' in the 1960s, a conceit hard to reconcile with Eugen Kogon's 1949 *Der SS-Staat* or Karl Dietrich Bracher's monumental 1950–4 *Die Auflösung der Weimarer Republik.*

At the time, religious thinkers, such as Gustav Grundlach or Romano Guardini, who had gone into exile or semi-retirement in the preceding twelve years emerged to find that they had a wider audience than they ever imagined despite the abstruseness of their work or the considerable limitations on what could be published. They were joined by a substantial number of conservative writers who applied two of their principal past complaints about Western modernity – namely 'massification' and 'de-Christianisation' – to National Socialism (and Communism), discovering in democracy a new form of defence of Western civilisation against what they were learning to call 'totalitarianism'.[41] Similar notions were popular among more exclusively religious thinkers who emphasised the evils of secularisation, and mankind's abandonment of a divinely decreed moral order, which had been replaced by an amoral world in which demonic forces used ruthless demagogues and crooked simulacra of religion to propel the masses towards ever darker moral degradation.[42]

This was the view taken by Konrad Adenauer in his first major speech as provisional chairman of the Rhenish Christian Democratic Union before a large audience at Cologne university. Like de Gasperi, who was sixty-four when he became Italian prime minister, Adenauer – who was sixty-nine in 1945 – benefited from the discredit which totalitarianism had wrought upon the cult of youth. Although not without dry humour, Adenauer had a face of almost oriental impassivity, as if carved from some exotic hardwood. He had become lord mayor of the Rhineland metropolis in 1917 and had already negotiated a seven-year period of Allied military occupation after the First World War. In late 1945 the British military authorities inadvertently facilitated the coolly impassive elderly gentleman's nationwide political ascendancy by rudely dismissing him from the post the Americans had appointed him to a few months earlier. His Cologne speech was a masterly mixture of shame regarding the crimes of the recent past with pride in the steadfastness of the German spirit. Although unsparing in blaming such groups as the generals and industrialists, Adenauer acknowledged a more pervasive German tendency to treat the state as an idol, sacrificing on its altar 'the dignity and value of the individual person'.[43]

Both Churches also made public pronouncements on the broader subject of guilt, something they had ostentatiously denied after the 1914–18 war when German 'guilt' became part of the Versailles peace treaty. The Catholic hierarchy were the first to make a solemn statement on the subject, in the joint pastoral letter issued by the Fulda Conference of Bishops in August 1945. This began with a rousing declaration regarding the impermeability of the faithful, or their refusal to bow the knee to Baal, as the bishops put it, before a measured acknowledgement of both crimes and moral complicity for which *some* of their fellow Catholics and Germans had been responsible:

> Terrible things were done in Germany before the war and by Germans during the war in occupied countries. We lament this deeply: many Germans, including those from our own ranks, allowed themselves to be deceived by the false teachings of National Socialism, remaining indifferent towards crimes against human freedom and human dignity; the conduct of many facilitated these crimes, many others became criminals themselves. A heavy responsibility applies to those who could have used their influence to prevent such crimes, and who did not do so, but rather made these crimes possible and thereby declared their solidarity with the criminals.[44]

Among those who failed to be impressed by the conduct of the Catholic bishops was Adenauer, who in early 1946 commented on an essay titled 'The Silence of the German People' by Max Pribilla, a Jesuit schoolfriend, and editor of the periodical *Stimmen der Zeit*. Giving the lie to the contemporary conceit that conservatives somehow conspired in a form of public amnesia regarding Nazi criminality, Adenauer said that everyone was aware of the illegality of 'the pogroms against the Jews in 1933 and 1938' and the 'unparalleled barbarities' in Poland and Russia. He argued that the bishops should have agreed among themselves jointly to denounce these things from their pulpits on a particular day, and that if they had done so many crimes could have been prevented: 'That did not happen, and for that there is no excuse.' If the bishops had been imprisoned or sent to concentration camps so much the better.

In October 1945 the Protestant Church leadership weighed in with the 'Stuttgart Declaration of Guilt', which was signed by eleven prominent Church leaders, including Otto Dibelius and Martin Niemöller, as an indispensable precondition for German Protestantism's readmission to the ecumenical community symbolised by the newly founded World Council of Churches, a delegation of which visited the Germans. It was a potentially difficult moment since countries that had fought the Nazis were being exposed to the full, post-facto shock of the extent of Nazi criminality, while the Germans felt deeply aggrieved at such injustices as 'de-Nazification', mass expulsions of population and being treated as international pariahs. The key passage of the declaration read:

> We are the more grateful for this visit [from the World Council of Churches], as we with our people know that not only are we in a great company of suffering, but also in a solidarity of guilt. With great pain do we say: through us endless sufferings have been brought to many peoples and nations. What we have often borne witness to before our congregations, we now declare in the name of Jesus Christ against the spirit which found a terrible expression in the National Socialist regime of tyranny, but we accuse ourselves for not witnessing more courageously, for not praying more faithfully, for not believing more joyously and for not loving more ardently. Now a new beginning can be made in our churches. Grounded on the Holy Scriptures, directed with all earnestness towards the only Lord of the Church, they now proceed to cleanse themselves from influences alien to the faith and to set themselves in order. Our hope is in the God of grace and mercy that He will use our churches as His instruments and will give them authority to proclaim His word, and in obedience to His will to work creatively among ourselves and among our whole people.[45]

Perhaps the most astonishing feat of the immediate post-war period was the creation of two avowedly interdenominational Christian political parties, which the far-sighted few – such as Adam Stegerwald and Konrad Adenauer – had advocated in the inter-war era and whose attractions had multiplied under the conditions of the Nazi regime. The presence of Stalin's legions on the Elbe further concentrated minds. Ecumenical activity was sanctioned at the highest ecclesiastical levels, since the Catholic Church quickly realised that it alone was not strong enough to combat a left to which Protestants might defect unless there were a powerful interdenominational conservative alternative. In Bavaria, where a nostalgic monarchism initially weakened the right, one of the key supporters of this development was Josef Müller, whom we encountered earlier as the key intermediary between the pope, the German conservative resistance and the British in 1940. In June 1945 Pius XII implicitly gave Müller the green light for interconfessional political activity, when he said that as Catholics and Protestants had stood together against Hitler they should work together against Marxism.[46] Both the Catholic hierarchy in Germany and such leading Protestant figures as bishop Otto Dibelius supported the new Christian Democrat parties. Although the Protestants were more circumspect, the Catholics were never going to support the SPD, whose leader Schumacher's embittered outbursts included calling the Catholic hierarchy the 'fifth occupying power', and who in 1945 said: 'It is precisely the Nazis and reactionaries who for better or worse want to keep what they have in hand and who would as gladly camouflage this under the term "Christian" as they previously did under the term "national".'

While the Social Democrats simply picked up where they left off in 1933, failing to adjust either their Marxist dogma or over-centralised organisation to evolving realities, the Nazis' virtual obliteration of the rest of the party-political landscape, and the Allies' limitation of what was acceptable on the right, provided a crucial opening which a remarkable generation of German politicians took creative advantage of. Occupation conditions meant that the new Party was inherently multi-centred and had subtle regional emphases. A witty French observer once described the Christian Democrats as 'socialist and radical in Berlin, clerical and conservative in Cologne, capitalist and reactionary in Hamburg, and counter-revolutionary and particularistic in Bavaria'. Actually, a better way of describing the CSU would be to imagine that Scotland had undergone a Catholic Counter-Reformation.[47]

Political activity in Germany resumed about six weeks after the end of the war. Separate Allied occupation zones, the disruption and restriction of communications, and large numbers of robust individuals with a strong local following entailed party-political initiatives of a highly localised character, with the Christian Democrats only coalescing into the Christian Democratic Union or Christian Social Union (the more particularist Bavarian branch of the Party) in the course of 1947, a development that roughly paralleled the Western Allies' decisions to merge the various western occupied zones. Andreas Hermes and Jacob Kaiser founded the first 'German Christian Democratic Union' (CDUD) in wartorn Berlin in June 1945. They sought to nationalise heavy industry, to afford workers a huge say in the running of businesses, and, last but not least, to establish a neutral, Christian socialist Germany that would mediate between West and East. Both a programme that seemed more socialist than Christian and the machinations of the totalitarian

Socialist Unity Party ensured that the former capital exerted less influence than some of the western regions.

Initially, the Catholic culture of the westerly Rhineland had more enduring significance for the CDU than the former Prussian–German capital, although over time the CDU has taken on the confessional colouration of whichever area it seeks votes in. Adenauer made strenuous efforts to recruit Protestants to the new party, winning over such distinguished figures as Eugen Gerstenmaier, Gustav Heinemann, Hermann Ehlers, Friedrich Holzapfel, Ludwig Erhard, Robert Pferdmenges, Gerhard Schröder and Otto Schmidt. In return for a fair share of influence within what was a heavily Catholic party, in key areas Protestants were allowed to set the policy agenda. This was despite the fact that a Dominican priory at Walberberg had been the setting for the earliest discussions of Party policy in which the 'social' aspects of Catholicism seemed dominant. This tendency was short lived. The spirit of Thomas Aquinas may have hovered over the Party's 1947 Ahlen Programme, but two years later that of Adam Smith prevailed in the CDU's Düsseldorf Programme.[48]

Compared with the Centre Party, a party of confessional beleaguerment, which limped along until the Catholic hierarchy deliberately killed it, and despite its high levels of support among practising Catholics, Christian Democracy seemed much more attractive to Protestants, to whom it made important concessions. It proclaimed that doctrinal differences between Christians were less important than the chasm that separated them from atheist materialists. Its emphasis upon social justice, largely derived from the tradition of Social Catholicism, would smooth the rougher edges of free-market capitalism, while stopping short of collectivist state socialism, thereby attracting some workers to what was otherwise a party of bourgeois self-assertion. A powerful Protestant component, deliberately rewarded with a fairer proportion of posts than a Protestant-dominated German state had ever conceded Catholics, ensured that social justice would not stifle individual enterprise. Ludwig Erhard's 'social market economy' did not camouflage the fact that this was a conservative pro-enterprise party, with the 'social' concerns increasingly hived off to committees of ecclesiastics and lay experts. As the official definition said: 'The "social market economy" means that the market is regulated by the needs of society – i.e. the activity of free and competent agents is directed to the highest possible degree towards the economic benefit and social justice of all.' By the late

1950s Catholic thinkers were worried that the interconfessional political experiment had been too successful, with the Protestants acting as a Trojan horse for liberalism and secularism by another name. These fears were increased by the CDU's coalitions with predominantly Protestant parties.

The incipient Federal Republic was always more than an 'economic miracle'. It is easy to forget that post-war Germany's new rulers came from a generation that did fourteen hours of Latin and Greek each week at school, and that many of them were more than conventionally devout. In broad cultural terms, the CDU reflected Adenauer's conviction that a fault line ran through Germany itself. Germany west of the Elbe and south of the Weser had been Christianised a millennium before the rest of the country, whose eastern regions had been pagan as late as the fourteenth century. The Prussian cult of the state had grown in this thin soil, itself providing the root stock upon which Nazism had thrived.[49] The CDU would be aggressively pro-Western, eliding Schumacher's socialism with an 'alien' Prussianism as well as the 'oriental' Kremlin, while its enthusiasm for Europe meant that for the first time a centrist conservative party could appear more internationalist than the parties of the left which persistently played the nationalist card. The Rhinelander Adenauer joined his fellow 'frontiersmen', de Gasperi from the former Habsburg Trentino and the Alsatian Robert Schuman, in discussing the future of Europe in a German that all three men spoke fluently.[50] The Western, Catholic orientation of the CDU, or what one distinguished scholar has called 'a German and European policy, with Cologne cathedral as centre', appalled a number of prominent Protestants who combined anti-Americanism with anti-Catholicism. The theologian Karl Barth routinely denounced the US while finding every conceivable excuse for Stalin's Soviet Union. Germans and Europeans, he felt, should opt out of the Cold War, pursuing what he claimed was Jesus Christ's neutralist 'third way'. Martin Niemöller similarly attacked the newly founded West German state as a Catholic confection, and denounced Adenauer for appearing to be in no hurry to forfeit Catholic dominance through reunification with a larger number of east German Protestants: 'the present form of the West German state', he declared in 1949, 'was conceived in Rome and born in Washington'. While the Catholic Church was concerned about the fate of the much smaller numbers of German Catholics marooned in the east, it generally supported Adenauer's pro-Western line, and was cooler towards the notion of a reunified Reich that historically had been dominated by its confessional opponents.

The broader relationship between the CDU–CSU and the Churches was far from straightforward, particularly in the case of the Protestant Churches, which sometimes viewed the Christian Democrats as a Catholic cabal. The crypto-constitutional Basic Law guaranteed religious freedom and the generous flow of Church taxes, while the immense range of institutions involved in health and welfare were financed by the state but run by the Catholic and Lutheran Churches. Protestant clergy generally kept a healthy distance from the new party, although the anticlerical stridency of Schumacher's Social Democrats propelled many of them to abandon their neutrality. By contrast, the Catholic Church was more forthcoming in its support, with cardinal Frings ostentatiously joining the CDU in December 1948, and many of his episcopal colleagues openly supporting it. The Church helped organise local Party groups and allowed them to use its halls in the absence of a political infrastructure. As in Italy, the Church actively encouraged the laity to obey the dictates of conscience during polls, although perhaps not so unashamedly as the Bavarian priest who said: 'It is not for me to tell you how to vote. But I do say: Vote Christian! Vote Social!' Others were more subtle: 'Everyone must vote according to his conscience. But it is clear that every true Catholic's voice of conscience recommends giving his vote to the candidate or the list which offers really adequate guarantees for the protection of the rights of God and the soul, for the true good of the individual, the family and society, in keeping with the law of God and Christian ethical teachings.'

In return for its support, the Catholic hierarchy expected to influence policy, seeing it as a vehicle for a Social Catholicism that seemed increasingly outmoded. It established a liaison office in Bonn, under monsignor Wilhelm Böhler, who in turn created a series of committees and working parties designed to represent Catholic views among politicians. Böhler thought that his remit included influencing appointments within the civil service, his main interlocutor being Adenauer's staunchly Catholic éminence grise Hans Globke, whose political influence had not been diminished by his co-authorship of the legal commentary to the 1935 Nuremberg Laws. Protestants in the Party responded with their own working group to tilt appointments the other way. In practice, the wily Adenauer gave the Catholic Church the illusion of influence, affording it a limited say in the formation of social policy, but that influence stopped whenever it threatened his broader political calculations. He thought that clergy of any stripe had no particular political competence. Not even the intervention by Pius XII in 1949 in favour of parents' rights to confessional schools, as guaranteed under the still operative Reich Concordat, could persuade Adenauer that this issue was worth permanently alienating the Free Democrats, and Social Democrats who no less vehemently opposed it.

One issue that fundamentally divided Catholics and Protestants was German rearmament, something the US military increasingly desired to counter both the might of the Red Army in central Europe and the 1948 decision of the German Democratic Republic to create paramilitary police units stationed in barracks. Events in East Asia raised the temperature in Europe. North Korea's invasion of the South in the summer of 1950 prompted the East German leader Walter Ulbricht to issue rhetorical threats linking the 'puppet regimes' in Seoul and Bonn. With the encouragement of Churchill, Adenauer suggested to the Americans that West Germany might supply 150,000 volunteers to a future European army. The cabinet retroactively sanctioned this recommendation, with one notable dissenting voice, the minister of the interior, Gustav Heinemann, who immediately resigned.

The leading Protestant within the CDU, and *Präses* (chairman) of the Protestant Synod, Heinemann believed that on two occasions God had justly removed weapons from the German people and that rearmament was morally wrong. He also thought that such a momentous decision should be subjected to a popular mandate, including the views of the 'brothers in the East'. Adenauer countered that a passive stance towards the Russians would only invite aggression, past experience having taught him that only strong defences, and the prospect of annihilation, would deter a totalitarian power from expansionist aims.[51] Heinemann was supported by the increasingly hysterical Niemöller, whom Adenauer began to characterise as 'an enemy of the state', after the pastor accused the chancellor of secretly manufacturing weapons and using former members of the Wehrmacht to organise an army against the wishes of the majority of the German people. Niemöller also seems to have imagined that he was entitled to pursue what amounted to a separate Protestant foreign policy, thereby straying into territory that Adenauer regarded as peculiarly his own. While Heinemann and Niemöller represented a nationalist and neutralist left-wing trend in German Protestantism, which chimed with the 'ohne mich' (count me out) mentality of many in the SPD, other conservative Protestants, such as Hermann Ehlers and Eugen Gerstenmaier, declined to support Adenauer's critics, and backed German rearmament. The Catholic response was even more extraordinary.

Cardinal Frings cited the authority of Pius XII for the view that 'it would be reprehensible sentimentality and falsely directed humanitarianism if, out of fear of the suffering of war, one permits every kind of injustice to occur. If, in the opinion of the Holy Father, going to war can be not only a right but also an obligation of states, so it follows that propaganda for an unlimited and absolute conscientious objection to military service is not compatible with Christian thinking.'

The Catholic Church opposed pacifism, neutralism and the granting of legal protection to conscientious objectors, reminding believers of such saintly warriors as St Sebastian in the remote past. In describing Charlemagne's western empire, Frings rather pointedly remarked that its eastern border was at Magdeburg. In March 1952 the Soviets seemed to dangle the prospect of German reunification as a reward for German non-participation in any coalition or military alliance directed at one of the victors of the Second World War. Adenauer rightly sensed a plot by Stalin to use a neutralised Germany to provoke a US retreat from western Europe. To that end, Stalin was prepared to write off the German Democratic Republic, hoping that with appropriate guarantees Communism would triumph in the end. Similarly those Germans, such as Heinemann, who claimed that Adenauer had missed a crucial chance to reunite Germany were ready enough to abandon the claims of the eight and a half million ethnic German refugees, half of whom hailed from east of the Oder–Neisse frontier that a reunited Germany would relinquish for all time.[52] The Catholic Church mobilised its lay organisations, with the Federation of German Catholic Youth and the Association of Catholic Men's Organisations in supporting government policy on rearmament. Clergy who dissented from this view found themselves banished to remote parishes, and leading Catholic intellectuals who advocated neutrality, similarly found themselves the object of the hierarchy's froideur.

The Protestant disarmers, meanwhile, founded a party – the All-German People's Party – which they hoped would be a Protestant neutralist alternative to the Western, rearming Christian Democrats. Unfortunately for them, few Protestants appeared to sympathise with their stance – which was permeated with vulgar anti-Catholicism and guilt-ridden naivety towards the Soviet Union. In elections that came a few months after the June 1953 popular uprising in East Berlin, the Party was wiped out by Adenauer's CDU, which became the first party to achieve an absolute majority in a German election, a result that was possible only because it attracted a very high number of Protestant supporters. The majority of Protestant Church leaders thenceforth neither endorsed nor opposed German rearmament, although Niemöller and his admirers continued their opposition for some years. In 1955 they did, however, strongly oppose the introduction of military conscription, arguing that each citizen should decide whether or not to take up arms. They largely got their way, especially after the equipping of the Bundeswehr with tactical nuclear weapons brought into question older notions of what constituted a just war when war seemed to promise indiscriminate annihilation. As a direct result of their interventions on behalf of reluctant conscripts, Germany introduced some of the most extensive exemptions for conscripts in the world.

Once again, the Catholic Church adopted an entirely contrary position, claiming that a professional army would be an updated version of the predominantly Protestant Prussian army of old, and opposing any exemptions for conscientious objectors. The Catholic Church also supported the deployment of nuclear weapons, with cardinal Frings choosing a visit to Japan, of all places, to declare: 'The Catholic Church does not advocate the outlawing of atomic and hydrogen weapons at the present time.' Christian Democrat politicians required more than this cryptic utterance when the Social Democrats went on the attack against the prospect of 'nuclear death' in the 1958 elections in North Rhine–Westphalia. The Catholic Church issued a lengthy justification of nuclear weapons, albeit within the desired context of controlled disarmament, against an enemy bent on destroying 'all contrary beliefs and life'. It categorically rejected the facile quip 'better Red than dead': 'If, then, a state belongs to such a defence alliance and carries out all the implicit obligations for defence, including acquisitions of the appropriate weapons, it is only fulfilling the obligation to its own citizens and to the international community.' While German Protestants in many respects anticipated what subsequently was called 'Ostpolitik', notably claims on what had become Polish or Soviet territory, just as its Stuttgart declaration on guilt was a harbinger of Willy Brandt kneeling in the Warsaw ghetto, the German Catholic Church remained silent.

While Christian Democracy helped return Italy and Germany to Western liberal democratic civilisation, the victory of Franco's forces in Spain inaugurated a reactionary regime whose preferred models were not Mussolini or Hitler, but the Catholic monarchs Ferdinand and Isabella. The symbolism of the Crusades was omnipresent at the Festival of Victory in Madrid which culminated in a five-hour march-past of Nationalist forces, with aircraft tracing 'VIVA FRANCO' in the leaden skies. At the royal basilica of Santa Bárbara, Franco presented his sword of victory to cardinal Gomá, in a setting heavy with relics of Spain's crusading past. Franco, whose bluff soldierly Christianity had deepened under the influence of his wife, vowed: 'Lord God, in whose hands is right and all power, lend me thy assistance to lead this people to the full glory of empire, for thy glory and that of the Church. Lord: may all men know Jesus, who is Christ son of the Living God.'[53] Unlike Croatia or Slovakia, or for that matter the Basque region or Ireland, where religion was integral to the national self-consciousness of a marginalised and repressed people, Franco's 'national Catholicism' was an attempt to recover past glories that had only been achieved in the first place through the total identification of Church and nation.[54]

Wherever the right triumphed, the Republic's anticlerical and secularising legislation was nullified in an atmosphere of distasteful ecclesiastical triumphalism. The measures that were reversed included civil marriage and divorce, the prohibition of the Jesuits, and the exclusion of monks and nuns from education, while intimate human affairs were resubjected to ecclesiastical courts. As part of the desired 'resacralisation' of Spanish life, crucifixes became compulsory adornments of classroom walls and religion part of the curriculum from elementary schools to universities. The quarter of a million Republican inmates in Francoist prisons were not spared the attentions of their erstwhile victims, since clergy were given extensive powers to bring about their conversion through compulsory mass and catechisms. Churches were rebuilt on a lavish scale with the aid of public subsidies and every area of public life was suffused with an outward show of piety through evangelistic rallies and processions. Their symbols were dwarfed by the largest cross of all time. In April 1940 Franco embarked on the megalomaniac Valley of the Fallen north-east of Madrid, a huge granite monument, with a five-hundred-foot cross, built by penal battalions of Republican prisoners who expiated their sins with blood smeared in the granite. Clerics from modest rural backgrounds manifested an embarrassing obsequiousness in the company of the rich and powerful. A degree of outward religious conformity was indispensable to getting and keeping a job, while men whose intellectual horizons were limited by their seminaries exercised censorship over films and books they knew nothing about. The only area in which the Church experienced minor setbacks was when confessional trades unions were absorbed into state syndicates and the Falange insisted on the prohibition of Catholic boy scouts.

While the Vatican had been careful to distance itself from the wartime effusions of most of the Spanish hierarchy, by May 1938 so many governments had recognised Franco that it duly followed suit. The regime may have heaped privileges on the Catholic Church, but it did not reciprocate with unqualified approval, for the more avant-garde or nationalist elements in the Falange included anticlericals, or were otherwise wary of the Church's internationalism at a time when Christian Democracy was ascendant elsewhere. The Church's concern for the losers of wars (in this case former Republican sympathisers rather than Fascists and Nazis) met with a stony-hearted response in Franco's Spain. When in April 1939 Pius XII sent the new government congratulations on its victory in the Civil War, passages in which he urged magnanimity and moderation upon the victors were cut from the Spanish transmission of his thoughts. Cardinal Gomá found that a similarly irenic pastoral letter *Lessons of the War and the Duties of Peace* was suppressed, as the regime did not care for his talk of respect for human rights or his strictures on the growing power of the state. In 1942 a pastoral letter from the bishop of Calahorra y La Calzada condemning National Socialism was similarly proscribed. In Seville, archbishop Pedro Segura had a number of run-ins with the local Falange when he refused to allow them to inscribe the names of their dead, and that of José Antonio Primo de Rivera, on the walls of his cathedral while declining to hold open field masses to round off Falangist rallies. When they tried to bring a giant rally up to his cathedral to make their point, Segura threatened to excommunicate them. Pointing out that the term *caudillo* meant the head of a band of thieves also took considerable nerve, as did his concern for the plight of imprisoned Basque republican priests in his diocese. Wherever he went, Segura was shadowed by armed Falangists.

Bishop António Pildain Zapiaín of the Canaries did not endear himself to the Caudillo when a pastoral letter condemning dancing coincided with Franco's attendance at a splendid military ball. The bishop also regarded the state's revival of monarchical influence on episcopal appointments with distaste.[55]

These critical voices were atypical, and did not address themselves to the black heart of the regime. Having made so many institutional advances, the Catholic Church was satisfied with Franco's vague assurances and the adoption of the outward trappings of a *Rechtsstaat*, while such Fascist provocations as the outstretched *saludo nacional* were quietly dropped. A repressive dictatorship seemed a small price to pay for what seemed to be an upsurge of religious enthusiasm, albeit in traditionally Catholic regions in the north, rather than among industrial workers in the cities or the de-Christianised helots of the south. The new primate, Enrique Pla y Deniel, co-operated in defining what critics called a 'national Catholic' identity to replace a Fascist Falangism that had fallen into disrepute. The Civil War, he claimed, had been a legitimate rebellion against the tyrannical Popular Front.[56] The primate became one of three members of the Regency Council, and, together with another prelate, also sat on the Council of State. Appointments in the universities began to be influenced by the National Catholic Association of Propagandists and the more secretive Opus Dei. Although only three of the thirteen members of Franco's first post-Fascist cabinet were identifiably Catholic politicians, it was striking that in addition to education and public works the head of Catholic Action, Alberto Martín Artajo, became minister of foreign affairs, the key figure in presenting an image of a 'post-Fascist' Spain on the wider world stage.

The defeat of the Axis powers in 1945 represented a perilous moment for the Franco regime. The Baptist freemason Harry Truman had no time for Franco, exclaiming: 'He wouldn't let a Baptist be buried in daylight. That's the truth. He had to be buried at night in plowed ground.' His administration duly struck Spain off the list of potential recipients of Marshall's largesse. In 1946 the UN Security Council described the Franco government as 'Fascist', agreeing to deny it diplomatic recognition if it did not quickly establish a more representative government. The border with France was closed and most of western Europe was deeply hostile. In addition to international isolation, there was domestic trouble too. Anarchists and Communists conducted assassinations and bank robberies, while there were large-scale strikes in the Basque country and Catalonia. Although it served to divide the opposition, there was trouble from the monarchists too. In March 1945 the pretender Don Juan issued the Lausanne Manifesto inviting Franco to step aside for a moderate monarchist regime. In response to this challenge, Franco appeared to leave the door ajar to restoration of the monarchy while he remained head of state for life. The regime began to cultivate Juan Carlos, the boy Bourbon heir, who from 1948 onwards was educated in Spain, while Franco's doling out of titles of nobility to various old cronies warned the pretender of where things might tend if he did not play ball. Franco's return to international semi-respectability was achieved through 'Hispanic' connections in Latin America, especially involving Perón's Argentina, and his claim to have been first into the anti-Communist lists, a theme that resonated as the US contemplated events in Greece, Italy and eastern Europe in the late 1940s. A powerful Spanish lobby operated in Washington – with huge sums going to the law firms involved – the most vocal supporters of Franco being a group of senators and congressmen from Nevada who in 1950 pushed through US government loans to Madrid.[57] Although the US declined Franco's offer of Spanish troops for Korea – which may have been a reflection of the fact that on ceremonial occasions the army minister sported a German Iron Cross – international tensions resulted in over two billion dollars of US aid, and executive agreements that established US military bases in Spain. Ironically, Franco whipped up anti-imperialist hysteria over Queen Elizabeth II's visit to Gibraltar at precisely the moment he was conceding Spanish sovereignty to the USA. Only some of the Catholic hierarchy, notably cardinal Segura, were unhappy about this closer association with the (Protestant) 'dollars of heresy', being partly compensated by ruthless repression of any public manifestations of 'heresy'.

While re-establishing amicable relations with the US was a paramount concern, Franco's government was also keen to wring a new concordat from a Vatican that was distinctly cool towards the regime. There was a preliminary treaty in 1941, but a concordat was not concluded until August 1953. This ratified many changes that had already taken place, including Franco's usurpation of the royal right to choose from the three names recommended for each vacant bishopric. The Church gained a powerful voice in education and social morality, while the Catholic nature of the Spanish state was expressly proclaimed. Pius XII appointed Franco a member of the Supreme Order of Christ, while

Spain itself was reconsecrated to the Sacred Heart. There were other more worrying developments beneath the outward pieties of open-air masses and processions. During the 1950s the number of religious vocations may have reached record numbers with a thousand priests ordained each year, but the Spanish working class was as hostile or indifferent to religion as ever, with a mere 5 per cent of a Catalan textile town's population attending Sunday mass. Large lay organisations, such as Catholic Worker Youth or Worker Brotherhoods of Catholic Action, became involved in wage disputes and strikes that brought them into conflict with the regime. In both the Basque country and Catalonia, radical priests became closely involved with regional nationalist movements, with ETA – the Basque separatist terrorist organisation – evolving out of a Catholic youth organisation. More worryingly for the regime, while the 'development society' of the 1950s acquainted Spanish people with enhanced demands, fundamental shifts in the international Church began to filter and then flood back into Spain itself.

CHAPTER 6

The Road to Unfreedom: The Imposition of Communism after 1945

I HATE NOW, IN ORDER TO LOVE LATER

Within a couple of years after 1945, half of Europe was returned to single-party totalitarian rule. The destruction of one totalitarian state, Nazi Germany, reinforced the omnipotence of the other, the Soviet Union of Joseph Stalin. While dazed people picked their way through cities reduced to archaeological rubble, purposive minorities – who often returned in the baggage train of the Red Army – scurried about eager to implement their ideological certainties. In every country where the Red Army and NKVD security police established a presence, minority Communist parties – in Romania amounting to fewer than a thousand people, in Hungary two thousand, among many millions – used a limited but effective repertoire of techniques to achieve the dominance of a single totalitarian party obeisant to an alien Asiatic power: the grim truth behind Stalin's public subscription to democratic principles. These emollient democratic noises were made to dampen the suspicions of his allies in the free world, although a wartime coalition solely based on the object of defeating Nazism rapidly came apart in the form of a Cold War that lasted for the next forty years.

In consonance with many European intellectuals in the 1940s, prominent historians, beyond the usual Western apologists for Stalinism, still endeavour to put a positive gloss on the imposition of totalitarianism, by failing to investigate the fraudulent ambiguities in the Communists' use of such terms as 'democracy' and their total contempt for law, truth and morality. They also insist on a moral equivalence between the electoral defeat of Communist parties in Western societies governed by the rule of law and the prior suppression, often by force and fraud, of all opposition in the Communist East, events that chronologically preceded the onset of the Cold War and which hence did not ensue from it.[1]

The word 'democracy', especially when 'people's' preceded it, was meaningless. As a naive young German Communist was once informed: 'It's got to look democratic, but we must have everything in our control.' Referring to the November 1945 'elections' in Yugoslavia, Milovan Djilas was no less forthcoming: 'We Communists did not want any opposition, none whatsoever. In the summer of 1945 when the draft of the elections was discussed, we deliberately included provisions that rendered it impossible for the opposition to participate.'[2] That November the Yugoslav Communists recorded a remarkable 95 per cent of the vote, while their scrutineers watched as voters dropped audible rubber ballots into one of two urns; only one of the parties they were voting for, the Communist-dominated Popular Front, actually existed. Elsewhere, coalition governments were formed, not as an act of Communist magnanimity, but usually because, as in Hungary, local elections had indicated the exiguous levels of support for the Communists, and the latter calculated that they could subvert or suppress larger democratic parties later. The Communists ultimately took their orders from the Soviets. When a leading Hungarian Communist, Zoltán Vas, protested that the Soviets were dismantling and hauling off industrial plant, he was summoned by marshal Klementi Voroshilov, the head of the Allied Control Commission in Hungary, with whom Vas was friendly: 'You see, Comrade Vas, don't be so stubborn! We trust you, you must agree with what we ask you to do.' When the Czech coalition government showed interest in the Marshall Plan, they were summoned to Moscow to hear Stalin inveigh against it as a capitalist plot intended to isolate the Soviet Union. The Czech foreign minister Jan Masaryk notoriously commented: 'I went to Moscow as the foreign minister of an independent sovereign state; I returned as a lackey of the Soviet government.' The Czechs declined to attend the Paris conference where the Marshall Plan was aired. Jan Masaryk would subsequently be found dead and broken beneath a high window.

For many people in central and eastern Europe, the arrival of the Red Army inaugurated a period of terror, which, in the nature of Stalinism, encompassed many members of that army itself. Suspicion coursed like adrenaline through the veins of its political officers, who were the objects of suspicion themselves. Former prisoners of war, or those such as Alexander Solzhenitsyn, naive enough to imagine that wartime sacrifices might usher in reforms in the motherland, were despatched to the gulag, including those who were repatriated to the Soviets by the Western Allies, in what many regard as a major post-war crime in which the West colluded. It was also ominous that, the day after Budapest fell to the Russians, a euphemistically phrased decree licensed cost-free abortion, on the ground that women were too weakened by wartime privations to give birth safely. In fact, many of them had been serially raped by Red Army soldiers, a theme that the historians Norman Naimark and Antony Beevor have highlighted in the case of Soviet-occupied Germany, where as many as two million women suffered this ordeal. Of course, it was not merely women who suffered, as can be understood from the terrible fate of the Hungarian retired bishop count János Mikes or his colleague bishop baron Vilmos Apor of Györ, who were shot dead when they interceded on behalf of village girls menaced by drunken Russian soldiers.[3]

The Red Army was a political tool as well as a fighting force, with its officers shadowed by NKVD secret policemen. It offered crucial logistical support to infinitesimal Communist parties through assignment of meeting halls, permissions to travel or supplies of ink, paper and petrol. In Hungary, the Russian occupiers permitted two Catholic weekly papers to appear, in reduced format and after being heavily censored, while encouraging twenty-four Communist dailies, together with five weeklies and several magazines.[4] The Red Army may have cut its occupation forces (although their presence was guaranteed in the peace treaty with Hungary to guard supply routes to Austria), but a Soviet general, Rokossovsky, was made Poland's minister of war and commander in chief in 1949, establishing the Felix Dzerzhinsky Military Academy to train officers to replace those purged. The Soviet NKVD was also omnipresent to intimidate and terrorise opponents, sometimes reopening Nazi concentration camps to imprison them.[5]

Even such notorious hell-holes as Auschwitz and Majdanek were used to incarcerate prisoners from the former Home Army who had spent four years fighting Nazism. These prisoners were then shipped eastwards in long freight trains. South-east of Moscow there were 'about 25,000 Polish political prisoners ... imprisoned there. This group includes a number of camps, separated from each other by 70–80 kilometres, each containing between 600 and 800 persons. The deportees live in barracks

... There is insufficient food: 120 grams of bread daily, coffee in the morning, soup at noon, coffee or soup in the evening.'[6] In eastern Germany alone, the Soviets interned 122,000 people, of who 43,000 died in detention and 736 were executed.[7] Under the October 1944 Decree Concerning the Defence of the State, a fearsome range of offences were reserved for military courts which liberally passed death sentences as well as lengthy terms of imprisonment on farmers who resisted land reform, railway workers who delivered faulty goods, or former members of the Home Army and of non-Communist political parties. A Special Commission to Combat Economic Abuses and Sabotage sentenced people to forced labour, there being one hundred labour camps which never housed fewer than 150,000 people in total. Until 1950 the Polish and Soviet security apparatus was involved in a bloody war against partisans of the former Home Army.

In Hungary, seven hundred thousand people were deported to Soviet gulags, either as prisoners of war or as civilians whom the Soviets decided to abduct as forced labour.[8] With their keen nose for the levers of power, Communists would insist on the interior ministry in coalition governments, which, as in Bulgaria or Hungary, gave them control of the incipient national security services. National Communist leaders were schooled in the tactics that had enabled them to rise to power within their own parties through the process of 'democratic centralism', and to survive the murderous rigours of exile in wartime Moscow where a knock on the door of the Luxus Hotel did not announce the night porter. As indefatigable committeemen, who would battle through a storm to get to the top table, there was not a committee or meeting these operators could not rig to give the illusion that decisions had been consensually arrived at. The practice of infiltration also meant that some Communists were covertly positioned within opposed parties, as was the case with the Hungarian Social Democrat and the Polish Socialist parties, just as they were concealed within the armed forces, police and judiciary. Obviously, as clients of Moscow they were also well placed to invoke the menacing spectre of their Russian 'advisers' to blackmail and bully opponents into compromises with the devil they thought they knew.[9]

The Communists benefited from the immense contribution of the Red Army in defeating Hitler, and from the dislocation and disarray of their opponents. Surviving democrats had already failed to halt Nazism or various indigenous Fascisms, and so confronted the triumphant Communists burdened with a record of abject defeat. Many, notably the

Czechs, had also had the experience of abandonment and betrayal by the Western Allies; others, such as the Poles, invested the latter with unreal expectations.

The war had wrought enormous transformations across the region. In many countries the aristocracy had been ruined, or, in the Prussian case, decimated by Hitler, a process compounded by land reform intended to win over populations largely consisting of peasant farmers. Industrialists were liable to the charge of collaboration under the Nazi occupation, a charge never levelled at their workforces. The property of the former was expropriated, as was that of Jews who had perished in the Holocaust, the thrust of Nazi crimes having been exclusively refocused on those committed against 'anti-Fascist' forces. Pre-war democratic politicians had been driven into exile or slaughtered by the Nazis, who did much of the Communists' work for them. Except for instances where either the Communists or the Nazis had killed the local intelligentsia – which both totalitarian regimes successively and successfully managed in Poland, where a third of such people perished – the survivors were often susceptible to the naive belief that only socialist central planning would be capable of hauling these countries into the radiant future exemplified by a Soviet Union whose realities few had witnessed.

Hope sprang eternal, from a supposedly scientific doctrine that resembled medieval chiliasm in its monumental unreality. Not a few of them, like the people loosely disguised in Czesław Miłosz's *Captive Mind*, were also drunk with the prospect of power and vengeance, the inevitable accompaniments of a developed sense of victimhood every-where. There was even hope for many on the extreme right. Inter-war right radicals could be converted into left-wing radicals, through both blackmail and a combined desire to sweep away the old elites whom both political extremes bitterly resented (it is striking how many former Fascists cropped up as Communists in the events described below). They were also useful when it was necessary to smear opponents with charges of antisemitism, since some of them had enthusiastically espoused this already.[10]

Selective memories played a part too. The nightmare of the Depression still compromised liberal capitalism, which Marxism blamed for the rise of Fascism, while the enormous aid the Western Allies had given the Soviets was conveniently forgotten, even as the Russians forbade the Czechs and Poles from benefiting from the Marshall Plan. Finally, the rhetoric of class struggle was not without resonance, organised rallies of 'workers' being the main ritual of the new Marxist political religion across central and eastern Europe. These were designed to give physical weight to the claim that events were historically inevitable. Like many intellectuals, trades unionists were tantalised by the raw power of the Marxist parties; thuggish demonstrations by miners and factory workers in urban centres were crucial to the intimidation of residual democratic parties representing more dispersed rural constituencies. When Hungary's cardinal Jozsef Mindszenty addressed huge crowds of penitents in Budapest in February 1946, agents provocateurs in the crowds cried out the name of the former Fascist leader 'Szálasi, Szálasi', while 'workers' were encouraged to hold 'anti-Fascist' counter-demonstrations with the slogan 'Work and bread – the rope for Mindszenty', despite Mindszenty having been imprisoned for urging surrender on Ferenc Szálasi's Arrow Cross tyranny.

Only the Churches remained as a potential source of opposition, and even here it was possible to identify and exploit weak points. Before turning to their fate, it is necessary to say something about a Communist modus operandi that exhibited several generic features, regardless of differences between national contexts, the tactics used to destroy political opponents then being transferable to the war on the Churches.

What the Hungarian Communist leader Mátyas Rákosi called 'salami tactics' were employed to fracture opposed majorities into isolable fragments, which could be absorbed or destroyed as opportunity arose. A slice at a time avoided indigestion. Since the 1930s the left has always been adept at using the charge of 'Fascism' to marginalise and destroy a wide range of opponents. The still-raw events of the war and occupation were used to discredit genuine, and merely putative, Fascists and collaborators, while nationalist passions were incited against the teeming German diaspora, millions of whom – men, women and children – were driven from territories they had settled for hundreds of years in the largest population transfer in modern history. The vacated territories were used as Communist colonies, from which non-Communist parties were excluded. The return of Transylvania from Hungary to Romania was similarly used to appeal to nationalists there, ironically in view of the fact that left-wing dominance of that minority had enabled the Romanian Communists to come to power in the first place.

Organised 'judicial' vengeance was a convenient and morally unimpeachable cover for the elimination of a wide range of opponents. Although Bulgaria had never sent troops to fight in Russia, and had saved most of its Jews from the Nazis, some fifty thousand Bulgarians were charged as 'war criminals'. Trials in Sofia of major political figures and parliamentarians meted out twice as many death sentences as the prosecution had requested. In Czechoslovakia, trials of former collaborators, notably former president Tiso, were used to divide and weaken the non-Communist Democratic Party in Catholic rural Slovakia, where the Communists lacked the support they enjoyed in the more industrialised, Czech half of the country. The Czech and Slovak Communist parties specifically tried to shape the trial to bring about this political result.[11] Another effective tactic derived from the multi-party inter-war popular fronts and wartime resistance movements. Various 'democratic' or 'anti-Fascist' blocs and fronts were formed, in which the Communists enjoyed the advantage of being part of government but also the main opposition. In addition to infiltrating various sectoral organisations for women, youth and so on, Communists artificially bolstered their rank and file through compulsory mergers with other socialist parties, even being prepared to sacrifice their name in favour of 'Socialist Unity' or 'United Workers'. These then levered themselves into power, routinely through fraud and intimidation, the reality of all the 'people's democracies'. The only goal behind such shotgun marriages was to liquidate the non-Communist political opponent.

In numerical terms, the Polish Peasant Party vastly outnumbered the Communists, yet it was the latter that won both a referendum and election in the years 1946–7. The methods were redolent of a banana republic. During the June 1946 referendum, designed to abolish the Senate and ratify the new borders, innumerable dead people appeared on electoral roles, while the town of Slupck recorded five thousand more 'yes' votes than there were inhabitants. Of the five thousand Peasant Party activists assigned to monitor the polling booths, only six hundred turned up, since the remainder had been arrested. During the January 1947 elections, Peasant Party candidates were arrested on charges of association with 'bandits' active in the Tatra mountains, who, when captured or killed, were miraculously found to have Peasant Party membership cards about their persons. Ninety-four members of the Party vanished without trace. Taking no chances even with government personnel, civil servants were obliged to sign certificates saying that they would vote for the 'Anti-Fascist Democratic Bloc', after which they were informed that it was not necessary for them physically to vote since this would be done for them. In this fashion the Bloc achieved a remarkable 95 per cent vote at the election. Shortly afterwards, the Peasant Party leader Mikołajczyk fled the country disguised as a British naval officer, as he was about to be charged with having arranged the murder of general Sikorski and with involvement in fantastical plots to restore the Habsburgs.[12]

II HOW COMMUNISM HELPED REVIVE CHRISTIANITY

With the political parties crushed out of existence or into subservience, attention focused on what limited manifestations of civil society had survived war and a 'liberation' that to many felt remarkably like another round of alien occupation. Not only were the communist regimes animated by an atheistic desire to eradicate religion, the Churches re-presented a constituency outside the totalitarian state, and in many cases were the main surviving repository of a sense of national independence, despite Communist attempts to hijack this sentiment.

The assault on the Churches had a different trajectory in each national context, and began where Stalin was confident that he had an important ecclesiastical ally, a tactic repeated in eastern Europe, where Protestant denominations, weakened by the mass expulsions of the Germans, and more easily controlled because of the indigenous nature of their hierarchies, were co-opted into attacks on the international Roman Catholic Church.

The first indication that Stalin's wartime professions of respect for religion were temporary and expedient was evident from his assault on the seven million or so Uniate Catholics who lived mainly in the western Ukraine, Czechoslovakia, Bulgaria, Hungary and Romania. Uniates were Slavs who had been converted to Catholicism by Jesuit missionaries during the Counter-Reformation. Under the terms of the Union with Rome, agreed at Brest-Litovsk in 1596, they acknowledged the primacy of the pope, while retaining the Glagolitic alphabet and Orthodox rites. In the Ukraine, the Uniate faith was an important vehicle for anti-Soviet nationalism, one of whose manifestations was a partisan war that raged against the Red Army and NKVD until 1952.

Stalin used every means at his disposal to bring about the dissolution of the Uniate Church, in consonance with the Orthodox hierarchy in Russia, which regarded the Uniates as schismatics. In April 1945, Alexei, the Orthodox patriarch, urged the Uniates to revert to their 'ancient attachment', claiming, 'Now Divine Providence has restored Russia to her ancient frontiers, and you are henceforth with us for ever.' In response, Gabriel Kostelnyk formed a collaborationist committee to effect such a union, although Ukrainian nationalists subsequently killed him outside a church in Lwów in 1948. These appeals were backed with the brute force of the NKVD. The Uniate archbishop of Lwów, Joseph Schlypi, was taken to Kiev and subjected to intensive interrogation, while being offered the Orthodox metropolitanate of Kiev should he renounce the Roman allegiance. When he refused to renounce his faith, a search of his residence produced 'evidence' of 'criminal complicity during the war with the German Fascist occupiers, and with the Gestapo', on which dubious basis he was sentenced to life imprisonment with hard labour – dubious because he had recently donated a hundred thousand rubles to aid Soviet war wounded. He was later observed, at a camp somewhere in the Urals, felling trees or excavating a canal, in the most appalling conditions, a martyrdom he endured until his release seventeen years later. Schlypi was relatively fortunate. The bishop of Przemyśl, who also found himself being interrogated in Kiev, had his ribs broken and his beard pulled out by the roots, an improvement on the fate of the Uniate bishop of Trans-Carpathia, for whom the Soviets arranged a fatal car crash. The allegiances of Uniate priests were further undermined by threats to their wives and families. Force was accompanied by a policy of divide and rule. Three renegade bishops were encouraged to declare the union with Rome at an end, thanking Stalin for 'your great deed in helping to unite us with Mother Russia'.

By the end of 1946 the public face of the Uniate Church in the western Ukraine had been wiped out, or at least forced to operate underground, while its churches and property were transferred to Orthodox clergy. The same fate befell the Uniate Churches in Czechoslovakia and Romania. The Romanian Orthodox patriarch, Justinian, and the Romanian Communist authorities colluded in a phoney synod – at which the government wrote the chairman's speech and the police were present – that requested readmission to the Orthodox Church. In Czechoslovakia, the Stalinist regime organised a synod at Prešov, which voted to abolish ties with Rome. In addition to expropriating the Uniate Church, the Czechoslovak authorities imprisoned its bishops, turning on to the streets priests who refused to become Orthodox, labour camps being the destination of the unemployed in all the emerging workers' paradises.

The Soviets and their local accomplices had to move more warily where they did not have the equivalent of the Orthodox Church urging the destruction of a hated rival. The worst case was probably the one we know least about. In Albania the hierarchy were arrested or shot, together with about a hundred priests and nuns and a Muslim lawyer who had tried to aid persecuted Franciscans. Elsewhere, many of the tactics used to destroy rival political parties were evident in Communist policy towards the Churches. Indiscriminate charges of collaboration and Fascism, divide and rule, and outright repression were all employed at various times to weaken the hold of the Churches. As in the case of Nazism, the matter of controlling the minds of the young became a crucial battlefield.

The first major trial of strength between a Communist regime and the Catholic Church occurred in Yugoslavia, the scene of one of the most rapid Communist takeovers. Several issues proved incendiary. Intimately bound up with their respective nations' sense of identity, the Croatian Catholic and Serb Orthodox Churches stood in the way of the creation of a federal Yugoslavia. The Vatican's refusal to countenance the extension of Yugoslav rule over Istria and Trieste's half a million Italian Catholics was a further source of tension between state and Church. While the self-proclaimed Catholic Croat Tito initially sought to detach the Croat hierarchy from the Vatican with vague assurances of greater independence within the emerging Yugoslavia, the actions of the Communists belied this apparent offer of compromise. The war in Yugoslavia ended in a bloodbath during which, with Allied connivance, all sorts of opponents of the Communists were repatriated and killed. Religious leaders were among those shot out of hand, including the Muslim mufti of Zagreb, the Orthodox bishop of Sarajevo, and the bishop of Dubrovnik. In one village in Herzogovina, partisans doused fourteen Franciscan friars with petrol before setting them alight.[13] Lesser provocations included the destruction of rural shrines, introduction of civil marriage and the active promotion of atheism in schools. In a pastoral letter issued in September 1945, the Croat bishops, under the leadership of archbishop Stepinac of Zagreb, pointed out that 243 clergy had been killed, 169 were imprisoned, and 89 were missing. A further nineteen theology students, monks and nuns had been summarily executed. The Catholic press had been suppressed. The Church's schools and seminaries had been taken over by the state. Plans were afoot to reduce the land available to each church to a paltry five hectares.

Tito indignantly responded by asking why the Church had never issued such a forthright condemnation of the Ustashe, warning that 'laws existed which forbade sowing discord and treachery, and anyone who wished his country well must honour these laws'.[14] The regime pursued a twin-track strategy by resuming diplomatic relations with the Vatican – an American bishop was sent as chargé d'affaires in Belgrade – the aim of which was to get the Vatican to recall Stepinac, against whom the secret police organised 'popular' demonstrations, in which rocks were thrown at his car, the second strand of their strategy. The government press attacked the Catholic Church, accusing it of harbouring pro-Ustashe conspirators. All of this was reported back to the Vatican, which with the appointment of four American cardinals – notably the charismatic Francis Spellman of New York – was much more dextrous in using the media and in making its concerns known at the highest political level.

The Vatican's alleged support for the Italians at the Paris peace conferences led to the arrest and trial of several clergy, who were indicted with the former Ustashe police chief so as to conflate the two in the public mind. After months of imprisonment, these clergy then implicated Stepinac in their 'conspiracy'. One senior Party figure recalled that 'He would certainly not have been brought to trial for his conduct in the war ... had he not continued to oppose the new Communist regime.' Stepinac was charged with having blessed and supported the Ustashe, of being responsible for the conversion of Orthodox Serbs 'with knives at their throats', and of conspiring to overthrow Yugoslavia's present government. The hearing exhibited many of the characteristics of a show trial. Stepinac's defence lawyer, Ivo Politeo, was appointed a week before the hearings began, and was denied the right to call various witnesses or to cross-examine key witnesses for the prosecution. His not implausible arguments were systematically ignored. Stepinac was found guilty on all counts and sentenced to sixteen years' hard labour. The Holy See excommunicated all those involved in his trial. Thanks to the public relations skills of the US Catholic bishops, the Stepinac trial became an international cause célèbre, one of the defining themes of the evolving Cold War being the hostility of Communism to the free practice of religion. Pius XII made the imprisoned Stepinac a cardinal in 1952; the 'ex-archbishop', as the regime referred to him, died still under house arrest in 1960.[15]

In Hungary, the programme of the second Congress of the

Communist Party in September 1946 focused on the need for a struggle against the Churches, on the ground that 'from the onset of the new era they were against democracy'.[16] This set the regime on a collision course with cardinal Jozsef Mindszenty, the sort of prelate who was the bane of 'anti-anti-Communists' (that is self-repudiating Westerners who imagine that the Cold War was a manipulative wickedness perpetuated by their own governments) everywhere. Mindszenty had the distinction of having been imprisoned by Béla Kun's murderous hundred-day Soviet in 1919, and then by the Arrow Cross Fascists after he tried to stop them turning Hungary into a battlefield. He was a blunt-speaking Hungarian patriot who regarded Communism with apocalyptic dread. His politics were those of a Catholic Habsburg-minded monarchist, for whom he saw himself as a sacred steward.[17] Initially, he engaged reluctantly in politics:

> I myself wanted simply to remain a pastor. I regarded politics as a necessary evil in the life of a priest. But because politics can overturn the altar and imperil immortal souls, I have always felt it necessary for a minister to keep himself well informed about the realm of party politics. Knowledge alone enables the priest to give those entrusted to his care some political guidance, and to combat political movements hostile to the Church. It would certainly be a sign of great weakness if a priest were to leave vital political and moral decisions solely to the often misled consciences of the laity.[18]

The more temperate Italian diplomats in the Vatican Secretariat of State sometimes quaked at his forthright manner of speaking. This was why, paradoxically, the Hungarian Communists were so insistent on the return of the papal nuncio, in the hope that they could use the power of Rome to divide the local hierarchy from the cardinal.

After nationalising the Church's lands, in June 1948 the Communists struck at religious schools, which constituted virtually the only education available in a country that was 75 per cent Roman Catholic. Marxist textbooks intended to replace those current in the Horthy era were the first bone of contention. When the Catholic Church refused to use these books in their schools, which accounted for all primary and three-quarters of secondary education, the government accused the schools of being responsible for 'Fascist plots', and resolved to ban priests from teaching and to nationalise the schools. Some 3,000 Catholic, 1,000

Calvinist and 375 Lutheran establishments were taken over by the state, in flagrant violation of the wishes of parents.

Mindszenty realised that 'The result of the nationalisation of the schools will be: First, religious instruction will become optional; then, after a suitable delay, religious lessons will be suppressed outright; finally, we shall have lessons in Marxist philosophy in their place.'[19] He immediately excommunicated Catholic teachers who agreed to take the state's salaries, thereby encouraging 4,500 others to refuse to teach. He also forbade Catholics to read government educational literature or to listen to government broadcasts. Finally, he excommunicated all those members of the Communist-packed parliament who had voted to take over the schools. Inevitably, there was a response.

The Communists launched a campaign of denigration against Mindszenty, beginning with the absurd charge that he had been imprisoned by the Nazis for refusing to hand over his illicit stocks of woollen underwear which they needed for their troops on the Russian Front. They also raked up the fact that during the war he had changed his surname from the Swabian Pehm to the Magyar Mindszenty – as the Mountbattens and Windsors know, a common enough phenomenon across Europe whenever Germans were unpopular. Next, they used the proven Stalinist tactic of sowing dissension among their opponents, in this case by imagining that the clergy were divided into 'progressives' or 'reactionaries', and sponsoring a group of 'Progressive Catholics' that included the historian Gyula Szefir and the composer Zoltán Kodály. Finally, in June 1948 they found a pretext to attack Mindszenty in person. This occurred in the wider context of the suppression of the democratic opposition.

Beginning in April 1946, the Communist secret police undertook searches of Catholic intermediary schools, in which they discovered weapons that they themselves had planted. A Budapest newspaper managed to report on a conspiracy centred on a Cistercian school at Baja before the search that revealed the alleged evidence had even been conducted. Each arrest offered further evidence of a Catholic conspiracy against the government. In conjunction with parents' associations, the Church organised a skilful campaign to resist the Communist assault on religious schools. This campaign was supported by the Smallholders Party, which was the dominant partner with the Communists in government. The Communist-dominated secret police arrested elected representatives of the Smallholders Party, while the Communist deputies refused demands for a parliamentary investigation into the role of those arrested representatives in an alleged conspiracy. The Communists next lifted the parliamentary immunity of Smallholder deputies, many of whom were arrested on the steps of the parliament building. The Party's secretary-general, Béla Kovács, was invited to attend a meeting with the Hungarian secret police; he reluctantly attended and was arrested by the NKVD, and disappeared, but not before he had made a confession implicating the prime minister.

Two months later, in April 1947, premier Ferenc Nagy was alleged to have engaged in a conspiracy against his own government. He was on holiday in Switzerland at the time. Threats to the life of his son ensured that he resigned and remained abroad, where many members of the Smallholders Party fled too. The Smallholders Party had been deprived of its majority in parliament. In June 1947 the Communist-dominated parliament revised the electoral laws. One consequence of this was a new electoral roll, from which the names of a million known opponents of the regime disappeared. When opposition decamped from the disintegrating Smallholders Party to a new Freedom Party, backed by the Church, the printers union refused to print its newspaper, and its leaders were arrested. The Communists then encouraged the formation of six new parties to fragment the anti-Marxist vote. When the election came, the Communists availed themselves of new provisions that enabled people to vote despite being non-residents. Buses and trucks took large groups of Communist supporters back and forth to vote, often with the aid of false registration cards. Should the resulting vote be larger than the number of eligible electors, the votes that had gone to non-Communist parties were simply discarded to even up the numbers. When the result still indicated a 40 per cent vote for the opposition, the Communists organised fraud charges against the Hungarian Independence Party, whose votes were duly discounted by a Communist-dominated electoral court. In October, 106 of the 109 parliamentary deputies of the Independence and Democratic parties were arrested.

Having rigged the outcome of the 1947 elections, the Communists felt confident enough to revert to the assault on religious schools. Employees in offices and factories, organised as 'local national committees', were encouraged to petition the government calling for the nationalisation of schools. On 3 June 1948 villagers held a meeting in the town hall at Pócspetri to protest the secularisation of the schools. When the police tried to break up the meeting, a policeman dropped his rifle and was fatally shot. The town clerk was charged with murder, while the local priest, János Asztalos, who had not killed anyone and had been present to calm the crowd, was accused of incitement to murder, his initial protestations of innocence being superseded after three days by confessions not only of his own, but of the Church's corporate guilt. Newspaper headlines proclaimed 'Murder at the Instigation of the Church'. The clerk was executed; the priest's death sentence was commuted to life imprisonment, and the village schoolmaster was sent to jail for eleven years. Riding a wave of manufactured indignation, 'parliament' duly voted to secularise the schools on the same day that Hungary became a one-party state through the fusion of the Communists with the Social Democrats.

The attack on Mindszenty was not simply designed to destroy the primate or to intimidate the Catholic Church, but to show ordinary Hungarians that if a cardinal was not safe, nor were they. Loudspeakers in streets, squares and factories continually broadcast lies about the primate: 'The hostile, antidemocratic attitude of the primate is the reason for the disunity and misery of our people. He is demanding the return of the confiscated estates, refuses to recognise the republic, is organising counter-revolution, and is blocking compromise between Church and state.'[20] In fact, the Hungarian bishops had offered to negotiate these relations, but the Communists insisted that they first recognise the Hungarian Republic, as had the Protestant Church leadership, after the Communists had had them all replaced. A menacing new phenomenon, dubbed 'Mindszentyism', became current among the cadres, the conversion of a person's name into an 'ism' being a sure sign that arrests were mooted. The Communist leader, Mátyas Rákosi, declared that 'the tolerant policy which has donned kid gloves for dealing with traitors, spies and smugglers clad in clerical garb is over. The policy of punishing only the small clerical criminal, and not the big fish, is over. The time has come when we have to defend our land against a group of reactionaries behind Mindszenty.' While the Hungarian ambassador to Rome tried to pressure the Holy See into having Mindszenty recalled, pseudo-meetings led to pseudo-resolutions in which the comrades bayed for the cardinal's blood. The presence in Budapest of the Soviet prosecutor Andrei Vyshinsky indicated whence the impulse to persecute a prince of the Church came. Refusing to flee, and evidently resigned to his fate, Mindszenty warned his clergy that if they heard he had signed a confession or had resigned, they should regard this as having been extorted.

After weeks of police surveillance, Mindszenty was arrested on 23 December 1948 and charged with high treason, espionage and currency violations. This was the first time a prince of the Church had faced capital charges since the Reformation.

No show trial is complete without sensational confessions and evidentiary revelations. A mysterious box was found by the police at Mindszenty's palace at Estergom, containing faked handwritten communications in which he had allegedly urged the West to overthrow the Communist government and the US not to return the much venerated Crown of St Stephen after they had liberated it from the retreating Germans. He was alleged to have conspired with Otto von Habsburg with a view to restoring the dynasty after the imminent Third World War. Shortly after his arrest, Mindszenty was taken to the AVO secret police headquarters in the former Arrow Cross building in Budapest, the local equivalent of the Lubyanka in Moscow or Berlin's Prinz Albrechtstrasse. Stripped of his cassock, he was forced to wear an oriental clown's suit, a wicked indignity for a man who had rarely ever worn civilian clothes since his ordination. Denied proper sleep for nearly forty days, Mindszenty was interrogated through the nights, an experience interspersed with extended torture sessions in which a police major assaulted him with a rubber truncheon. Mysterious doctors were on hand to administer drugs, which Mindszenty was certain were also being mixed in his food. Now and then, one of his co-accused would be brought in, in an unrecognisable state, to confirm the interrogators' view of things. After a month of this mistreatment, Mindszenty was prepared to sign documents, the contents of which he was scarcely cognisant of, and whose dates and facts were altered to suit the case the Communists were preparing. He also wrote to the minister of justice, confessing his illicit involvements with the British and Americans in order to bring about a federal monarchy in central Europe under the Habsburgs, and offering to resign to pre-empt the need for a trial.

Mindszenty and his co-accused appeared in February 1949 before a People's Court, whose judges represented the Communist Party. The president of the Court was especially zealous, as befitted a former member of the Fascist Arrow Cross who had become a Communist. A co-operative Italian Communist senator was wheeled in to aver that the cardinal had not been tortured. Mindszenty's elderly defence lawyer was coerced into taking the brief, by the expedient of threatening to cut off his pension, and then given two days to study a case the regime had prepared for him. Whenever he deviated from the script, the judge told him off. Three further conspirators were added to the accused, although Mindszenty had not even been asked to confess conspiracy with them. Although both the main accused and some of the others in the dock with him were renowned as articulate men, all spoke in a halting, disconnected fashion. When the president of the tribunal asked, 'Are you mentally tired?', Mindszenty replied: 'Yes ... Mr President ... Mr President ... I am [long silence] a man broken in mind [long silence] ... and in body ... ' Surreally, many of the questions from the defence lawyer Kiczkó were designed to elicit admissions of his clients' guilt. Although the charges involved capital crimes, and a conspiracy spanning continents, the trial was concluded inside three days. Mindszenty admitted many of the charges against him, as well as such absurdities as bribing the head of Vatican Radio with a car paid with US money. He was found guilty and sentenced to life imprisonment. A group of visiting British clergy and the Italian Communist organ *L'Unità* pronounced the trial fair; more accurately the Italian premier de Gasperi called it 'a trial which would be inadmissible under any Western government, a sentence that would be unthinkable in any country governed by equitable laws, a challenge to the civil conscience of the world'. Mindszenty was shuttled between several prisons, until the 1956 Hungarian Uprising afforded him brief respite. After fleeing into the US embassy in Budapest, he remained there for fifteen years until a deal resulted in his transfer to Vienna and freedom.

The case of Mindszenty caused such an international furore that it was not repeated. In both Czechoslovakia and Poland the Communist authorities attempted to divide as well as selectively harass the national clergy, while trying to isolate them from Rome. In the case of Poland this was relatively easy, since the Vatican refused to recognise the new Polish state and objected to its brutal expulsion of ethnic Germans. The government retaliated by accusing the Holy See of imposing German bishops upon Polish dioceses during the war, putting on trial monsignor Splett, the German bishop of Danzig and wartime administrator of Chelm, who was imprisoned for eight years as 'a Gestapo collaborator imposed on a Polish diocese'.

In both countries, the regimes singled out priests who believed it was possible to reconcile Catholicism with Marxism, or who were credulous to Communist professions of a desire for international peace. In Poland, the Communists alighted upon the former leader of the inter-war Polish 'falanga' Fascist organisation, Bolesław Piasecki, as the founder of a new group of 'progressive Catholic' intellectuals with their own publishing house called PAX. About two hundred clergy also joined the so-called Patriotic Priests, whom the Communists encouraged in their willingness to disobey Rome and their own bishops in matters other than faith and morals. Whereas every subterfuge was used to make the lives of bishops difficult, notably forbidding them to operate collectively, or to take up appointments in the Regained Western Territories of Pomerania, Prussia and Silesia, the regime positively encouraged the Patriotic Priests to participate in political activities such as meetings where they signed 'peace declarations'. Patriotic Priests endorsed the Communists' sequestration of the Caritas charitable network, which included crèches, libraries, homes and hospitals, on the spurious grounds that there had been 'abuses' in their administration. They became untouchable. When bishop Kowalski disciplined one of the priests involved, he was arrested and treated so violently that he subsequently endorsed the sequestration himself in a well-publicised letter to Bolesław Bierut, the president of the Polish Republic.

The Polish hierarchy was also in a confused condition at the end of the war. The primate, cardinal Augustus Hlond, had gone into exile in France and Italy in 1939, and then been arrested and imprisoned by the Gestapo in 1944. In Hlond's absence, de-facto leadership devolved upon archbishop Adam Sapieha of Cracow, a seventy-five-year-old whose stock had risen because of his noble conduct during the German occupation. When Hlond was sent to Poland in August 1945, he ostentatiously refused any contact with Poland's new rulers, who responded by abrogating the 1925 Concordat, breaking off diplomatic relations with a Vatican that continued to recognise the exiled government in London. The abrogation of the Concordat removed internationally recognised guarantees of the Church's status. In addition to expressing sympathy to the German Catholic hierarchy over the fate of ethnic German expellees, the Holy See further alienated the Poles by refusing to recognise the Oder–Neisse frontiers before the conclusion of a peace treaty. Although the Communists began with such gestures as allowing crucifixes to remain in schools and courtrooms, or the ostentatious attendance of leading Communists at Church feastdays, the reality was revealed by the introduction of civil marriage and divorce; the removal of Catholic books and periodicals from state libraries, and the expropriation of around 375,000 hectares of Church lands. Ironically, the latter measure served only to increase popular respect for the clergy, who could not be accused of defending any corporate material interest.[21]

A further issue was assiduously exploited to complicate the reputation of the Polish Catholic Church abroad. The approximately eighty thousand Polish Jews who survived the Holocaust were joined by a further hundred thousand who were repatriated by the Soviet Union. Slight privileging of these people, some of whom were disproportionately represented in the Communist Party and UB secret police, was accompanied by the wholesale suppression of the massive contribution that the Home Army, rather than the Communist Lublin Poles, had made to resistance against the Nazis. While a monument commemorated the uprising in Warsaw's Jewish ghetto, there was no memorial to the subsequent uprising by the Home Army and its supporters. Rather in the way that present-day anti-Americanism is often bound up with antisemitism, so the latter often permeated anti-Communism – the common factor being the belief that it was, and is, the Jews who were 'really' in control. These factors encouraged the myth of the 'Zydokomuna', that is the belief that Poland was being run by Jewish Communists. Of course, the opposed leftist demonology involved caricaturing anti-Communists as 'antisemites', 'bandits', 'criminals', 'Fascists', 'nationalists' and 'reactionaries', although we hear far less about that process of demonisation. Whether or not encouraged by the NKVD, during the July 1946 referendum pogroms occurred in a dozen Polish cities and towns, of which the worst was at Kielce, where more than forty Jews were killed and a hundred injured. Conveniently, at a time when it was trying to defeat the Peasant Party, the government claimed that these outrages, committed by political delinquents in an anarchic atmosphere skilfully captured in Andrei Wajda's film *Ashes and Diamonds*, had been inspired by the Peasant Party and its supporter the Catholic Church, the tenuous evidence for that connection being the folkloric justifications some of the perpetrators used for their despicable actions. Ironically, one of the Communist's main anticlerical propagandists, Wojciech Pomykało, would be prominent in 1968 for accusing dissident intellectuals and students of being alien 'Zionists'.[22] The Polish Catholic hierarchy demonstrated the grudgingly leaden touch it invariably revealed on an issue where, for some, it was an article of faith that all Catholic Poles were antisemites. Although primate Hlond had publicly condemned antisemitism, on this occasion he and the rest of the hierarchy refused to do so. When the US ambassador forced him to hold a press conference, Hlond managed to make a hash of it, with remarks like 'the Jews occupying leading positions in Poland in state life are to a great extent responsible for the deterioration of these good relations', which sat ill with the fact that two thousand Jews had been killed in Poland between 1944 and 1947. Such infelicities isolated the Polish Catholic Church from its supporters in the US and elsewhere, one of the principal aims of Communist policy.[23]

Meanwhile, relations between the Vatican and the Polish government deteriorated. In March 1948 the Polish press got their hands on a letter from Pius XII to the German bishops, in which the pope deplored the circumstances of the expellees and appeared to cast doubt on the wisdom of Poland's land-grab in the west. The Polish government tried to use this letter to drive a wedge between the Polish hierarchy and the Vatican. Skilfully, Hlond extracted papal endorsement for a pastoral letter recognising 'all former German territory in Polish hands as definitely Polish'. This accommodating tone smoothed the way for a settlement between Church and state in Poland. The sixty-seven-year-old Hlond died in October 1948 and was replaced by the forty-seven-year-old bishop Stefan Wyszyński of Lublin. Wyszyński had a background in Catholic trade unions and youth movements, and was considered to be on the left of the Polish hierarchy. He had spent the war on the run from the Gestapo, using the code-name 'Sister Cecilia'.[24]

In July 1949 the Holy Office issued a decree excommunicating members and supporters of the Communist Party, together with those who published or read its materials. Although this decree had been conceived in advance of the 1948 Italian elections, it was now given general applicability. In response, the Polish government nationalised the Caritas welfare organisation. Chaplains were forbidden to minister in state institutions such as hospitals and prisons. The regime managed to attract fifteen hundred priests to meetings to denounce the Church's administration of welfare. Many of them were coerced into attending.[25] In March 1950 the government took over the Church's larger landed estates, claiming that the revenue would be diverted into social reform. After this battering, the hierarchy thought an accord the better part of valour, having already initiated informal discussions of Church–state relations which resulted in four meetings in the second half of 1949. To the surprise of Pius XII, on 14 April 1950 Wyszyński and his fellow bishops signed an accord whose nineteen clauses recognised the Communist president, the government's 'socialist programme', including agricultural collectivisation, and – in opposition to the Vatican – the inalienably Polish character of the western territories acquired after Yalta whose ethnic composition the Communists were altering by transferring people from the east of the country. Several articles of the agreement were designed to prohibit the Church from making any public criticism of the regime or its wider socio-economic programme, let alone the daily illegalities of the system. In return, the Communist authorities agreed to respect religious freedom and to guarantee the continuance of religious education, including the world-renowned Catholic University of Lublin.[26]

This accord shocked the Vatican authorities, who regarded it as effectively worthless, because, as under-secretary of state Tardini commented: 'These people do not attach the same meaning to words and phrases as other people do.' Having placed a wedge between the episcopate and the Patriotic Priests, the Communists had now magnified existing tensions between the Holy See and the Polish bishops. The latter appeared to have adopted the line of the Patriotic Priests, to the effect that Rome's spiritual authority was separable from purely temporal political issues, a division that the pope unequivocally rejected. The accord raised the ominous prospect of national Churches entering into separate deals with the Communists.

The Communists had no intention of respecting this accord, especially after an agreement with the German Democratic Republic in July 1950 conferred recognition of the so-called 'Peace Frontier' between the two Communist countries. Although the regime had guaranteed religious education, it made sure that this was to occur solely in the hour before children were due to return home, for which the authorities simultaneously scheduled sporting activities so as further to encourage non-attendance. They created a range of secular schools, offering free meals and new textbooks, which were alluring at a time of post-war austerity. Threats to the jobs of parents ensured that children were enrolled in secular schools. The Marxist organisation Fighting Youth distributed anti-Christian propaganda among the young which contained lurid images of how Christianity had been introduced to Poland a thousand years before. Further petty harassments included attempts to marginalise Christian festivals by celebrating the anniversary of the (Russian) October Revolution or Stalin's birthday, while Christmas Day became 'New Year's Day' and Christmas trees 'Trees of Light'. Taxes were imposed on church collections, and priests were expected to account for every trifling sum they received for marriages, baptisms and funerals. This accounting measure helped the security police identify those who still dared to support the Church.

These stratagems fired Wyszyński into responding forthrightly; he repeatedly compared the onset of Communist rule to the tyrannies of the Dark Ages, although he confidently (and accurately) predicted that 'this evil ideology which seeks to destroy man's belief in God will not last'. The primate instructed parents to ignore possible reprisals by educating their children at home, rather than subject them to the godless state system. The Communists responded by accusing the primate of politicising the pulpit and of reneging on the 1950 accord. They also took the opportunity to attack Pius XII, not only for allegedly being pro-German – he had recently interceded for the life of the former gauleiter Artur Greiser, whom the Poles had sentenced to death – but for being 'in league' with German and US militarists. It was at this time that the Soviet secret police commenced their black propaganda regarding Pius XII's alleged 'silence' during the Holocaust, as well as accusing him of being the chaplain to the incipient North Atlantic alliance. They used such preposterous headlines as 'The Pope Receives High USA Militarists' or 'Bishops behind the Scenes at General Staff Conference'. They also availed themselves of the death in May 1951 of the enormously popular cardinal archbishop Adam Sapieha of Cracow, whose role in resisting the Nazis had rendered him untouchable, to settle accounts with that highly conservative Catholic city by stepping up censorship of the prestigious newspaper *Tygodnik Powszechny*. A few months later they closed the paper, reopening it only when its editorial board had been restocked with reliable Patriotic Priests. The next step was to strike directly at six of the bishops and nine hundred priests by arresting them. Bishop Kaczarek of Kielce was accused both of homosexual activity and of working for the CIA, while spying too for Polish émigrés in Munich. He was jailed for twelve years after a classic show trial. Further trials were conducted against the archbishop of Cracow, who was charged with hiding valuable artworks and furniture on behalf of exiled aristocrats. In the spring of 1953, the regime awarded itself the right to appoint and dismiss priests and bishops, and insisted that all clergy were to take an oath of loyalty to the Polish state. This made it clear that the Communists had never been satisfied with a Western-style separation of Church and state, but sought the total subordination of the Church to the state, in line with the totalitarian goals they pursued towards society as a whole.

Wyszyński replied to this outrageous assertion of state power with a sermon in St John's cathedral in Warsaw in which he said: 'We teach that it is proper to render unto Caesar the things that are Caesar's and to God that which is God's. But when Caesar seats himself upon the altar, we respond curtly: he may not.' The Polish bishops issued a joint memorandum condemning the new decrees and announced a moratorium on ecclesiastical appointments. The regime responded by accusing them of high treason. The security police struck at Wyszyński himself; he was arrested, deposed from his see, and secreted in a remote monastery. Careful to avoid a messy trial of Wyszyński, the authorities shuffled him from one monastery to another, while tempting him with exile abroad. He refused to abandon his flock. In 1955 the Communist state prohibited religious education in schools. By that time, two thousand Catholic activists were in jail. They alone had spoken out for freedom and human rights during the most intense phase of Polish Stalinism. As Czesław Miłosz has written: 'Churches were the only places that could not be penetrated by official lies, and church Latin allowed one to believe in the value of human speech, which elsewhere was being degraded and used for the basest tasks.'

After the agreements at Košice, recreating a Czechoslovak state in which the Slovaks were recognised as an independent nation, a National Front government was formed with the Communist Klement Gottwald as prime minister. The Communists in Czechoslovakia initially moved relatively cautiously, over-confident that securing 38 per cent of the vote in the first post-war elections in May 1946 would translate into durable success. A period of drift brought a collapse in probable Communist electoral support and stern criticism from foreign Communist parties. Stung by these criticisms, the Communists stepped up their infiltration of rival parties, while in Slovakia they used the trial of Tiso in 1946 as an opportunity to smear the Slovak Democratic Party and the Catholic Church as crypto-Fascists, even though Rome had refused to intercede on behalf of a priest it had already disowned. Although the Democrat Party had initially been based in the Lutheran community, after the banning of the Slovak People's Party it became the home for many of Slovakia's Catholics. The fact that it achieved 62 per cent of the Slovak vote, as against 30.3 per cent for the Communists in March 1946, made it an urgent object of Communist attentions. Despite Tiso being regarded as a national hero by many Slovak Catholics in the US, the Holy See comprehensively distanced itself from him, pointing out that it had urged him not to assume high political office and had condemned his regime's antisemitic measures.[27]

Instead of weakening the Democratic Party, or its democratic counterparts in the Czech parts of the country, Communist talk of plots stiffened their resistance. In February 1948 they demanded that the Communists cease packing the ranks of the provincial police with their own supporters. The Communists formed (armed) 'action committees' from the trade unions, as well as a fifteen-thousand-man 'people's militia'. When the non-Communist ministers in the government resigned, in the vain expectation that president Beneš would invite them to form a non-Communist government, armed trade unionists attacked the buildings of these parties. The massing of the Red Army on the borders indicated that the Soviets were not prepared to contemplate a non-Communist Czechoslovakia. President Beneš invited Gottwald to re-form a National Front government. The foreign minister Jan Masaryk was found dead beneath the Foreign Ministry window. Three months later Gottwald replaced Beneš as president.

The Czech Communists endeavoured to conciliate the Catholic Church in order to win support in elections designed to legitimise their February 1948 coup. Gottwald encouraged clergy to stand on the government slate in the elections, although the Czech primate, archbishop Beran, specifically instructed them not to do so. Only monsignor Josef Plojhar defied this prohibition, which resulted in his being suspended as a Catholic priest, although it opened a new political career.

Simultaneously, the Czechoslovak government sought to isolate the Catholic Church from fellow Christians. The most important Protestant denomination, the Hussite Evangelical Church of the Czech Brethren, was prevailed upon to declare that Communist policy was not anti-religious. Usefully, the Communists also alighted upon the Erastian Czech National Catholic Church, whose hostility towards a pro-Habsburg papacy was compounded by memories of Western betrayal at Munich, a betrayal which they imagined the Vatican had been a party to. The government endeavoured to attract dissident Catholic clergy to this national Catholic Church through the simple expedient of offering them higher stipends and pensions and improved living quarters. Its leader, Plojhar, an advocate of reconciliation between Christianity and Communism, was appointed minister of health, which resulted in his suspension from the priesthood. Further priests were suspended when they became commissars for engineering or posts and telecommunications. These Peace Priests were rolled out to demand such things as the dissolution of the monasteries or to condemn US policy in Korea. They participated in staged-managed occasions, where they could always count on the support of such foreign useful idiots as Hewlett Johnson, the 'Red' dean of Canterbury.

Confident that they had successfully divided the Catholic Church, in 1950 the Czech government opted for a show trial. The target was carefully chosen. Ten monks selected from the different orders alleged to be conspiring on behalf of Rome indicated that the regime had had difficulty in subverting the loyalty of regular clergy to either Rome or their bishops. State radio accused the monasteries of being nests of 'Germanism', their cellars brimming with arms, radios, spies and assassins. Such old anticlerical standbys as the charge that monasteries harboured not just the idle but dangerous pederasts were thrown into the list of accusations against these 'Vatican monkish agents'. Before the cassocked accused stood a table upon which crucifixes and monstrances lay alongside dollar bills, pistols and machine guns. Three of them had been in Nazi concentration camps, and put up a better defence than expected. Despite this, all of the monks received sentences ranging from life to long terms of imprisonment. No sooner had this show trial concluded than the Peace Priests called for the dissolution of the monasteries. These were converted into museums or social clubs, schools and hospitals, while the remaining monks (and nuns) were concentrated in two institutions, one for abbots and superiors, the other for the ordinary brethren. These were known as 'concentration monasteries or nunneries'. The daily regimen was a grotesque parody of what the monks were used to, except that they worked in docks and uranium mines rather than gardens, and their reading materials were reduced to the customarily leaden fare of dialectical materialism.

When archbishop Beran attempted to read a pastoral letter denouncing these policies, a 'workers' militia' contingent drowned out his words in his own cathedral. People who tried to stop this hooliganism were arrested by the police. A month later, Beran was arrested and confined in his residence, before being sent to a remote 'concentration monastery'. During the war, the Nazis had imprisoned him in Pankrac jail in Prague, before sending him to Dachau. The Communists had themselves awarded the archbishop Czechoslovakia's highest military honour in recognition of his wartime bravery. Other bishops were banished to the countryside, while priests were prohibited from leaving their parishes, unless they were among those forcibly transferred in the dead of night, to disrupt the continuity of their ministry. To round off these measures, the Communist regime introduced state salaries for all clergy, who had to be approved by the state, and who were obliged to swear an oath of loyalty to the 'people's popular democracy'. Rather than face extinction, the bishops enjoined their clergy to swear this oath, the alternative being the mass arrest of the orthodox Catholic clergy and their replacement by the heterodox Peace Priests.

Within a remarkable short time totalitarian rule had been reimposed on half a continent using a combination of force and fraud. Democratic political life was brutally extinguished in favour of single-party states with a monopoly of opinion. Although they were subjected to relentless assault from state-sponsored atheism, the Christian Churches remained the only licensed sanctuaries from the prevailing world of brutality and lies. Appropriately enough, as we shall see, they played an important role in the overthrow of Communism forty years later. So, it has to be said, did the diffusion eastwards of what in the 1960s became a homogeneous youth culture consisting of conforming nonconformity. For that we have to visit what was known as 'swinging London', one of the epicentres of generational revolt born of unparalleled affluence.

CHAPTER 7

Time of the Toy Trumpets

I WERE THE BEATLES BIGGER THAN JESUS?

The most elegant statement of a chronicle of a death foretold was 'Church Going', the absence of a hyphen being significant, by the English poet Philip Larkin. He wrote the poem in early 1954:

> . . . wondering, too,
> When churches fall completely out of use
> What shall we turn them into, if we shall keep
> A few cathedrals chronically on show,
> Their parchment, plate and pyx in locked cases,
> And let the rest rent-free to rain and sheep.
> Shall we avoid them as unlucky places?
>
> . . .
>
> Power of some sort or other will go on
> In games, in riddles, seemingly at random;
> But superstition, like belief, must die,
> And what remains when disbelief has gone?
> Grass, weedy pavement, brambles, buttress, sky,
>
> A shape less recognisable each week,
> A purpose more obscure.[1]

If Larkin's poem seems prescient with hindsight, at the time Britain's Protestant Churches were basking in a post-war religious revival, as reflected in peak memberships in the years 1955–9.[2] In 1954 they received a major boost with the US Evangelist reverend Dr Billy Graham's 'sweep for God through Britain'. In three months, some 1,300,000 people flocked to the greyhound track at Harringay, as part of a crusade that culminated with nearly two hundred thousand people packed into Wembley and White City stadiums. Some 1,200,000 people, or nearly three-quarters of the city's population, also attended Graham's rallies in Glasgow. In the following year, the Jehovah's Witnesses attracted forty-two thousand people to rugby's Mecca at Twickenham.[3] According to the leading British Church historian Hugh McLeod, this revival can be generalised across the West:

> In most parts of the Western world these were years when organised Christianity had a high profile, whether because of the size of congregations, the numbers of new churches being built, the huge participation in evangelistic rallies, Christian influence in the fields of sexual morality, family life and gender-roles, the role of the churches in education and welfare, or the political strength of Christian Democratic parties, then at the height of their power.[4]

This was about to change.

This revival proved evanescent in England as well as elsewhere. After a long period of constancy between 1890 and 1960, all the major indices of formal involvement with the Churches went into a sharp decline in the 1960s. Ordinations to the clergy fell by a quarter, Anglican confirmations by a third; baptisms fell below 50 per cent of live births, and less than 40 per cent of marriages were celebrated in church. Attendances at Sunday Schools, which had grown in the 1950s, plummeted a decade later. In 1900 over 50 per cent of children had attended these schools; by 2000 this was true of only 4 per cent of them, which indicates that one of the major means for transmitting the Christian faith had virtually been extinguished, although in the meantime others have developed and proved highly popular, such as the 'Alpha' courses.[5]

The 1960s were the crucial turning point. It is a decade that still uniquely polarises opinion, especially among the middle aged and elderly, who are divided for and against. This is either because posterity lives with that decade's real (and imagined) consequences, or, on the contrary, because those nostalgic for those times regard them as a golden age of energy, exuberance and irreverence before the present 'age of anxiety'. Since novelty has its limits, many of today's teenagers have revisited forms of expression that seem remarkably like those of the 1960s – notably four young people with three guitars and a set of drums.

It is exceptionally difficult for anyone who lived through that decade to disentangle what happened from the illusions of hindsight and tricks of perception. Film, with its rapid cutting techniques, lent itself to the illusion of life speeded up, while patches of garish colour brightened the grime and grey of the 'austere' 1950s, one symbol being the brightly coloured Mini Coopers darting along 'Swinging' London streets. The films of the period have paradoxically achieved greater longevity than much of the art output, especially that which did not strive for the deliberately ephemeral. Who now remembers Peter Blake? A few square miles of Soho streets, with their strip joints, coffee bars and music dives, with outposts on Chelsea's psychedelic Kings Road, and grittier Liverpool or Newcastle, were the modest epicentres of this 'cultural revolution'.

Even Britain's allegedly calcified class system seemed to become miraculously fluid as Bermondsey boys, like the actor Sir Michael Caine, began their rapid ascent to wealth and fame, through their portrayal of 'Jack-the-lad' cockneys in such movies as *The Ipcress File* or *Get Carter*. Chirpy girls like Cilla Black, Lulu and Twiggy set forth on their forty years of stardom, while the more middle-class Marianne Faithfull became the decade's symbolic victim, a role she has settled into. Few might have imagined that by about 2000 the proletariat would have become culturally hegemonic, with their accents, tastes and manners sufficiently dominant for the rest of society to feel obliged to adopt them to assert street credibility.

At the time, many people felt that they were experiencing a revolution of attitudes and values, and that revolution ultimately revolved around liberated sexual mores, the development whose effects have been most enduring, because of its long-term impact on women. In 'Annus Mirabilis' Larkin identified the major enthusiasms, although it is relevant that his poem was the work of a provincial librarian who had been in a sexually liberated relationship for the previous fifteen years and was at that time acquiring a supplementary long-term mistress:

> Sexual intercourse began
> In nineteen sixty-three
> (Which was rather late for me) –
> Between the end of the *Chatterley* ban
> And the Beatles' first LP.
>
> Up till then there'd only been
> A sort of bargaining,

A wrangle for a ring,
A shame that started at sixteen
And spread to everything.

Then all at once the quarrel sank:
Everyone felt the same,
And every life became
A brilliant breaking of the bank,
A quite unlosable game.[6]

As the finest new British historians, like Dominic Sandbrook, are revealing, some of the changes associated with the 1960s were evident in the preceding decade, while many young people remained conservative, with a small 'c', in their tastes and opinions. They were as likely to be train spotters, shivering enthusiastically at the sight of a rare locomotive on a platform in Crewe, as 'Mods' and 'Rockers' brawling in seaside Brighton, let alone student revolutionaries occupying the LSE. They thought that trad jazz or skiffle was more 'authentic' than the twangy electrified sounds of The Beatles or the pseudo-blues of the Rolling Stones. Their lives were centred on home, school and work rather than on events in the US or south-east Asia. Despite the reality of relative consumer affluence, and the appearance of increased social mobility, much of the working class still passed from secondary modern schools to dead-end jobs, subject along the way to the cycle of boom and bust, while their middle-class grammar school fellows were more likely to end up among the living dead on the living hell of British commuter trains than among the advertisers, designers, hairdressers and 'snappers' of Chelsea.

Religion plays little part in even the best accounts of that decade, with British social historian Arthur Marwick more tantalised by teenage sexuality than by Vatican II, to which he makes no reference. A number of trends came together in ways that were catastrophic, rather than merely detrimental, to the Christian Churches, and ultimately to Christian belief, although various surrogates – above all those based on New Age mumbo-jumbo – received a minor boost in a decade that prided itself on its enhanced global awareness as symbolised by hashish, kaftans and joss-sticks. The ensuing 'cultural revolution' was not a state-decreed 'Year One or Year Zero', of the sort that had tantalised the Jacobins or the Sorbonne's home-grown dictator Pol Pot, although some conservative moralists may have regarded BBC director-general Hugh Carlton Greene in that capacity. Television indeed became the main vehicle for the diffusion of sensational images, with a capacity to analyse and evaluate editorial bias lagging way behind the flickering pictures.[7]

The decade also witnessed the expansion and 'massification' of higher education, although in Britain this took longer to engineer than on the continent. Socialism enjoyed its Indian summer, as state planning experienced its last gasp before being thoroughly discredited in the following decade. A relaxation of high Stalinism in eastern Europe and the Soviet Union, and the emergence of distant Marxist hybrids that spoke to a certain agrarian romanticism, served to relegitimise more local forms of Marxist heterodoxy, which became hegemonic in the Left university. Entire courses were given over to the theological scrutiny of various incomprehensible theorists who are now largely forgotten. The publication in 1968 of such landmark books on Communism as Robert Conquest's *The Great Terror* passed unregistered in circles that spoke unselfconsciously of 'democratic totalitarianism'. A series of earnest but sinisterly silly gurus of revolution, from Jürgen Habermas to Herbert Marcuse, supplied such sophistical notions as 'repressive tolerance' and 'structural violence' to square the circle of one's own coupling of demands for freedom from violence with resort to violence up to and including the terrorism that would derange the 1970s. The West German extreme left (covertly encouraged by the post-Stalinist GDR) con- tributed a peculiarly masochistic and self-repudiating strand to this process, culminating in the spectacle of German terrorists waving machine guns over Israelis they had hijacked.

In various European centres, juvenile revolutionary sectarians turned forests into a flurry of leaflets, while throwing up a few toy barricades and behaving boorishly to eminent professors, viscerally reminding some of the latter of the antics of Nazi students in the 1930s. Other professors tried to curry favour with the insurgent young or in the case of some gurus incited them; junior-ranking academics often behaved with the customary amoralism of the desperate. Across Europe and the US a series of inconsequential confrontations took place, whether in protest against university overcrowding or against the tons of bombs raining down in south-east Asia.

The immediate effects of these gestures, memory of which still excites tenured radicals, were as nothing compared with the subsequent march of these formerly militant students through major institutions, notably the universities themselves, which thenceforth were dominated by people with little or no experience of what is popularly (and properly) called 'the real world'. Unlike their predecessors, future generations of academics had no experience of code-breaking, being parachuted into France or Greece, or commanding a tank squadron on the Normandy beaches. Instead they inhabited a peculiarly trans-temporal space where the quest for vicarious rejuvenation often meant remaining juvenile into one's retirement, sometimes manifested through vampiric interest in female students.[8]

Change came to be fetishised for change's sake. The transformations inaugurated during the 1960s reflected enhanced individual choice, albeit shaped by powerful commercial forces or the disproportionate suasion of moral avant-gardes, which sought to lead the masses into a promised land of chiliastic indeterminacy. They have since camouflaged the fact that they themselves are a highly nepotistic and unchallengeable elite, zealously tending various sacrosanct liberal pieties while sniggering at anything other than their own ascent to money, power and influence.

By identifying themselves as *the* future, whose deleterious long-term potential consequences they rarely thought through, the 'innovators' and self-styled 'revolutionaries' of the 1960s had an easy time of it with culturally conservative opponents, who readily lent themselves to caricature and satire that was rarely applied to the revolutionaries themselves. What plausible defence could one mount for the prosecution barrister Mervyn Griffith-Jones, who during the 1960 trial of Penguin Books over the alleged obscenity of D. H. Lawrence's *Lady Chatterley's Lover*, pompously invited the jury to ponder whether it was 'a book that you would have lying around in your house . . . a book you would even wish your wife or your servants to read?'[9]

A senior Anglican cleric, bishop John Robinson, appeared as a defence witness for Penguin Books, claiming that it had been Lawrence's intention 'to portray the sex relationship as something essentially sacred'. Archbishop Fisher of Canterbury publicly rebuked Robinson for apparently condoning adultery.[10] Legal publication of *Lady Chatterley* unleashed a torrent of soft-core pornographic magazines, some of which, like *Penthouse* which appeared from 1965 onwards, claimed to be leading 'the struggle for moral and intellectual freedom', although it seems doubtful whether that noble cause, rather than girls with big breasts, was uppermost in the minds of its readers. A similar relaxation of what appeared on television – which came into people's homes rather than sitting on a shop top shelf – led the teacher Mrs Mary Whitehouse to found the 'Clean Up TV' campaign in 1964, which a year later metamorphosed into the National Viewers' and Listeners' Association. Resembling the frowsy straight-woman playing minor fiddle to the baroque Dame Edna Everage, Mrs Whitehouse became the (willing) butt of innumerable media smart-alecs. This is, of course, not to deny that television was capable of powerful documentaries or drama, some of which linger in the mind forty years later, as an object of nostalgia amid the ambient squalor of modern television.

Rather in the way that some of the major cultural characteristics of the 1960s were anticipated in the 1950s, so the latter witnessed the first cautious stirrings against traditional moral thinking from within the Churches. As early as 1953, the Church of England Moral Welfare Council invited the home secretary to initiate an official inquiry into the nature of homosexuality and the laws regarding it. At that time the annual rate of prosecutions for homosexual activities ranged between four and six thousand. Several Churches contributed to the 1957 report by Sir John Wolfenden, along the lines that already governed attitudes towards marital infidelity, namely that although homosexuality was a sin, sexual relations between consenting adult males in private should not be treated as a crime.[11] In the following year, the Lambeth Conference's Committee on 'The Family in Contemporary Society' relativised the notion that the sole purpose of Christian marriage was procreation by claiming that contraception – hitherto regarded as legitimate in dire medical or social emergency – could enhance the 'sacramental' human values implicit in sexual congress *within marriage*. Various Christian thinkers elaborated what by the mid-1960s were known as 'situation ethics', in which 'love' became a constant imperative while 'law' was demoted into something relative to past societies that no sane modern would wish to emulate. What seems nowadays like an ongoing Christian obsession with sex (an obsession now shared with television commissioning editors) was heralded by such curiosities as *The Quaker View of Sex* (1963).

A slew of legislation in the late 1960s, much of it sponsored by the Labour MP Leo Abse or the Liberal David Steel as private member's bills, swept away many of the legal prohibitions (if not the taboos) that were still pervasive in the 1950s, creating what the Labour politician Roy Jenkins advertised as the 'Civilised Society'. Jenkins was by some accounts a decent man, although his hagiographers provide more details of his love of the liquid lunch than of what principles, if any, drove his own reforming impulses beyond vague reference to 'expert' opinion.[12]

His autobiography is similarly unenlightening as to why the 'liberal hour' struck on his Home Office watch.[13] By contrast, his prime minister, Harold Wilson, who came from a Congregationalist family background, once tellingly remarked: 'What I mean is that I don't care for religious attitudes and ideas of morality which seem to depend on intolerance of one kind or another.'[14] A lengthy Criminal Justice Act made it harder to send people to prison through resort to such alternatives as suspended sentences. The 1967 Abortion Act permitted medical terminations of pregnancies before twenty-eight weeks on specific grounds and subject to the recommendation of two doctors. It was also made available, along with contraceptives, on the National Health Service. Only the Free Church of Scotland and the Roman Catholic Church battled steadfastly against the legalisation of abortion; the latter stood alone in continuing to oppose artificial contraception in line with the 1968 encyclical *Humanae Vitae*. In comparison with the other Churches, the Roman Catholics were relatively liberal on the theme of gambling, requiring as they did bingo and raffles to cross-subsidise their role in education.

In 1967 private homosexual relations between consenting adults were legalised by the Sexual Offences Act, while two years later the divorce laws were revised to the fundament of irretrievable marital breakdown.[15] A united clerical front against the latter bill broke down when the Church of England moderated its opposition to 'breakdown of marriage' and dropped its insistence on 'matrimonial offence'. In general, the response of the Churches to such legislation was to join progressives in highlighting the evils of legislative inanition – notably the blackmailing of homosexuals, or the gruesomeness of back-street abortions – while continuing to hope that the assertion of Christian ideals would inhibit any potential slide into limitless moral nihilism. That may have been a realistic expectation as long as society included significant numbers of Christians, and before the 'new morality' spread from the pot-smoking middle and upper classes to teenage heroin addicts on 'sink' public-housing estates in Cardiff or Edinburgh.

Aided by mafia-like local authorities, modernist architects built their brave new worlds – at Park Hill in Sheffield, or Ernö Goldfinger's projects in London's Camberwell or Poplar. Glamorised visual projections showed happy housewives exchanging the time of day on their disconnected twelfth-floor walkways; the current reality of these shoddy constructions, like the Balfron Tower in Poplar, being condensation stains on concrete, lifts reeking of urine, rotting windows and infestations of crack addicts. Christians struggled vainly against the deleterious consequences of the 1960 Betting and Gaming Act, which licensed off-course betting, bingo halls and casinos, and more successfully against attempts to desacralise the sabbath through the extension of normal patterns of trade and consumption. They had some success in holding up the introduction of a national lottery.[16]

The 'cultural revolution' of the 1960s has been described as 'demotic', a fancy word for plebeian, and 'antinomian', the Christian term for those who think that grace releases them from observance of the moral law.[17] As we shall see, these were not the only characteristics of that decade, but they are a serviceable place to start, even if both were already evident in the later 1950s. Britain is important because in many respects it was in the vanguard of the changes inaugurated in the late 1950s and early 1960s, although thanks to modern communications – pirate radio, television, imported records and so forth – by the end of the decade, after the advent of satellite television transmission in 1963, the US West Coast, the Chinese 'cultural revolution', race riots and the grinding war in Vietnam suddenly became vividly present on a day-to-day basis through the masochistic medium of television. Endless images of the grunt's-eye view (usually of village huts going up in flames) inevitably skewed the broader strategic picture. Everything in the US seemed like a more intense version of things Europeans seemed to be playing at, partly because of the searing effects of slavery and segregation, but also because of the wider availability of guns. If the dominant image of the US in the 1950s had been an idyll of suburban tranquillity united in resistance to the Red menace, now its underlying urban chaos stood revealed as young and old, white and black shouted past one another. This was relentlessly exhibited by Europeans who hated America.

The affluence of the long Macmillan era between 1957 and 1963 was the breeding ground for much diffuse anger and resentment. Kingsley Amis and Philip Larkin railed against the dominance of a desiccated metro-politan modernism, before becoming self-consciously sour reactionaries exchanging unfunny thoughts about black people. Less lastingly, others whose social chip was the man raged against anyone ignorant of the fact that it was 'grimly authentic up north', a reality hammered home in endless accounts of struggle and strife in places that are often pretty grim nearly fifty years later. The actor cum playwright John Osborne ranted against everyone in the time-honoured tradition of the drunken nasty piece of work. Youthful satirists self-righteously assailed such easy targets as Macmillan's Etonian-dominated and, in the end, scandal-ridden government, which had been one of the most impressive reforming administrations of the century. Ministers were accused of hanging around too long, although some of the satirists – like Alan Bennett or Jonathan Miller – seem to have lingered a remarkably long time at the summit of the media Establishment, the former becoming a national treasure, the latter a roving polymath with a grievance. The 'revolution' threw up a few celebrities such as the Pakistani activist Tariq Ali (nowadays a wealthy television producer) and the Australian Richard Neville who in 1967 founded the hippie paper *Oz*. The Beatles were the first British pop group to make a global impact, eventually occupying all top five slots in the US charts, and famously inciting crowds of teenage girls into hysterics as they cooed 'Love, love me do, I'll always be true'. Although The Beatles were ordinary young Liverpudlians who combined conventional ambitions with a native Scouse cheek, John Lennon developed messianic delusions, imagining that his pop group was 'bigger' than Jesus, and going on to try to improve the world's collective karma from a bed. Trends that seemed daring at the time hid a squalid reality, in which people such as Cynthia and Julian Lennon got hurt. As the son of the man who co-wrote 'All You Need is Love' had it: 'Dad's always telling people to love each other but how come he doesn't love me?'[18]

The most significant changes were among the screaming young girls to whom film-makers obligatorily cut away in their star-focused films. Changes that were infinitely subtle can only sound rather mechanical. Because of the progressive feminisation of piety in the preceding hundred years, mothers and grandmothers were primarily responsible for instilling religious values in their children. Now the cycle was broken, in the sense that, as a contemporary investigation has shown, religious parents only have a fifty–fifty chance of replicating their beliefs among their children, thereby giving those beliefs a 'half-life', whereas non-believing parents are successful in transmitting their own non-belief. This process did not arise overnight.[19] During the 1960s a distinctive youth culture fuelled by recent relative affluence displaced older ideals of domesticity, one consequence being that traditional religion lost the main site for its transmission through the generations, based as that was on the binary stereotypes of pious and respectable women corralling more wayward men into the traditional Christian home and family. According to the historian Calum Brown, author of the most innovative work in this field, almost overnight girls' and women's magazines that celebrated a traditional range of domestic feminine virtues were swept aside by such products as *Jackie* in which 'Stories focused on the words "you", "love" and "happiness".' Out went the good-deed-doing Four Marys of its predecessor *Bunty*, and in came making oneself appealing to The Monkees, a US pop group manufactured to subtract market share from the British Beatles. Much of the content of *Jackie* seems incredibly innocent, from a contemporary vantage point where magazines aimed at very young teenagers speak to concerns hitherto confined to younger adults. A new range of magazines for young women, notably *She* and *Cosmopolitan*, the former a product of the 1950s and the latter of the early 1970s, provided more adult versions of the same shift in moral discourse, highlighting women as people with careers or as consumers, while even the more staid *Woman's Own*, which was primarily aimed at housewives, witnessed an expansion of the range of problems countenanced by their agony-aunt columnists, so as to breach such taboos as female sexual satisfaction, while references to religion disappeared.[20]

The Churches and the more semi-detached tribe of theologians often responded to rapid changes in the wider world by trying to assimilate secular cultural and social enthusiasms, while jettisoning anything that still smacked of 'superstition', sometimes including God as well as the devil. Both tendencies had been evident for a hundred years. Radical German and US theology was sensationally vulgarised for British audiences by bishop John Robinson, beginning with his 1963 book *Honest to God*. Centuries-old liturgies were abandoned in favour of 'happy clappy' church services, although few ventured as far as the (Catholic) college students whose antics in chapel were fictionalised by David Lodge:

> Each week the students chose their own readings, bearing on some topical theme, and sometimes these were not taken from Scripture at all, but might be articles from *The Guardian* about racial discrimination or poems by the Liverpool poets about teenage promiscuity or some blank verse effusion of their own composition. The music at mass was similarly eclectic . . . They sang Negro spirituals and gospel songs, Sidney Carter's modern folk hymns, the calypso-setting of the 'Our Father', Protestant favourites like 'Amazing Grace' and 'Onward Christian Soldiers', and pop classics like Simon and Garfunkel's

'Mrs Robinson'. . . or the Beatles' 'All You Need Is Love'. At the bidding prayers anyone was free to chip in with a petition, and the congregation might find itself praying for the success of the Viet Cong, or for the recovery of someone's missing tortoise, as well as for the more conventional intentions.[21]

While this seems silly and shambolic, during these years Christians of all hues were important to the success of many of the newer 'niche' charities of the period, such as Amnesty International, the Cyrenians, Oxfam, the Samaritans, Shelter and the hospice movement, whose work for the distressed, dying and desperate has been an impressive manifestation of Christian charity. Organisations like Alcoholics Anonymous synthesised Christian spirituality with secular forms of self-improvement to wean people off the demon drink. By contrast, the Campaign for Nuclear Disarmament, which briefly flourished in the early 1960s, almost invites a more dismissive response, since it was largely a stage upon which Christian and humanist radical eggheads could flaunt their woolly moralism and vulgar anti-Americanism, in the conceited delusion that the hard-headed realists in Moscow or Washington would take notice of them.[22] The introduction of apartheid by the Nationalist government of South Africa provided a further ongoing focus for moral fervour, with the reasoned advice of such informed figures as the Anglican bishop of Johannesburg about how to combat doctrines that violated fundamental Christian tenets, trumped by those whose solutions were progressively indistinguishable from those of secular radicals. The Churches (other than those Protestants who espoused apartheid) had to negotiate between the evil of making Christians worship in racially segregated churches and the consequences of a Marxist-dominated ANC coming to power. Marxism created desolation wherever it was essayed, not least in Africa, where the forces of 'liberation' in Angola, Ethiopia or Mozambique presided over decades of civil war whose direct and collateral casualties dwarfed those in South Africa or what became Zimbabwe. Post-colonial guilt was responsible for the terrible condescension visited upon Africa by a racist liberalism.

While British Christians – or at least those who wished to find favour with the shapers of secular elite opinion – took up the causes of 'the World', what was happening to the beliefs of ordinary Britons in the same period? The religious could console themselves by arguing that, despite their own desperate trendiness, declining involvements with the

Churches merely reflected waning memberships of political parties, trade unions and other institutions, and that traditional beliefs broadly held up among large swathes of the population. And so it seems today. On the face of it, surveys like the 2000 'Soul of Britain' investigation reveal that 31 per cent of people describe themselves as 'a spiritual person' while a further 27 per cent claim to be 'religious', with 'convinced atheists' and agnostics languishing at respectively 8 and 10 per cent. As the sociologist Steve Bruce has cogently argued, the questions posed virtually guarantee that in order to avoid the relatively well-defined pigeonhole of 'convinced atheist', with its unBritish connotations of ideological militancy or fanaticised Darwinian scientism, large numbers of people who are actually indifferent to, and wholly ignorant of, religion will respond with the truthfulness of the smoker who tells the doctor that his consumption of twenty cigarettes is really five a day. It is more self-flattering to claim to be a 'spiritual' or 'religious' person than to concede that one's universe is bounded by supermarket aisles, the antics of Posh 'n' Becks or television programmes devoted to Hitler, sharks and the 'reality' of a woman masturbating with a bottle. Opinion monitors no more press people on what they mean by 'spirituality' than doctors dispute prodigies of willpower involving one of the most addictive substances known to man.[23]

Ironically, in view of events across Europe, one major change in the field of religion attracted almost no attention at the time, namely the translation of people from countries where religion was all-pervasive to a developed society where the dominant creed was secular liberalism with Christian remnants. While the arrival of many migrants who were devout Muslims, Hindus and Sikhs (not to speak of Pentecostalist and Seventh Day Adventist Christians from the Caribbean) gave a timely boost to the numbers of religious believers in an otherwise secularising society, the need to acknowledge their faiths (as well as that of the existing Jewish minority) dealt a lasting body-blow to the exclusively Christian Constitution of Britain. The fact that there was no apparent or conceivable challenge to the verities of Western liberalism (except from a lunatic far right and its analogues on the extreme left) meant that the religious implications of mass immigration went unattended. The idea that Britain is a 'multi-faith' society has become so ingrained, often with the explicit encouragement of the Establishment, that it is easy to forget how this development happened. This is mysterious, because what seemed a promising celebration of difference has turned out to be highly divisive.[24]

The experience of imperialism had long encouraged the British to take an intelligent and sympathetic interest in the religions of indigenous peoples, although in the case of missionaries these faiths were regarded as preliminary foundations for the higher Christian truth. Initially, in an era of rapid decolonisation and immigration, most Christians who thought about the religious implications of Commonwealth immigration confined themselves to generalised expressions of respect for other faiths, albeit as less-developed paths to the truth of God, although Calvinists and some Evangelicals were not so indulgent and actively proselytise to this day.

The first substantial attempt by a theologian to address these issues was by John Hick, a Presbyterian minister and philosopher at Birmingham, in his 1973 book *God and the Universe of Faiths*. This rejected the traditional belief in *extra ecclesiam nulla salus* or 'no salvation outside the Church' and called for a 'Copernican revolution' which would recognise all major religions as 'valid' routes to God. The various creeds were parallel 'ways through time to eternity'. Hick resolved the problem represented by the divinity of Jesus by claiming that the doctrine of the incarnation was a necessary myth.[25] While the majority of British people remained directly unaffected by immigration, in such centres as Birmingham the larger question of race relations led to Christian initiatives to defuse tensions, which had briefly exploded into violence, through dialogue with people of other faiths. These meetings were of a very informal kind, and were as much about exchanging knowledge, with a few cautious experiments in common worship. It took time for the development of forums of the kind that had long existed to facilitate dialogue between Christians and Jews, even though the latter had in the interim been eclipsed as Britain's main religious minority.

In many respects, the monarchy took the lead. Unlike the US, where public holidays such as Memorial Day or Thanksgiving, and the four-year cycle of presidential inaugurations and annual state of the Union addresses, provide frequent occasions for the affirmation of 'civil religion', in Britain, where royal coronations happen at most once or twice in a lifetime, explicit declarations of British values are much more anodyne and episodic, restricted as they are to such things as the Queen's Christmas Day broadcast or major events in the royal family's life cycle which as yet did not include serial marital breakdown and divorce.[26] The fact that these broadcasts, which were televised from 1957 onwards, were also directed to the Commonwealth ensured that from the start they eschewed Christian exclusivity in favour of the Queen's references to 'whatever your religion may be'. Official recognition that Britain was a 'multi-faith' society was symbolised by the service held in 1966 in St Martin-in-the Fields, in which a congregation led by the Queen heard affirmations drawn from the sacred books of the major religions. Significant numbers of Christians objected to the politique omission of references to Christ, and to the implication that they should bury very real differences of belief in the interests of an inoffensive religious syncretism. In subsequent decades, other members of the royal family, with Prince Charles most prominent, have been among the most enthusiastic promoters of 'inter-faith' dialogue, while the presence of representatives of the plural faiths has become obligatory whenever the British state celebrates its civil religion, especially at the annual Remembrance Sunday services. While very few indigenous British people converted to such alternative faiths as Islam or Hinduism, the 1960s saw an opening up of the mind to a range of eastern religions, or rather to a dilettantish smattering of what elsewhere were extraordinarily elaborate religions. In 1980 a speaker at the Anglican General Synod lamented that young people 'go to India in search rather than to the parish church – to the transcendental meditation, to the Divine Light, to Hare Krishna, to the Moonies, to find enlightenment and joy. If only we were equipped and obviously ready to offer instruction in meditation and prayer, some of them, I believe, would turn to us.'

The numbers of young Britons making pilgrimages to India in the 1960s was small, and the ranks of those joining the Moonies and similar fringe sects even smaller. In the light of successive moral panics, it is salutary to remember that less than a hundred people in Britain are currently estimated to be engaged in the worship of Satan, though one can never tell.[27] There were very few conversions to the rigours of Islam or Orthodox Judaism, although a few celebrities flirt with such mystical strands as the Kabbalah, the faith of choice for Madonna, partly because this would involve taking on the outward characteristics of visible minorities. Ironically, the greatest attraction was to the religions with which Western people had the least direct contact, perhaps suggesting that there were no disillusioning moments such as liberals began to experience with Islam, when the tremors of upheavals thousands of miles away began to rock Europe's shores during the late 1970s.

The late 1960s prepared much of the ground for what have come to be called 'New Age' beliefs, which fused snippets of Eastern mysticism, astrology and occultism, environmentalism and psychotherapy, and whose eclectic philosophies nowadays adorn entire walls in bookstores. New Age beliefs are generally eclectic, holistic and self-centred. They appeal overwhelmingly to white middle-class people – a Briton of West Indian origin dismissed a New Age event as 'something for poofs' – including a disproportionate number of women, who believe that they will discover their true inner self, suppressed by the combined forces of male-dominated Western scientific rationality and a consumer economy. In a sense, it represents a spiritualised form of the Marxist quest for an end to alienation, although New Age is less coherent than the study of economics. Its powerful ecological component reflected the concerns first articulated by the American biologist Rachel Carson in her 1963 warning against indiscriminate use of DDT and other pesticides, *Silent Spring*, as well as more heady attempts to fuse science and religion, notably the British chemist James Lovelock's notion that life on earth has a collective consciousness symbolised by the earth-goddess 'Gaia'.[28]

New Age religions often reach backwards to pre-modern (or utterly fantastical) cultures and times – the Native Americans and King Arthur are favourites – or reach outwards to less developed societies. Viewed superficially, the New Age religions seem little more than an updated form of the romantic belief in *ex oriente lux*, a post-imperial cultural cringe that has replaced the alleged arrogance of Western imperialism with limitless credulity in response to the spiritual beliefs of the under-developed world. New Age religions share much of the Western culture of self-repudiation that is evident in other more supposedly rational areas of life. But this is to miss how the Western New Agers have subtly transmogrified these beliefs, leaving out anything that does not chime with their own existing views and desire for Western comforts. Out goes anything resembling stoical acceptance of the insignificance and transience of our lives on earth, for otherwise what would be the point of religions based on self-development? Out goes any notion that re-incarnation may involve a judgement on the moral character of one's past or present life, or, put crudely, the prospect that one might accordingly be reborn as a rat or cockroach. Out too goes any notion that the heights of spiritual 'awareness' might require exercises of a kind that might once have taxed an Ignatius Loyola. Instead, spiritual wisdom can be acquired through a weekend at a Scottish or Welsh meditation centre or on a two-week holiday in Thailand. If time is pressing, the distilled experience, can be made available through the crash course as video or DVD package, paid for with a credit card over the internet, in ultimate obeisance to the Western rationalisation of time that the New Age deplores. Ancillary services include such things as feng-shui consultants to check out the presence of gremlins and hobgoblins in a new house.

Although the therapeutic culture is probably here to stay, it seems doubtful whether New Age religions will have much greater longevity than the hippies, a few of whom still linger on in tepees in Welsh valleys that time forgot. Some of it will undoubtedly be absorbed into the dominant commercial culture, whether as management technique or as a branch of healing. Pity all those (graduate) corporate employees who have to play motivational games that might excite five-year-olds. The eclectic nature of New Age beliefs also militates against their being easily communicable to fresh generations of adherents, who, as products of conventional education, may react to the beliefs of their middle-aged parents with incredulity. Although Christians may deplore what could be called a soft recrudescence of European paganism with orientalised accretions, it is salutary to remind ourselves that even the most generous estimate of the numbers currently involved in organisations catering to New Age spirituality amounts to a mere third of the worshippers lost to one Christian denomination (the Methodists) over a forty-year period.[29]

II FORTRESS OR SIEVE? VATICAN II AND AFTER

Under Pius XII, one institution at least seemed impervious to cultural change, his stance being a source of reassurance to many in an increasingly bewildering world. Five years after his death, Pius would become the object of Communist-inspired denigration in the form of Rolf Hochhuth's historically fanciful 1963 play *Der Stellvertreter*. Suddenly the aquiline hollow-eyed face that had long adorned postcards in millions of Catholic homes was changed for that of a jowlier man of robust peasant stock who while a priest had served as a sergeant in the First World War. In January 1959 the newly minted John XXIII announced his intention to summon a general council, part of his wider desire to bring the Church into greater conformity with modern times, a process known in Italian as *aggiornamento*. He knew it would be like trying to steer an oil tanker, a task requiring great skill, with many potentialities for utter disaster. His summons was received with consternation among some of the more reactionary members of the Roman curia, whose leader – cardinal Ottaviani of the Holy Office – moved to retain control of the key doctrinal commission, the area where least change was effected by Vatican II. In 1960, John XXIII established a Secretariat of Unity under a German Jesuit cardinal Augustine Bea. As the secretary of the World Council of Churches, Visser 't Hooft, exclaimed, 'We finally have a friendly address in Rome.' This office invited leading Anglicans and Orthodox to Rome, with archbishops Fisher and Ramsey calling on the popes in 1960 and 1966, and patriarch Athenagoras in between. In 1965 pope (Paul VI for John had died in the interim) and patriarch would rescind mutual excommunications first promulgated in AD 1054.[30] A further by-product of the Council was the Council for Inter-Religious Dialogue, established in 1964, which was charged with promoting a more respectful relationship with the Jews and Islam, a task of some delicacy given the mistrust and racism that characterises relations between these monotheisms, both of which include people who refer to what postmodernists call 'the Other' as respectively 'goyim' and 'kufar', terms of abuse for which Christianity has lost an equivalent vocabulary.

The Council opened in October 1962. In addition to 2,500 bishops and heads of religious orders, four hundred theological experts (or *pertiti* in the language of conciliar communication) and about forty observers from non-Catholic communions attended. Like many meetings, this one was shaped by vocal minorities, since only two hundred people are recorded as having spoken over all four sessions.[31] The experts included a number of distinguished theologians upon whom the curia had cast a beady eye under the pontificate of Pius XII, notably Karl Rahner and Henri de Lubac. The presence of so many brilliant minds, from Rahner to Ratzinger, gave the Latin deliberations an academic and theoretical flavour, whose vagaries would lead to confusion and conflict when attempts were made to translate them into concrete measures. Feelings of betrayal abounded. This was doubly the case because the theologians themselves naturally differed in terms of their receptivity towards an historical or contextual approach focused on the world and its ideologies in the present, or one which spiralled after truth within the Church's own deep experience. If Hans Urs von Balthasar represented the first tendency, then Edward Schillebeeckx and the liberation theologian

Leonardo Boff embodied the latter, albeit from respectively a European and a Brazilian perspective. By the opening of the third session the rules were changed so that the experts expressed a view only when asked. Furthermore, talk of collegiality was all very well, but when the bishops and theologians decamped, the Church's structure remained hierarchical, with scope to make conservative appointments from the centre to stymie a more reformist, synodal – or micro-conciliar – spirit on a national or local level. Of course, it was left to such conservatives as Ratzinger to remind people that during the Nazi era the collective statements of the German bishops were far more tepid in tone than those of bold individuals.[32]

The Council ended its fourth and final session in December 1965, under the new pope Paul VI who faced the difficult task of ensuring that change did not result in the helter-skelter of revolution, a position urged on him by conservatives who suspected him of liberal sympathies. They feared that a too uncritical opening to the world would transform a fortress into a sieve.[33] His first encyclical, *Ecclesiam Suum*, dated 6 August 1964, effectively gave the green light for the Council to engage with a number of problems that as pope he himself appeared to eschew:

> We have no intention here of dealing with all the serious and pressing problems affecting humanity no less than the Church at this present time: such questions as peace among nations and among social classes, the advance of new nations towards independence and civilization, the current of modern thought over against Christian culture, the difficulties experienced by so many nations and by the Church in those parts of the world where the rights of free citizens and of human beings are being denied, the moral problems concerning the population explosion, and so on.[34]

Debates on an impossibly wide range of huge subjects – including 'war', 'peace' and 'women' – were condensed into conciliar documents inaugurating a dialogue within world Catholicism that still reverberates today. This was when many current battle lines were formed; the migration of the thirty-five-year-old Ratzinger, the current pope, Benedict XVI, from a liberal to a more conservative stance being symptomatic of the tensions within one individual. The Council differed from Vatican I, which had never formally concluded its work amid the disarray caused by the Franco-Prussian War, in the sense of being less concerned with the papacy than with the Church and Christianity in general. The Council itself reflected the insistence on enhanced collegiality in the running of the Church, while both the notion of the 'People of God' and various liturgical innovations brought changes that could be summed up as popular and participatory, reducing the obvious distinctions with Anglican or Lutheran worship, to the regret of many devotees of tradition who watched as what had been normative for millions became an object of intense nostalgia for the equivalent of devotees of medieval costume or madrigals. These liturgical reforms coincidentally smoothed the ecumenical conversations that have done much to allay crude mutual suspicions stemming from the priestcraft aspects of the Roman tradition. These conversations extended beyond the community of Christians into relations with the Abrahamic faiths. The 1965 conciliar document *Nostra aetate* categorically condemned antisemitism 'at any time and from any source' and disowned the ascription of collective guilt to the Jews for the Romans' crucifixion of Christ. A dialogue with Islam and eastern religions was more tentative. Flowing on from John XXIII's position, there was a very slight relaxation of Pius XII's implacable anti-Communism based on a new-found recognition of the distinction between a materialist creed and the common humanity of its believers. This meant that in some circles, Paul VI would be caricatured as a 'Communist'.

A flying visit to U Thant's United Nations in New York established the pope's claim that the Church was 'an expert in humanity' with a complementary role to that of the UN:

> We are happy to note the natural sympathy existing between these two universalities, and to bear to your terrestrial city of peace the greetings and good wishes of our spiritual city of peace. One is a peace which rises from the earth, the other a peace which descends from heaven, and their meeting is most marvellous: justice and peace have kissed one another. May God grant that this be for mankind's good.[35]

World problems, whether the prospect of total destruction by nuclear weapons or hunger afflicting entire continents, seemed to require global solutions from an organisation whose international credibility was as yet largely unsullied by corruption, inertia or rampant anti-Americanism. Speaking in French the pope connected disarmament and dialogue with what would later be known as the 'peace dividend' through which increased aid would be disbursed to developing nations in an age almost innocent of the fact that it might be embezzled. Paul also reminded his auditors that beyond the urgent plight of the starving was the need to maintain human dignity, to which the right to religious liberty was integral. The answer to rampant population growth was not artificial contraception, or 'cutting down the number of guests at the banquet of life', but a concerted effort to improve world food production. The second part of this agenda was drowned out by those who were outraged by a corollary that seemed to fly in the face of the liberated mood of the times.

If acknowledgement of the complementary role of the UN was one consequence of Vatican II, so was an unequivocal condemnation of resort to area bombing or the use of nuclear weapons against major population centres, although some clerical Cold Warriors felt there should be scope left for nuclear strikes on remote military targets such as radar systems, airbases and weapons silos. On the political front, the Church continued to recognise the diversity of governmental forms in the modern world, arguing: 'In the face of such widely varying situations it is difficult for us to utter a unified message and to put forward a solution which has universal validity. Such is not our ambition, nor is it our mission. It is up to the Christian communities to analyse with objectivity the situation which is proper to their own country.' In fact, the Church was burying its inherited hostility to the legacies of the French Revolution, and radically scaling down some of the larger claims that had once been made on its behalf. The word 'monarchy' ceased to be a part of its constitution. The Council acknowledged that several models of Church–state relations were no longer appropriate to modern times, including theocracy, Erastianism and Caesaropapism. Regimes were acceptable as long as they respected the human right of religious liberty, including the right to conduct missions or to proselytise, to minister to dissident or oppressed groups, and to proclaim the Church's moral and social teaching. By drawing together senior ecclesiastics from different political settings – including Poles whose Catholic orthodoxy was steeled by the fight against Communism, and Central and Latin Americans who were internally divided by how they lined up for or against authoritarian regimes and ruling classes that were not shy of asserting their self-interest by massacring as well as oppressing poor people – the Council itself became an important source of plural opinions that seeped back into national contexts where debate was hitherto highly restricted.

Predictably enough, the decrepit contingent from Franco's Spain distinguished themselves with highly reactionary opinions; at one point the assembled fathers laughed as the elderly bishop Gómez warned that hellfire awaited non-believers. The microphone was removed from him. By contrast, the abbot of Montserrat gave an interview to *Le Monde* in which he said that 'Colectivamente neustros hombres políticos no son cristianos,' a remark that cost him his job. The Council sent shockwaves through the Spanish Church, while Franco thought that it represented a conspiracy by freemasons. Any regime whose claim to legitimacy rests upon an international Church risks shipwreck when currents within that Church wash back through the mind's permeable borders. For the first time Spanish clergy, who were the youngest in the world even as their bishops were among the oldest, learned of French novelties in pastoral work or the convolutions of celebrity German theologians.[36] During the later 1960s, Paul VI made a number of crucial appointments to auxiliary bishoprics (which did not have to have the regime's stamp of approval) before he appointed the pro-Vatican II archbishop Vincente Enrique y Taracón to the see of Toledo and the primacy.

Independently of the impact of the Council, many Spanish Catholics chafed under the deadening orthodoxies of 'crusading Catholicism', that is the militant 'throne and altar' ideology that had been deployed during the Civil War and in the campaign to petrify and purify Spanish society thereafter. It seemed all too appropriate that the regime's most lasting edifice was the Valle de los Caídos, the monument to the Nationalist fallen, which took twenty years, and much blood, to excavate from the granite outcrops north of Madrid. Discontent was fed by the fact that the regime's strident Catholic rhetoric had not advanced the self-proclaimed goal of 're-Christianising' Spanish society, but seemed rather a gambit to win international recognition, as reflected in the August 1953 Concordat and the economic and military pacts with the US that followed. Paradoxically, the ascendancy of what many feared as a secretive and reactionary 'secular institute' – Opus Dei – founded by the Aragonese priest José María Escrivá de Balaguer in 1928 – may have contributed to the secularisation of Spanish society. It combined the authoritarianism of Pius IX with a concern for the bottom line that would not have embarrassed Henry Ford. Originally founded to counteract the influence of freemasons, Opus inserted adherents into academia, banking, business, publishing and journalism, culminating in a 1957 cabinet that included eight Opus Dei members.[37] Its high-placed members were represented

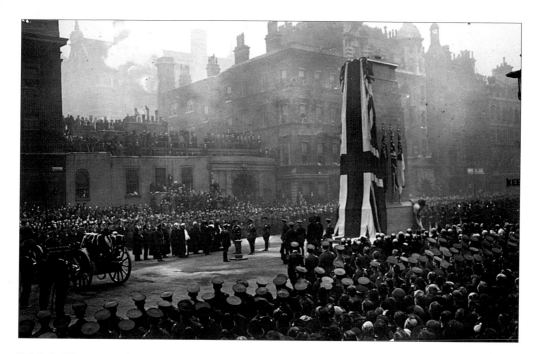

Whitehall's Cenotaph, unveiled by George V in November 1920, and used to commemorate the dead of the world wars, has become the epicentre of a British civil religion.

Above: Georges Roualt's monumental series of etchings *Miserere* are a moving example of a Christian attempt to comprehend the losses of the Great War.

Right: Louis Haeusser was one of the so-called 'inflation prophets' abroad in the chaotic and disturbed early years of the Weimar Republic.

Pope Pius XI (1857–1939) was a forceful opponent of the totalitarian political religions of the inter-war period.

The Bolshevik League of the Militant Godless held ceremonies and processions mockin Orthodox Christianity and other religions, unconscious that their own symbols, including th stars, hammer and sickle shown here, themselves were part of a political cult

A Bolshevik attempt to create a 'visible god', in this case a constructivist machine.

The detonation of the remaining shell of Moscow's Cathedral of Christ the Saviour. Other churches were converted into museums of atheism

A crude Republican caricature of the key forces on the Nationalist side in the Civil War.

Republican troops shooting up the statue of Christ on the Cerro de los Angeles outside Madrid on 20 August 1936.

Like the Bolsheviks, Spanish Republicans went in for the exposure of the remains of religious figures, in this case the corpse of a Carmelite nun. The object seems to have been to disprove the popular superstition that such bodies were not subject to decomposition.

Post-War Christian Democratic poster in which the Party claimed to be defending the nation, family and freedom against Communism.

The Communist threat to the integrity of the family shown in the form of a Russian tank about to crush a child.

The totalitarian challenge to the churches in the form of the image of Mussolini projected onto the facade of Milan's cathedral.

Homeless refugees sheltered in the summer residence at Castelgandolfo of Pope Pius XII.

Hitler frequently incorporated Christian texts into his speeches, a process depicted in a painting entitled *In the beginning was the word.*

The Italian artist Signorelli anticipated the siren-voiced dictators of the twentieth century when he painted the Antichrist preaching several hundred years before.

It is often overlooked that the Nazis launched a vicious campaign against the Catholic Church, in this case SS cartoons alleging sexual misconduct.

LA LÉGION FRANÇAISE
VEUT FAIRE
LA RÉVOLUTION
POUR LA
FAMILLE

Sections of the Catholic Church backed the Vichy regime's claims to be remoralising French society after the decadence of the inter-war years.

Pius XII has routinely attracted criticism for his conduct during the Second World War, and for having a hieratic conception of his office. Here his human side is more evident at an audience for children.

The totalitarian dictators saturated the public sphere with their own graven images, in this case banners of Stalin at a May Day parade in Bucharest, Romania.

The League of the Militant Godless outrage heavenly hosts with their continuous working week.

The Roman Catholic Church played a complex role in Latin America, both legitimising and opposing authoritarian and military regimes. Archbishop Oscar Romero was assassinated in his own cathedral in San Salvador on 24 March 1980.

The Western Left found it inconvenient that highly religious industrial workers were among the staunchest opponents of an alien Marxism in Poland.

Sniggering at any form of authority was one of the cultural trends of Britain during the 'swinging sixties', although in the interim, the sniggerers have largely gone on to occupy much of the academic, cultural and entertainment Establishment.

During the 1960s rioting students were briefly in the vanguard representing various sacred causes, few of which were shared by the majority of their tax-paying fellow countrymen, who mysteriously felt no enthusiasm for Mao's Little Red Book.

John Lennon and Yoko Ono personifying much of the silliness of the sixties at a 'bed-in' in a hotel in Amsterdam, the Netherlands being where the spirit of the 1960s became institutionalised in a country known for high levels of complacent conformity.

Ireland's matriarchal culture played a key role in keeping the sentimental flames of Republican nostalgia alive throughout the Troubles. Funerals were one of the central features of this sinister cult.

The world marvelled as the ancient right to march to an ugly Protestant church at Drumcree was aggressively asserted.

The reality of political violence in Northern Ireland. Two British soldiers were lynched by a Republican mob in 1988.

United Airlines flight 175 strikes the south tower of the World Trade Center on 11 September 2001. This act of mass murder announced the onset of unlimited Islamist aggression against western civilisation.

The far enemy. Captured members of Al Qaeda and the Afghan Taliban detained in August 2003. Failed states are swamps that terrorists find congenial.

The near enemy. In 2004 the controversial Dutch filmmaker Theo van Gogh was murdered on an Amsterdam street by a second-generation Moroccan immigrant.

Scenes of anarchy in the suburbs of Paris among youths with 'rights' rather than 'responsibilities' – one of the main disorders of modern western civilisation.

among the technocrats responsible for neo-liberal economic policies, which, by bringing prosperity, the regime hoped would compensate for the absence of political freedom. Terms like 'development dictatorship' or even 'modernisation' did not augur well for the preservation of a traditionally religious worldview that relied on an unchanging socio-economic order.[38] Simultaneously, the horizons of this traditional society were being expanded by the exodus of many young Spanish people to work in more liberal parts of Europe, and the influx of enormous numbers of foreign tourists, whose holiday Deutschmarks and pounds Sterling were essential to the nation's economy. Even as they purchased their leather wine flasks and toy toreadors they also helped diffuse more relaxed mores, if only by displaying expanding acreages of tanning flesh.

The Church itself was in crisis. Religious vocations plummeted. In 1963 some 167 priests had opted for the secular life; by 1965 that had reached 1,189, and then an all-time high of 3,700 four years later. The Jesuits lost a third of their members in a decade. In the Basque country and Catalonia, many priests made common cause with local auton-omists, inexcusably including the Marxist terrorist organisation ETA, while others exchanged their spiritual vocation for various forms of social radicalism, making fiery sermons and taking part in protests that often ended in violence. According to a survey conducted in 1977, apart from an intransigent minority, the majority of Catholic clergy espoused political positions that were congruent with those of the semi-Marxist intelligentsia. The Church's espousal of social radicalism meant such novelties as a 'Concordat jail' at Zamora, to hold priests who had stepped too far out of line, and the growth of extreme right-wing anticlericalism directed against the 'Communist pope' and the 'Red clergy'. By the late 1960s financial scandals involving Opus Dei enabled the diehard Falangists to revenge themselves on the organisation, a further sign of growing alienation between the Church and the regime.[39]

Each time one of the elderly reactionary bishops died, he was invariably replaced by a 'conciliar'-minded successor. The bishops of Andalucía denounced the fact that casually employed labourers were suffering from malnutrition and living in caves; the newly appointed bishop of Bilbao contrived to excommunicate policemen who had beaten up one of the clergy in his diocese. By 1970 some 187 Basque priests were in prison while across Spain priests were being assaulted by the regime's tougher supporters. A year later, a clerical assembly narrowly failed to support a motion condemning the idea that the Civil

War was a crusade, a motion that would have carried comfortably had the Basque and Catalan clergy been represented. In 1972 the Bishops' Conference issued an epochal statement renouncing the Church's political privileges and calling for political pluralism.

The funeral of prime minister admiral Carrero Blanco in 1973 was a symbolic moment. He had been murdered by a huge ETA bomb in a tunnel below a Madrid street as he travelled from mass to his office. The blast left his car, with his body and those of his guards and driver, perched four floors up on the roof of a Jesuit church. When the Spanish primate attended the funeral, he was greeted by loyalist cries of 'Taracón to the firing squad'.[40] In the regime's dying years, the Vatican sought to renegotiate the 1953 Concordat, while nominating only one person to each vacant see so as to frustrate the regime's exercise of its residual regalian rights. In 1975 Paul VI led the world in protesting against the retroactive death sentences imposed on ETA (and FRAP – a Madrid-based Marxist grouplet) terrorists who had killed three policemen. The Church played a significant role in removing any religious justification for what was au fond a brutal and thuggish military regime. Under the skilful guidance of king Juan Carlos, the best argument for monarchy in the late twentieth century, and the transitional centre-right governments, the Church was able to find a new place for itself as a free Church in a free Spain. It helped that the Church renounced any intention of backing a Christian Democratic party, while generational changes meant that the leaders of both the Socialist PSOE and the Communist PCE were less afflicted by a mindless anticlericalism than their predecessors.[41] This role was consolidated by the 1978 Constitution that codified the relationship between Church and state, and resolved the thorny question of state subsidies to religious schools. Separate bilateral agreements with the Vatican terminated the state's rights of ecclesiastical presentation, while the Church agreed that state subsidies for the Church should be phased out. The Spanish confessional state had been dismantled with the co-operation of a Church that was integral to this most successful transition to democracy.

Similar strains, albeit in a minor key, between Church and state also became evident in Portugal during the 1960s. Although opposition thinking had spread to some lay Catholic organisations, until the late 1950s the highly conservative hierarchy was immune to it. Then in 1958 the hitherto unremarkable bishop of Oporto, António Ferreira Gomes, drew upon himself the dictator's ire by criticising the lack of freedom, the pervasive rural poverty and the plight of indigenous peoples in Portugal's corruptly run colonies. In 1959 the bishop was prevented from returning to the country after a holiday abroad and spent the next decade in exile. In the same year the secret police uncovered the 'Cathedral plot', in which a few hundred civilians (many of them Catholics) and junior army officers were conspiring to overthrow the regime, having built up weapons in the crypt of Lisbon cathedral. The pope's condemnations of colonialism and imperialism also created tensions between the Salazar regime and the hierarchies in Angola and Mozambique, where since 1961 the colonial power had been facing armed revolutionary insurgencies. Paul VI angered the Portuguese with his 1964 visit to India, which had recently annexed Goa, and then again six years later when he cordially received the leaders of anti-Portuguese liberation movements in the Vatican. Despite such accommodations, the latter tended to be suspicious of the Catholic Church as a long-term support of the colonial power, a suspicion compounded by the fact that many of the leaders of FRELIMO were Protestants. Matters in Portugal did not evolve in conformity with events in neighbouring Spain, partly because the local impact of Vatican II was minimal and the Church could not imagine a role for itself outside the conceptual framework of the 'New State'. While there were priests and laymen who entered into dialogue with the opposition, the hierarchy remained on the side of the ailing dictator and his successor Marcelo Caetano, who was overthrown in a coup in April 1974.

Arguably the most significant repercussions of Vatican II were felt in Central and Latin America, home to half of the world's billion Catholics, and where Brazil and Mexico were its two largest national Churches. Like the Middle East, the continent was not for the squeamish. Privilege and poverty were starkly represented, and the privileged were not hesitant in defending their interests in a brutal manner, through kidnapping, torture and murder, sometimes aided and abetted by the CIA and other organs of the US government, which left behind the gentlemanly rules of the Cold War when they crossed the Mexican border. For historical reasons, the Church was itself part of the privileged, although in fairness, unlike professors of sociology, it also operated large-scale charities for the disadvantaged. Ironically, it had been Pius XII who in 1955 urged the hierarchies there to adapt themselves to the challenges of the present, one by-product of which was the creation of the Latin American Council of Bishops (CELAM). In line with secular intellectuals, some clergy had grown sceptical of the idea that Latin America's problems could be solved by 'development', arguing instead that the continent's disadvantaged were the consequence of structural 'dependency' upon the rich 'North' and entrenched domestic oligarchies within. Decolonisation in Africa and Asia helped popularise the notion of 'liberation' as the favoured solution to the continent's problems, although the example of the French 'worker priests' also played a part. The Cuban Revolution seemed to promise more success than the more cautious reforms of Latin America's few Christian Democrats, although in reality a 'Sovietised' Cuba would acquire a grim notoriety, and not just for religious persecution, although something like eight hundred priests soon became eighty in a population of around six million. In parts of Latin America individual clerics sought to give away Church lands, campaigned against pervasive illiteracy or endeavoured to make Catholicism less Hispanic to appeal to the Amerindian populations, who were the poorest of the poor. One defrocked Colombian priest, Camilo Torres Restrepo, joined the National Liberation Army and in 1965 met the fate of Che Guevara in a fire-fight with government forces. In Guatemala, fathers Arthur and Thomas Melville of the Maryknoll Congregation were ejected from the country for advocating armed uprisings. The 1968 Latin American Bishops' Conference in Medellín noted these developments, and issued a detailed call for social justice for the poor. It even warned that 'institutionalised violence' was taxing the patience of the oppressed, a claim that the US government over-interpreted as the Church opting for violent revolution. One result of this conference was the creation of 'base communities', in which clerics tried to recentre the Church within the world that the poor inhabited. Although estimates of the extent of these communities vary wildly, in fact they only embraced about 5 per cent of the continent's huge population of Catholics.

Both Catholic and Protestant theologians, above all Clodovis and Leonardo Boff, Gustavo Gutiérrez, Juan Luis Segundo and the Protestant Rubem Alves, elaborated the controversial notion of 'liberation theology', which was doubtless exciting at the time. These largely European-educated scholars repudiated the 'Eurocentrism' of the old continent in favour of theologies that centred on the practical needs of the 'young' societies they hailed from. Doctrinal orthodoxy was downplayed in favour of what is called orthopraxy – or what Marxist–Leninists call propaganda of the deed. Their concern with practice did not preclude a naive ingestion of many of the analytical premises of Marxism, which in some cases led to the wholly ahistorical notion that Jesus Christ had been a proto-Marxist revolutionary as well as the dubious concept of 'structural sin'. Ill-digested economics and sociology flooded into the minds of theologians for whom the Gospels were not sexy enough unless flavoured with a heavy shake of Marxism. Other gestural sillinesses included trying to exclude the rich from the eucharist.

While conservatives, especially those with dismal experience of adolescent student Marxists in western Europe or real existing socialism in its eastern half, regarded accommodation with an atheist doctrine with horror, some argued that liberation theologians in reality repre-sented a highly conservative view of things despite the obeisance paid to what was modish in university sociology departments. Like any nineteenth-century ultramontane, they were opposed to the atomised circumstances produced by capitalism and individualism; they rejected 'liberal' attempts to confine religion to a private sphere; and like Pius IX they thought they would achieve the re-Christianisation of society as well as the realisation of the Kingdom of God on earth.[42] Though that argument seems sophisticated, it fails to comprehend that it is heretical to collapse eschatology into the expectation of this-worldly redemption by identifying salvation with the merely political. Although in some countries, notably Brazil, liberation theology enjoyed high-level spon-sorship, calls to democratise the Church's own hierarchical structures alienated many Latin American bishops who were already exercising a de-facto 'preference for the poor'. The election of the Polish pope John Paul II, and more particularly the ascendancy of Josef Ratzinger, who had bitter experience of the totalitarian mindset abroad among the would-be revolutionaries at Tübingen university, as prefect of the Congregation for the Faith, led to reminders that the Church itself disposed of entirely adequate teachings regarding human dignity and justice for the poor. It did not require dilettantish borrowings from secular creeds that had resulted in hecatombs of corpses and mass material and spiritual immiseration throughout eastern Europe, Russia and China, as well as the pollution of Western universities with agitprop masquerading as scholarship. There were personal animosities on all sides, some of them the result of liberation theologians who courted media celebrity more than the retiring but remorseless prefect in Rome.[43] Leonardo Boff, whose first book Ratzinger had helped publish, was even-tually silenced, while a combination of censorship and self-censorship stemmed the flow of liberation theological writings. In truth, the Church in Latin America would continue to exhibit a plurality of responses to economic inequality and military repression. In some countries, Argentina most of all, it would condone a brutal military junta under which people disappeared out of helicopters; elsewhere, in Brazil or Chile, it played a pre-eminent role in the opposition. If the presence of four priests in the Marxist Sandinista regime in Nicaragua caused consternation in the Vatican, it is also worth mentioning that archbishop Oscar Arnulfo Romero of El Salvador, who was assassinated by a death squad while celebrating mass in his own cathedral, was a highly conservative cleric with close connections to Opus Dei.[44] These darknesses were worlds away from the bright lights of the decade we began with. And then there was the grey zone of eastern Europe.

The 1960s will always be associated with its pop culture and a student elite waving banners emblazoned with Marx, Lenin and Mao. The former resonated with young people across the Iron Curtain, in societies where listening to rock music was a subversive offence. By contrast, Marx, Lenin and Mao had no purchase there, as the first two were the philosophers of the presiding Communist Establishment. That was why eastern Europe and the Soviet Union were like the elephant in the room for Western student revolutionaries. To the horror of the Western liberal-left, the Christian Churches also had tremendous authority in some Communist societies, and it is to that distinctive combination of circumstances that we will turn, after a visit to one place where an atavistic conflict may, in fact, illustrate our future.

CHAPTER 8

'The Curse of Ulster': The Northern Ireland Troubles c. 1968–2005

I AN UNSENTIMENTAL VIEW FROM ENGLAND

Readers in the US, especially the forty million people with remote Irish ancestry, albeit 56 per cent derived from the Scots-Irish, may subscribe to a republican fantasy view of Ireland, much as the English have an idea of Provence or Tuscany derived from cookbooks, novels and a villa holiday. Imaginary Ireland has led some Irish-Americans (not least those organised as the Irish Northern Aid Committee or NORAID) to pour money into Sinn Fein and the Irish Republican Army, or IRA, although the largest source of weapons in recent decades was the Libyan dictator Colonel Ghaddafi, memorably dubbed 'loony tunes' by the Irish-American president Ronald Reagan. Ghaddafi is only one of the international friends of Sinn Fein–IRA (which also include Cuba and Syria, the PLO and ETA) who are regarded with intense suspicion by those who conduct US foreign policy.

Even some American Jews are not immune to the obscure romance of the IRA. In 2005 no less a personage than Elie Wiesel, the living conscience of the Holocaust, attended a New York junket organised by Bill and Hillary Clinton on religion, conflict and reconciliation. One of Wiesel's fellow guests was Sinn Fein president, and IRA Army Council member, Gerry Adams, whose organisation is a long-standing supporter of Palestinian (and Basque) terrorists. Perhaps Wiesel was ignorant of the fact that Adams's father lit bonfires on the Black Mountain to guide Luftwaffe bombers towards Belfast, where they killed over a thousand people in a devastating series of raids that wiped out 50 per cent of the housing stock. Adams's party, Sinn Fein, also annually celebrates around a statue of Sean Russell, an IRA terrorist whose organisation declared war on the British in January 1939, putting the Nationalist community under the protection of Nazi Germany, to where he went to train as a spy.[1]

Modern England exhibits many signs of cultural decline, which amuses or saddens the quick and witty. However, the decline of English culture is at least matched by what has happened across the Irish Sea, which despite the lingering flutey-voiced sentimentality has become a vulgarised version of Essex, a rather beautiful English county unfairly traduced as the epicentre of a crass materialism symbolised by 'Essex man', whom like greed Margaret Thatcher allegedly invented.

A post-imperialist cringe means that less attention has been paid to the inhabitants of an island once associated with Joyce and Yeats, or C. S. Lewis and Louis MacNeice if one is so minded, for Ulster had a culture too. Due to its affluent American diaspora and the European Union, the Irish Republic has become much richer, while Northern Ireland is kept afloat by an inflated public sector providing outdoor relief to its middle class. Since 1997 the Irish have overtaken the British in per-capita GDP, and in a few years are projected to be richer than the citizens of affluent Luxembourg.[2] There is a price. Some of Ireland's most prominent businessmen have a, doubtless ill-deserved, reputation for ruthlessness.[3] Fans regard such figures as genially piratical; others think they are greedy and mean-spirited, a description that might also apply to large swathes of the Irish in the English building trades, although competently reliable young Poles are displacing this horde of bodgers and shysters. Bizarrely, English local government enables Irish travellers to defy – for they are not ignorant of them – planning laws that apply to everyone else living in these islands. Taste inhibits one from dwelling on the predominant ethnic identity of Catholic clergy involved in the paedophile scandals, who along with the bishops responsible for the cover-ups have disgraced and embarrassed the Church. They are not Chinese, Germans or Filipinos. Although many Irish people fought for Britain in the two world wars, one might not know it, for the number of memorials to the dead of those conflicts is dwarfed by those to the heroes of the Easter Rising and the Civil War. The last world war has other embarrassments. In addition to president de Valera's unfortunate condolences to the German people on the suicide of their Führer, there is the less-known matter of how the Catholic Church provided sanctuary for Croatian Catholic war criminals.[4]

Then there is 'the culture', which should not be confused with the occasional minor Irish poet winning the Nobel Prize for literature. Various provincial cliques and coteries, whether eccentrically Anglo-Irish, or just plain Irish, are inflated out of all proportion to their actual significance by their admiring fellows in the metropolitan British media. It is also depressing that the only celebrated visual art is the political graffiti – known as Muriels in Belfast – that adorns the ends of terraced housing. Hollywood contributes its quotient of surreal movies about nobly moody Irish terrorists allegedly facing agonising moral dilemmas, rather than the reality of intimidating drunks cutting (Republican Catholic) people's throats in Belfast bars for such grave offences as spilling their drink, a practice that assumed global notoriety after the slaying of Robert McCartney (1971–2005). It can depict Irish-American cops as crooked or psychopathic in such movies as *LA Confidential* or *Internal Affairs*, but realism departs once the movies are about the emerald isle. Drink plays a large role in what is a deeply unattractive fusion of sentimentality and violence, where people are quick to take offence as Robert McCartney and Brendan Devine discovered (senior Belfast IRA figures stabbed and battered McCartney to death in Magennis's bar after Devine had made an observation about one of the females in their company). Speaking of bars, dingy Irish theme pubs are ubiquitous in Europe, with their fake swirling Celtic tat and Guinness, and giant monitors for football and rugby, Gaelic or otherwise, which only partially drowns out the relentless, mindless gabbling known as 'craik'. Some evenings these places are given over to interminable fiddle and jiggy music, or to tear-jerking rebel songs, although a truly weird cultural format, consisting of boys and girls hopping up and down with their arms rigid at their sides, has even made it on to the West End stage in London.

England has undergone the reverse cultural colonisation of the erstwhile oppressed. As fluent talkers, the Irish have colonised entire areas of British television, with the benignly unctuous Terry Wogan succeeded by the vulgarly queer Graham Norton, whose sexually obsessive innuendo even managed to fall below the (very) low standards of British television comedy. The skill of the late Dave Allen or Dermot Morgan (whose capacity to speak home truths about dipsomaniacs and crazy clergy made them both very unpopular in Ireland) has become but a memory.

Membership of the European Economic Community in 1973

broadened the Irish people's horizons away from their perplexing love affair with Franco's Spain or their curious penchant for living in England while muttering about the 'fookin' British' in the queue for handouts in English post offices. The Irish have skilfully dipped into the bottom-less trough of EU funds, securing £14 billion (or roughly US$25 billion) between 1973 and 1991, mainly in the form of agricultural subsidies. Some border areas have never progressed from a long history of banditry and smuggling. Criminal jiggery-pokery with differential duties and Value Added Tax rates between the Irish Republic and Northern Ireland mean that smuggled diesel fuel, pigs and tobacco are a major source of income for the entrepreneurial gangsters of the Provisional IRA, who then recycle their illicit profits into Bulgarian, English and Italian property holdings as well as arms or drugs. A good movie, *The General*, about the murder of Veronica Guerin, an investigative reporter who probed that sordid milieu once too often, conveys the brute realities of the Dublin criminal–terrorist scene rather better than the Irish-American cinematic kitsch that merely flirts with the subject of terrorist violence. A significant exception to this sentimentalising trash is the amusing, and explicitly trashy, *Halloween V*, with its gleeful demolition of Irish-American Celtic fantasies, as a mad entrepreneur uses insect-riddled pumpkins to destroy America's children following a trigger signal hidden in a subliminal TV advert.

A leading role among the European Union's smaller nations has been followed by participation in other manifestations of soft power, such as international organisations and NGOs, where the Irish have found forums for their impassioned moralising self-assertion. Any cook or pop star can become a celebrity seer nowadays in a culture where other forms of authority have withered.[5] Superannuated rock musicians have boarded this bandwagon, with Bono and saint cum sir Bob Geldof in the van of vulgarly formulated attempts to strong-arm governments seeking the youth vote into giving away more money that by and large finds its way into the Swiss bank accounts of African kleptocrats. It is startling to watch British politicians lapping up abuse from this mouthy sloven, until one notes that knowledge of pop music is nowadays a crucial part of obtaining high office. Ireland's professional moralists are represented, at most disasters and 'tragedies', by Irish television news reporters, again omnipresent on British TV, with a nice line in emoting about the world's starving, a sight that makes many of a cooler dis-position long for the old days of stiff upper lips.

As such fine commentators as Kevin Myers have said for decades, wallowing in victimhood is an essential element in the Irish problem – as of so many other problems – providing as it does the emotional and moral 'justification' for bullying, intimidating and killing others, whether they belong to one's own 'tradition' (the euphemism for tribe) or that of the opposing group. The republican version of History, with its roll-call of martyrs of British (and Irish) perfidy, and its assurances that the Celtic warriors will triumph in the end, is integral to this conflict, as it is to so many other conflicts around the world. The Celtic warriors are as risible as Islamist militants who depict themselves on horseback with swirling sabres when in fact they go about in Toyota SUVs. Although it should know better, the Catholic Church, in the shape of cardinal Tomás O'Fiaich, colluded in giving such convicted terrorists as Bobby Sands, the lead IRA H-Block hunger striker in the Maze prison, a spurious christological air, although that is not the sum of its relationship with terrorists.[6]

Words, we are told by the inside group that writes about nothing else, matter in Northern Ireland. A writer can choose to call the province Ulster or the Six Counties, or use either Derry or Londonderry, to describe that grim little town. While it is well known that some Protestants call nationalists 'Croppies', 'Fenians' or 'Taigs', it may be less appreciated that some nationalists describe Protestants as 'West Britons' or 'Orange bastards'. For reasons that will be explored below, nationalists and republicans successfully erased awareness of the fact that their ranks include many Catholic bigots. Detached outsiders will also have noticed something else about the use of language, which may be less apparent to insiders for whom the jargon is second nature. Like armed robbers, who while studying law or sociology in prison adopt a professional vocabulary that sits unnaturally with their tattoos and stony faces, so many of those closer than a detonator or trigger's length to colossal violence, have become plausible (at least to themselves) in an argot that they stream forth: 'identity', 'tradition', 'the situation' being favourites among terrorist–politicians who regularly bring their little frisson of violence (and smirking evasiveness) to British television studios.[7]

Among senior journalists, other than those who have gone native, the mere mention of Northern Ireland, which many have covered for decades, produces an effect on their faces akin to watching a sunlit field darkened by a passing cloud. British statesmen have often reacted thus, Shortly after the Great War Winston Churchill observed:

> Every institution, almost, in the world was strained. Great empires have been overturned. The whole map of Europe has been changed. The position of countries has been violently altered. The modes of thought of men, the whole outlook on affairs, the grouping of parties, all have encountered violent and tremendous change in the deluge of the world. But as the deluge subsides and the waters fall short we see the dreary steeples of Fermanagh and Tyrone emerging once again. The integrity of their quarrel is one of the few institutions that have been unaltered in the cataclysm which has swept the world.[8]

Others expressed their view of Ireland more acerbically. They had heartfelt reason to, because any political involvement with the problem meant a lifetime of armoured limousines and bodyguards, not to speak of colleagues and friends murdered by terrorists, the fate of such leading Conservatives as Airey Neave and Ian Gow. The place drove many to the bottle. In 1970, Reginald Maudling, home secretary in the newly elected Edward Heath government, is said to have despaired to a stewardess on a flight back from Belfast: 'What a bloody awful country. For God's sake bring me a large Scotch.' Margaret Thatcher was eloquent on what it was like to deal with Ireland: 'In the history of Ireland – both North and South . . . reality and myth from the seventeenth century to the 1920s take on an almost Balkan immediacy. Distrust mounting to hatred and revenge is never far beneath the political surface. And those who step onto it must do so gingerly.'[9] Her highly astute chancellor of the exchequer, Nigel Lawson, wrote succinctly of 'the curse of Ulster'.[10] One did not need to be a British minister to feel bleak about Northern Ireland. A young southern Irish Catholic priest who arrived in a parish on Belfast's Lower Falls area was told by a fellow cleric: 'Look! This place is hopeless. The people are hopeless. They're as thick here as bottled pig-shit. You'll be wasting your time getting involved with them.'[11]

However much one may sympathise with these views, discussion of the last thirty years of Northern Ireland's 'Troubles' is unavoidable, although many argue that this was not an exclusively religious quarrel, but one about questions of sovereignty or economic and political power that played out in communities whose respective nationalisms were heavily tinged by their remarkably high degree of religious observance.[12] Actually, the long-term origins of the problem are obviously religious. In a fit of absent mind, the Tudors had left Ireland predominantly Catholic and ruled by chieftains they were allied with. Fearing that Ireland was England's Achilles heel, the Stuarts settled large number of Scots Calvinists in the north-east. The beleaguered intransigence of these frontiersmen was matched by a heavy Counter-Reformation Catholicism. These antipathies were etched into the physical landscape. Many villages and towns in Europe still reveal their historic topography, most obviously London where the names of some tube stations like Aldgate or Moorgate dimly recall old perimeters, or for that matter Madrid where the medieval Moorish city is still just about evident. In Northern Ireland the stones and streets are frozen reminders of ethno-religious battlelines, with the Catholics of the Bogside and Creggan below and beyond Protestant Londonderry's bastions and walls, and villages where Protestants and Catholics live at opposite ends of the main street. Even Belfast's roads and pavements testify to riots past; cobblestones were replaced with granite setts, and then by massive flagstones too cumbersome to be thrown fluently.[13]

At the beginning of the twentieth century, Ireland was part of the United Kingdom of Great Britain and Ireland, with a population consisting of three and a quarter million Catholics and a million Protestants. Although there was a distinctive Anglo-Irish civilisation in the South, which went back to the Middle Ages, the majority of Protestants had been 'planted' in the north-east in early modern times to counteract Catholic rebellions which were sometimes linked, in a fifth-column sort of way, with the imperial ambitions of France and Spain, or indeed Germany in the twentieth century. While a three or four hundred years' presence might indicate a right to be counted as indigenous, rather in the way that an illegally built shed would be considered legal after about five years, their enemies regard these Protestants as an alien infestation – or as the lickspittles of English imperialism.

In modern times, Irish nationalists sought Home Rule, that is an autonomous parliament in Dublin but under Westminster's overall aegis vis-à-vis the highest affairs of state, a goal that horrified Unionists, that is people who believed they were British and regarded themselves as a bulwark of Protestant liberty against an authoritarian 'popish' tide. Real differences between the religions should not be glossed over; the point about Northern Ireland is that these are visceral today in ways that are not true of the British mainland or anywhere in Europe beyond the Balkans.

Protestants had radically differing views about authority, ecclesiology and the transferability of spiritual merit through the mediation of an elite priesthood that were irreconcilable with the beliefs of the Catholic Church. They were intensely suspicious of what they regarded as the theocratic nature of Catholicism, and of the ways in which priestcraft seemed to control the minds and bodies of their adherents. While it would be wrong to claim that religion is the only source of conflict, it renders everything into apocalyptic and absolute terms, to the incomprehension of English people whose dominant religion is based on so many hard-fought, judicious compromises.[14] Religious endogamy also fixes people within their respective communities. Between 1943 and 1982 only 6 per cent of marriages in Northern Ireland were of mixed confession, in contrast to England and Wales where over the same four decades, 67 per cent of Roman Catholic marriages were of a bi-confessional nature. Religious polarisation was also evident in education, since even today only 2 per cent of children in the province attend mixed-confessional schools. Attempts to mix a primary school resulted in the spectacle of politically fanatical adults intimidating small children. Regardless of whether one regards religion as a cause of Northern Ireland's problems, it is certainly high among the enduring effects.

It is not necessary to retrace the problem back to the battle of the Boyne, as that is refought on an annual basis. By 1914, the Unionists had armed themselves with weapons from imperial Germany to defend their rights against what they saw as an imminent sell-out by the Liberal government at Westminster, betrayal by the perfidious English (or Welsh in this case) being an underlying pathology among these descendants of predominantly Scots settlers of Dissenting stock, who themselves had been victims of Church of England discrimination and persecution. The fact that both antagonistic communities in Northern Ireland have a very developed sense of their own victimhood partly explains why they have such difficulty in understanding the victim status claimed by their opponents, a pathology that bedevils the Israeli–Palestinian conflict. Both communities also have an extraordinary capacity to find excuses and mitigating circumstances for what to anyone else seems like psychopathic violence, whether committed drunk or sober, a mania that has a long history on the island.[15]

Although many Irish Catholics fought in the Great War, they were quickly expunged from the public memory of what would become an independent South. Five thousand Ulstermen in the 36th Ulster Infantry Division perished on the first day's action on the River Somme, a loss that Unionists have never allowed to be forgotten. Communal myth ensured that these men shouted 'No surrender' (the Protestant war cry from the 1970s) as they went over the top of their trenches.[16] Unionists have also never forgotten the abortive Easter Rising in 1916, which they regarded as an act of betrayal in wartime. The Rising had the effect of swelling support for an independent Irish Republic, which in 1921 was only partially assuaged by the creation of the Irish Free State, consisting of twenty-six southern counties, although that soon degenerated into a vicious civil war. In the new Free State, Protestants were driven out or forced into quiescence, in one of the twentieth century's stealthier instances of 'ethnic cleansing'. As the southern Presbyterian Church reported in 1921: 'In more than one congregation members have received threatening notices and have been compelled to abandon their homes . . . church property has been stolen, burned, or otherwise destroyed. A very large number of Protestants was compelled to leave the country, in some cases, nothing being left to them but their lives and the smoking ruins of their homes.' A separate British polity, called Northern Ireland, based on six counties in the north-east, was rapidly and successfully ramified, with its own devolved parliament, eventually situated at Stormont, and dual representation at Westminster. Following the 1949 Ireland Act, Northern Ireland's constitutional status could be changed only with the explicit consent of the (Unionist) majority, a democratically sensitive arrangement that endures (just about) to this day.

Unlike in the Free State, where the post-Tridentine Roman Catholic Church was grimly hegemonic, Northern Ireland Protestantism consisted of a bewildering array of denominations, as well as the Anglican Church of Ireland, which lingered on in dilapidated splendour south of the border. Irish Protestantism was as fissiparous as Protestantism elsewhere, with divisions between liberal modernists and believers in scriptural inerrancy within the Presbyterian camp, which is not to be confused with the maverick reverend Ian Paisley. Since many southern Protestants relocated to the north after being made to feel unwelcome in the Free State, they developed an even more visceral siege mentality than that community already possessed as part of its memory of being on the frontier. Betrayal from within or without was a constant possibility, and typically it was given historical expression. The Loyalist term for a traitor – a 'Lundy' – harks back to the prudent lieutenant-colonel Robert Lundy who was less than resolute during the 1688 siege of Londonderry by Catholic Jacobites. By contrast, the Apprentice Boys stood firm.[17]

To an outsider, the modest stage on which these conflicts have unfolded is as incredible as the incapacity to forget historic grievances and slights in a society when 'they' did in 1690 is parried by what 'we' did in 1691. This is hell around a parish pump. The entire population of Northern Ireland is about a seventh of that of Greater London, and roughly equal to those of Birmingham or Glasgow. Initially, Northern Ireland, or Ulster, had one and a half million people, a third of whom were Catholics. Only two counties, Antrim and Down, were over-whelmingly Protestant in composition; 320,000 Protestants in Belfast outnumbered the city's 95,000 Catholics in the west of the city. Much of that urban Protestant majority felt imaginatively closer to people living in similar circumstances and doing similar jobs in mainland Leeds or Glasgow than they did to the backward rural sea beyond. At its shortest point, the distance between Ulster and Scotland across the North Channel is a mere twelve miles. By contrast, in Armagh, fifty thousand Catholics faced sixty thousand Protestants, in a rural borderland county that Northern Ireland might have been better off without when the boundaries were established, a sentiment many English people share about the entire province, which over the last thirty years has brought them nothing but international embarrassment, financial loss and grief. Although they do not often remark on it, the Troubles have cost the English, Welsh and Scots a great deal of treasure. Even now, a huge per-centage of the province's population are employed in a grossly inflated public sector, giving every second cousin of a terrorist a job, although that could be deflated with the arrival of peace.[18]

Religiously inspired discrimination was endemic on both sides of the border, although the northern Protestant strain appears better known among those credulous enough to elide the likes of Gerry Adams or Martin 'Pacelli' McGuiness with Martin Luther King or Nelson Mandela. It was very difficult for any Protestant to get state employment (as distinct from entrance into a liberal or scientific profession) in the Free State, which also connived at the suppression of the Protestant Orange Order through tacitly supported popular violence. Compulsory Irish-language tests thinned out the ranks of Protestant civil servants and members of the professions, who invariably emigrated. The Catholic Church contributed its own form of exclusionism by insisting in the 1908 document *Ne Temere* that the children of mixed marriages be raised as Catholics, an injunction that became part of Irish law, although the Church eventually relaxed this in 1970 with *Matrimonia Mixta*. Just as the Orange Order was held to exert a malign influence north of the border, so Catholics had the Ancient Order of Hibernians – nowadays best known as the organisers of New York's St Patrick's Day parade.

In the North, Protestants enjoyed the lion's share of jobs in the engineering, linen and shipbuilding industries, which were located in 'their' territories in Belfast. It is vital to emphasise that they regarded this as their entitlement for being loyal to the British Crown. Why should potential traitors and fifth columnists prosper? When some nationalist-dominated local councils ostentatiously professed allegiance to the Irish Free State, and refused it to the government of Northern Ireland, the Unionists responded by abolishing proportional representation, gerrymandering electoral boundaries and disfranchising those who paid no rates, the British term for local property-based taxes. Protestant businesses and householders enjoyed an inbuilt advantage over poorer Catholics who lived cheek by jowl in tenements with lodgers and tenants. Local councils controlled by nationalists fell from twenty-three to eleven, notably Londonderry where 7,500 Protestants were more amply represented in the city's government than 10,000 Catholics. Such invidious arrangements, which were not universal across Northern Ireland, had knock-on effects in terms of the construction and allocation of rented public housing and employment in the local government sector, as each tribe sought to benefit its clientele in an era when there were no anti-discrimination laws and jobs were filled on the basis of family connections and word of mouth. The wrong address, identifying as it did religious confession to those alert to such things, would simply guarantee a life of low pay, marginalisation and unemployment. Again, it is important to qualify these broad assertions since there are respectable studies that indicate that discrimination, for example in housing, was less pervasive than is often suggested. Catholic entrance into higher education was also growing – the percentage of Catholic students at

Queen's university Belfast grew from 22 to 32 per cent between 1961 and 1971, but that only exacerbated matters as they did not progress into a society with equal opportunities.[19]

Access to political power was uneven. In 1943 there were no Catholics in the top fifty-five jobs in the provincial civil service and only thirty-seven represented among the six hundred in the next rung down. By contrast, seven years later, 40 per cent of council labourers were Catholics, as their restricted educational opportunities had equipped them for little else. In 1934 when the home affairs minister learned that a Catholic was working as a telephonist at Stormont, which two years earlier had become the home of the Ulster parliament, he ceased using the telephone until she was transferred. Although many Westminster politicians disapproved of these developments, which were accompanied by fierce bouts of sectarian cleansing in which people were persuaded to move elsewhere by having their homes firebombed, the ultimate effect was the creation of monocultural working-class ghettos, which were often ominously proximate to one another.

Religion played a vital part in bridging social divisions within the Protestant camp. Strident Protestant rhetoric secured working-class Protestant support for an elite class of rural landowners, many ostentatious in their wartime military ranks as 'captain' this or that, a form of cap-doffing deference that took decades to disappear, as it eventually did with the emergence of Democratic Unionism. An Orange Order civil religion consisting of celebrations of royal occasions, the 12 July and 12 August commemorations of the 1689 siege of Londonderry and the 1690 battle of the Boyne, and the more universal cult of the dead in the Great War cemented Unionist or Loyalist solidarities across divides of social class or town and countryside. The Orange Order, a Protestant quasi-masonic self-defence organisation that came into being in the late eighteenth century to combat the depredations of the Catholic 'Defenders', exercised as much influence on Unionist politicians as the trades unions used to hold over the British Labour Party. Rather than anything necessarily sinister, that may simply reflect their background in committee work and public speaking, although in Catholic eyes sinister it certainly was. Some find its rituals quaint in our otherwise homogenised and globalised metropolitan cultures – a sentiment too far that this metropolitan author does not share.[20]

Unashamedly, the Unionists monopolised the judiciary, police and civil service, in addition to what was briefly a flourishing local economy, although, again, it is worth pointing out that the province's first chief justice was a Roman Catholic. They were not shy in expressing this 'ascendancy', to give it an emotionally freighted name, as when prime minister Sir Basil Brooke told a Unionist gathering in 1933: 'There were a great number of Protestants and Orangemen who employed Roman Catholics. He felt he could speak freely on this subject as he had not a Roman Catholic about his own place . . . Roman Catholics were endeavouring to get in everywhere. He would appeal to Loyalists, therefore, wherever possible to employ good Protestant lads and lassies.'[21]

Throughout his tenure of office, Sir Basil Brooke – or viscount Brookeborough as he inevitably became under Britain's Ruritanian system of organised deference – refused to visit a single Catholic school. Although it began with a limited Catholic presence, well below the third of places reserved for them, over time the Royal Ulster Constabulary (RUC) became ever more Protestant in composition. The continued activities of the IRA ensured that fearful (and aggressively militant) Protestants flocked to the ranks of the Ulster Special Constabulary, a part-time paramilitary police force, notorious as the 'B-Specials', equipped with extensive powers and weaponry unknown in the rest of the UK, powers they used not only to harass Catholics in spiteful ways but to ensure that the Civil War anarchy in the South did not spread northwards. The fact that the IRA tended to shoot Catholics rash enough to join the police reinforced the RUC's sectarian character, the republican contribution to indirectly fostering the institutionalised discrimination Catholics complained of being a relatively under-explored aspect of this saga.

Ironically, the early 1960s seemed to presage an end to Ireland's unholy wars of religion, if that is what they were. Even in these benighted parts, where the Enlightenment never happened, the winds of change were felt, as faiths were modernised and hands outstretched. The reforms instituted by the Second Vatican Council promised a new respect on the part of the Catholic Church for those it had hitherto regarded as heretics, in that it up-graded them as 'separated brethren', although it could not quite bring itself to refer to the brethren's Church as such. The mainstream Protestant response was encouraging. The Church of Ireland, Methodists and Presbyterians issued statements welcoming the Council's spirit, and admitting that there had been 'uncharitable' discrimination against Catholics. By contrast, the Irish Roman Catholic hierarchy resembled that of Franco's Spain in dragging its feet in adopting the new guidelines from the Vatican. Predictably, the sight of Protestant clergy happily flitting back and forth to exchange platitudes with the pope, let alone such worrying signs of the time as pronouncements that 'God was dead', increased the feeling among the hotter sort of Ulster Presbyterian that the forces of Antichrist were rallying for the kill. Maybe all the 1960s did was to ensure that the terrorist gunmen tended to have long hair.[22]

In 1965, Sean Lemass, the taoiseach of what in 1949 had become the Irish Republic, paid the first official visit of a southern Irish leader to the new Northern Ireland premier Terence O'Neill. The latter was an Eton-educated army officer with an aloofly English manner that grated as much on the more fiery Protestants as his patronising remarks about slothful child-rich welfare dependency outraged Catholics. The two Irish leaders were propelled into each other's arms by chronic economic problems in both parts of Ireland, which in the North consisted of a worrying decline in its outmoded heavy industries, and the related need to attract foreign investment in more advanced sectors such as car manufacture. Artificial fibres and Third World production ravaged the linen industry, while ships became ten a penny because of Far Eastern and Polish competition. Between 1961 and 1964 the workforce at Belfast's Harland and Wolff yards fell by 40 per cent (or 11,500 men) and the Troubles themselves would ensure that it, and other jobs dependent on inward investment, would fall much further until youth unemployment alone would touch 50 per cent.[23]

Other elements of the geometry shifted as for the first time in Northern Ireland's history English politicians opened their eyes to Unionist dominance of a province that was annually costing the British Exchequer £45 million by the early 1960s. The advent of a British Labour government under Harold Wilson added urgency to O'Neill's desire for economic modernisation and political reform. Wilson represented a constituency in Liverpool with many Catholics, while some ninety Labour MPs would join the Campaign for Democracy in Ulster, a powerful pro-republican caucus within his party. An O'Neill era aptly described by the journalist T.E. Utley as 'government by gesture' ensued. His reforms were never enough for Catholics, especially university graduates whose chief career path was emigration, while reform as such terrified both rural Evangelical Protestants and those working-class Unionists who imagined that O'Neill was selling them out to pushy papists.

The Lemass visit was the final straw for the keener sort of Protestant already bothered by other O'Neill gestures. As a mark of respect, O'Neill had ordered flags on government buildings to be flown at half-mast when the internationally regarded pope John XXIII died in 1963. Worse, he had sent a telegram of condolence to archbishop William Conway, commiserating on the death of a pope who in more extreme Protestant minds was already perspiring in hell. Ian Paisley, the firebrand head of the Free Presbyterian Church, led a march on City Hall against 'the lying eulogies now being paid to the Roman antichrist by non-Romanist Church leaders in defiance of their own historic creeds'. In his world-view, one does not send commiserations to a Church that persecuted most of one's own religion's founding fathers. The past was so real that one could almost smell Latimer and Ridley still burning in the air. In 1964 Paisley provoked days of riots when his agitations forced the more pragmatically inclined police to enter an IRA-dominated building to remove the Irish tricolour flag, which it was illegal to fly in Northern Ireland. When Lemass arrived at Stormont in 1965, Paisley and his supporters demonstrated with placards reading 'NO MASS NO LEMASS'. Who was this troublesome cleric, currently the leader of the largest political party in Northern Ireland and the member of the European parliament with the entire continent's highest total of votes?

As many have remarked, Paisley was the religious equivalent of a Trotskyite, positively revelling in his capacity to divide a Church until he was in charge of a purer version of his own (the Free Presbyterian Church), although it has affiliates nowadays in many parts of the world. Up in the chill solitudes, the pure air is said to go to one's head. An acute sense of political theatre, and a vivid and sometimes humorous turn of phrase, ensured Paisley's instant notoriety, as did his willingness to court fines and time in jail for his beliefs. On his earliest political outings, he seemed like a deranged US Evangelical preacher as he threw his imposing black-garbed bulk – sometimes rounded off with a black fedora or a jaunty Russian fur – into unseemly mêlées. His moving mass was accompanied by baritone shouts of what in local argot sounds like 'Tach yoor hunds off that mon' or the more comprehensible 'No surrender', always barked out with a snarl. In person, Paisley is said to be congenial, in a monomaniacal sort of way. As in the case of republicans, the monocultural background is not conducive to flexibility of mind. That he comes from a highly religious family and has founded a political dynasty, as well as a successful Church, contributed to his singularity of purpose, since there was no nagging domestic voice to damp things down, either from Mrs Paisley or from their many children, some of whom (like Ian Paisley Junior) have become politicians in their own right.

It is important to note that the emergence of a more implacable strain in Unionism antedated the emergence of the 'civil rights' movement, and that its chief cause included the perceived betrayal by the Unionist political caste, whose reforms signified that religion could be confined to a separate compartment, a view that was anathema in these circles where creed is all. Paisley's supporters regarded O'Neill as a patrician, English-sounding traitor. Although dissident Unionist leaders could mobilise people through such quasi-Fascistic phenomena as William Craig's Vanguard Movement, by the late 1960s 'Official' Unionist candidates for Stormont and Westminster were being challenged by what were initially called 'Protestant' Unionists, the forerunner of today's Democratic Unionist Party, founded in 1971, the word 'democratic' signifying an end to the politics of deference towards what the populist Paisley memorably dismissed as 'the fur-coat brigade'. In other words there was a crisis within Unionism as well as within nationalism. Meanwhile, the contemporary phase of terrorist violence gained momentum and would result in over three thousand people being killed over the following thirty years. The point of no return was approaching.[24]

In 1966, republican celebrations of the fiftieth anniversary of the Easter Rising witnessed the emergence of an Ulster Volunteer Force, or UVF, which in a brief drunken spree managed to kill three people, including an elderly Protestant widow, a Catholic man who shouted 'Up the Republic' within their earshot, and a young Catholic barman who strayed into the wrong pub in the ultra-Loyalist Shankill Road area.

While these killers were quickly apprehended, the increasingly well-educated Catholic middle class discovered the US civil rights style of politics, which focused on the simple slogan of 'one man – one vote' and sectarian discrimination in employment and housing. A Northern Ireland Civil Rights Association was founded in 1967, an umbrella organisation that included the student-based People's Democracy, whose Trotskyite leaders – above all Bernadette Devlin – would promote a campaign of civil disobedience against Unionist 'supremacy'. That term, and others like it, often strays into writing about Northern Ireland. It reflects an agenda. White Irish Catholics managed the public relations trick of appearing to be African-American or South African blacks, with the Orange Order standing in for the Broederbond or Ku Klux Klan, and the RUC for Sheriff 'Bull' Connor or the Afrikaner police in the era of apartheid. Ironically, of course, a hundred years before, the Catholic Whiteboys had been the ones gallivanting around on dark nights in white sheets. It is also worth remarking that while there was discrimination in Northern Ireland, this did not inhibit inter-confessional dating, which was certainly so with inter-racial dating in either South Africa or the southern USA.

This was an important sleight of hand, as all shades of US liberal opinion bought it, extending far beyond the narrow band of diehard Irish-American republicans with their mischievous folkloric view of the English, who as we all know invariably play the baddies in Hollywood movies. A broad coalition emerged, which together with neighbour-hood defence groups, militant students and civil rights enthusiasts included elements of the IRA, who regarded the civil rights movement as a useful front for provoking street confrontations with the authorities in order to subvert the Unionist government at Stormont. Their Wolfe Tone Societies had actually envisaged this movement as the latest form of struggle. As the future Sinn Fein president Gerry Adams would assert, albeit with an element of self-aggrandisement, the civil rights movement was 'the creation of the republican leadership', although Adams was not a leader at the time.[25]

While Presbyterians and Roman Catholic clergy made cautious declarations of irenic intent, on the streets the politics of rage took over and the province's government lost control of events. What were clearly examples of systematic discrimination were instrumentalised in ways which – rather than provoking a cross-confessional 'class war' – resulted in sectarian violence as the two northern tribes slid into outright war. The O'Neill government's enthusiasm for dirigiste planning prompted complaints that the sprawling new housing estates like those at Lurgan or Portadown, or for that matter a new university provocatively named Craigavon, were being deliberately sited within Protestant territory. Corruption in the allocation of a public council house to the secretary of a councillor's lawyer, who was also a Unionist candidate, drove a Nationalist MP to squat symbolically in the house until he was ejected by policemen, who included the official tenant's brother.[26] This incident led to demonstrations by the civil rights movement, which in turn attracted counter demonstrations by loyalists. The civil rights movement organised a march from Belfast to Londonderry, as if these were Montgomery or Selma, which some participants hoped would lead to confrontation with the authorities. In Northern Ireland, marching is not a neutral stroll down the street, but a way of claiming territory by walking on what Unionists archaically call the 'Queen's Highway'. That is why so many of the marches look like military processions with each tribe sporting its uniform – bowlers and sashes here, balaclavas and parka jackets at republican funerals there, and guns everywhere. In January 1969 at Burntollet Bridge Paisleyites ambushed the marchers, who had already been set upon by the RUC and the B-Specials, amid scenes of primal violence.

These incidents appalled international opinion, while forcing the Westminster government to lean hard on the Unionist authorities in ways to which they were not accustomed in their role as Britain's shock absorber. O'Neill was called to London to be dressed down by Wilson and home secretary James Callaghan, who threatened to cut the flow of subsidy from English taxpayers to Northern Ireland. Regional indignation over the fact that the Northern Irish, Welsh and Scots pay taxes too does not wash, since the British revenue system is wholly posited on transferring money from the rich south-east (and in particular London, where a quarter of the country's money is made) to the less affluent regions. O'Neill's desire for reforms was counterbalanced by those Unionists, like Brian Faulkner, who regarded the civil rights movement as a front for the IRA. His home affairs minister William Craig resigned over O'Neill's weakness and the bullying tone of a British government the Unionists were professing aggressive loyalty to. Mysterious bombings of electricity sub-stations were attributed to the IRA but in fact carried out by the UVF so as to destabilise the O'Neill government. They were successful since O'Neill was forced to resign, an event which many loyalists marked with bonfires.

The new premier was a relative of O'Neill's, James Chichester-Clarke, an Etonian former Irish Guards officer turned gentleman farmer. Under him Ulster went to hell in a handcart. Permission to allow a march by the Protestant Apprentice Boys of Londonderry, who included the usual quotient of hooligans, sparked an uprising among the rival hooligans in the Catholic 'Bogside', the Catholic settlement beyond the historic walls. When the RUC smashed down barricades, Protestants stormed into the area like a marauding army. Violence spread to the working-class districts of Belfast, and evolved from stone throwing to a shooting war in streets as lamps were shot out and the only light came from raging fires and the arcing spin of Molotov cocktails. Eight people were killed and

750 injured, including 150 with gunshot wounds. Families living on the wrong side of the sectarian divide quickly packed their possessions on to carts and trucks and moved elsewhere. The Irish Republic briefly considered military intervention, rapidly realising that this was inadvisable, but settling instead for field hospitals situated near the border to treat the injured, and, in the case of some members of the government, covert shipment of weapons to IRA terrorists whom they had hitherto suppressed in the Republic. Although the Republic under de Valera and Lemass had itself interned Official IRA men between 1957 and 1962 for waging war north of the border – thereby contesting the fundamental right of the Dublin government to declare war or peace – their successors seemed to have struck a tacit deal with the IRA, whereby the latter would confine their activities to the North, refraining from attacks in the Republic, a dirty deal reminiscent of how some French governments deflected French-based Basques back into Spain.[27]

Because the three-thousand-strong RUC had been stretched to exhaustion, Chichester-Clark asked London to despatch regular army soldiers, who began arriving in Northern Ireland from August 1969. There was a price for this assistance. Westminster insisted on the disarming of the controversial B-Specials as a prelude to their disbandment (they were replaced by the Ulster Defence Regiment) and reform of the RUC under an English chief constable. Had the British army not intervened, Londonderry's Catholics would have been massacred. There was no one else to defend them, given that – as the graffiti in Catholic areas had it – 'IRA' meant 'I ran away'.[28]

This demoralising experience accelerated a split within the IRA, between its southern Marxist leadership under Cathal Goulding, who clung to the illusion of a unified Catholic and Protestant working-class struggle, and the newly founded Provisional IRA and Provisional Sinn Fein, which saw their mission as one of community defence, sectarian retaliation and the total rejection of parliamentary institutions, whether north or south of the border, which they regarded as the illegitimate offspring of Partition. They were often pre-political and bigoted fundamentalists. One of their leaders, the half-English John Stephenson – his Irish half winning out with Sean MacStiofain – had views on Communism that would have warmed the heart of Pius XII.[29] The split occurred in 1969 and signalled a general downgrading of the political struggle for a united Ireland in favour of terrorist violence. By 1972 the Official IRA had formally renounced violence – on the ground that this increased sectarian mayhem – although they did not relinquish their weapons. Thenceforth most IRA violence was perpetrated by the Provisionals, or 'Provos'. A further breakaway faction from the Officials, called the Irish National Liberation Army, INLA, managed to combine Marxism with psychopathic violence, which was erratically directed at the two larger factions of the IRA as well as at the security services and loyalists.

Although many of its leaders and activists were drawn from the South, the Provisional IRA tilted towards tough northern Catholics, from both city and countryside, animated by a desire to avenge themselves on Protestants and clan mythologies in which many of their relatives had lengthy involvements with the IRA. The former barman Gerry Adams, who despite never having fired a shot rose to the summit of the IRA, came from two families in which his great-grandfather, grandfather, father, and mother had histories of involvement with republican organisations. His comrade Martin McGuinness had once worked in a butcher's shop before devoting himself to armed struggle. Mighty matriarchs, some members of the IRA women's organisation Cumann na mBann, kept the home fires of sectarian hatred burning, while younger women helped move weapons or lure victims to their deaths through sexual entrapment. Children were recruited to the Fianna, the IRA's youth wing.

Eventually, the Northern Command would effectively take over the organisation, putting the older southern godfathers out to grass on their farms in Kerry, one of the South's hotbeds of republican extremism. Training camps situated in remote areas of the South taught northern city boys how to use rifles and to handle explosives, most of which were mixed on southern farms. IRA members had their own argot and culture, which included 'nutting' or 'stiffing' people – that is, shooting them – balaclava hoods, combat jackets, high-velocity rifles and US-manufactured machine guns. As the memoirs of convicted terrorists and informers amply illustrate, there was a hierarchical command structure, in which there were many chiefs and few Indians, fancy military titles, and much admired specialists such as bomb-makers, snipers, interrogators and torturers, roles that brought added kudos to those involved. Soubriquets like 'Dr Death', 'Geronimo', 'Hack Saw', 'Slab' and 'Tonto' were used not solely to discriminate among too many people called Murphy, but to convey a specific air of menace in the way that Americans will be familiar with gentlemen called 'Fats' or 'Fingers'. The parallels can be developed further. Two of the Belfast IRA figures suspected of killing Robert McCartney are regarded locally as 'made men', who in the run-up to the 1994 ceasefire had murdered various Protestant paramilitaries.

Speaking of gangsters, among Boston's criminal fraternity, shipping weapons to the IRA seems to have been a way of enhancing the local status of gangs like the Murrays, who were so untrustworthy that the IRA kept two of them hostage to ensure that an arms shipment was completed.[30] The IRA's constant search for revenue ineluctably meant armed robbery, the supply of rigged slot machines, forgery and money-laundering, drug trafficking, motor-insurance fraud, and various scams connected to the construction industry in England, which are indistinguishable from the methods used by the mafia. It was and is a criminal organisation, whatever its political rhetoric, the only ascertainable difference being that the profits go to the organisation rather than to individual gang members, most of whom live modestly – often on British welfare – and appear to pay no taxes.

One fact about these people needs to be emphasised. Violence was glamorous in inner-city working-class areas and small rural towns that were largely deprived of it; every hick or urchin could play the role of 'romantic' rebel. Some, like Gerry Adams, whose abilities had got him into a grammar school run by the Christian Brothers, effectively wrecked their own education and career prospects when the IRA alternative path to the top beckoned. Adams and many others made up for this during spells in prison, which acted as universities for republicans and loyalists alike, although men like Adams prided themselves on their autodidactic achievements, to distinguish them from those of the educated Catholic middle class, like John Hume, a French teacher who rose to lead the moderate nationalist SDLP. Others, like 'Slab' Murphy, a Gaelic-football-loving bachelor farmer, could play the local Mr Big, building a fearsome crime empire under the guise of humble pig farmer.[31] Below that august level are the usual quotient of loquacious dullards or stony-eyed psychopaths whose reputation is dependent on their skill in the kill.

The demi-educated leadership talked a good class struggle, but visceral sectarian hatreds were involved that are invariably presented in a one-sided fashion. A Belfast Provisional observed as he surveyed the Protestant areas of the city: 'that's my dream for Ireland. I would like to see those Orange [Protestant] bastards just wiped out.'[32] A spiral of violence ensued, in which the militarisation of searches and arrests by the security services led to incidents of heavy-handedness by soldiers who discovered this was not Wiltshire with more rainfall, while IRA shooting of British soldiers – some from cities, such as Glasgow or Liverpool, with their own sectarian history – resulted in the latter's inclination to deploy their considerable firepower regardless of any rules of engagement. The introduction of internment without trial in August 1971 ratcheted up the tension – without effectively rounding up terrorists – many of whom sneaked across the border of the complaisant Republic, where the mythology of 'rebels' had some sway. The impression of British injustice was compounded with the introduction of non-jury Diplock courts in 1972 – an inevitable solution to the fact that potential jurors were too terrified to sit on cases involving terrorists since they had to make their way home at night.

On 30 January 1972 a civil rights march in Londonderry against the recently introduced policy of internment became a major tragedy. Soldiers of the Parachute Regiment, whom other regiments of the British army regarded as 'thugs in uniform', were despatched into the Catholic Bogside to contain the disturbances and *arrest* rioters. For this task they were dependent on biased and inaccurate intelligence from the RUC. Arresting people was not a task paratroopers were suited to, so arguably those in London responsible for their deployment were at fault. The republicans were not blameless. The local Provo leader in waiting, allegedly Martin McGuinness, flitted and skulked about in the shadows, with guns and pipe bombs.[33] Since the Parachute Regiment soldiers had (or claim to have) been shot at, they opened fire on the crowd, shooting dead thirteen unarmed people, in scenes that became a propaganda coup for the republican movement. 'Bloody Sunday' (although the IRA were responsible for bloodshed on every day of the week) attracted so many potential recruits to the IRA that the organisation was incapable of absorbing them. Republican tempers were further provoked when in April 1972 lord chief justice Widgery, who was predisposed to the forces of law and order, exonerated the actions of the Parachute Regiment. A re-run of the Widgery investigation (the Saville Inquiry) is still ongoing, which for the lawyers involved has turned into the most lucrative case in British legal history, with their fees totalling £85 million out of net costs of £163 million. This does not seem to embarrass the lawyers, but to many people it is a disgrace, especially because the peace process enabled McGuinness to use his own appearance/non-appearance as theatre.[34]

It is also the most egregious instance of how British soldiers, who have been responsible for 8.2 per cent of all deaths in Northern Ireland, have been constantly subject to politicised inquiries, while republican terrorists, responsible for 58.3 per cent of fatalities, evidently do not excite the imaginations of lawyers or the human rights industry.[35] It is worth noting, as a sort of glaring parenthesis, the callous treatment of relatives of people murdered by the IRA when they sought explanations from its senior figures. In 1991 a dissident republican, Eoin Ta'Morley, was shot twice in the back with a rifle when he defected from the IRA to the INLA. His father, the former head of republican prisoners in the Maze, asked Martin McGuinness, a logical port of call in such situations, to investigate whether inter alia his son's murderers had been drunk. The 'investigation' took place in a bungalow with the murderers present. 'Were youse drinking?' asked McGuinness, who presumably got his legal training while packing bacon in James Doherty's butchers shop. 'No, we don't drink,' replied the murderers, one of whom made to leave. 'Sit down, Patrick, I am finished, I'm quite satisfied,' said the scrupulous investigator. He reported to the parents of the dead man that this (ten-minute) 'court of inquiry' had found no wrongdoing. At least such investigations are cheap and don't involve lawyers.[36]

In the wake of ever more killings and following the failure of the internment policy advocated by prime minister Brian Faulkner, the Heath government prorogued Stormont and opted for direct rule by the secretary of state (William Whitelaw being the first to venture into the political graveyard) with a Northern Ireland Office as the local administrative apparatus. This imposition of political tutelage without any regard to the wishes of the majority outraged Protestant opinion. Craig and Paisley organised a two-day strike that paralysed the province, while a hundred thousand Protestants marched on Stormont. One significant effect of the suspension of Stormont was a wholesale exodus of aristocratic and middle-class Protestants from Unionist politics, which left the field wide open for lower-middle-class demagogues and sectarian toughs from the Protestant ghettos. The number of British troops stationed in Northern Ireland climbed from seventeen thousand to nearly thirty thousand that year.[37]

Although there were covert discussions between Whitelaw and the IRA in London, these brought temporary ceasefires rather than a cessation of IRA violence. Both sides were clearly also testing the wills of their interlocutors for the serious military conflict that was not long in coming. Twenty-one car bombs that detonated simultaneously in Belfast on 'Bloody Friday', 21 July 1972, killed nine people and injured dozens more. An eyewitness described the scene at the Oxford Street bus depot where four bus drivers were slain:

> You could hear people screaming and crying and moaning. The first thing that caught my eye was a torso of a human being lying in the middle of the street. It was recognisable as a torso because the clothes had been blown off and you could actually see parts of the human anatomy. One victim had his arms and legs blown off and some of his body had been blown through the railings. One of the most horrendous memories for me was seeing a head stuck to a wall. A couple of days later we found vertebrae and a ribcage on the roof of a nearby building. The reason we found it was because the seagulls were diving on to it. I've tried to put it at the back of my mind for 25 years.[38]

The violence the IRA meted out to anyone failing to conform to their way of thinking within what they regarded as 'their' own violently 'greened' communities was terrifying. Jean McConville was a Belfast Protestant who converted to Catholicism when she married a Catholic builder, who died of cancer a year before his wife's disappearance. Menaced out of her home in a Protestant area, she and her family, which included ten children, moved to Catholic West Belfast. In 1972 the widowed McConville rashly tried to comfort a British soldier who had been shot virtually on her doorstep. In December, four republican women burst into her house, dragged Jean McConville from her bath, and abducted her in front of her brood of children. She was never seen alive again, although the IRA did eventually admit that it had killed her as a suspected informer. Her remains were discovered on a beach in 2003; she had been shot in the head. Eight of her children were put into care after her murder, as each of these killings has ramifications for many more than the victims.[39]

As the RUC retreated from what became Catholic ghettos, the IRA assumed the role of surrogate police force, delivering rough justice to delinquents, who, if they refused to emigrate to England, were treated with baseball bats, concrete blocks, electric drills, all applied to their arms, knees or ankle joints, as well as the ultimate sanction of death by shooting. Such vigilantism caught up with twenty-eight-year-old Hugh O'Halloran in West Belfast on 10 September 1979. A Catholic with five children, O'Halloran had allegedly knocked a girl over in his car. A group of men connected to the IRA beat him to death with hurley sticks and a pickaxe handle as he returned home late at night. The attackers were all drunk.[40]

Endemic violence also brought massive job losses. According to one of the most realistic Labour Northern Ireland ministers – the former coalminer Roy Mason, who occupied that office in 1976–9 – the number of jobs created by inward investment fell from three thousand to three hundred per annum during his term in office. Seventy-two government-sponsored firms folded and sixteen factories were destroyed. Business costs were inflated by political extortion. One in five homes in Belfast were rendered uninhabitable.[41]

IRA violence was also directed against England, which had become dulled by the chronicle of death across the water. Scotland and Wales were exempted, less because of pan-Celtic sentiment than because the ferry routes from Northern Ireland to Scotland were used as what the terrorists dubbed their 'Ho Chi Minh trail'. In 1972–3 the IRA bombed London's criminal court, the Old Bailey, and the Protestant UVF killed thirty-three people with bomb attacks in Dublin and Monaghan in the Irish Republic. In incidents that are etched into the mind of most English people of my generation, a small IRA cell conducted devastating attacks on pubs in Guildford, Woolwich and Birmingham. These places were targeted on the notional grounds that off-duty soldiers frequented these establishments, but the wider expectation was that killing English civilians would attract enhanced news coverage, undermining English support for British government policy in Northern Ireland. So did the campaigns to free the Guildford Four and Birmingham Six, that is those Irishmen who were convicted of two of these attacks, campaigns that in the left imagination eclipsed memory of the carnage the IRA had been responsible for.

In rural South Armagh, violence was savagely sectarian. Gunmen from nationalist and loyalist terror organisations simply burst into bars and the like to spray the patrons with bullets. Only murders in double figures attracted the big publicity. In one of the foulest incidents, twelve masked IRA men flagged down a red minibus containing a party of workmen on a lonely road at Kingsmills. The men had been chatting about English football. The IRA separated the sole Catholic from the eleven Protestants, whom they lined up and murdered in a hail of automatic gunfire – although one victim would survive despite being hit eighteen times. When the emergency services arrived on the scene, it was awash with blood, as well as littered with boxes of sodden sandwiches. This was a blatant sectarian killing, as it was certainly not part of the class struggle. Armagh, with its IRA roadsigns warning of 'Snipers at Work', became so dangerous that it soon bristled with military watchtowers while troops moved around by helicopter.

In the mid-1970s the British government adopted a twofold strategy of deploying Special Air Service troops to apprehend or kill IRA men as they perpetrated acts of murder, and 'Ulsterising' the public face of security through the RUC and part-time Ulster Defence Regiment. While this meant that part-time policemen (and prison officers) bore the brunt of IRA attacks, whether shootings or bombs wired into their cars, it did not immunise the British army. In 1979 two trucks filled with Parachute Regiment troops were blown up by an 800lb bomb as they passed a hay trailer. The survivors, and soldiers who had come to their rescue, were decimated by a second 800lb bomb concealed in milk churns, which had been deliberately positioned in anticipation of their probable defensive position. Eighteen young soldiers died at Warren-point that day; a surviving soldier was killed by the IRA a year later. The day also saw the murder of the seventy-nine-year-old earl Mountbatten, prince Philip's uncle, his grandson and a teenage helper, when a 50lb bomb exploded in the *Shadow V* as they pulled up lobster pots. The earl's daughter Lady Brabourne died of her injuries the next day. To the IRA's warped mindset, Mountbatten was nothing more than a symbol of the British Establishment. As the *Republican News* explained: 'We will tear out their sentimental heart. The execution was a way of bringing home to the English ruling class and its working-class slaves that their government's war on us is going to cost them as well.'

These crimes had a complex impact on the major Churches. Violence between republicans, loyalists and the British armed forces and their local auxiliaries polarised communities, which in turn expected their respective clergy to clarify their own stance. The leaders of the flock were often the led. Internment was supported by many Protestant clerics, whose instinct was to support the forces of law and order, even as the Catholic hierarchy vociferously opposed it. While the priest Edward Daly was caught on film desperately waving a blood-soaked handkerchief after tending a victim of 'Bloody Sunday', both the Church of Ireland and the Presbyterians regarded the rioters as the cat's paw of attempts to coerce Protestants into an all-Ireland republic. The Roman Catholic position was complicated by centuries of anti-Catholicism on the part of a Britain for whom militant Protestantism is a residual part of its identity, albeit a sentiment tempered by increasing tolerance of the Catholic minority in England, a minority that includes a huge Irish diaspora. Sections of that clergy had also imbibed the usual Gaelic cum Celtic mythology, and the republican ideology of the martyred rebel.

Most Catholic clergy north and south of the border felt strongly about the need for social justice for the northern Catholic community, as did the majority of citizens of the Republic, so long as they did not have to take on the fiscal burden of the North's far more extensive welfare arrangements. They condemned harsh army search tactics and the use of mental or physical coercion in police interrogation centres, although the Republic's Gardai were not known for their gentle approach to offenders. Like any normal person, they took grave exception to such instances as an ambulance being deliberately kept waiting at an army roadblock so that the wounded terrorist inside could bleed to death. 'That's the point, mate,' an English soldier explained when a priest objected. Emotionally clergy supported the goal of a united Ireland, and sympathised with the anti-British outlook of their parishioners. There was another reason for the clergy to become involved in the civil rights movement, namely the concern of their bishops that it might otherwise be dominated by secular Marxists in the IRA.[42] Some priests went further in more or less overtly supporting the so-called 'armed struggle', by hiding weapons or ferrying terrorists about and taking them to safe houses that they themselves provided. Only one priest, father James Chesney, seems to have been actively involved in terrorism – the 1972 bombing of the village of Claudy, which killed nine people, including nine-year-old Rose McLaughlin. Although it is difficult to get at the truth of the matter, memoirs of former terrorists frequently allude to the bigotry of Catholic clergy and their uncritical espousal of an un-reconstructed republicanism. While making his getaway after murdering Peter Flanagan, a Catholic RUC Special Branch officer, in a pub, IRA operative Sean O'Callaghan – the future head of its southern command – stayed in a priests' house. The prospect of an over-cooked fried chop was enlivened by the TV and radio:

> The IRA had regularly used this house for meetings, for the induction of recruits and as a general safe house and base in the area. The priest was an active IRA sympathiser with influence at

the highest levels of the republican movement. He was as good as regarded as a senior IRA activist.

There was another priest in the house. Home on extended leave from a stint in the foreign missions, he was nowhere near as shrewd as the first priest and was regarded locally as a friendly, irresponsible simpleton. He was a 'groupie' who liked to spend time in local republican safe houses where he would try to get people like me to talk abut operations. 'That was a good job,' he would say with a sly, cunning look on his round, simple face. Once he was there he was happy. I never regarded him as more than an idiot, useful at times but mostly tiresome and irritating . . . Over dinner and more holy water, having listened to more radio and TV reports of the day's events, the senior priest said to me, 'Flanagan was an abominable man who sold his soul to the devil.'[43]

It is important to recall, however, that some priests had unique insights into the evils of republican violence and were clear-minded about this. On some occasions, the IRA allowed those it had tortured as alleged informers the consolation of a last confession, an act which sheds light on the presence of committed Catholics within the movement. After the arrival of armed and hooded figures at the parish house, priests were taken to secret locations where they were confronted with men who had been drowned in baths or burned with cigarettes to extract information. A priest recalled:

I froze when the bathroom door closed. I was suddenly dealing with evil and not just talking about it. The man in the chair was one of my parishioners. I remember looking at the bath filled with water wondering what they had done to him. He was stripped to a pair of wet underpants. His hair and body were wet, so they'd obviously been holding him under the water. Looking back I observed so many things in a matter of seconds or perhaps because I now just imagine that was so . . . He was badly bruised and his eyes were so swollen that he could hardly see me. My first thought was whether I could get him out of there when the bathroom door opened. 'Remember, Father,' one of the gunmen told me, 'any funny business and you're both for it. Anyway, there's somebody out the back even if you could get him out the window.'

Unable to rise, the victim murmured 'Please help me, Father.' The man made his confession. 'I'm going to die. Isn't that right, Father?' The priest put his arm around him. On the way out, the priest remonstrated for the man's life: 'This is against the law of God.' The stony response was: 'You look after the law of God, and we'll look after our business.'[44]

That last remark had echoes of the term 'cosa nostra' – or 'our thing'. Some criticise the Catholic clergy for not refusing IRA murderers absolution or for failing to excommunicate them. In reality, although the subject is necessarily opaque, few IRA gunmen were likely to confide their crises de conscience to a mere cleric, while excommunication would have had as many pitfalls as internment. If practised in Ireland, it would have had to be universalised, including those cases of extreme repression where political violence might have been morally justified, as it probably is in hellholes like Guatemala or El Salvador. More importantly, the cultural Catholics and secular leftists in the IRA would have simply ignored it, as would Catholic Provos who regard murder as a venial sin, and the clergy would have forfeited all influence in such circles. Discreet influence may not be as glamorous as impassioned statements, but if it saved even one life, it was probably justified.

The Catholic hierarchy were never prepared to sanction indis-criminate terrorist violence. As the bishops said in September 1971:

In Northern Ireland at present there is a small group of people who are trying to secure a united Ireland by the use of force. One has only to state this fact in all its stark simplicity to see the absurdity of the idea. Who in his sane senses wants to bomb a million Protestants into a united Ireland? At times, the people behind this campaign will talk of defence. But . . . their bombs have killed innocent people, including women and girls. Their campaign is bringing shame and disgrace on noble and just causes . . . This is the way to postpone a really united Ireland until long after all Irish men and women living are dead.

That does not quite exculpate such senior clergy as cardinal Tomás O'Fiaich, who was appointed archbishop of Armagh in 1977 by an Italian papal nuncio whom both Fine Gael and the Irish Labour Party had wanted recalled because of his republican sympathies. A fanatical Irish folklorist and supporter of Gaelic football, O'Fiaich made a grotesque comparison between people living in sewer pipes in Calcutta and convicted Irish terrorists who chose to cover themselves and their prison cells in their own shit. The Calcutta poor do not choose to live in sewers. He was heavily criticised by such Irish politicians as Jack Lynch and Garret FitzGerald for his republican enthusiasms, which became evident in his responses to the Maze hunger strikes. Roy Mason tersely remarked that O'Fiaich's 'words could have been written by any propagandist from Sinn Fein'.[45] One area where priests have played a vital political role has been in brokering contacts between elements of the republican leadership and other sections of the broader nationalist movement north and south of the border, an indispensable contribution to persuading the former back on to the tracks of constitutional politics. The Belfast Redemptorist priest Alec Reid was highly active in arranging such contacts for the likes of Adams, his vision being of a pan-nationalist front.

Under prime minister Margaret Thatcher, who instinctually sympathised with the Unionists until they managed to alienate her, IRA terrorists rediscovered the virtues of 'martyrdom' or what psychiatrists call passive aggression. In the spring of 1981, convicted IRA terrorists, including Bobby Sands, who was serving fourteen years for possession of a gun while on an IRA mission, went on a hunger strike within the Maze prison. 'Geronimo' Sands, to give him his sinister IRA soubriquet, was the officer commanding IRA prisoners, a powerful role belied by his long-haired bearded image, which among the credulous suggested an innocent drummer in a rock band.

On the most parochial level, these men were engaged in a familiar struggle with the prison authorities regarding whether they or the prisoners were running the prison. They were also determined to avenge an earlier hunger strike in December 1980 which had collapsed after one of the strikers went blind and the event was called off. Since this was Northern Ireland, they were also participants in a war of nerves with the government of Margaret Thatcher, over what the convicted IRA terrorists saw as attempts to 'criminalise' them through the wearing of prison-issue rather than personal clothing, a struggle that had already resulted in them covering themselves and their cells with their own excrement in what was called the 'dirty protest'. Bearded naked troglodytes flitted about in cells smeared with primitive brown markings. The hunger strike was the next stage of the struggle. Although Margaret Thatcher undoubtedly won, unfairly contributing to her image as the 'Iron Lady',

Britain's image gained little from the interventions of the Red Cross and the Vatican, from the pictures of starving men beamed across the world, including Bobby Sands who had got himself elected as an MP at a Westminster parliament that the republican movement has never acknowledged. Sands died after sixty-six days.

Those engaged in the second strike, ten of whom died of their own volition, knew that their emaciated images would be mentally blended with that of the crucified Saviour, and that their funerals and wakes could be turned into IRA recruiting demonstrations. The clerical response to acts of suicide is crucial. Prison chaplains, notably Denis Faul, were in the unenviable position of being confronted with men fully prepared to die for their beliefs, an act they had learned about at several removes in the tales of missionaries they were taught to admire at theological seminaries. As Faul recalled: 'Here were these men doing for a temporal cause, a doubtful, disputable temporal cause in many ways . . . they were making the very sacrifices that Jesus had done, and that Catholic priests and Catholic people were called upon to do. They were doing it . . . and there was a religious motif to it . . . they were doing it for the people.' The Catholic hierarchy, with the conspicuous exception of Tomás O'Fiaich, condemned the hunger strikes on the ground that suicide was sinful, a line endorsed by England's cardinal Basil Hume and the papal pro-nuncio to London Bruno Heim.[46] Father Faul, who got to know the dying very well, was sceptical of their motives, seeing that the men were bent on death (and conspicuous funerals) as a political statement, acts of self-immolation with a long history within the republican movement. The strikers assumed the exterior mantle of Christian martyrdom without much sense of its spirit. The first four to die had achieved concessions on the subject of clothing, so it seemed unjustifiable to expect others to starve in order to achieve the rectification of further grievances. Towards the end, Faul succeeded in getting the remaining men's relatives to take the opportunity of the strikers lapsing into comas to insist on their being fed intravenously, although other republican prisoners tried to combat this inexplicable 'weakness' on the part of relatives by producing fabricated letters of support from other members of their families.

Both the international sympathy that the hunger strikers generated and the mass grief manifested at their funerals suggested to the more sinuous leaders of Sinn Fein–IRA that the ballot box had as much potential for achieving power as the bomb and bullet, especially since the British army (and covert police units) had inflicted serious damage on republican ranks. Much of this was due to improved intelligence, with virtually total surveillance possible in such a small society, not to speak of informers and supergrasses who, if nothing else, sent the paranoia of terrorist bosses into overdrive. Cases may have collapsed or convictions been overturned, but much of the energy of the IRA was turned upon itself. A reassessment of the political track was the key lesson of Sands himself, who was elected to Westminster by an impressive margin. Between 1982 and 1985 Sinn Fein contested four elections, averaging 12 per cent of the vote, but 40 per cent among nationalist supporters. They threatened to eclipse the moderate SDLP in the foreseeable future, while demographic trends promised a longer-term Catholic victory. In 1983 Adams was elected to Westminster. While he refused to take up his seat (without relinquishing the considerable parliamentary expenses to bolster his British benefit payments), which would have involved swearing the oath of allegiance, he nonetheless used his visits to London to establish amicable relations with such figures as the Greater London Council leader Ken Livingstone, an ultra-leftist who did not even have the usual pro-republican excuse of Irish ancestry which seems to have conditioned the sympathies of such Labour figures as Clare Short and Kevin McNamara. Sinn Fein–IRA joined a diffuse range of 'causes' which the 'loony leftist' Livingstone, a caricature radical, vicariously dabbled in before Mrs Thatcher sent the GLC packing. The prospect of Sinn Fein holding the finely balanced politics of the electoral ring in the Irish Republic helped concentrate moderate opinion north and south of the border. Both Mrs Thatcher, who in October 1984 narrowly escaped an IRA assassination bid in Brighton, which paralysed the wife of a close political ally Norman Tebbit, and the taoiseach Garret FitzGerald realised the urgent necessity of bolstering constitutional (that is unarmed) nationalism to stymie the rise of Sinn Fein in both parts of Ireland. British and Irish civil servants held a productive series of meetings out of which came the November

1985 Anglo-Irish Agreement. Maybe their experience of the European Union made it easier to contemplate the notion of pooled sovereignty in the case of Northern Ireland.

The Agreement solemnly repeated the view that no change in Northern Ireland was possible without the consent of the majority of its people. It institutionalised inter-governmental discussions and 'structures' that gave the Republic (and constitutional nationalism) a say in the affairs of the province. Unionists were horrified by an apparent British (and Irish) shift in stance to a benign neutrality towards this most dysfunctional of places that was costing both governments prodigious sums of money. The Protestant response to this agreement was predictable. In addition to incendiary speeches by the likes of Paisley, they organised massive demonstrations, while turning their own paramilitary forces against the police. The homes of five hundred RUC men were firebombed, and 150 of them were forced to move house. The ferocity of the Unionist response shocked British politicians, who from then onwards – in the eyes of nationalists – failed to follow through with the reforms that the Agreement seemed to herald. The republican side also consistently demonstrated bad faith. Even as it appeared to pursue electoral politics, the IRA availed itself of the generous arms shipments from Libya that we began with, to wreak havoc both in Northern Ireland and on the mainland. The security forces proved vulnerable, even within fortified police stations and watchtowers, as the IRA proved itself skilled in the improvisation of mortars. Nothing was sacred to them either. In November 1987 they exploded a bomb that killed eleven people, injuring a further sixty, at a Remembrance Day ceremony in Inniskillen. Violence spiralled out of control in a sequence of bizarrely interlinked events. In 1988, SAS soldiers shot dead three IRA terrorists, including a female, on the island of Gibraltar, as the latter reconnoitred the route of a British army band which they planned to blow up. Apologists for the IRA claim that the three terrorists were unarmed and that the SAS used excessive force, but the soldiers insisted that the three had made suspicious movements. Although these shootings were very popular with the man and woman in the English street, among nationalists and their various fellow travellers they were regarded as acts of murder, blithely ignoring what would have been the fate of the army bandsmen. Worse followed.

The funerals of the dead terrorists were attended by thousands of nationalists who flocked to west Belfast's Milltown cemetery. A lone loyalist gunman, Michael Stone, ran amok in the crowd, firing indiscriminately with a handgun and throwing grenades, until he was cornered and almost beaten to death. One of Stone's victims was an IRA man. During his funeral at Milltown, two British army undercover operatives took a wrong turn, and inadvertently drove into the cortège. Surrounded by hostile mourners, who mistook them for further loyalist terrorists, corporals David Howes and Derek Wood produced weapons and fired warning shots. Armed with such things as wheel braces, the crowd of republican sympathisers, dragged the two corporals from their car and beat them semi-conscious. They were then hauled, in their socks and underpants, into a taxi which took them to a wasteground, where IRA gunmen shot them in scenes that were recorded by army surveillance helicopters. The reason for their murder was an identity card with the word 'Hereford' on it – location of the SAS headquarters. In fact it said 'Herford', a small town in Germany, where one of the men had been based.

The IRA developed a variety of new tactics, including kidnapping people who were used as involuntary human bombs by being chained into trucks that were blown up after being despatched towards army bases. In February 1991 I obliged a visiting German professor, on his first trip to Britain, who wanted to see London from a taxi on what was a snowy afternoon. We left the LSE where the students were throwing snowballs, and, after heading along the Strand and around Trafalgar Square, turned into Whitehall, where pandemonium broke out. The IRA had fired several mortars through the roof of a van parked behind the Ministry of Defence which landed in the garden of the Downing Street complex where prime minister John Major was chairing a meeting on the Gulf War. In early 1992 the IRA struck at the financial heart of the British economy (that is, the part that produces 25 per cent of its GDP) when they detonated two gigantic fertiliser-based bombs at the Baltic Exchange in the City of London, causing £700 million of damage. A 'ring of steel' consisting of police boxes and CCTV cameras appeared around the entrances to the City, where the extreme proximity of towering modern buildings along quaintly named medieval lanes and alleys was almost ideal for maximising material damage. The physical shabbiness of many British cities was not unconnected to the removal of all wastebins lest the IRA put bombs in them. The effects of these bombs had no appreciable impact on the morale of the British people, nor did the British confuse Irish immigrants with those allegedly bombing on their behalf.

The 1990s saw the beginnings of what has become known as the 'peace process'. Two affable conservative Northern Ireland secretaries, Brooke and Mayhew, under prime minister John Major, signalled that Britain had no 'imperialist' agenda in the province and that they would not rule out talks with anyone. Paradoxically, relations with the lower-middle-class Unionist political class were strained by Mayhew's patrician manner and Anglo-Norman-Irish ancestry, stretching back to the thirteenth century, an improbable source of tension that would not count anywhere else on the planet.[47] From 1990 they authorised MI5, Britain's domestic intelligence agency, to reopen contacts with militant republicans. Although Major's slender parliamentary majority might have increased his dependence upon the Ulster Unionists, in fact he was genuinely committed to resolving the Northern Ireland problem, and lucky in the warm relations he enjoyed with the Republic's Albert Reynolds. In a further departure from recent tradition, in 1988 the constitutional nationalist John Hume, who was enormously influential in Washington, held secret talks with the Sinn Fein–IRA leader Gerry Adams. These talks were arranged by father Alec Reid, the Redemptorist priest, who also forged contacts between Hume and the Unionists. These were so sensitive that they were held in Germany.[48] Hume's outlook was remarkable, even if he had something of the droning pedagogue about him. He regarded the IRA as a species of Fascism, observing that if he were to re-establish the civil rights movement in the 1990s the IRA would be the principle object of criticism, since they had murdered and tortured thousands; dehoused people; killed people on campuses, in schools and hospitals; wrecked the economy and transport infrastructure; and caused massive unemployment, with robberies of post offices depriving the unemployed of state benefits. He memorably said: 'They [the IRA] are more Irish than the rest of us, they believe. They are the pure master race of Ireland. They are the keepers of the holy grail of the nation. That deep-seated attitude, married to their method, has all the hallmarks of undiluted Fascism.'[49] Yet Hume was also willing to talk to anyone in the cause of peace, regardless of whatever criticism, or worse, this brought upon him, for as in the case of all politicians in Northern Ireland violence is never far away.

Paradoxically, a sudden surge in the incidence of violence added fresh impetus to the quest for normality. The IRA struck at a fish shop in the Shankill Road, managing to kill nine ordinary people rather than the Ulster Defence Association leadership. The IRA commented laconically:

'There is a thin line between disaster and success in any military operation.' Retaliation came very fast from terrorists on the other side. At Halloween, loyalist paramilitaries – one of whom shouted 'Trick or treat' – burst into a village bar at Greysteel where two hundred people were listening to country and western music. Eight Catholics were shot dead, with a further nineteen injured. It may be that such killings were so viciously senseless that the perpetrators forfeited any residual legitimacy even among their own supporters. A palpable atmosphere of fear spread over the province, with taxi drivers only visiting areas of the same religion as their own, and people scurrying home as quickly as possible after going to church lest a crowd be visited by IRA or UDA–UFF gunmen.

In December 1993 the British and Irish governments issued the Downing Street Declaration, which seemed to reconcile the conflicting agendas of consent and self-determination, while the British ventured further down the path of neutrality between the warring parties. The US played an ever larger role, especially when a president was elected who was prepared to spend up to 40 per cent of his time on the tiny troublesome province. The leaders of both tribes began to log up the airmiles. David Trimble had the intelligence to see that the Unionists, after several decades of being outmanoeuvred by the rich and influential Irish-American republican lobby, needed to remind many Americans of their 'Scots-Irish' ancestry so as to counteract republican propaganda. Much effort was put into elaborating a distinctive Protestant Ulster identity. The fact that Major had allegedly been partial to George H. Bush in the presidential election campaign, by raking through Clinton's harmless Oxford past, probably inclined Trimble not to pal up with US conservative opponents of the victor, although the Ulsterman's proverbial refusal to charm probably played a part in his thinking. Clinton helped the peace process along by controversially rescinding the prohibition on granting a visa to Adams, who was soon duly lionised by the US liberal media and the Irish-American Catholic Establishment. This reversal of policy was deemed a form of payback for Major's earlier partiality for the Republican Party. The US ambassador to London, Raymond Seitz – one of America's most highly regarded envoys – was outflanked by the US ambassador to Dublin, who thenceforth was known to Unionists as Nancy 'Sodabread' rather than Soderberg. Adams had to tread carefully. While he relished his newfound international celebrity, by carrying the coffin of Thomas 'Bootsie' Begley, the Shankill Road IRA bomber, who had managed to blow himself up, the Sinn Fein–IRA leader counteracted resentments among his IRA comrades that he was becoming over-fond of fancy limousines, fine dining and expensive suits, for in Ireland resentments tend to be lethal.

On 31 August 1993 the IRA declared a 'ceasefire'. Two months later, the leading godfather of loyalist paramilitary violence, Augustus 'Gusty' Spence, also declared a ceasefire, one of the most significant developments of the preceding years being the emergence of a more politically astute leadership within the ranks of imprisoned Protestant paramilitaries. Any hopes that these ceasefires would result in political talks were frustrated by the IRA's stubborn insistence that a ceasefire did not include their day-to-day criminal activities – least of all their idea of rough communal justice – and by their refusal to surrender their massive arsenal of weapons. Clinton appointed the Lebanese-Irish senator George Mitchell as head of an external team charged with assessing the size of the IRA's arsenal and working out how to get rid of it, a task Mitchell performed with aplomb. The annual cycle of Protestant street marches brought further tensions, notably regarding the right of twelve hundred Orangemen to march to the church at Drumcree and back via the Garvaghy Road along which many Catholics live, which became a trial of strength between the loyalest of the loyal and Sinn Fein–IRA 'community' activists. The march, which had already been re-routed in the 1980s out of deference to Catholic sensitivities, was to be led by the Royal British Legion lodge. The Portadown Orange Lodge was the oldest in the province, Portadown being known as the 'Citadel' or 'Vatican' of Orangeism. They had marched out for a service on the Sunday before the commemoration of the battle of the Boyne since the early nineteenth century. One of the Catholic Garvaghy Road residents' group members had convictions relating to blowing up of the Portadown Legion Hall by the IRA in 1981. Nothing straightforward here then, for once again Sinn Fein–IRA were using their passive-aggressive tactic, and the Unionists duly obliged them. The world focused on men in bowler hats and sashes wishing to bang big lambeg drums on a few flyblown streets in a British province, or rather on the bizarre cat-and-mouse game that the would-be marchers played with the RUC and British army, who improvised defensive wire to frustrate them. The UVF murderer in chief, Billy 'King Rat' Wright – a celebrity terrorist responsible for killing a dozen people – turned up with his aura of shaven-headed belligerency in the midst of the trouble with a view to using a mechanical digger as a primitive tank.[50]

In February 1996 the IRA detonated a powerful bomb near the Canary Wharf complex on London's Isle of Dogs, killing two newspaper sellers, one a twenty-nine-year-old Muslim, and causing millions of pounds' worth of damage to this prestige project. In June another massive blast ripped up the commercial heart of Manchester. The following summer brought renewed confrontations focused on a churchyard in Drumcree so embittered that the province teetered on the brink of a sectarian meltdown. In October 1996 the IRA blew up the headquarters of the British army at Lisburn, killing one soldier and effectively ruling themselves out of any further talks with the British government of John Major, who was personally affronted by the obvious discrepancies between the rhetoric and reality of Gerry Adams. Major's mounting difficulties with his scandal-ridden party diminished the likelihood of a Northern Ireland settlement while he was in office, although that does not detract from the contribution he and the Conservatives made to one.

The landslide election victory of New Labour's Tony Blair in May 1997 brought greater authority to the British government position vis-à-vis the Unionists, for Labour's huge majority required no alliances of convenience, and a leader who was prepared to take the bold step of negotiating with Sinn Fein–IRA without insisting on prior decommissioning of IRA weapons. An Ulster Protestant mother and a Liverpool Roman Catholic wife seemed to leave no trace upon how Blair approached this problem, which was with his characteristic pragmatic steeliness. His relative youth and habit of knowing where the train of History (as well as Clio's hand) was headed lent new momentum to the peace process. Obvious republican sympathisers within Labour ranks were sidelined, it being helpful to Blair that they belonged to a left-wing of his party that he regarded as akin to a lingering odour. Although the faux uncouthness of the new Northern Ireland secretary, a former academic called Marjorie 'Mo' Mowlam, managed to offend the more old-fashioned manners of the Unionists, the prime minister injected authority and realism into talks, just managing to keep Adams, McGuinness and the rising Unionist star David Trimble in the same building with one another. Communal meals were designed to engender glimmers of humanity amid the unpolitical talk about fly-fishing. On one occasion, a dish of porcini mushrooms prompted the thought that the diners had been slipped 'magic' mushrooms, so improbable did it seem that these men would be eating with one another.[51]

These talks produced the April 1998 Belfast or Good Friday

Agreement, for even its name is contentious. The Republic formally renounced its constitutional claims on the North in return for a continued 'all Ireland' framework, together with a new 108-seat assembly, and the devolution of decision-making in such fields as agriculture and education to a local administration based on the strength of the respective parties. The Agreement was subjected to referenda in both parts of Ireland, although the Unionists were much less enthusiastic in their support. After much agonising, David Trimble became 'first minister' of the new devolved Northern Ireland executive. A bomb attack by the so-called Real IRA at Omagh in August 1998, which killed twenty-nine people, increased Unionist dismay that Trimble was willing to preside over an executive that included former republican terrorists. Many found it hard to stomach Martin McGuinness, whose hands they thought were steeped in blood, as minister of education with influence over the lives of their children. Another republican headed up the health service, so that republicans dominated the two areas with the lion's share of the budget. While the British government tinkered with the Royal Ulster Constabulary – which had born the brunt of terrorist violence for three decades and many members of which were plagued by post-traumatic stress disorders – the IRA was allowed to drag its feet on the matter of surrendering its arsenal.

The asymmetrical nature of the peace process rightly outraged a large number of conservative British journalists, who, since Unionist politicians did not translate well to the mainland media, became the most articulate spokesmen of a cause that fashionable opinion deemed antediluvian or atavistic. That anyone holds the view that both sides are as bad as each other is something of a public relations achievement for the Unionists. Because of the IRA's failure to disarm, which Unionists rightly insisted upon, the British government decided to suspend the Northern Ireland executive, returning power to Westminster after what had been a mere seventy-two days of limited devolved government. Elections in June 2001 indicated that only the extremes grew stronger, as Sinn Fein–IRA crept up on Hume's SDLP, and the Democratic Unionists began to eclipse Trimble's Unionists. Frustrated by the ongoing jiggery-pokery of the IRA, Trimble resigned as first minister. Protestant terrorists decided that, as IRA–Sinn Fein violence had won them rather a lot, they would adopt the same tactic.

The events of 9/11 initially confirmed the US Republican administration of George W. Bush in its implacable hostility to all forms of terrorism. An Arctic wind blew towards Adams, McGuinness and the rest from the new Republican White House, especially since three Irish republicans, including Jim 'Mortar' Monaghan, the IRA's head of 'engineering', had been detained in Colombia a month earlier on an alleged training mission with FARC narco-terrorists, their defence being one of 'bird-watching' rather than franchising murder. While the Bush administration's hard line on terrorism has weakened the position of Sinn Fein–IRA (and an Irish Republic neutral about Iraq) vis-à-vis the US's most loyal ally, the loyal ally has paradoxically insisted that the IRA should not be conflated with Al Qaeda, which presumably explains such things as amnesties for convicted terrorists and such dubious innovations as 'community restorative justice', the first step to a dual or federal legal system and an ominous precedent. Incredibly the Blair government is now proposing to rake through every instance of killings by the British security forces in Northern Ireland.

An ambiguous peace, rather than goodwill, has come to Northern Ireland, although it is anyone's guess whether this will hold. For the time being, the most recent burst of creative energy among those who deal with the province has been exhausted, especially since Blair will soon leave office, while the ball has passed into Adams's and Paisley's court on the assumption that they can 'de-fang' the men of violence. This is the modern analogue of handing considerable local power to tribal chieftains for the sake of a quiet life in the imperial metropolis, a deeply worrying development in Europe's response to aggrieved minorities, where governments surrender power to leaders of so-called communities on the presumption that these figures are 'moderate' and that they control the 'communities' they claim to speak for. In this manner, entire cities or parts of them are being subtracted from the purview of the democratically elected government to create what amount to 'no-go' areas.

It seems unlikely that the presence of some thousands of amnestied terrorists will readily allow the province to slip into the regional decline that would otherwise be its fate were anyone to reduce the lavish monies that the Troubles attracted towards it. The killing of Robert McCartney suggests the high price being paid for the 'peace process'. Women supporters of the IRA expertly cleaned the murder scene, Magennis's bar, with bleach, while CCTV film disappeared. Many of the seventy-two eyewitnesses claimed to have been otherwise engaged in two tiny lavatories, cynically known nowadays as the 'Tardis' after the police telephone kiosk – with a capacious time machine within – used by the television time lord Dr Who. Once out of the spotlight of the world's media, the five McCartney sisters have been forced out of their homes in the Short Strand Catholic enclave, where their families had lived for five generations. The IRA offer on 8 March 2005 to shoot the unknown perpetrators of the murder was a dismal insight into their conception of justice, while Martin McGuinness's warning that they should not allow themselves to be politically manipulated was sinister in the extreme.

The manner and rhetoric of adult terrorists has seeped downwards into the minds of every hooligan and petty criminal, many of whom in Northern Ireland are viciously violent. Mainland Britain has plenty of juvenile delinquents who always 'know their rights'. In Northern Ireland they have the paramilitary-dominated ghettos to flee to – where the newly minted Police Service of Northern Ireland is reluctant to enter, lest gunfire accompany the bottles and bricks, the fate of officers trying to investigate the McCartney murder. Even teenage suspects seem to have memorised the Provo handbook's sections on counter-interrogation. Respect for lawful authority has virtually disappeared, as it has in much of the mainland. That tendency may become more widespread, in Britain and elsewhere, as police forces already fearful of charges of 'Islamophobia' or 'racism' surrender local power to com-munal vigilantes and strongmen, in a manner vaguely reminiscent of the late Romans watching as power leached away to the barbarians.

No one can foresee the future of a precarious peace, which involves turning a blind eye to extraordinary explosions of communal violence and to the mafia grip of paramilitary armies on entire communities. Other countries pay for the place, while no one really wants it. Not the thriving Irish Republic, because the amount of British government subsidy to the province – which has British levels of health and welfare – equals the Republic's entire revenue from taxation. Why would it wish to assume responsibility for a population twisted by decades of war? Not the British, who either wish to be rid of the place or hope it will sink into provincial quiescence like any other disadvantaged region that the EU may eventually raise from the dead on a raft of taxpayers' money. The 'stakeholders', to use the meaningless jargon of New Labour Britain, have renounced 'ownership'. We are horribly wrong in imagining that Northern Ireland is some atavistic throwback to the religious wars of the sixteenth or seventeenth centuries. Its model of the state surrendering 'communities' to the tender mercies of their so-called leaders may presage the future, except it will involve minorities who worship another God. The gloomy spires of Fermanagh and Tyrone will continue to haunt us, despite such epochal events as the collapse of Communism in eastern Europe and the Soviet Union, but they may well be outnumbered by the gleaming domes of Europe's proliferating mosques, in areas from which the state has quietly retreated.[52]

CHAPTER 9

'We Want God, We Want God':
The Churches and the Collapse of
European Marxist-Leninism 1970–1990

I SLUM CLEARANCE

A persuasive way of understanding the collapse of Communism in Europe and the Soviet Union is to think of nineteenth- or twentieth-century slum clearance. For in many respects the Soviet Empire was a slum of continental proportions. Beyond the grotesque architectural assertions of an alien ideology, public housing – almost all housing – consisted of anomic and primitive concrete barracks where the smells of cabbage, damp and low-grade tobacco combined. Rivers and lakes were polluted by chemicals, with the Pleisse river in East Germany alternately turning first red then yellow. Other waters mysteriously dried up because of dams and developments elsewhere. The air reeked of sickly lignite fuel, which in Leipzig was strip-mined on the edge of the city. Local wits argued that in Leipzig one could see what one was breathing. At Bitterfeld the groundwater had a chemical reading mid-way between vinegar and a car battery. In Cracow, the sun disappeared on hot afternoons behind a veil of industrial fumes.

Shortages of basic foodstuffs, as well as consumer goods, meant the exhaustion and ill-temper of interminable queues. People seemed grey and shabbily dressed, especially whenever their garb echoed some long-forgotten Western fashion. Pervasive alcoholism was reflected not in hooligans having a carnival, for that would have been illegal, but in rheumy-eyed figures morosely clutching a drink in grim station bars. What British cultural critic Jonathan Meades called the 'pissocracy' was not confined to the drunks in the Kremlin, but reached via the workplace to park benches. Food poisoning was routine in canteens and restaurants. Uniquely in the industrialised world, the average age of mortality *decreased*, not just because people were prematurely worn out, but because of dangerous and dirty working conditions and substandard health services. All of this may suggest nothing strange to a housing estate in Cardiff or life in the decaying suburbs of Paris. But far more than indifferent living conditions was at stake.

The countries of the Communist bloc were ruled by unelected gerontocracies, and their younger clients, who lived in hermetically sealed government quarters, like Wandlitz in East Berlin, venturing out in motorcades with curtains obscuring the passengers from curious eyes. Although their living conditions did not aspire to the opulent vulgarity of Western nouveaux riches, they had private hunting parks and access to shops filled with Western luxuries. But, again, many politicians in Western democracies treat high office as pigs regard their troughs. The most striking evils of the Communist regimes were hidden away in jails, camps, asylums and orphanages, while the police state had listening devices and shadowy watchers to remind people of their existence, whether through blackmail and intimidation or through the ubiquitous men with cameras. Notoriously, East Germany had to build an immense wall to prevent its own citizens fleeing the worker–peasant paradise. It also arbitrarily expatriated people, or sold them to West Germany for large sums of money in what amounted to a form of human trafficking.

Historically, slum clearance was never solely an exercise in replacing insalubrious dwellings with improvements, but also involved reform of the moral and social evils that slums engendered. That is where the analogy takes off. East European dissidents reversed this process by deciding to eradicate moral disorders before watching as the vast slum created by Marxist–Leninism crashed down as a result of factors intrinsic as well as extrinsic to the system. That approach involved standing Marxist materialism, as well as other progressive delusions, on their head in favour of such intangibles as mind, values and spirit.

Naturally there were major external actors who contributed to the success of these popular revolutions, but this should not detract from the courage of less well-known figures within the societies concerned. The history of the revolutions in 1989–90 is also that of dissidents, many of whom were from the working class, it being academic whether they or intellectuals played the more important part. Some of the workers were highly intelligent, if that means they thought deeply about things, rather than possessing one of the regimes' qualifications for mindless conformism. In what meaningful sense was the wily dissident electrician Lech Wałęsa less 'intelligent' than some conforming dullard with a history or philosophy PhD written according to the spurious 'laws' of Marxist–Leninism? In some countries, 'intelligence' almost correlated with failure to oppose the regime, it being remarkable how few East German students, for example, participated in the popular demonstrations that brought the regime down.

Sometimes major events have very small beginnings that at the time few notice. Paradoxically, just as the post-Stalinist Soviet leadership thought it had secured long-term legitimacy for its outer Empire, it conceded what it cynically regarded as a small ancillary cost that could be subsequently ignored with impunity.

The background to this development lay in the heyday of détente in the 1970s, when Western leaders lined up to find permanency and virtue in Marxist tyrannies. The European Conference on Security and Co-operation's Final Act, signed in Helsinki in August 1975, turned out to be a pyrrhic victory for Soviet leader Leonid Brezhnev, who imagined it had consolidated all that Stalin had gained at Teheran and Yalta by persuading the West to renounce military intervention in the affairs of signatory states, even though it was the Soviets themselves who had violently intervened in East Germany (1953), Hungary (1956) and Czechoslovakia (1968).

What were called 'Basket Three' of these deliberations contained a number of provisions regarding human rights, together with monitoring mechanisms to police them. Principle VII committed signatories 'to respect human rights and fundamental freedoms, including the freedom of thought, conscience, religion or belief, for all without distinction as to race, sex, language or religion'.

This provided a cloak of legitimacy to a number of human rights groups, which in the teeth of Communist repression could claim that the regimes themselves had twice signed up to these values, not only at Helsinki, but in their own constitutions. Since the constitutions notionally guaranteed various rights, why not insist that these regimes observe their own laws? That was one of the chief considerations for the Czech signatories of Charter 77, named after the 'year of the political prisoner' in 1977. Its three spokesmen included the playwright Václav Havel and the philosopher Jan Patočka, who would die after an eleven-hour police interrogation worsened an already bad case of flu.[1]

Realists on either side of the Iron Curtain may have preferred to regard the Cold War as a game of chess, played out by experts versed in such arcana as arms controls or nostalgic for the era of the post-Napoleonic Congress System when ordinary mortals did not count. But the Helsinki Accords ensured that questions of freedom and morality would continue to matter. In the long term, they were the only product of the era of détente to yield results, because through such provocations as the invasion of Afghanistan and the stationing of intermediate SS20 missiles in the Ukraine the Soviet Union certainly failed to observe its spirit.

At the time, détente had such widespread purchase that even the Holy See was not immune, a worrying example of the Churches' general permeability to evanescent secular ideologies, as represented by clergy joining in hysterical clamour against nuclear weapons that had ensured that neither superpower risked direct confrontation.

Breaking with the implacable and principled anti-Communism of Pius XII, Paul VI encouraged dialogue with the Communist regimes, granting many of their leaders private audiences and acting as if the arbitrary rigidities of Yalta and beyond were past repair or recovery. There was even talk of guiding Marxism back to its 'Christian' roots. While such uncompromising figures as Mindszenty were replaced with younger moderates, the maverick Yugoslav dictator Tito was received by the pope in 1971, the first Communist leader to be accorded this honour, followed in the next four years by Nicolae Ceausescu, Todor Zhivkov and György Lázár, encounters that would have had Pius XII whirring in his grave. In the eyes of Vatican diplomats, the need to avert thermonuclear war was paramount, as was a naive belief in the gradual convergence of the two antagonistic political systems, something they had picked up from the wisdom of social 'science'.

II SPIRITUAL VOICE OF THE WESTERN WORLD

It has become fashionable to deprecate the role of ethics, religion and people power in the anti-Communist European revolutions.[2] Actually, the development and diffusion of a highly subtle way of thinking about, and living within, totalitarian regimes was at the heart of things, and could not have been otherwise, once Karol Wojtyła, the cardinal archbishop of Cracow, was elected the first Polish pope in October 1978. Some people like to downplay his contribution to Communism's collapse. The KGB and Bulgarian secret service did not agree since in 1981 they recruited a fanatic Turk to kill him.

Wojtyła brought his nation's sophisticated Catholic traditions of moral philosophy, as well as an absolute abhorrence of Communism, to the Vatican, together with what proved to be a highly useful aptitude for showmanship, as he was an accomplished actor. Wojtyła's words and writings resonated in a society where everyone faced explicit moral dilemmas every day. As the future Solidarity leader Lech Wałęsa would comment: 'The invocation of a moral order was the most revolutionary response that could be made to the increasingly dogmatic socialism practised in Poland, and people were caught up in this wave of moral reawakening – each expressing it in his or her own way, at work or in the home, in professional and in personal relationships.'[3]

A record in cultural activism is one important clue to the subsequent effectiveness of the pope in dealing with Communism. His early adulthood was spent under Nazi occupation, where Wojtyła was part of the non-violent Christian Resistance that tried to sustain an independent Polish culture that the Nazis had sought to eradicate by reducing the Poles to illiterate helots. Having suffered so much death, the surviving Polish Catholic clergy emerged with enormous popular credibility, in a country that was 96 per cent Roman Catholic as a result of the war-time (and immediate post-war) loss of Germans, Jews and Ukrainians. Catholic ranks also extended well into the Communist Party, which, however appallingly it acted, was never entirely hardened to appeals of conscience.[4]

Wojtyła was a charismatic, practical man, who spent the war working in a limestone quarry and a chemical plant, and a gifted scholar with deep reserves of spirituality. His doctorate was on philosophical aspects of moral choice, the very area that would be so crucial to later opponents of the Communist totalitarianism that succeeded Hitler.[5] As archbishop of Cracow from 1964 onwards, cardinal Wojtyła, as he became three years later, intensified contacts with the intellectual milieu he came from, including representatives of the secular non-Communist left, but also with the industrial workers of the new suburb of Nowa Huta around the Lenin Steelworks. This concrete monstrosity was a deliberate act of social engineering designed to swamp the old city's conservative Catholics with the 'new' socialist man.[6]

Unfortunately for the Marxist regime, the workers were as devoutly Catholic as the peasants whom industrialisation and modernisation were intended to render obsolete. Their migration from the countryside had been too precipitous for them to be effectively reconstructed as socialist 'new man' overnight. The workers' desire to erect a church in the midst of this Marxist–Leninist concrete paradise became a bone of contention between Church and Party for nearly twenty years. Wojtyła defiantly conducted open-air masses until what was called the Ark Church was dedicated in 1977.

His election to the papacy in October 1978, following the unconscionably brief pontificate of Albino Luciani, or John Paul I, culminated in a four-hour installation mass, deliberately drawn out to stop the Polish Communist Party's media arm from giving it their own negative gloss. John Paul II's final words were 'Be not afraid,' one key to understanding the impact of his papacy upon those who fought for liberty under totalitarianism. Another was his constant insistence that it was not enough to be against Communism; one had to think in terms of the moral renewal that would accompany this. Parallel criticism of Western materialism, and espousal of the dignity and rights of labour, made him difficult to position in conventional political terms.

The diplomat Paul VI's pursuit of what in German is called 'Ostpolitik' was quietly discarded. The difference John Paul II's election made in eastern Europe is rather tellingly illustrated by the evolution of cardinal František Tomášek of Czechoslovakia, who had formally denounced Charter 77; by 1984 the same figure blessed its spokesmen.[7] John Paul II spoke in terms that resonated with many dissidents, regardless of their ethnic, political or religious background, for as the former archbishop of where Auschwitz is situated he was acutely conscious of the need for repair in the Church's relations with the Jews. They were ready for the message in the sense that someone like Adam Michnik had transcended the visceral anti-Catholicism of many Jews and among the secular left intelligentsia. In that respect, Michnik's book, *The Church, the Left and Dialogue*, published in France in 1977, represented a landmark, for forces that the regime had managed to keep inimical coalesced in an alarming fashion.[8]

John Paul II constantly reiterated the importance of human rights, pressing governments to enforce the Helsinki Accords. Coming from a city with a Christian history of nine centuries, he emphasised both Europe's common Christian culture – using the metaphor of two lungs without which East and West could not breathe – without neglecting the national distinctiveness with which that manifested itself. As this highly cultured man remarked, Shakespeare was both essentially English and profoundly universal, something he knew from his days as a keen amateur actor. This implied that Marxist–Leninism was an alien and evanescent doctrine dealing in vapid universalising generalities that bore no resemblance to its grim reality. The pope knew that to challenge it one had to stress what was more rooted and satisfying.

Wojtyła had noted the effectiveness of this approach between 1957 and 1966, when cardinal Wyszyński's Great Novena, in the long run-up to celebration of a millennium of Polish Catholicism in 1966, effectively denied the Communists' version of historic time, while focusing minds on a rival range of visual symbols like the Black Madonna of Czestochowa and the Church's own calendar, feasts and processions. As an actor, Wojtyła had a keen feeling for the theatrical.

The Polish Catholic Church, and not the Polish People's Party, was the popularly acknowledged guardian and repository of the nation's identity, just as the Church had been during the era when Poland had no statehood between 1794 and 1918. By the 1970s, this defiance had become so worrying to the Party's cultural functionaries that they deliberately manufactured secular ceremonies that parodied the much more popular Christian exemplars. State officials were paid monthly bonuses to drum up takers for 'name-giving' (baptismal), 'honorary guardianship' (godparents) and 'personal identity-card awarding' (confirmation) ceremonies to augment compulsory civil marriages.[9] With mounting desperation, Edward Gierek's government essayed a thirtieth-anniversary celebration of the Communist regime (in 1974) highlighting the guest of honour Leonid Brezhnev, and then a thirty-fifth-anniversary celebration (in 1979) which omitted the international (big) brotherly element in favour of the Communists' version of patriotism.

By the late 1970s there were three further actors on the international scene. After decades of centrist drift, the conservative right had come to power in the US and Great Britain, under Ronald Reagan and Margaret Thatcher. These were highly imaginative thinkers, whose outlook had been respectively shaped by domestic experience of equivocation and soul-searching in the US of the Vietnam era and Britain's decades of managed decline. They took on a whole corpus of 'progressive' assumptions and shibboleths in both domestic and foreign policy terms, with Thatcher earning the undying hatred of Britain's left-wing Establishment in the universities and the BBC. Rejecting much of what passed for academic as well as political wisdom, Reagan wittily remarked that détente was 'what farmers have with turkeys before Thanksgiving Day'. He totally rejected the inevitability and permanence of Communism. In a major speech at Notre Dame in 1981, he said: 'The West won't contain Communism, it will transcend it. It won't bother . . . to denounce it, it will dismiss it as some bizarre chapter in human history whose last pages are even now being written.' That proved prescient.

Both leaders were highly informed about eastern Europe and the Soviet Union, relying on the knowledge of Robert Conquest and Richard Pipes, rather than the dazzling political insights of Noam Chomsky, Eric Hobsbawm, Harold Pinter and the entire field of academic international relations. Western sophisticates, including Helmut Schmidt and Valéry Giscard d'Estaing, were snobbish about the gun-slinging 'cowboy actor' Reagan and the handbag-bearing 'housewife' Thatcher, thereby underrating not just their intelligence and single-mindedness, but the significance of their long record of political activism – in Reagan's case as a labour leader and industrial motivational speaker. Neither leader was conventionally religious, but both had a Churchillian sense of right and wrong, and when they spoke of 'freedom' they meant it, even if that moral clarity was not always evident in Reagan's dealings with Central America and Iran. Both signalled a readiness to use military force, whether by bombing the absurd Colonel Ghaddafi or sending a battle fleet thousands of miles to defend a few miserable South Atlantic islands.

They also proved sympathetic to the networks in the West which ensured that the little flame of freedom was never entirely extinguished in the Soviet Empire. Magazines such as *Encounter* and *Index on Censorship* made it their business to follow events in the Communist world. Individual writers of great stature ensured that there was no excuse not to know, from Victor Kravchenko's *I Chose Freedom*, via Arthur Koestler, to Robert Conquest's *The Great Terror*, and above all Alexander Solzhenitsyn's novels and his factual *Gulag Archipelago* with its unforgettable opening about hungry gulag *zeks* horrifying Soviet archaeologists as they fried up fossilised fish. Leszek Kołakowski, the exiled former professor of Marxism at Warsaw university, learnedly demolished the high texts of the dogma in his path-breaking *Main*

Currents of Marxism. Why, he asked, bother with a substitute religion when Christianity provided a real one?

In addition to talking frankly about freedom, Reagan restored a moral tone to international affairs, most memorably when in March 1983 he referred to the USSR as the 'evil empire' – against the advice, as it happens, of Robert Conquest. While that led the Soviets to imagine that they were dealing with a US president crazed enough to launch the bomb, paradoxically Reagan had a horror of nuclear weapons, and consistently urged on the Soviets the need to eradicate them through effective anti-ballistic missile defences. That idiosyncratic offer in the form of the Strategic Defence Initiative – for deterrence had relied on the absence of just these systems – unlocked the frozen cage of the Cold War in the twofold sense that it denied its permanence while forcing the Russians to realise that they could never compete with America in the most advanced computer and laser technologies. It did not matter whether or not such a system was feasible; after all, for the first twenty years of the Cold War the Soviets had bluffed the West into imagining that they had a much greater nuclear arsenal than was the case.[10]

There was one further significant individual. After 1985 Reagan and Thatcher found themselves dealing with a new, charismatic Soviet leader possessed of a relatively open mind as well as a pulse. Realising that to Europeanise Russia he would have to de-Sovietise eastern Europe, the fifty-four-year-old general secretary Mikhail Gorbachev signalled that the rulers of the outer Empire could no longer rely on Red Army tanks as their trump card in their dealings with their own peoples. It is worth stressing that the abandonment of the Brezhnev Doctrine was effective from 1981 onwards, when the Soviet Politburo ruefully acknowledged that it could not send troops into Poland without having to fight the Polish armed forces as well as the civilian population. Gorbachev made this explicit.[11]

The rest of Gorbachev's vision of a humanised and reformed Marxism–Leninism was hopeless: a slackening of the reins of Party control over managers and technocrats; a liberalised private small-business sector; co-option of the more biddable elements of the opposition into a reformist front without surrendering the hegemony of the Communist Party. He was a tragic victim of an illusion. As he beguiled and bewitched the world stage, a submarine went down with all hands; the Ukraine and much of northern Europe were hit by the toxic clouds of Chernobyl; and the Red Army disintegrated into a drunken or drugged rabble in the vast strangeness of Afghanistan, in graphic illustration of the cost to the USSR of imposing Communism on people who rejected it. In desperation, from 1985 onwards, Gorbachev took private economics lessons from the US secretary of state, the Stanford economist George Shultz, an appalling indictment of real existing socialism's total systemic failure. Of course, it was not just about a mere lack of competitiveness, as if the USSR was like some factory wedded to older ways that could be changed. Communism was morally as well as economically bankrupt. During the twentieth century over a hundred million people lost their lives around the world in the course of this monumental failed and futile experiment with human nature.

III BEING A DISSIDENT

Several of John Paul II's concerns with cultural and spiritual transformation also preoccupied many of the future leaders of eastern European (and Soviet) dissent. It was vital to have a keen sense of good and evil, to 'shake that evil off, escape its power and to seek the truth' as the Czech Václav Benda had it. That involved calling things by their proper names. While leftist anti-anti-Communist Western scholars split hairs over what to call Communist systems, dissidents who had to live under them eagerly embraced the Western concept of totalitarianism. So did Gorbachev for that matter. Rather than finding some relativising explanation for inhumanity, why not attach a perfectly serviceable name to it, while also acknowledging the existence of dark forces in human affairs? This led to a much more fine-grained analysis of the corrupting impact of Marxist–Leninism than any number of social 'scientific' studies, most preoccupied with attaching meaningless neologistic labels to things that European Christian culture had already given names for. Self-knowledge helped too. By acknowledging that Communism was capable even of corrupting its opponents, dissidents were more fully able to combat it.[12]

This quiet moral transformation involved living life as if the oppressive cope of Marxist–Leninism did not exist, or was moribund, while creating and expanding spaces so that 'civil society' could function within a system that – having failed to politicise every aspect of human affairs – had settled for docile acquiescence. First isolated individuals, followed by larger groups, began to stand up, straightening their spines, until the day when the Communists became an isolated clique whose primary loyalty was to an alien power. No one not involved should underrate the heroism of those involved, least of all Western intellectuals who in a fit of self-dramatising conceit created Charter 88 and the like in free societies.

A major individual effort was involved in shrugging off the quotidian moral complicities, and the easy acceptance of major lies about the past, present and future, that such a regime required to shore up its illegitimacy, for like a tumour Marxist–Leninism had insinuated itself not only into such concepts as peace and internationalism, but into nationalism and patriotism too. To dissent was to have secret policemen on one's tail; searching one's home and rooting through the pages of each book; or being hauled in for hours of interrogation in the middle of the night. All personal relationships could, potentially, result in betrayal by people one loved, as many in East Germany – where the Stasi had perfected technologies of control – were shocked to discover.

People had to keep the political equivalent of a Bach keyboard variation ringing clearly in their heads, to blot out the ambient ideological Muzak with its bogus messages of happiness, goodwill and progress. The reality was of a privileged Party elite, with its own shops and marks of favour, with a two-tier system of shops for everyone else, such luxuries as coffee only being available in PEWEX shops that took hard currencies. Most shops had lengthy queues snaking around the block, including those organised by committees, whose members took it in turns to hang around on the off-chance that rolls of coarse brown lavatory paper or a refrigerator might turn up, despite the long trail of loss stretching from the distributors to the factory backdoor. Even if people managed to get that refrigerator, should it break down there were no guarantees, no one to complain to, no repairs or spare parts, no consumer watchdogs, and no competitors to turn to for a new one. As well as no choice, there was frequently just nothing. Or rather one week there would be a glut of shaving cream, but no razor blades; the next week, razor blades but no shaving cream. More generally, any social mobility that the system had encouraged, mainly through the huge post-war population transfers after the ethnic cleansing of the Germans, had ground to a halt. Young people, who were both better educated and more curious about the wider world than their parents, found their upward trajectories blocked by those the regime had already privileged.

Significantly, a third of the workforce that would flock to Solidarity were under twenty-five when the Revolution happened. They knew of the existence of a wider world, with their wages docked in support of Cuba or Vietnam, but they could never visit it.

Industrial conditions were generally appalling; the point of the official trades unions was to communicate the wishes of the Party to the workforce, rather than to represent the workers' grievances to the Party employer. Marxist historians who write about these things peculiarly avoid such matters as conditions at work, housing and welfare, but then the only workers they know are in the abstract.[13] After a day's gruelling labour in unhygienic and unsafe conditions, those workers not crammed together in hostels caught the single bus that wheezed up to the suburban housing barracks. Despite a high divorce rate, families lived cheek by jowl in cramped conditions, with parents and unwanted partners crammed together. The one telephone kiosk for thousands of people rarely functioned, and was rigged so as to make only local calls, if you could find it at night in the absence of street lighting, and there was no lateral communication between individual apartments except by revisiting the ground and then working back upwards. It was telling that flats that had been deliberately constructed with every inconvenience for the Nazi occupiers (such as bathrooms that were poky and airless) by wartime Polish builders were considered highly desirable in the 1980s. Admittedly there were small oases of comfort amid the ambient grey of societies without advertising. Party officials either lived in large pre-war apartments, in purpose-built quarters, with communal gardens and swimming pools, or if they were really important in large villas in areas cordoned off with signs saying 'Military Area: Entry Strictly Forbidden'. First secretary Gierek, who had the state build a house for him for twenty million złotys, which he promptly purchased for four million, also had children from a neighbouring orphanage relocated to cut out unwelcome noise. The state picked up the tab for maintenance of elite housing, it being impossible for any ordinary citizen simply to summon an electrician, carpenter or plumber. The Party elite could also avail itself of special clinics, pharmacies, sanatoria and reserved wards in public hospitals, while the rest of the population had to make do with dirty, poorly equipped facilities, where one had to bribe a doctor to be treated, always assuming that he or she could lay their hands on drugs or hip-replacement joints. Despite having two hundred thousand inhabitants, there was only one hospital with a thousand beds in Nowa Huta.[14]

Newspapers, magazines and books had to be read through an ideological filter – assuming that they actually contained some coded references to the truth – or better yet, not read at all, which necessitated alternative sources of uncontaminated information. Radio Free Europe, Radio Vatican and the World Service foreign language broadcasts of the BBC provided channels of uncorrupted information, although strenuous efforts were made to jam their signals, or to assassinate broadcasters whose criticisms delved too deeply into the miasma of dynastic Communist corruption. The Bulgarian dissident Georgi Markov was killed with a lethal injection from an umbrella in the middle of London.

The recreation of an autonomous culture was a major achievement of intellectual dissenters. Since the universities were in the hands of the embourgeoised ideological soulmates of the West's tenured radicals – that is, corrupt, conformist mediocrities armed with their Party lapel badges and cards – dissidents created 'flying' alternatives in which 'heretical' thoughts could be aired in people's flats, sometimes with visits from such Westerners as the conservative philosopher Roger Scruton. Theatre people like Kenneth Tynan or Tom Stoppard kept up contact with Václav Havel. Samizdat publishing provided an alternative to the flood of cheap official books, enabling small circles of people to become acquainted with the thought of such writers as Havel and Michnik. In Poland, where censorship was lighter than elsewhere, *Tygodnik Powszechny* became the paper of record for dissidents. But there were myriad underground newspapers, such as KOR's *Bulletin*, the name a conscious echo of the wartime Home Army paper, or *Robotnik* aimed at the workers, all produced by very brave people operating in basements and attics. Two things are worth adding at this juncture about dissidence under Communist regimes in eastern Europe.

Firstly, the concentration on moral, cultural and indeed environmental issues denied Communists the ground on which their skills in coercion and manipulation would have operated to their advantage. In their worldview, culture was a secondary by-product of profounder economic and social forces, so to treat it as the priority was tantamount to deranging their thought processes. Acting as if Communism did not exist, or could be ignored, proved a more effective tactic than a head-on-collision with these regimes, as had been tried and had failed in 1953, 1956 and 1968. Why pick an open fight with a dying man? Secondly, the renunciation of violence not only recognised the asymmetrical balance of power, but in itself delegitimised Marxist–Leninist fantasies of heroic revolution, which further confused these regimes' responses. As they had become heavily dependent on Western loans to help their ailing economies, they were taking risks whenever they bludgeoned or murdered obviously non-violent dissenters. The election of the first Slav pope, with its inevitable refocusing of world attention on eastern Europe, ensured that any repression in that half of the continent would attract the full glare of publicity, especially because few attempts were made to curtail the movements of representatives of the Western media.

Most crucially, the creation of enduring contacts and viable coalitions between intellectuals and manual workers not only denied Communist regimes the gambit of divide and rule (especially if they could point to dissident intellectual Jews so as to push the buzzer of latent antisemitism) but also enabled opposition movements – which deliberately kept their organisations loose or nebulous – to transcend classical divisions between right and left. There clearly were major differences of opinion, but these were muted in favour of the more pressing struggle against Marxist tyranny. The presence of manual workers in coalitions of dissidents who talked about human rights and religious freedom delegitimised regimes whose public propaganda was ostentatiously 'workerist'; the sturdy wielders of axe, drill, hammer or shovel were not supposed to kneel in prayer or go into raptures about the Polish pope. These enthusiasms caused consternation in some perplexed foreign circles. German and French leftists, together with the British Communist-dominated National Union of Mineworkers, rushed to dub the worker activists of Poland's Solidarity movement 'Fascists', their catch-all term for anyone who inexplicably rejected their ultimately economistic view of the world in which workers were supposed to be concerned with bread-and-butter issues. Certainly, dissident workers were concerned about prices, wages, working conditions and pensions, but they also insisted on an impressive range of basic freedoms for which cheap refrigerators were not a worthy pay-off.

IV WAR OF THE SYMBOLS: SOLIDARITY

The route to this historic alliance between workers and intellectuals was stained with bloodshed. Bread-and-butter issues may have triggered the initial uprisings, but they soon evolved in new directions. Shortly before

Christmas 1970, the Gomułka regime in Poland hiked food and fuel prices without a commensurate rise in wages. The workers in what had recently been renamed (to save a manager's job) the Lenin Yard at Gdańsk went on strike, which after police intervention led to violent confrontations. Lech Wałęsa, a young electrician, made his debut mediating between striking workers and the militia, who eventually deployed machine guns and tanks. On 16 December the Polish army shot down striking workers, killing twenty-eight (that being the lowest estimate of fatalities) and wounding twelve hundred more. Thousands of people were arrested and interned. Priests helped trace people who had disappeared, and recorded burials carried out by the secret police at night. Trouble spread along the coast to Gdynia, Sopot and Szczecin. In Szczecin the workers burned down the Party district offices, the militia headquarters and the District Council of Trade Unions buildings.[15] As a result of these uprisings, Gomułka was dismissed and replaced as Party first secretary by the younger Edward Gierek whose temporarily effective pitch was all hands to the pump to stop the ship sinking. While workers grudgingly returned to work, a further strike by women textile workers in Łódź forced the government to rescind the price increases. Having begun with promises of a little Fiat and housing for everyman, Gierek's honeymoon gradually turned into a fractious divorce from the Polish people whose name he so readily evoked.

Gierek sought to raise loans on the international capital market, so as to modernise the economy and pay back the loans through exports. Recycled petrodollars would support wage rises and price controls. This resulted in disaster, since the inefficient Polish economy was incapable of producing goods of a standard world markets required. By the late 1970s Poland had levels of debt rivalling Latin America at US$23 billion. The cost of serving this debt mountain rose from 27 per cent of export income in 1974, to 43 per cent in 1975 (and 70 per cent in 1980). More loans, at punitive interest rates, were incurred just to pay the interest on the original debt.[16] Six years after the abortive price rises, Gierek raised them again in the summer of 1976. The price of meat went up by 100 per cent. Riots occurred at the Ursus Tractor Factory in the capital, while in Radom armament workers burned down the local Communist Party headquarters. Although these price increases were revoked, this time the police, Security Service and Party militia pursued a vindictive campaign against those involved that led to many detainees being physically assaulted. These brutalities led a group of intellectuals, including Jacek

Kuroń, Bronisław Geremek and Adam Michnik, many of them social democrats by political avocation, to found the Committee for the Defence of the Workers (or KOR by its Polish acronym) which pursued the cause of workers being persecuted by their own Party-state.

In the summer of 1979 John Paul II returned for a triumphal nine-day visit to his homeland. Thirteen million people turned out to hear him, many no doubt agreeing with the miner who said he had come 'To praise the Mother of God and to spite those bastards.' There were other encounters of a less agreeable kind. At receptions with members of the regime, John Paul categorically rejected their insistence that the Church had merely cultic functions within society:

> Given that [the temporal dimension of human life] is realized through people's membership of various communities, national and state, and is therefore at the same time political, economic, and cultural, the Church continually rediscovers its own mission in relationship to these sectors of human life and activity. By establishing a religious relationship with man, the Church consolidates him in his natural social bonds.[17]

At a mass on Warsaw's Victory Square, the crowd responded to John Paul's sonorous classical Polish with chants of 'We want God, we want God, we want God in the family circle, we want God in books, in schools, we want God in government orders, we want God, we want God.'

Following this gigantic anti-Communist plebiscite, the Church was never far removed from the final cycle of unrest to hit Poland. On August 1980, workers at the Gdańsk Lenin Yard went on strike following the dismissal of a crane operator who was a labour activist. Lech Wałęsa, who had been sacked earlier, climbed back into the yard and took over leadership of the strike committee. The local bishop managed to calm things, by going into this prestige Communist project to say an open-air mass for the strikers below a giant wooden cross which the workforce had made to commemorate the victims of government repression a decade earlier. Since it became obvious that no local deal was going to prevent this wave of unrest from spreading across the country's entire workforce, the regime concluded the Gdańsk Agreement on 31 August 1980, which recognised the rights to strike and of association, conceded construction of a permanent memorial to those workers shot down in 1970, and a relaxation of censorship. Wałęsa signed the agreement with a huge pen capped off with a picture of the pope.

A National Co-Ordinating Committee of a New Self-Governing Independent Trades Union came into being, called Solidarity for short, under Wałęsa as its chairman. One of his first acts was to have himself photographed beneath a large cross. A young Gdańsk designer, Jerzy Janiszewski, provided the logo in which red letters on a white ground not only echo Poland's national colours, but seemed to lean on each other for support.

Government concession of the right to independent trade unions was followed by wildcat strikes and the gradual disintegration of the Party even as it conducted a fitful dialogue with the Solidarity leadership. Three million lowly members of the Party joined Solidarity, while the abandonment of 'democratic centralism' meant that at the 1981 Extraordinary Party Congress 90 per cent of the old-guard leaders were rejected by an 'electorate' that had hitherto known when to put its hand up. The Russians commenced ostentatious military manoeuvres, code-named Soyuz 81, including landing marines on Poland's Baltic beaches. In early December 1981, the Solidarity Executive discussed free elections and a referendum regarding Poland's main external alliance. Wałęsa demurred. The discussions were heard through intelligence-service listening devices. Fearful that the Russians would intervene militarily, which the grim East German leader Erich Honecker was urging them to do, General Wojciech Jaruzelski, who had become prime minister in February, and first secretary of the Party in September, began making ominous dispositions. Soldiers sent to the countryside to help distribute food also used the opportunity to record the addresses of Solidarity activists. Jaruzelski met both Wałęsa and the new primate, cardinal Józef Glemp, in November 1980, to give the appearance that a negotiated settlement was possible. That month Brezhnev warned the general that 'there was no way to save socialism in Poland' unless 'a decisive battle with the class enemy' was fought. He was probably bluffing since the Soviet Politburo ruled out intervention even as US intelligence was confirming its imminence. As the Kremlin's chief ideologist Mikhail Suzlov expressed it: 'If troops are introduced, that will mean a catastrophe. I think we have reached a unanimous view here on this matter, and there can be no consideration at all of introducing troops.'[18]

Among those who live by sensation, it is often said that US president Ronald Reagan – who came to office in January 1981 – initiated intelligence sharing with John Paul II, although the Vatican possessed one of the best information networks in the world. In fact, his predecessor Jimmy Carter had initiated this practice. Secretary of state Zbigniew Brzezinski, himself of Polish extraction, showed the pope US satellite imagery, while Carter warned the Russians that there would be 'very grave' consequences for the superpower relationship if they intervened in Poland. At a minimum, the US would prompt world trades unions to impose a total boycott on Soviet air and sea traffic. John Paul II sent a strongly worded letter to Brezhnev, reminding him of the consequences of the violation of Poland's sovereignty in September 1939, and of obligations that the Soviet Union had solemnly committed itself to at Helsinki. Three months later, Stanisław Kania and Jaruzelski were summoned to Moscow to learn that the Soviets would not intervene.[19]

Jaruzelski reassured the Soviets with the prospect of imposing martial law, the details of which he doubtless thrashed out with his opposite numbers in Soviet intelligence agencies. Towards midnight on 12 December 1980 the nation's three and a half million private telephones went dead and the army occupied the streets, positioning armoured vehicles at major intersections. Ten thousand people were detained and put in internment camps. The leaders of Solidarity, including Wałęsa, were arrested in what was a cross between a coup and an invasion, and was called 'a state of war' in official pronouncements. Wałęsa was shunted around various Party villas to maintain the pretence that the regime was negotiating with him. The only major resistance was in a mine in Silesia where twelve hundred miners barricaded themselves in the pit, and had to be forcibly extracted by Security Service and ZOMO riot police, at the cost of nine miners killed.

Jaruzelski imagined he could detach the 'extremists' among Solidarity's leaders from ten million followers, who would then be satisfied with economic rewards which the regime could not deliver. Black propaganda was used to discredit the interned Wałęsa, who the combination of inaction and the Party's well-stocked food and drinks cabinets had made portly. Hidden cameras recorded him in private conversation with his brother. Snippets were re-edited with the voice of an actor added, which 'revealed him' obsessed with the rate of interest his 'fortune' would accrue in Vatican banks. This was crude stuff and wholly ineffective.[20]

The junta tried to conceal its brutal demonstration of police power with a claque called the Patriotic Movement for National Rebirth. In fact, the junta relied on curfews, the militarisation of the workforce, the abolition of unions for journalists and film-makers and the suspension of Solidarity to stifle dissent. There was really no coherent strategy beyond that. Jaruzelski thought he could isolate the Church by forcing it to distance itself from opposition 'violence', and with the promise that it would have a powerful voice once the old order was restored, perhaps by allowing it to set the moral tone against various manifestations of decadent Western secularism such as pornography. He thought he could pacify the workers with what is known as Kadarisation, that is a Hungarian-style liberalisation of the consumer economy.

He imagined that the West, concerned about Poland's mountain of external debt, would make ritualised protests before seeing how best it could recoup its money. After all, the West German government said little about martial law – Helmut Schmidt was throwing snowballs from a balcony with Erich Honecker when it was declared – lest it impact on its relations with the GDR or on ethnic Germans throughout the Soviet Union. Germany's left-'liberal' intelligentsia also maintained its customary solipsistic provinciality, or bored on about US policy in Central America, for the purposes of doing down Reagan rather than by virtue of knowing or caring much about people in whose oppression the US colluded there.

All of Jaruzelski's hopes proved illusory, not least the conviction that the Church would cynically cut a deal to defend its institutional interests. Glemp was not a widely admired figure – some called him 'comrade Glemp' – partly because people were unaware that, by inaugurating a more collegial style of Church governance than his predecessor Wyszyński, he had allowed different voices on the episcopal bench to be heard, some of whom notoriously put more trust in Jaruzelski than in Wałęsa. However, a Polish pope, whose sympathies were with Solidarity, could always be relied upon to trump Glemp's authority, as he did when he indicated that the Church was not to play the role of neutral arbiter.

The primate also discovered that lower clergy found the vivid reality of the popular movement more compelling than ecclesiastical hierarchy. Indeed, in 1982 Glemp sat stony-faced as two hundred of them attacked his stance in the harshest terms at a meeting of the Warsaw curia.[21] One priest began to attract the attentions of the Party-state. Father Jerzy Popiełuszko was a young priest in a parish in Warsaw's Żoliborz district. He was ordered to establish relations with workers in a major steel plant. Typically, this slight and uncharismatic figure wondered why they applauded or wept as he walked in, thinking someone more important must be behind him. After the imposition of martial law, he held monthly masses for the fatherland in his church of St Stanisław Kostka; these were attended by workers from the capital's proletariat. Popiełuszko addressed crowds of ten or fifteen thousands about the need to resist the evils of the regime. Glemp reminded him of the rather spurious distinction between being patriotic and political. Pressure from the Soviets may have induced the Interior Ministry to do something about him. The regime's chief mouthpiece, Jerzy Urban, described Popiełuszko as 'the Savanarola of anti-Communism'. If Communist regimes could conspire to assassinate the pope, a troublesome priest was hardly a major challenge. In 1981 the Bulgarian secret police – and probably the KGB and Stasi – orchestrated Mehmet Ali Ağca's attempt to shoot John Paul II as he toured St Peter's Square in his 'Popemobile'.[22] A first conspiracy to kill Popiełuszko in a faked car crash evidently failed. But a week later, on the night of 19 October 1984, his car was flagged down by three security service officers; despite being handcuffed, the priest's driver managed to escape, which shows how confident the authorities were of getting away with it. Popiełuszko was repeatedly beaten up each time the car stopped. He died and was dumped, still bound, in a Warsaw reservoir, his body weighed down with stones. Thousands of people flocked to his church, sceptical of the government's claims that he had been kidnapped. When news that his corpse had been found came to the crowds in his church, there was a real risk of major public disorder. This was averted. Hundreds of thousands of people turned out for his funeral and his grave became a Solidarity shrine. Popular pressure and international outrage forced the regime to hold a trial of the perpetrators, who included a captain in the security service. The prosecution tried to find extenuating circumstances, insinuating that Popiełuszko had brought about his own death by making provocative statements.

Father Popiełuszko became part of the ad-hoc symbolic arsenal with which the opposition confronted the regime. Its symbols radiated more power than those used by the Party. For example, dissident workers wished to commemorate their friends and colleagues who had been shot down in 1970, a process begun with the laying of simple wreaths – which the security service endeavoured to clear away or obstruct – followed by crosses marking the places where they had been slain. The northern shipyards became the unlikely sites for an explosion of popular poetry, theatre and religious folk art. It is difficult to convey that hour when it was blissful to be alive. The most powerful symbols were the giant steel crosses that replaced earlier wooden efforts in the Lenin Yard. The regime used every conceivable form of chicanery, including calling for a national competition, so as to delay the inevitable, or building a wall, to frustrate the memorial being built or subsequently seen. Workers appropriated the wall's bricks as souvenirs. The steel crosses, consisting of long tubes welded together in a triangle to evoke Golgotha, symbolised faith, sacrifice and solidarity, while the anchors (symbolising the professions of the sea as well as the wartime emblem of the Home Army) welded on to the top signified hope in the future. Around the base were lines from the Psalms, by the poet Czesław Miłosz and by the pope. The opening ceremony was designed to marginalise Communism, not only in the sense that it was a religious service, but in the sense that the people and Solidarity's leaders all faced the monument from the same level, in marked contrast to the Communist practice of having the Party leadership gazing down from a monument upon the serried masses beneath them. Similar monuments to the fallen were erected in Gdynia and Szczecin. They give the lie to the idea that Polish workers were solely concerned with bread and meat prices.

Symbols are no substitute for political victory. This seemed distant. In the summer of 1983, John Paul II made a second visit to his depressed and fearful homeland. His presence encouraged crowds in their chants of 'Solidarity'. In private sessions with Jaruzelski, the temperature grew heated as the pope insisted that the general resume a dialogue with Solidarity. That, after a delay of five years in which the regime further demonstrated its inability to master Poland's chronic economic problems, was what eventually occurred in early 1989. Jaruzelski was persuaded of this course by his slippery new prime minister, Mieczysław Rakowski, who thought that by bringing Solidarity into government – and especially by handing them economic portfolios – the union would share the blame for the country's parlous state. Following the round-table talks, the regime relegalised Solidarity, conceded that the Soviets were responsible for the wartime Katyn massacres nearly fifty years earlier, and granted elections in which half the seats in the Sejm were to be freely contested. In these polls, Solidarity candidates won 99 per cent of the seats in what was a rout for the Polish Workers Party. Although Jaruzelski remained as president and commander in chief, having resigned from the Party, a new administration was formed with the Catholic Solidarity activist Tadeusz Mazowiecki as prime minister of the

'Polish Republic'. He fainted at his own inauguration ceremony in shock at this personal turn of events. To all intents and purposes, the Party was dead and Poland enjoyed its freedom for the first time since the Second World War. Of course, the defeat of the enemy that had concentrated minds inevitably led to bitter disputes among the victors, who had very different ideas about Poland's political future. But this should never detract from the way in which this remarkable nation threw off foreign tutelage, and within a relatively brief period established itself as one of the most important states in contemporary Europe.

V A VERY PROTESTANT REVOLUTION?

What became the German Democratic Republic in 1949 was the only Communist state to have a Protestant majority, numbering four-fifths of the population in 1946. By the time the regime collapsed in 1989, it had created the least religious society in the entire Communist system, with only about 10 per cent of the population acknowledging any religious affiliation. But paradoxically the Churches played a significant role in the regime's downfall.

The two major Protestant Churches consisted of eight regional Churches. Five adhered to the Evangelical Church of the Union – forged by the Prussian state in the nineteenth century between Lutheran and Reformed Churches – the remaining three being combined in the United Evangelical Lutheran Church. Each of these territorial Churches elected its own bishops and synods, and reflected subtle differences in both ecclesiology and theology. Bishops shared power with synods of clergy. Below them were district superintendents with oversight of individual parishes, each of which had an elected parish council. There were many lesser denominations, free Presbyterian Churches and sects.

There were also one million Roman Catholics in the GDR, mostly in the south where many expellees and refugees from Catholic regions of eastern Europe had settled. From the start, the Catholic Church simply refused to accept or co-operate with state socialism, but it also decided not to oppose it actively. As bishop Otto Spulbeck of Meissen put it in 1956: 'We live in a house, whose structure we have not built, whose basic foundations we even consider false. We gladly contribute, living worthy and Christian lives. But we cannot build a new storey on this house, since we consider its foundations false. We thus live in a diaspora not only in terms of our Church, but also in terms of our state.' The authoritarian and centralised nature of the Catholic Church meant that it never dabbled with such dubious concepts as 'the Church in socialism'. The fact that most of the East German Church belonged to larger dioceses in West Germany or Poland helped maintain its independence.[23]

The Protestant churches were in a much more complicated position. Initially, the majority of clergy rejected the state's demands for total identification with socialism, but there were significant differences in how the Churches responded to insistent demands that such an identification occur. Some clergy vividly recalled their role under the Nazis, and opted for the stance of being 'watchmen', alert to every violation of human rights by the totalitarian state. Others were attracted to the notion of 'the Church within socialism' as a way of avoiding marginalised irrelevance. A third group followed traditional Lutheran teaching on the two kingdoms by bowing to the state as 'the force of order' and practising political quietism. A further group, who had been influenced by Dietrich Bonhoeffer, thought the Churches should be a refuge for the alcoholic, the old, the weak, the imprisoned and the politically persecuted. And finally there were those who believed in 'critical solidarity' with a regime whose overall vision they approved. To make matters as complicated as in reality they were, relations between the regime and the Churches were also affected by generational differences, as people who were already old when the Nazis fell were succeeded by younger leaders who had come to adulthood in the GDR, a process with mostly negative effects on the capacity of the Churches to resist Communism.[24]

Throughout its existence there was no formal separation of Church and state in East Germany. Indeed, until the mid-1950s, an atheist regime compelled citizens to pay dedicated taxes that it redistributed to the Churches. In addition, the regime paid fees and rents for the Church properties it had expropriated. There was also support for the Protestant theological faculties attached to six universities, and an impressive health and welfare nexus. The largest Protestant Church, for example, controlled 44 hospitals, 105 homes for the disabled and mentally ill, 19 orphanages, 310 community services centres, and 278 kindergartens and nurseries. The state allowed the Protestant Churches to print five major regional newspapers on its own presses, an arrangement that made censorship easier.

Since the Communist regime was virtually flown into Germany by the

Russians, incidentally giving the lie to it having 'resisted' the Nazis, it began by handling the Churches with tactical restraint. Partly because of contacts established between dissident pastors and Communists in Hitler's camps and prisons, many of whom were subsequently purged from the ruling Socialist Unity Party (SED), the SED initially downplayed the antagonisms between Marxism and Christianity. The Soviet Military Administration, which really ruled East Germany, allowed the Churches to carry out their own de-Nazification procedures to weed out former adherents of the German Christians. This honeymoon period continued from 1945 to 1948.

In that year, the state began to interfere in religious instruction and to insist that people work on Sundays. In 1949 it foisted two 'progressive' pastors on the Church's weekly radio broadcasts, which resulted in the Church withdrawing from the programme. Next, schools were forbidden to celebrate Christmas; Stalin's birthday on 21 December became the obligatory alternative. Christmas became the 'winter vacation' and Jesus 'the Solidarity child'. The money disbursed to Churches was drastically cut, while permission was denied to make up the shortfall through collections. With historical materialism marching into education, by 1952 Bible study groups were banned. At tertiary level both Marxist–Leninism and the Russian language became obligatory for all students, the latter a means of culturally isolating people from England and France. In 1952 the regime closed the borders with West Germany, while bundling 8,300 suspect citizens over the border. More than seventy clergy and laity were arrested as 'agents' of Western intelligence services. Christians were subjected to insidiously systematic discrimination in education, employment and welfare. A campaign was launched to use the law to eradicate what remained of the commercial middle class and private farming. The conviction of 'economic criminals' allowed the state to take their property. It also resulted in chronic shortages of such basic foodstuffs as butter and margarine. By the end of March 1953 the number of people convicted for trivial offences against 'the people's property' had risen to ten thousand a month, and the number of prisoners more than doubled from thirty-one thousand to over sixty-six thousand within a year.

The chief effect of attempts to 'build socialism' in the GDR was that people upped and left. In the first half of 1952, seventy thousand people fled to the West, followed by a further 110,000 before Christmas. Another 330,000 fled in the following year. The trend was so worrying that even Lavrenti Beria, the rapist former head of the NKVD, and member of the post-Stalin Soviet leadership, thought that the GDR might be expendable if a united Germany could be kept neutral. This sobering message was passed on to Walter Ulbricht and others after they had been summoned to Moscow in early June for instruction on how to improve the running of their country. Although they instigated reforms on their return, they nonetheless insisted on a 10 per cent increase in productivity rates, a substantial pay cut that infuriated the very class in whose name they claimed to rule.

On 16 June 1953, construction workers at an East Berlin hospital downed tools. They were joined by more men working at a site on the Stalinallee, who imagined that the first group were being held captive, and then by the huge workforces at three major plants in the south of the capital. They marched on government ministries, tearing down propaganda posters and overturning official cars en route. They demanded the rescission of the new productivity norms, free elections and the resignation of the government, whose representatives – with one exception – were too cowardly to meet the protesters. One hundred and fifty thousand people subsequently went on to the streets in what amounted to a workers' revolution that soon spread to seven hundred other places. There were calls for free elections and national unification. They attacked security service buildings and prisons, the former task made easier by the fact that the Stasi had been sent into the factories. Western intelligence agencies in West Berlin were taken totally by surprise, although the GDR leadership would subsequently blame them for the uprising.

Since the police and Security Service were in no position to deal with a mass uprising, Soviet tanks appeared on the streets of East Berlin from 17 June onwards. Fifty people were shot – twenty of them by summary firing squads; forty Red Army soldiers lost their lives – most of them executed for refusing orders to shoot German civilians. Three thousand demonstrators were arrested, together with a further thirteen thousand people picked up after the event.[25]

This revolt, the first against any Communist regime since the war, had two important consequences. First, the head of the Ministry of State Security (Stasi for short), the veteran Communist Erich Mielke – who had carried out political murders in the Weimar Republic before fleeing to Russia – determined that the Stasi would never stare into the abyss again. It would be the sword and shield of the Party, a metaphor it owed to the Polish founder of the Cheka, the Bolshevik secret police. Mielke constructed an enormous secret police apparatus, including what became some two hundred thousand spies recruited from among the population who reported to controllers within the one hundred thousand permanent Stasi officers. It kept files on four million citizens in the GDR and a further two million on people in the Federal Republic, files which ran to a hundred kilometres in East Berlin, with a further eighty kilometres on shelves in the provinces. In Leipzig, the Stasi maintained card entries on two-thirds of the city's half-million inhabitants. What this meant for dissidents was epitomised by the case of the man who discovered that twenty-two of his close acquaintances, including a cousin, had regularly informed on him.[26]

The Stasi files were administered by three hundred full-time archivists, with the technology constantly being upgraded. There was even a bottled collection of the personal scents of dissidents – derived from their clothing – whom one might have to send the dogs after. The Stasi was lavishly funded with an operational budget of four billion marks a year. In addition to its imposing headquarters on the Normannenstrasse in East Berlin, it had two thousand safe houses and covert installations from which to photograph people, together with regional offices in each district. Riot police squads – or rather a domestic army, with armoured cars and cannon – were augmented by the four hundred thousand men organised in the Party's industrial militias which could strike down any future worker protests at source. Among the population at large, the bloodbath in June 1953 suggested the inadvisability of any further insurrections.[27]

With very few exceptions, clergy kept their distance from the 1953 worker uprising, which became an official holiday in the Federal Republic. Ironically, Edgar Mitzenheim – the brother of bishop Moritz Mitzenheim of Thuringia – was sentenced to six years' imprisonment for taking part in what his brother denounced as 'Fascist provocations'. While few clergy went that far, their tacit support for the regime reflected the fact that at their meeting with the SED trio on 2 June the Russians had instructed the rulers of the GDR to liberalise their policies towards the Churches with a view to turning the latter into pliable political instruments.

While the Party-state could do little to diminish the faith of older Christians, it could affect the young, especially those raised on vast suburban housing estates where building churches had been overlooked. The regime introduced, dropped and then reintroduced the requirement that all schoolteachers had to be Marxists. It was responsible for remorseless atheist propaganda as well as crude polemics against the manifestly superior Federal Republic. It created a massive youth organisation – the FDJ – to rival Christian youth organisations. In virtually every respect this was a copy of the earlier Hitler Youth, although its creed was 'anti-Fascism', the public ideology of the GDR.[28] The very young were encouraged to join the Pioneers, where they were inducted into the religious cult of Ernst 'Teddy' Thälmann, the Weimar Communist Party leader shot by the SS in 1945, who was presented to children as the Red equivalent of the Protestant Sweet Jesus.

In 1954 the regime inaugurated youth dedication ceremonies – the Jugendweihe – an idea they took over from nineteenth-century secularists. The socialists and Communists of the Weimar Republic had used similar ceremonies, as did the Nazis who introduced them for children joining the Hitler Youth. In the GDR these ceremonies were preceded by the secular equivalent of catechetical classes in which atheism was aggressively propagated. Fourteen-year-olds were informed that religion was a tool 'for holding down the masses and oppressing them'. They received such books as Nikolai Ostrovsky's turgid *How Steel is Hardened*, set in the Russian Civil War, or *The Universe, the World, and Mankind*, whose title alone must have daunted fourteen-year-olds. This atheist alternative to confirmation ceremonies was supposed to be voluntary, but by 1958 it had become general, so many advantages did it mysteriously confer. There were other pseudo-religious aspects to becoming an adult socialist. A visit to the memorial within the former concentration camp at Buchenwald became a pilgrimage for millions of FDJ members. There they learned about the former prisoners' 'Oath of Buchenwald' and dedicated themselves to the anti-Fascist struggle that lay at the heart of the GDR's self-understanding. They presumably did not learn that the Soviets had continued to use Buchenwald until 1951 to house prisoners, who included former Nazis as well as many of their former opponents.

By contrast, it was made harder for Christians to bring their children up in the faith of their choice. The 1956 Fechner Decree, introduced by East Berlin's municipal government, banned religious instruction in the hours before school commenced, and insisted on a statutory two-hour interval after a child had returned home, before evening instruction might begin. Parents and children who were still prepared to learn about Christianity later at night had to get written permission, which was renewable on a tri-monthly basis. The payment of Church taxes became voluntary, while the churches themselves fell into dilapidation, and pastors had to make do with meagre stipends that barely kept them above the breadline.

Up to this point, relations with the Churches had been left to the deputy prime minister, Otto Nuschke, the head of the CDU-East Bloc Party, one of the licensed transmission belts to non-Communist constituencies under the overarching dictatorship of the SED. But in March 1957 Nuschke's Office for Church Relations was transformed into a State Secretariat for Church Affairs under Werner Eggerath, East Germany's former ambassador to Romania. His opening communication to the bishops invited them to use Easter services to denounce the nuclear bomb. East German clergy were told to sever their ties with their West German co-religionists, while bishop Otto Dibelius was suddenly banned from the eastern parts of his own diocese. Smear posters linked Dibelius with Heinrich Himmler and a sex fiend called Balluseck. In April 1957 the regime arrested a popular pastor, Georg-Siegfried Schmutzler, who was jailed for five years for 'agitation to boycott the Republic', for supporting the Hungarian Uprising, and for supporting the Evangelical Church's agreement to appoint military chaplains to the West German Bundeswehr. While he languished in prison, bishop Moritz Mitzenheim of Thuringia took the lead in finding a modus vivendi with the regime. The private meeting became the normal mode of communication between Church and Party figures.

In July 1958 Mitzenheim and the SED first secretary Walter Ulbricht issued a joint statement which claimed that 'the Churches ... are in fundamental agreement with the peace efforts of the GDR and its regime'. Ulbricht also averred that 'Christianity and the humanistic ideals of socialism are not in contradiction'. There were other spectacular betrayals. In October 1958, the eminent Swiss theologian Karl Barth – one of the few Protestant theologians to have opposed Nazism root and branch – wrote an extraordinary letter to Protestant pastors in the GDR, claiming that since West Germany was in the grip of former Nazis and NATO warmongers, they should have no hesitation in giving their loyalty to the East German Communist regime.[29]

With the progress of time, and the retreating prospect of reunification, several smaller Churches, as well as the Lutherans, formed autonomous organisations within the Communist Republic. The main Evangelical Church resisted pressure to follow, but by 1969 even it had formed the

Federation of Evangelical Churches in the GDR. Once detached from Churches in the West, the Evangelical Church found ideological concessions easier. By 1971 its leaders talked about the Church 'in socialism' rather than either against or alongside it. Ironically, although the regime seemed to have closed the front door to the Churches in the Federal Republic, it was in reality allowing those same Western Protestant Churches to export huge amounts of goods and services, which in addition to building or repairing churches went to alleviate the plight of political prisoners or to purchase people out of the workers' paradise.[30]

Ulbricht's successor, Erich Honecker, rejected the confrontational approach to the Churches that had been pursued until the late 1950s. Honecker was a former roofer whose involvement with Communism began early. In 1937 the Nazis sentenced him to ten years' imprisonment, from which he emerged in 1945. He was conspicuously inarticulate, and looked and sounded like a sanctimonious schoolmaster, listening to reports from his Politburo colleagues on issues that he and a handful of cronies had rigged privately prior to each meeting. However, he had a keen sense of where ultimate power lay. He established close relations with Brezhnev, and was selected by him as Ulbricht's replacement when the Russians decided it was time for the old man to go. Honecker began by permitting such minor subversions as pop music, jeans, beards and long hair, as well as allowing people to watch West German television and to use the Deutschmark as a second currency without fear of prosecution. In March 1978 he made various concessions to the Churches, including giving them quarterly access to state television, pension rights for clergy, compensation for expropriated property and permission to build new churches provided this was financed by the Federal Republic.

The subtler approach towards the Churches reflected Honecker's desire to defuse potential clashes over the introduction of 'pre-military' training for fourteen- to sixteen-year-olds – the reality behind the constant exhortations to peace; it also reflected the regime's reversal of its blanket condemnation of the German past, so as to invent a spurious legitimacy for itself. This was part of a broader emphasis upon the GDR as a separate socialist nation, as reflected in such adjustments as the German Academy of Sciences becoming the Academy of Sciences of the GDR. By allowing selected clergy to travel abroad to meetings of the World Council of Churches and the Lutheran World Federation, Honecker also hoped to win international recognition for the GDR while having the clergy represent GDR policies.

Paradoxically, at the same time as the regime realised that religion had its practical and symbolic uses, the Churches became magnets for dissent. The quincentenary of Martin Luther's birth in 1983 was a key moment in the regime's politicised revision of history, since hitherto the Party had disdained Luther (and much of the historic German past) as an enemy of 'the people', at the same time lavishing praise on the Anabaptist 'revolutionary' lunatic Thomas Müntzer, who had turned sixteenth-century Münster into a vision of hell. Now, in its efforts to graft itself on to the root stem of Prusso-German history, as well as to cash in on Western tourism, the GDR leadership discovered positive value in the patriotism, and passive political theology, of the great Protestant reformer. But the Churches had moved on from the high-level deals of the late 1970s.

During the early 1980s the Churches became key sites where heterogeneous ecological and peace activists could meet, for any other gatherings of more than half a dozen people required the state's permission. Churches also had the advantage of possessing telephones with long-distance facilities, enabling contacts to be forged across the GDR, although there was always the risk of there being three callers on the same line. Courageous individual pastors, such as Christian Führer or Christoph Wonneberger, allowed their churches to become shelters for myriad oppositional groups. Some of these people were Christians, others not, but the key point was that the Churches helped them all overcome the intense atomisation which the regime had deliberately fostered, be it isolating and persecuting active dissidents or encouraging individuals in harmless private pursuits. Now they came together in candlelit vigils and prayer, a mode of organisation that was difficult to combat with police dogs and water cannon as the moral balance was so blatantly asymmetrical, while the peaceful forms nullified the entire Communist mythology of violent revolutionary upheaval.

Multiple ambivalences characterised the relationships between the Party-state, the Churches and opponents of the regime. Activists effectively carried out a laicisation of the Evangelical Church, a process called organising 'the Church from Below'. But at the same time the Party-state regarded this as the 'theologisation' of 'hostile–negative activities'. While some clergy were sympathetic to critics and opponents of the regime, others, often in the hierarchy, worried about the repercussions on the Churches or resented the lay tail wagging the clerical dog. All Churches in East Germany also faced the problem of the progressive secularisation of society, and wondered whether sheltering dissenters might reverse what accommodation with an atheist regime had conspicuously failed to achieve.[31]

At the St Nicholas Church in Leipzig, where prayer meetings held every Monday evening at 5 p.m. from 1980 onwards became a major focus of opposition, pastors allowed opponents of the regime to camouflage themselves as 'church groups', which then co-determined the increasingly politicised content of the prayer services. The strictures of Old Testament prophets against sinful kings developed into free exchanges of information and opinion during discussion periods, then to the posting of lists of people who had been arrested, and finally to confrontations with the Stasi as people debouched from the church into the main square.[32]

By 1987, the regime was actively concerned about how the 'temple police' they had been confident in managing had been suborned by activists and dissidents whom they generically stereotyped as 'rowdies'. As a member of the Politburo explained to a bishop: 'At official church offices, people answer the phone saying "contact office", "Solidarity office" or "co-ordination centre".' The forcible closure of the underground Environmental Library within Berlin's Zion Church in November that year was symptomatic of the regime's discomfort.

In June 1989 the Ministry of State Security estimated than there were 2,500 hardcore opponents of the regime who met in as many as 160 groups. All but ten of these (the chief exception being the Initiative for Peace and Human Rights) met under the aegis of the Churches. That was why the Stasi went to such lengths to recruit informers throughout the churches, notoriously including senior administrators, and why it planted three listening devices in the home of a single pastor, Rainer Eppelmann.[33] If the proximity and ubiquity of the Western media made the use of police force unadvisable, the Stasi hoped to use strategically placed informers to influence the political choices of opposition groups, many of whom wished to create some version of socialism with a human face, a project with more appeal to artists, writers and intellectuals than to the ordinary Manfred.[34]

Compared with the Poles' heroic emphasis on morality and liberty, there is something depressingly provincial about the causes espoused by the East German opposition. Maybe that prosaic, predictable quality lies at the heart of revolutions that negated the entire mythology and 'pathos' of revolution. Opposition groups harboured by the Churches alighted upon the creeping militarisation of GDR society, which stood in glaring contradiction to the emphasis upon peace in the state's foreign policy rhetoric. Of course in reality the GDR maintained a sizeable 'Afrika Korps' which dabbled in various sub-Saharan tragedies, while the National People's Army had plans for a quick thrust over the Rhine into France and Belgium. Dissidents called for non-military alternatives to national service that would not blight anyone's future career, as had been the case with those who had joined 'construction brigades', which the regime had conceded in 1964 as an alternative to military service, but which still involved labour on military bases. About four to five thousand people opted for this alternative, or had gone to prison for refusing it. They became one nucleus of opposition. This usually went together with opposition to the introduction of alarming civil defence drills, including the regular sound of sirens, or the appearance of mini-tanks in kindergartens. Those who wanted peace and opposed Communist militarism were vocal in such campaigns as 'swords into ploughshares', which the regime duly suppressed. Other opponents denounced the ecological devastation that Communism had inflicted through crash industrialisation, a cause that had become massively fashionable across the inner-German border with the rise of 'Green' politics. The outrageous denials of liberty within the GDR animated a third group. They campaigned for the right to emigration (or on behalf of those awaiting exit visas) from a state that was so popular among its citizens that it had surrounded itself with minefields, watchtowers and high concrete walls. The emigration issue introduced a potential fault line in the opposition between those who were desperate to get out and others who imagined that a reformed socialist state might eventuate that it would be worth remaining in. Finally, there were younger people who were fed up with the aged leadership and their middle-aged clones, and who sought the usual range of lifestyle freedoms enjoyed by their counterparts in the Western world. In this context, being a 'punk rocker' really was a political statement.

Although the GDR liked to tout its economic successes, claiming to be the world's tenth strongest economy, in reality, after the oil crisis of the early 1970s, it depended on economic subventions from its richer neighbour, some of which involving buying people freedom from the worker–peasants' paradise, a squalid trade vaguely reminiscent of human trafficking. The GDR was also fatefully and slavishly reliant on the Soviet Union, whose armies had ultimately created it. The Nazi experience largely accounted for the grovelling of East Germany's leaders towards the Russian big brother.

By the mid-1980s, however, in order to embarrass the East German regime, its domestic opponents merely needed to invoke Gorbachev, who, ironically enough, became an object of uncritical veneration among East Germans just as Stalin had been before him. The SED regime found itself in the novel position of censoring what appeared legally in Russia, as when a Berlin newspaper called *Die Kirche* was forbidden to republish an article on religion from *Moscow News*. In 1988 the paper was censored fifteen times. The media were prohibited from reporting on deliberations at synods, while Stasi personnel set upon a march by two hundred people led by a pastor who were protesting against government repression.

Despite the presence even at the highest levels of Stasi informers in their ranks, the Churches became a major force in the mounting opposition to the regime. In May 1989 they afforded shelter to groups monitoring the results of local elections, in which negative ballots were mysteriously under-represented. In Prenzlauer Berg, in East Berlin, the regime counted 1,998 negative votes, though in reality the figure was 2,659. When two hundred people demonstrated against this fraud outside the St Sophia Church, they were beaten up by the security service.

East Germany ultimately collapsed because it had no external supporters and faced a West German leader, Helmut Kohl, who was more nimble on the international stage than his provincial background or vast bulk suggested. The Hungarian decision not to maintain border fences with Austria – a decision covertly encouraged by aid to Budapest from the Federal Republic – led to a dramatic exodus of East Germans via that breach in the Iron Curtain. Thousands more claimed asylum within West German embassies in Prague and Warsaw, while hundreds managed to get inside West Germany's diplomatic outpost in East Berlin. When the East German regime organised trains to take these people from Czechoslovakia to West Germany so as formally to expatriate them, those fleeing threw their identity papers in the faces of officials. The police had to be deployed to stop less lucky East Germans from storming railway stations to join this licensed exodus.

The Evangelical Church demanded urgent reforms and the introduction of a multi-party system. The GDR's fortieth-anniversary celebrations turned into a public relations disaster as the Evangelical Church organised prayers for peace and vigils in Berlin, Leipzig and Dresden, which were attended by hundreds of thousands of people. An uncomfortable Gorbachev regarded an even less happy Honecker as if he were the spectre at his own feast. What could the regime do about crowds chanting 'Gorby, Gorby'? The rest of the East German Politburo got the message from the Russian leadership. Although Honecker briefly contemplated Deng Xiaoping's Square of Heavenly Peace option – in which thousands of protesting Chinese were shot down and crushed by tanks – and so took a sudden interest in Sino-German friendship, a fronde among his colleagues forced his resignation and flight to the Soviet Union. His Marxist comrades in Chile granted him asylum.

His successor, the Uriah Heep figure Egon Krenz, met privately with Church leaders, who were then represented at the round-table discussions to decide East Germany's future. For a brief interval the GDR was ruled by Hans Modrow, the last 'reform' Communist seeking to conserve the GDR as an independent entity – the hint of aspic or vinegar being intentional in the choice of verb in that sentence. A coalition government headed by the Christian Democrat Lothar de Maizière was elected, in which there were four Protestant pastors. There were fourteen pastors in the new democratically elected parliament.

Among the new government's first acts was the restoration of Christian holy days, including Christmas and Easter, as well as the release of documents showing Stasi penetration and subversion of the Churches. This dealt the Churches a body-blow at a time when they were temporarily riding high. The reunification of Germany meant the end of the independent 'East German Church', just as it dispelled the illusion of there being a 'third way' between Western liberalism and socialism, an illusion supported by some East German dissidents as well as many West German left-liberals. Reunification also precipitated a flight from the Churches when West German Church taxes – from which people have to opt out explicitly – began to be levied. Many women were also alienated by what appeared to be the West German Churches' role in extending the Federal Republic's stringent abortion laws to eastern *Länder*, where there had been abortion on demand. But this is to slide into the provincialities of German domestic politics. After a brief period of hysterical concern about the recrudescence of a 'Fourth Reich', a united Germany duly succumbed to its greedy ingestion of the Communist East and its own corporate-welfare sclerosis. Its moralising neutralism towards the war in Iraq and solipsistic self-preoccupation also ensured that by 2000 it counted for less than Poland in the esteem of the Anglo-Saxon world. For the first time in thirty years, no one was much interested in anything its left-liberal intelligentsia had to say, with even their hand-wringing ruminations on the Nazis becoming a bore to many sophisticated people elsewhere. For the platitudes about globalisation were to take on an entirely new meaning as an extreme version of religion reappeared as one of the major motive forces in human affairs. This takes us to an autumnal dawn in Manhattan when the world really did change.

CHAPTER 10

Cubes, Domes and Death Cults: Europe after 9/11

The eleventh of September 2001 dawned a bright late-summer day on the US east coast, with blue skies and light glinting off its tall buildings in the cities dotted about the plains. At airports it was the cusp of the day, when night shifts left for home, and those who had arrived for work were organising their papers and thinking about their first cup of Starbucks. Earlier that morning, nineteen men, including fifteen Saudis, rose in nondescript hotels to board four transcontinental flights at Boston, Newark and Washington Dulles airports. They briefly flit across grainy videos recording their pestilence-like passage through terminals.

They had evaded every security system designed to prevent potential weapons coming on board, every method of screening used to identify suspected terrorists. Since a belt clasp or metallic watch routinely triggers detector alarms, readers who find all airports an ordeal like running the gauntlet may find this difficult to fathom. The hijackers were like the bomb-laden terrorist professor in Joseph Conrad's *Secret Agent* who 'passed on unsuspected and deadly, like a pest in the street full of men', the word 'pest' being used in the original sense of a plague, the 'street' in September 2001 being the polished floors of terminals. Each team included a man who had learned to steady a large commercial jet in the air and to alter its course, although landing the plane was not deemed a priority when they had their lessons earlier in the year. The muscular bulk of each of the four teams was needed to intimidate and kill crew and passengers. Who were these people? We can start with their leader.[1]

Born in 1968 in Egypt, Mohammed Atta had a degree in architectural engineering and had worked as a town planner in Cairo. His technical background may or may not be significant, in terms of inculcating a cold, problem-solving mentality, although it could just as well be seen as an inevitably utilitarian attitude towards knowledge common to developing societies where the arts are a luxury. In 1992 Atta moved to Germany, where he studied urban planning at the Technical University in Hamburg, an ugly city ravaged by British bombers and post-war planners. His thesis was on architectural restoration in Aleppo, which may or may not afford insight into his hatred of the anomie and arrogance of New York, the metropolis of the Western world.[2] In 1995 he returned briefly to Cairo, where plans were afoot to prettify part of the old city and to fill it with actors in traditional costume to entertain Western tourists. The rage mounted. Back in Hamburg, his religious opinions became more pronounced, evidenced by a decision to grow a beard, and to communicate with his tight group of friends only in Arabic rather than his excellent German. The university, typical in its mindless multiculturalism, thoughtfully provided a hut for them to reinforce their hatreds of the Western world. Apparently they all stopped laughing in public so as to symbolise their newfound earnestness. Like others in his circle, Atta was pathologically antisemitic, regarding New York as the epicentre of Jewish world power, hatred of the coldly teeming cosmopolitan metropolis being one of many pathologies these men owed to a thoroughly European anti-modernism that would have been modish in France or Germany eighty years earlier. Atta's immediate circle in Hamburg included Ramzi Binalshibh, a Yemeni he had met in a radical mosque, and who moved into an apartment with him, along with another student, Marwan al-Shehi from the United Arab Emirates. A fourth member of the group was the Lebanese Ziad Jarrah, whose relationship with a Turkish girl complicated his dealings with the other three. Extraneous ideological influences included the London-based fanatic Abu Qatada, whose smiling rants were available on video or through the internet as well as through clandestine visits that went unremarked by British security agencies who in the eyes of their European equivalents seemed to be presiding over 'Londonistan'.

No movements or relationships have been better studied, after the fact, than those of the 9/11 suicide-murderers. In 1999 all four members of the Hamburg cell slipped away from Germany to visit Afghanistan;

various Moroccan members of their circle ensured, by paying outstanding bills for them, that nobody noticed their absence. While in Afghanistan, they met Osama bin Laden and those who initiated the coming 'spectacular' attacks against America. They returned to Hamburg in early 2000, shaving off their beards, and appearing to relax back into Western life. They acquired new passports, thereby erasing the visas for their trip to Afghanistan via Pakistan, and began making inquiries about US flight-training schools. Only the Yemeni Binalshibh was refused a US visa, since Yemenis were known to outstay their welcome, although he would be hyperactive as their link with the terrorist mastermind Khalid Sheikh Mohammed, who may have presented the 'planes plan' to bin Laden.

Shortly after arriving in the US, the three men enrolled at two different flight-training schools near Venice, Florida, spending the summer acquiring pilots' licences. While a frustrated Binalshibh wired funds to the three who had successfully entered the US, in Afghanistan bin Laden's attention was drawn to Hani Hanjour, a Saudi jihadist and trained pilot, who was soon despatched to join the others in America. After completing their courses, the four pilots graduated from light aircraft to using simulators to learn to fly large commercial jets, although none of them showed any proficiency in doing so.

Meanwhile, others had recruited thirteen fit men in their twenties, mainly unemployed Saudis, who had been sent to Afghanistan by radical clerics in their native country. During their military training they were personally selected for the 9/11 mission by bin Laden, who prided himself on being able to identify a fellow fanatic in ten minutes. After securing US entry visas in their new clean passports, from April 2001 pairs of these men began arriving in America. By 4 July all nineteen hijackers were in place, using safe houses in Florida and New Jersey. While the muscle men visited gyms, the pilots began making long surveillance flights – usually travelling business or first class – on the types of aircraft they would hijack, scouting out their targets and establishing how easy it was to bring box-cutters on to a plane. In July, Atta flew to Madrid to confer with Binalshibh, who relayed bin Laden's final instructions. In mid-August there was one close call when a replacement hijacker, the Frenchman Zacarias Moussaoui, was arrested for immigration violations after making a spectacle of himself at another flight school. By the third week of August all nineteen hijackers were booked on four transcontinental flights on 11 September, flights which require immense quantities of aviation fuel. The rate of intercepted terrorist telephone traffic intensified. On 9 September, in faraway Afghanistan, two Al Qaeda operatives, one posing as a television cameramen, blew up the Northern Alliance leader Ahmed Shah Massoud, an act designed to propitiate their Taliban hosts, and to allay their fears about probable repercussions after what was about to happen in the US.

On American Airlines Flight 11 from Boston to Los Angeles, four squat Saudi heavies used knives and disabling sprays to terrorise the cabin crew and passengers, who were forced to the rear of the plane. New hands and eyes pondered the banks of illuminated instruments, the screens and dials and the flap and thruster levers that fill a pilot's cabin. Mohammed Atta, who was now flying the aircraft, altered course and then flew southwards in 'an erratic fashion' towards New York. He may have said into the radio, 'We have some planes,' though it took many minutes for anyone to realise these words' full import. Simultaneously, United Airlines Flight 175 left Boston's Logan Airport for Los Angeles. As it climbed to its cruising altitude, the pilot and first officer reported disturbing radio transmissions from Flight 11. Hijackers struck soon afterwards on this aircraft, using sudden violence to commandeer the plane. Both pilots were probably killed. As desperate passengers and personnel used their mobile phones to alert family and friends to their plight, Flight 175 changed course for New York. To the south, at Dulles Airport, American Airlines Flight 77 left at 8.20 bound for Los Angeles. Thirty minutes later hijackers armed with box-cutters took over the aircraft. The plane altered course, as passengers, including Barbara Olson, the wife of the solicitor-general, made desperate calls to relatives. By 9.30 Flight 77 had descended to a much lower altitude and was heading towards the White House. The plane then abruptly banked to alter course for the Pentagon, the world's largest building, housing a department with a budget larger than Russia's GDP.

With the hijackers jabbering prayers in Arabic to stifle their own last-minute panic, for they used religious incantations as a combined stimulant and sedative, American Airlines Flight 11 flew into the North Tower of the World Trade Center in New York at 8.46, killing all on board and an unknown number within the building. United Flight 175 sliced into the South Tower fifteen minutes later, the aircraft almost emerging from the opposite side of the building, and its exploding fuel tanks sending out clouds of smoke from the point of impact. Half an hour later, Flight 77 slammed into the Pentagon at the full-throttle speed of 580 miles per hour, killing all 64 people on board and 125 Defense Department personnel. Defense secretary Donald Rumsfeld felt the shockwave of the impact in his office.

These three attacks, which collapsed the twin towers, resulted in the deaths of three and a half thousand people, although initially it was thought the casualties were double that number. These assaults appalled the world, although there was jubilation in some Palestinian refugee camps and in the Moroccan quarter of Rotterdam.[3]

By this time early-morning America, and then the world in its sequential time zones, was transfixed by what unfolded on television. I remember watching TV from early afternoon, when my sister-in-law called from Yorkshire to alert us to what was happening, and not switching off until thirty-six hours later. Ten days after that I was writing lengthy articles about it. Other horrors unfolded that day in one's peripheral vision, the main event being the heroism of New York's police and fire departments as they struggled to evacuate the World Trade Center before the towers collapsed.

The last aircraft, United Airlines 93 left Newark in New Jersey late, at 8.42, bound for San Francisco. An alert flight controller warned the sixteen flights on his watch, including Flight 93, two minutes before the plane was hijacked. At that moment the plane dropped 700 feet, and the captain began transmitting mayday distress signals. Four hijackers wearing red bandannas, rather than the five deployed elsewhere, overpowered the crew, leaving the passengers free to make mobile phone calls in which some learned of the fate of the other hijacked aircraft. Some of the thirty-seven passengers decided to regain control of the plane, although they knew less than the hijackers about how to fly it. They tried to break through the cockpit door. The hijacker–pilot rolled the plane sideways, and then raised and dipped the nose, to throw the insurgent passengers off balance. Recorders in the cockpit picked up the cries of the passengers outside the door and shouts of 'Allah is the greatest!' from the terrorists. Just after ten o'clock Flight 93 hit a field in Pennsylvania at 580 miles per hour, leaving no survivors as it ploughed its way to a dead end.

We know every detail of how the highest levels of US government reacted to these events. President George W. Bush was in a classroom at the Emma E. Booker Elementary School in Sarasota, Florida when a senior aide informed him that a small plane had hit the World Trade Center. Bush – like anyone watching this on TV – thought it must have been an accident. Maybe the pilot had suffered a heart attack. Just after 9 a.m. another aide whispered to Bush, 'A second plane hit the second tower. America is under attack.' Maintaining his outward composure, Bush remained listening to children read for a few minutes, seated before a blackboard that proclaimed, 'Reading makes a country great,' one of the domestic projects that he and his wife Laura hoped would define his narrowly won presidency. Much has been written about how Bush reacted to events that were designed to defy the imagination. The enormity of the attacks was reflected in his face, which does not conceal intense emotion well, and in the buzzing pagers and mobile phones of the journalists accompanying him. By 9.30, just as he boarded a plane for Washington, he learned of the attack on the Twin Towers. He concluded a call with vice-president Cheney with the words: 'We're at war . . . somebody's going to pay.' Since American Airways Flight 77 was at that point heading towards the White House, where Dick and Lynne Cheney were soon manhandled into a bunker guarded by agents with machine guns, the president's security staff ordered Air Force One to take off without a fixed destination. Laura Bush (or FLOTUS) and daughters, code-named 'Turquoise' and 'Twinkle', were secreted in secure locations in Washington, New Haven and Dallas. Practised emergency managers and counter-terrorism experts switched to their pared-down, time-saving use of language, interspersed with an occasional 'fuck' this or that. Because it seemed like an ongoing attempt to decapitate the US government, Air Force One was diverted to an airbase in Louisiana, where the folksy and far from articulate leader of the free world made a brief televised address, before the Secret Service whisked him off to a base in Nebraska equipped with better communications connecting him to the crisis teams assembling in the capital. Anyone with emotional intelligence, or taste, and whose mind was not corrupted by anti-Americanism, could see the enormity of the burden placed upon Bush, who in that hour had the sympathy of the world.

At his own insistence, by 6.30 that evening Bush was back in the Oval Office, having spent the day making such surreal decisions as authorising US combat jets to shoot down unexplained passenger aircraft. Rounds of meetings ensued as Bush and his team sought to shape a strategy that initially focused on Al Qaeda's presence in Afghanistan. It was going to be bloody. The CIA's counter-terrorism chief, Cofer Black, warned Bush that American agents and servicemen were going to die, a reminder of the nation's traumas about death in foreign parts, from Vietnam to

Somalia. Black chilled the cabinet when he added: 'When we're through with them, they'll have flies walking across their eyeballs.' Thereafter he was known as the 'flies-on-the-eyeballs guy', the man to summon when the mood was vengeful. This was war all right, but against whom? Who were 'they'?[4]

II 'THE KNIGHTS OF DEATH ARE HARD ON YOUR HEELS'

The 11 September attacks were not the first, or the last, terrorist assault nominally directed against Western interests and values, for the victims were not solely diplomats, spies or soldiers, nor always 'Western', but people engaged in such threatening activities as dancing in a tropical discotheque or shopping in Nairobi, Madrilenas reading on a train entering Atocha station, or writers and film-makers. Although the latter are single individuals, the ways in which religious fanaticism reached across legal systems to kill them is indicative of a broader trend as well as of the paramountcy of religion over localised jurisdictions. This is worrying as it also demonstrates that these ulterior loyalties are evident among second- and third-generation immigrants, suggesting a conspicuous failure to integrate them into the host societies. When young Muslims speak of their brothers and fellow countrymen, they sometimes mean not their neighbours in Barcelona or Bradford, but people in Chechnya, Iraq or Palestine. Bizarrely, many of them have managed to combine radical Islam with a street culture that owes more to Los Angeles than to Islamabad, with their booming stereos, hoods, sweatshirts and trainers.

Following a precedent set in 1989, when the Iranian regime issued a religious edict inciting the murder of the British writer Salman Rushdie, the forty-seven-year-old Dutch newspaper columnist and film-maker Theo van Gogh was killed on 2 November 2004 as he bicycled to work along the Linnaeusstraat in a mixed-nationality quarter of Amsterdam. It is a busy street lined with budget stores and with an average quotient of the city's pervasive squalor. Van Gogh was like a cross between the bumptious US film director Michael Moore and the more tough-minded British columnists Rod Liddle or Richard Littlejohn. He was a typical Dutch anticlerical, who had attacked Christianity and Judaism as well as Islam. His murderer, a twenty-six-year-old Dutch-born Moroccan, Mohammad Bouyeri, was a student drop-out who had drifted into minor criminality before joining a terrorist group called the Hofstad Network which planned to blow up Schiphol airport and to kill such figures as the conservative Dutch MP Geert Wilders.

A close friend of the victim talked me through Theo van Gogh's fate at the crime scene. He brought along van Gogh's wicker basket from his bicycle. Bouyeri ambushed the film-maker as he stopped at a zebra crossing in the cycle lane, shooting him once in the side. Van Gogh fell from his bike, but then raised himself from the ground and stumbled across the road. Bouyeri trailed him to a waste bin to which van Gogh clung, where he shot him twice more. He pulled out a butchers' knife to cut off van Gogh's head, settling for plunging a smaller weapon with a letter affixed into his victim's chest. This contained death threats against the Somali-born Dutch liberal MP Ayaan Hirsi Ali and Amsterdam's socialist Jewish mayor Job Cohen. Van Gogh's offence had been to help Ayaan Hirsi Ali make a film exposing Muslim mistreatment of women, an act that, of course, is not an offence in any Western society, although radical Islamists had murdered prominent writers in Egypt in the early 1990s. Bouyeri fled along the Linnaeusstraat to a park where he opened fire on a Dutch policeman and bystander who were both critically injured. Another policeman shot the perpetrator (whose flight was impeded by his traditional dress) in the leg and apprehended him. On Bouyeri's person they found a poem, 'Baptized in Blood':

> I also have a word to the enemy . . .
> You will certainly come off badly . . .
> Even if you go all over the world on Tour . . .
> Death will be on the look-out . . .
> The Knights of Death are hard on your heels . . .
> Who will colour the streets Red . . .
>
> To the hypocrites, I say in conclusion . . .
> Wish for death or else keep your mouth shut and . . . sit

It should be emphasised that until Khomeini announced open season on Salman Rushdie, no religious edict had ever been issued regarding a Muslim living in a non-Muslim country. No religious edict decided the fate of the non-Muslim van Gogh. Bouyeri has said nothing about his murderous actions.[5]

Van Gogh's murder, which has caused a sea change in Holland, up to and including the firebombing of the occasional mosque, was one event in a depressingly lengthy catalogue of atrocities occurring around the world. 9/11 was preceded by the 1993 ambush of US forces in Mogadishu (where imported terrorists brought down special-forces helicopters) and the truck bomb that exploded into the World Trade Center; the 1995–6 attacks on US and Pakistani personnel in Riyadh and Dharhan in Saudi Arabia; the 1998 bombing of US embassies in Kenya and Tanzania which killed hundreds and wounded five thousand Africans; the 2000 holing of the USS *Cole*; and unsuccessful conspiracies to kill former President George H. Bush and simultaneously to destroy airliners over the Pacific. 9/11 was followed by an attack on (mainly) Australian tourists in a night-club in Bali; the killing of two hundred commuters in Madrid's morning rush hour at Atocha station; and two suicide attacks, one horribly effective, the other a failure, on the transport system in London that killed fifty-two people, and then in October 2005 a further attack in Bali that claimed the lives of twenty-five people and further bombings of a Hindu festival in India. November 2005 saw bombs explode in three hotels in Amman, one of which wiped out a Muslim wedding reception. Realistically, in the six to nine months between this book being finished and its publication there will be more.

Over the same period, Israel has been subjected to a murderous campaign of suicide bombings by the Palestinian terrorist organisation Hamas, in which bus drivers have emerged as unexpected heroes of a society under siege. British-born suicide bombers were responsible for one such attack, on Mike's Bar in Tel Aviv, while another Briton – product of a minor private school in Essex and the LSE – killed the *Wall Street Journal* reporter Daniel Pearl for being a Jew. The Iranian-sponsored terrorist organisation Hizbollah regularly fires rockets into Israel too. Every day, Allied coalition forces, and vaster numbers of Iraqis, come under murderous assault from remnants of the Saddam regime and from foreign Islamist fighters drawn to Iraq by anarchy and bloodshed. The bombs get bigger as the addicts of orange light require ever greater explosions. At the time of writing, sixty Iraqis are being killed each day.

The young Jordanian Abu Musab al-Zarqawi eclipsed the elusive Osama bin Laden in terms of his violence and public notoriety, which began when he personally cut off the head of Nicholas Berg for the benefit of a video camera. Al-Zarqawi formed his organisational network in Iran, whose evil regime seems to be inciting events in Iraq (by the delivery of shaped explosives to the insurgents) so as to deflect any attack on their illicit quest for a nuclear capability. No word of condemnation comes from any Arab regime regarding the murder of innocent Iraqis on a daily basis, although Ayman al-Zawahiri bizarrely condemned these attacks from his cave residence in Afghanistan. One suspects resentment that al-Zarqawi was hogging the limelight, for competitive egos were at work as they are among professional gangsters. So-called Muslim community leaders (so-called because they merely represent other groups) in countries like Britain have also failed to condemn these killings of fellow Muslims, while finding every 'contextual' excuse for global Islamic violence.[6]

Leaders of Muslim countries trying to hold the ring against radical Islamism face regular assassination bids, the fate of Anwar Sadat hanging over Egypt's Hosni Mubarak and Pakistan's Pervez Musharraf, who (whatever one thinks of their regimes), are personally very courageous. In Algeria, an estimated 150,000 people have been killed in the dirty war that ensued when the army decided to ignore the Islamic Salvation Front victory in the 1997 parliamentary elections. Radical Islamist violence spread to Bosnia, Chechnya, the former Soviet republics of Central Asia, Indonesia, Malaysia, the Philippines and the Xinjiang province of China. We should not follow these countries' autocratic rulers in eagerly conflating each local terrorist threat with the network responsible for the anti-Western atrocities, to gain either Western assistance in crushing them or some form of aid that has hitherto been outstanding. Other claims should not be taken uncritically either. The terrorists', and their wider penumbra of passive supporters', rhetorical claims that they are moved to kill by the plight of their co-religionists in Bosnia, Chechnya, Iraq or Palestine should be queried more than is the case. There is little to stop, for example, a British-born Muslim from working in a Palestinian hospital or orphanage rather than blowing up a Tel Aviv pub or an underground train in London.[7]

Islamic terrorist atrocities are a fact, and not a figment infiltrated into our anxious imaginations by our rulers, a favourite trope of the superficially clever who regarded the Cold War in similarly domestic instrumental terms. Van Gogh was cut to pieces not by a phantom, but by a real man, who is currently sitting in a Dutch prison. Al-Zarqawi was not some Arab Robin Hood but a psychopathic murderer. These atrocities reflect an incapacity on the part of the perpetrators and their sympathisers to understand us – the Western 'Other' – who are reduced to a few crudely paranoid stereotypes that the otherwise outward 'gazing' postmodern 'Left university' in the West religiously ignores. Our murderers are inspired by hatreds of the occident that owe as much to the history of Western self-repudiation as to resentment or puritanical and politically radicalised versions of Islam.[8]

The inspiration (and the finance) for many of the terrorist atrocities catalogued above came from Osama bin Laden (1957-?), the tall and ascetic seventeenth child of a pious Saudi construction magnate who had a total of fifty-seven children. His father, Muhammed bin Laden, was a penniless immigrant from South Yemen who built up his Bin Laden Group on the back of lucrative construction projects connected with Saudi Arabia's holiest places Mecca and Medina, as well as with remote US bases built to defend the desert kingdom against Saddam. He died in a helicopter crash, the craft that had enabled him to pray at all three Saudi holy places in a single day. Osama bin Laden studied civil engineering and management, clearly to some effect, at the universities of Medina and Jeddah, although both his pious family background and events in the world around him ensured that religion and politics became his overriding obsessions. By all accounts he was an effective project manager, a skill that stood him in good stead in his chosen path. He had the money to indulge his beliefs because between 1970 and 1994 he received US$1 million a year as a legacy from his father. Several factors contributed to his worldview. Some were long range, others proximate. Like many indulged rich kids, bin Laden was susceptible to older gurus, a tendency still reflected in the Egyptian surgeon and Islamic Jihad leader Ayman al-Zawahari, who presumably still hovers behind his shoulder in whatever caves they inhabit if bin Laden is still alive. Who were these influences on the young Saudi?

At Jeddah university, bin Laden was taught by one of the Palestinian co-founders of Hamas, Dr Abdullah Azzam, subsequently blown up in Pakistan, and by the Egyptian Dr Muhammad Qutb, a brother of Sayyid Qutb, the Islamist ideologue who had been hanged in 1966. Azzam advocated implacable confrontation – 'Jihad and the rifle alone: no negotiations, no conferences, and no dialogue' – in order to restore the Caliphate as far as Spain's Andalucía. Azzam would play an important part in luring bin Laden to Afghanistan, where the former established an organisation to funnel through Arab fighters as a sort of Islamic international brigade. Bin Laden is presumed to have killed him subsequently.

Sayyid Qutb was a source of the moralising convolutions that appeal to impressionable minds such as bin Laden's. Qutb had grown up in a middle-class family in a poverty-stricken village in Upper Egypt, about which he wrote a book. Having memorised the Koran by the age of ten, he developed an aversion to Westernised women he had encountered as a student at a secular teacher-training college. His political involvements, while employed as a teacher and civil servant, led the Egyptian government to allow him to leave on an extended pedagogical fact-finding trip to the USA in 1949. An encounter with a drunken woman on board ship clouded his vision of that society even before he arrived there. In a Washington DC hospital the forty-three-year-old virgin, who was ill, was assailed by a lecherous nurse, whose 'thirsty lips . . . bulging breasts . . . smooth legs . . . and provocative laugh' simultaneously attracted and revolted him. In Greeley, originally a utopian settlement in eastern Colorado, Qutb hated the manicured lawns as symptomatic of Western individualism and materialism. He could not find anyone to cut his hair properly, although having a bad-hair day is surely a poor excuse for such fanatical hatreds. Worse, in the church halls and crypts he was horrified to discover Christian clergy aiding and abetting youngsters clasped to each other in the darkness of a sock hop as 'Baby It's Cold Outside' played on the gramophone. Clearly, Qutb had a problem with Western women, a theme common to many Islamic militants and puritans everywhere. In the big cities even the pigeons seemed to live joyless lives amid the promiscuous tumult. When he returned to Egypt in 1951, he saw signs of the same decadence and soullessness all around him, the result of Nasser's attempt to build Egypt on the Western creeds of nationalism and socialism, which had led the country into military defeat and systemic poverty. As with so many Islamist militants, the failure of these imported Western creeds prompted Qutb to intensify his religious convictions. As a member of the Muslim Brotherhood, he was arrested in 1954, tortured, tried and sentenced to twenty-five years' hard labour. In 1958 he witnessed a massacre of mutinying fellow prisoners. A year after his release in 1964, he was rearrested for further plots, and then hanged.

During his imprisonment Qutb wrote two works which were to become highly influential, despite his having no recognised authority as a religious teacher. One of them, *Milestones,* has the status of the *Communist Manifesto* in such circles. Faith was the guarantor of the innermost being of the true believer in a world of inauthentic otherness. The word *jahiliyya,* used by the Prophet to describe pre-Islamic pagan Arabia, and then in the thirteenth century applied to Mongols who adopted Islam but not the sharia law, was revived to describe the benighted chaos of modern unbelievers and those in the Islamic world who were contaminated by them. If the word originally signified naive idolators who worshipped several false gods as well as Allah, it came to mean those who consciously sought to replace Allah with the worship of things.[9] It was the duty of an enlightened and pious vanguard to reverse this, by reviving a lost purity that had existed until the eclipse of the four Rightly Guided Caliphs who followed the Prophet in AD 650. This was essentially an Islamist version of the Marxist–Leninist idea of the revolutionary vanguard whose role was to raise the consciousness of their more benighted potential followers.[10] This was a recipe for a war against virtually anyone who did not meet school inspector Qutb's exacting moral standards. Qutb's harrowing fate, mostly brought down on his own head, provided a stirring story of martyrdom for a noble cause in an Islamic world otherwise dominated by repulsive dynasts, madmen and military dictators whose wealth stemmed from selling oil and the return on their investments in the West.[11]

The tribal rulers of bin Laden's native Saudi Arabia derived their legitimacy from a puritanical Wahhabi strain of Islam, named after the eighteenth-century revivalist figure Muhammad ibn Abd al-Wahhab. The Kingdom's public face is that of a puritanical insistence on Islamic law, including absolute intolerance of other religious traditions, it being impossible to open churches or to practise other religions. Petrodollars have enabled the ruling elite to live a lavish lifestyle, largely thanks to hired Western technicians and armies of Third World helots who are miserably treated by masters with Latin American-style manners.[12] The contrast between public puritanism – with its strict code of sharia – and the hedonistic lifestyle of the Saudi ruling elite in the fleshpots of Addis Ababa, Mayfair or Monaco led to the incident in 1979 when four hundred Islamist militants stormed the Grand Mosque at Mecca and called for the overthrow of the ruling dynasty. French commandos had to be given a dispensation to help ten thousand Saudi troops eject them. Many of bin Laden's messages are exposures of the endemic corruption of the Saudi ruling dynasty, a view few would gainsay.

Two further events in 1979 were of profound significance. The Iranian Revolution that overthrew Shah Reza Pahlavi proved the viability of an Islamic government, which sponsored terrorism in the Lebanon and made martyrdom a central concept, notably in the case of the child soldiers mown down in Iraqi minefields during the long and bloody

Iran–Iraq war. The ayatollah Khomeini was vociferous in his hatred of a West that had sheltered him from persecution, a syndrome shared by many lesser mortals who have passed westwards since then to enjoy hospitality, tolerance and welfare payments. As Shia Iran became the focus of political Islam, the Sunni Saudis sought to boost their rival Islamist credentials, albeit while maintaining their traditional grasp on power. The war that followed the Soviet invasion and occupation of Afghanistan at Christmas 1979 must have seemed heaven-sent in the eyes of the Saudis, who, with US encouragement and support, were able to despatch militants to fight the atheist Marxist enemy.

Between 1979 and 1982, bin Laden was responsible for raising money and providing weapons for the Afghan mujaheddin, before venturing there himself in 1982 to team up with his former university teacher Azzam. He graduated from helping with strategic construction projects to hosting Arab fighters, and then commanding them in military operations, although his talents were as an organiser rather than combatant. He consolidated his local presence through acts of generosity to all and sundry, a 'Robin Hood' tactic he shared with, for example, the cocaine barons of Colombia who effectively provided an alternative welfare state in that similarly benighted country. He organised a global funding network called the Golden Chain to funnel cash and arms to the 'holy warriors' who were recruited through a related Bureau of Services. They were initially trained under the auspices of Pakistan's Inter-Services Intelligence Directorate, which was heavy with Islamic militant sympathisers. Bin Laden and his associates founded Al Qaeda – meaning 'the base' of activists in the vanguard of the struggle – in 1988 to maintain networks established to wage 'jihad', a word signifying not only self-overcoming, but also wars of defence and offence depending on how theologians and others choose to interpret it. Gradually a hard core emerged, numbering in the low hundreds, many of them relatively well-educated Egyptians like Ayman al-Zawahiri, although that brought the complication of wanting to concentrate terrorist attacks on Egypt or Israel, the line subsequently argued by Abu Musab al-Zarqawi, a former Jordanian petty criminal turned terrorist, who until killed by a US-led task force, had emerged as bin Laden's main competitor for the title of super-terrorist.

Relationships were developed through volunteers, with any number of groups engaged in militant struggles. Experts claim that the appropriate analogy for Al Qaeda is of commercial 'franchises' and 'venture capital groups' rather than anything resembling the rigidly hierarchical and villainous SPECTRE organisation in Bond movies. Again, the debt to the West is striking, the combination of borrowings from defunct Marxist–Leninism and ultra-capitalist modernity being extraordinary.

In 1989, following the withdrawal of Soviet forces, bin Laden returned to Saudi where he was invited to redeploy his Afghan veterans in resisting the Marxist regime in South Yemen. When Saddam Hussein invaded Kuwait in 1990, bin Laden offered the Saudi monarchy help should Saddam decide to go for broke by attacking Saudi Arabia. This offer was declined and the king turned to the US for more substantial military assistance. In bin Laden's eyes the presence of such US forces, however discretely corralled, represented the pollution of the most sacred sites in Islam, especially since America's new model army is blind to gender as well as race, a further offence among racist and sexist Arabs, aspects of the culture that we know about but rarely mention. He wrote an open letter chiding the elderly Saudi mufti bin Baz for sanctioning the stationing of such troops, and for welcoming the 1993 Oslo peace accords.[13] Bin Laden had a high regard for the (drug- and drink-sodden) Soviet troops he fought in Afghanistan, in contrast to his low opinion of the fighting capabilities of the clean-living Americans. If God had enabled the faithful to defeat the Soviets in Afghanistan, and to win the Cold War, for which bin Laden took sole credit, why shouldn't Allah direct his wrath towards the godless United States of America and its lesser confederates?

Sensing that his days in Saudi were numbered, bin Laden moved to Sudan, where, since 1990 the religious ideologue Hassan al-Turabi had encouraged him to settle to develop roads and to fight Sudanese Christians in the south. Bin Laden's wealth also enabled the regime to import wheat to feed its starving people. He developed a network of business enterprises based on selling Sudanese commodities to the European Union via Cyprus, and using the profits to sponsor Islamic mercenaries despatched to Bosnia or Chechnya, two conflicts that have been gradually 'Islamised'. It cost about US$1,500 (or £750) to put one such combatant in the field. In 1991 he became stateless when the Saudis removed his passport, although in that sinister world of dark mirrors few of their responses to bin Laden were so unambiguous. The radicalism of bin Laden's vision – and his focus on the US 'Crusader–Zionists' as the ultimate source of Islam's problems – meant that he became a magnet for the deracinated flotsam and jetsam that began to float free of specific conflicts and whose primary loyalty was to the 'emir' or 'sheikh', as he dubbed himself from 1996 onwards. His education, wealth and conspicuous height (six foot five) and white garb contributed to his charisma among simple souls, the most committed handful of whom took Fascist-style 'blood oaths' to obey his orders.[14] On countless videos we can see the sheikh delivering his words of death with a simpering smile in an Arabic monotone, the outward calmness belied by a revealing incident involving the BBC foreign editor John Simpson. In 1989 Simpson was filming a mujaheddin group in Afghanistan. A figure dressed in white appeared, with a Kalashnikov and expensive calfskin boots. He told the mujaheddin to shoot Simpson. One group of fighters regarded this as a dishonourable request and won a straw poll on the issue. Not giving up so easily, Osama bin Laden offered the driver of an ammunition truck US$500 to run Simpson over instead. The driver laughed and drove away. Simpson's crew came upon the Arab in white, lying on a camp bed and crying while he beat the pillow with his fists in infantile frustration.[15]

Al Qaeda was the first truly global terrorist organisation, consisting of alienated and displaced persons whose loyalties were to the organisation and who could emerge anywhere with new identities provided by the Sudanese authorities, who issued them with false passports. Al Qaeda also seemed to transcend the historic division between Sunni and Shia Muslims. Its operatives learned about bomb-making from Hezbollah in the Lebanon, and had extensive contacts with Iranian secret agents. They also had more fitful contacts with the intelligence services of Saddam Hussein, whose relationship with Islamic militants was characteristically opportunistic as this national socialist evolved into a would-be Saladin.

One or two words of caution are necessary to avoid giving Al Qaeda more importance or coherence than it possesses. Like the Securities and Exchange personnel who shaped war-crimes charges before Nuremberg in 1945–6, intelligence services and policemen who are engaged in hunting Al Qaeda operatives are primarily interested in organisational relationships to help prove the charge of conspiracy underlying assassinations and mass murder, not to speak of more shadowy linkages with rogue regimes which provide a more calculable target for the West's enormous military capabilities. Several informed journalistic investigators of Al Qaeda are sceptical whether the organisation is as effective as this suggests, pointing out its habit of claiming responsibility for atrocities with which it has no real connection as a means of magnifying its own global scope and magnitude.

Most of Al Qaeda's estimated US$30 million annual operating budget came from donations to dubious Islamic charities from oil profits and investments in the West racked up by rich Saudis. However, terrorist operations around the Middle East (in particular an attempt in 1995 to kill Hosni Mubarak) led to pressure from Egypt and Saudi Arabia on Sudan to expel bin Laden, although apparently the Saudis also tried to assassinate him. Fearing that the Sudanese might betray him to the highest bidder, in 1996 he returned to Afghanistan, or more specifically to the Pakistani-sponsored religious movement called the Taliban (students) who were fighting for control of that country. They routinely increased their limited local support by paying extremely poor peasants US$300 to join them in further wrecking that war-ravaged country. Wherever they imposed their rule, 'idols' were destroyed – notably the Bamiyan Buddhas in 2001 – and those guilty of such crimes as adultery were shot dead in the middle of Kabul's only football pitch with a bang and a collapsing burqa. Despite their ostentatious puritanism, the Taliban had amicable relations with the disintegrated nation's opium growers and drug traffickers. Any faint reservations their leaders may have had about the tall Arab were allayed by annual payments of up to US$20 million for their hospitality. The Saudis may have restricted bin Laden's access to his family's wealth, while the Sudanese looted his modest local assets, but the Golden Chain ensured that Saudi money soon flowed to him as he oscillated between Pakistan and Afghanistan. There he organised a network of training camps for international terrorists, including the Jordanian al-Zarqawi, who would carry out attacks either of his devising or proffered to him by other Islamic radicals who lacked means of their own. The camps grew more sophisticated in scale, including huge tunnels cut into mountains like the complex at Tora Bora.

The devastating bombings in Kenya and Tanzania in 1998 were the first obvious fruits of bin Laden's strategy of internationalising Islamist terrorism. Although US president Bill Clinton was acutely aware of the threat from this remote quarter, legal restrictions and a collective post-Vietnam fear of ground operations going wrong in inhospitable places meant that the US response was confined to firing (quite modestly laden) million-dollar Cruise missiles at camps and houses the terrorists had long vacated. No technology could bridge the time that elapsed between intelligence, which had to be evaluated, and firing a submarine-based missile over airspace that was nervy because of tensions between

India and Pakistan. Each possible response was also 'lawyered to death' by those anxious about collateral casualties. Advance warnings to Pakistani intelligence inevitably became warnings to bin Laden to move on in his little convoy of SUVs.

Beginning in late 1998 and extending into late 1999, bin Laden and his most senior associates turned their minds to what became known as the 'planes operation'. This was payback time for the Christian Crusader footsoldiers of Zionism, for bin Laden detects the Jews behind everything. It involved co-ordinated 'spectacular' attacks in which hijacked commercial airliners would be crashed into symbolic targets. These were carefully chosen. The twin towers, and by extension the US itself, were described by bin Laden in October 2001 as 'the Hubal of the age', a reference to the large stone moon god, one of the 360 idols worshipped by the Arabs in the period between Abraham and the coming of the Prophet. The wily rulers of Mecca had refused to destroy the idol lest this diminish pilgrim traffic. Nor would the holy hypocrites of Medina help the Prophet destroy the idol. Despite them, the Prophet returned to Mecca, defeated the pagans and tore down Hubal. This was a case of propaganda by the deed, in which by one imagination-defying act they gave the world a graphic demonstration of their total faith in God, merely passing through glass, flames and steel to the musky scents of paradise while leaving the smell of death lingering behind them. As Mohammad Bouyeri's poem had it: 'Accept the deal . . . And Allah will not stand in your way . . . He will give you the Garden in place of the earthly ruin.' In his various reflections on the event, bin Laden himself seems mesmerised by the financial loss 9/11 occasioned, totting up the cost of the 'successful and blessed attacks' in terms of lost business, employment and reconstruction to arrive at the figure of US$1 trillion, a big return on his US$500,000 investment. The loss of three thousand lives warranted no mention, for there was no defining line between taxpayer civilians and US soldiers.[16]

III US, THEM AND 'EURABIA'

Western culture is infinitely rich in resources for making sense of these murderous assaults on its values, although politicians, with horizons confined to the present, rarely avail themselves of them or shy away from giving offence. Orvieto cathedral in Umbria contains a masterly fresco cycle painted around 1500 by Luca Signorelli, an artist from nearby Cortona in Tuscany. One bay shows the Antichrist preaching with Satan whispering in his ear, a relatively rare theme in Western art. Although the allusion seems provocative in connection with contemporary Islamist terrorists, especially to anyone sceptical of George Bush's use of the (Islamic) term 'evil-doers' to describe them, in fact the simulacrum-like face of the Antichrist is useful for our purposes, and not just because the simpering smile resembles that of bin Laden. Not only does it force us to think about evil as an endemic presence in human affairs, but in this case the not-quite-right features of the Antichrist remind us that Islamist terrorism is in some respects like a cover version of ideas and movements that have occurred in modern Western societies, as well as a radicalised caricature of what over a billion Muslims, who simply wish to live out their lives in their full human complexity, believe. That is why the more intelligent commentators on 9/11 consulted their Joseph Conrad and Fyodor Dostoevsky novels for people similarly intoxicated by orange explosions and livid bloodshed. There were no answers to be had in the Greenwich plastic dome or the gleaming Parisian Grande Arche de la Défense.

The 'clash of civilisations' inaugurated by Islamist terrorists (for bin Laden uses that concept) has provoked a Western crisis of identity although most 'civilised' people are as affronted when someone is blown up in Nairobi as they are when this occurs in Madrid or London. Although some would like to see Europe as a utopian oasis insulated from a US hypostatised as 'Texas', in fact this is impossible, as soaring petrol prices and the flow of manufactured goods from China or India readily indicate. Certain European governments, above all the Zapatero socialist regime in Spain, and the always disappointing left in Germany, are under the illusion that they can appease militant Islam through an 'alliance of civilisations', or by willing a fusion culture based on the Mediterranean that will distinguish them in the eyes of Islam from their Atlanticist fellows in northern Europe. Of course, if the Spanish people think they have more in common with Libya or Tunisia than with England, Holland or Germany that is entirely their own affair, but one doubts whether they or the Italians share this liberal elite view.[17] Islamist terrorists are also linked to Western culture without evidently under-standing it, beyond its technological marvels or what they (and many Western commentators) regard as liberal decadence. All of the terrorist atrocities chronicled above were entirely reliant on sophisticated Western technology as well as box-cutters. These included laptop computers, mobile encryption phones, global positioning equipment, car and truck hire, credit cards, fast-flow international bank transfers, an Islamist press, video cameras, DVDs and satellites to transmit ideas and images, aircraft and sophisticated bombs, to say nothing of the repeated attempts to acquire bacteriological, chemical or nuclear weapons to cause a major catastrophe. Even the wealth that enables terrorist attacks is based on revenue that Saudi Arabia acquires from selling oil at vast profit to the insatiable West and other major markets. Much of this revenue is in turn squandered, not simply in supporting the ruling elite's extravagant lifestyle or buying weapons for armies whose record is dismal, but in converting the public face of the urban Arab world into a travesty of the travesties that already characterise Western modernism. Fake charities and phoney NGOs ensure that some of the surplus revenue flows not only to madrassas and mosques but to terrorist organisations.

The internet has become the broadband river whereby noxious ideologies (and the practicalities of terrorism) can be accessed in the privacy of bedroom or study in provincial towns and major cities of the West by young people, of whom significant numbers applaud the actions of Al Qaeda and other Islamist terrorists. It is curious that while security services can monitor internet paedophiles, they seem to be unlucky with people planning bomb attacks. The effect of attacks like 9/11 was primarily evaluated by the chief perpetrator in terms of its impact on Wall Street's financial markets, although as an engineer bin Laden also savoured the physical impact on the structures his planes collapsed. The fact that his outlay of US$500,000 caused what he estimated to be US$1 trillion worth of damage seems to have particularly excited his imagination.

Anyone who has visited a National Health hospital in Britain will have been struck by the dedication of the large number of migrants who work in them for a combination of long hours and low pay. The rapid ageing of Europe's population and its looming pensions crisis seem to many observers to mean that it has few alternatives to welcoming a much younger, and largely male, migrant workforce, which has added twenty million legal Muslim entrants throughout the continent since 1970. The illegal numbers would push that much higher. They are the ghost army of night cleaners in urban offices, or of those who toil in the hothouses of Andalucía for minimal wages. No one should mistake the human suffering involved, as wars intensify poverty and a tide of desperate humanity heads Europe's way. The distressing scenes at Sangat near the Calais channel-tunnel entrance, or amid the barbed-wire fences of Spain's Moroccan enclaves at Ceuta and Melilla, or on the heavily patrolled borders of the Ukraine indicate the scale of this human tide surging up from Africa and Arabia for the privilege of living in some rat-infested and rackety slum in Paris or toiling in the 100-degree heat of an Andalucian hothouse. Conditions in the banlieus of Paris have reached boiling point. One unfortunate by-product of this may be to surrender supervision of these troubled housing estates to self-appointed 'uncles' and Islamic clergy, probably supported by a form of militia, who will further detach these areas from modern French life, in pursuit of Islam's goal of creating extra-territorial moral and legal enclaves where the writ of the Western secular state no longer runs. Few intelligent observers were consoled when French imams decreed a 'fatwa' against young rioters, since it is not they who make laws in France.[18] If the stimulus this disorder has occasionally given to far-right parties, and to mavericks such as the flamboyant entrepreneur homosexualist Pim Fortuyn in Holland, is one over-studied phenomenon, an equally disturbing trend is the way in which Muslims have influenced those who represent them in Europe's parliaments. The real test of being British is not who one supports in cricket, but whether one accepts that Britain has autonomous national interests which are not subject to the veto of this or that minority. The price of domestic harmony, and of a seat representing part of Birmingham or Walsall, seems to be the sacrifice of Europe's staunchest ally. Like Britain's bumptious George Galloway, one can also get elected to parliament by playing to the Muslim gallery, punctuating one's discourse with crazed anti-American slanders and the occasional 'Shalom aleichem'.[19]

It would be wrong to imagine that the West solely imports, rather than exports, its present difficulties. It has been incautious about what face it blithely exports, and not only because its cultural diplomacy has collapsed since the days of *Encounter* during the Cold War.[20] Mass tourism has become the means whereby affluent Westerners, who are ignorantly indifferent to local sensibilities, have established outposts of their way of life on the coastal fringes of more traditional cultures. Instead of getting blind drunk in Birmingham, Benidorm or Bremen they do it in Eilat, Marrakech or the Maldive Islands. Moderate Muslims say this is no bad thing and that the natives gradually get used to it, but then they are part of a privileged elite that does not have to encounter such horrors on a daily basis. Satellite television enables people in the remotest societies to access such ghastlinesses as MTV where even wild animals are not safe from being stuffed down teenage trousers, in the mindless antics of those American teenagers whom Michael "Halloween" Myers has not yet murdered. Joking apart, international corporations, whose lack of local legal anchorage and arrogance appals as many on the right as on the left, leave their sordid traces on virtually every society on the planet, notwithstanding their commercials extolling cultural sensitivity. There is something wrong about the Gadarene rush of US companies and armies of private security contractors into the Iraqi war zone where robotic-seeming US troops already look, and often sound, like something that has strayed from a *Terminator* movie.[21]

Islamic terrorism also draws on a tradition of 'occidentalist' hatreds that are partly reliant upon the West's own tradition of repudiating modernism. The ideological indebtedness of Islamist terrorists to the West extends beyond the 'vanguardism' that links Al Qaeda and other groups to the Marxist–Leninist tradition of militant elites making up for persistent failures of popular consciousness. If some call these terrorists Islamo-Bolsheviks, in recognition of similarities not with only the Bolsheviks, but also with the Russian nihilist precursors who so appalled Dostoevsky, others have also semi-hit the mark by describing them as 'Islamo-Fascists', a term that resonates in the left-wing imagination and is employed by President Bush too. Now although almost nothing connects Al Qaeda with a movement like Nazism – apart from an antisemitic mania that sometimes draws upon such historical resources as the *Protocols of the Elders of Zion* – militant Islam shares something of the cultural pessimism of nineteenth-century *Western* critics of mass urban industrial society. It is not difficult to alight upon any number of Western (as well as Chinese or Japanese) critics of the comfortable soullessness of major cities, where innocent country folk were thrown into a witches' whirligig of deceit, deracination, vice and the richness of life reduced to sordid commerce. Everything and everyone had a price, in the great whore city filled with whores of both sexes. Only one type of being seemed entirely at ease in this environment: the Jews, who it was claimed, were at the cold dark heart of these unnatural arrangements, a theme that resonated in Europe's most profoundly provincial nation, namely Germany, where being provincial is celebrated as a virtue, and managers of hedge funds are compared with 'locusts' by politicians who ought to know better. Bin Laden's writings and messages are saturated with antisemitism, as well as with such bizarre claims as that AIDS originated in the US rather than sub-Saharan Africa, his claim that the US exports AIDS being undermined by the fact that only 0.3 to 0.6 per cent of the population there are carrying it.[22]

As in the case of earlier crises, whether the fin-de-siècle Dreyfus Affair which influenced Emil Durkheim to write about the socially constitutive role of religion, or the domestic repercussions of the Vietnam War which led the US sociologist Robert Bellah to write about civil religion, a clear and present external and internal threat has triggered waves of introspection on the subject of what we in the West stand for. Newspapers in Britain, to take one example, are filled with 'the fundamentals of law in this country' or 'what it means to be British', while traditional patriotic history books and a *Little Book of Patriotism* have become bestsellers.[23] These debates, which are replicated elsewhere in Europe, have overlapped with parallel attempts to build a supranational European 'moral' identity in contrast to the allegedly 'alien' identity of the USA. Clever noise about Martians and Venusians has been two-way, and it is assuming dangerous proportions since several vested interests are keen to open rifts within what have been effective alliances.

One fantasy that tantalises some Europeans is that of a decent, humane polity whose 'soft' moral power would rival the 'hard' power deployed around the world by the faltering Colossus across the Atlantic. Europe would be based on the explicit rejection of such practices as the death penalty, on subscription to multilateral institutions like the UN, and on high-maintenance social policies that are partly possible due to the US taxpayers' generous underwriting of Europe's ultimate security through NATO.[24]

The role of religion is crucial to these widening divisions, although many commentators have remarked that, while Americans may exaggerate their religiosity when questioned by telephone pollsters, Europeans may correspondingly over-egg their secularity in similar surveys since they imagine this response is fashionable. Moreover, it worth stressing that in a global perspective it is northern and central Europe that is 'exceptional' in this regard rather than America.[25] My own entirely subjective impressions of religion, gathered in a few years' experience of varied regions of the US, is that it adds a surprise dimension to people that is increasingly not met with in western Europe, that it provides a warm hearth for people in a vast and highly mobile society that can be cold beneath the superficial amiability, and that devout 'black- or yellow-necks' are just as evident as the Bible-bashing 'rednecks' of European legend.

In contrast to the US where, despite a formal separation of Church and state designed to preclude 'Establishment', religion has a significant impact on politics, many Europeans are determined to write Christianity out of the picture. They include British leftists, despite Evangelical Christianity being integral to British socialism, and aggressive secularists, in Belgium, France or Spain, who patrol battle lines established a century earlier over such issues as education. Religion in these circles signifies the Basques, Belfast, Bosnia and Bush, at any rate something horrid, like the 'national Catholicism' of Franco. Actually, the Democrat presidents John F. Kennedy, Jimmy Carter and Bill Clinton were not slow to invoke the Almighty, with genuine conviction in the case of Carter. The ultra-conservative and much maligned Ronald Reagan was not much of a churchgoer, even if rhetorically he appointed Him an honorary member of his cabinet.[26] The US religious right did not emerge from nowhere, and nor did it do so without liberal provocation. It is important to recall that the politicisation of conservative American religion began with the 1963 Supreme Court ban on prayers in public schools, and gained momentum through Roe v. Wade a decade later, the Supreme Court decision which struck down state laws against abortion, and that conservatives are as widely represented among America's largest de-nomination, Roman Catholics, as among the Evangelical Protestants who seem to attract the most media attention.

In similar fashion, the US right established an impressive array of think-tanks and autonomous centres, largely because they felt, with reason, that their views were excluded from the 'Left university' and much of the media, a process extended through maverick 'bloggers' seeking to balance liberal bias in America's established networks and newspapers.[27]

There are other cultural differences. Although the European media chooses to ignore it, the US has an extraordinary range of religious public intellectuals, such as William Buckley Jnr, Stephen Carter, Richard Neuhaus and George Weigel. By contrast, although Europe has such outstanding figures as Leszek Kołakowski, Hans Maier and Josef Ratzinger, its public culture is dominated by sneering secularists, who set the tone for the rest of the population and can make light work of the average bishop rolled out to confound them, especially in the case of Anglican bishops who share so much liberal common ground. Much of the European liberal elite regard religious people as if they come from Mars, except when they operate within such licensed liberal parameters as the Campaign for Nuclear Disarmament, the struggle against apartheid or the US civil rights movement in which Christians, notably Dr Martin Luther King, played a distinguished role.[28] 'Britain', we are loftily told, 'has not, since the 16th century, been ruled by bishops or mullahs and has been the better for it.' In fact, 'mullahs' have never ruled Britain except in that columnist's imagination. The last truly politically significant English cleric was the seventeenth-century archbishop William Laud. This line neglects the contribution that clergy have made to the public affairs of Britain, and, more worryingly, the fact that it is not only 'fundamentalist lobbies' who 'curse' modern politics, but professional lobbies representing animal rights, gays or the planet (causes that inspire sectarian devotions to the fox or Gaia, so to speak) which could equally be deemed a mixed blessing were it not politically suicidal to say so. Idiot and ignorant actors and playwrights are integral to all these causes.[29] Although many European politicians are highly religious, including the leaders of Britain's major political parties, notably Tony Blair, and many parliamentarians, it was thought expedient to let it be known that Downing Street does 'not do God' lest secularists make hay with it as they did when Blair announced that he felt accountable to God.

One European politician who did not dissemble his conservative religious convictions, the distinguished Italian philosopher Rocco Buttiglione, was the subject of a gay cum secularist media witch-hunt which refused to acknowledge that as EU justice commissioner he would be as capable of separating his private beliefs from his official brief as he had been in every earlier appointment. That the political thugs and gangsters of ETA, IRA–Sinn Fein and various neo-Fascists are represented in the European parliament is apparently deemed less shocking than the appointment of a single Catholic professor, but then the media seems to be fascinated by its brushes with people who commit shocking violence while smirking at them.[30]

Instead of religion, the liberal elites prefer their monopolistic mantra of 'diversity', 'human rights' and 'tolerance' as if they invented them, unaware of the extent to which these are products of a deeper Christian culture based on ideas and structures that are so deeply entrenched that most of us are hardly aware of them. As the great contemporary French philosopher Marcel Gauchet has written: 'Modern society is not a society without religion, but one whose major articulations were formed by metabolising the religious function.'[31]

That truth was suppressed in the draft 2004 European Union Constitution, which Dutch and French voters have since pushed into limbo. This document grandly traced Europe's ethos and telos from Thucydides to the Enlightenment. Vociferous objections from Italy, Poland, Spain and pope John Paul II forced the drafters to concede the scantest reference to the continent's fifteen centuries of Christianity. Among the most vociferous was Aleksander Kwásniewski, the atheist president of Poland, who said: 'There is no excuse for making references to ancient Greece and Rome, and to the Enlightenment, without making reference to the Christian values which are so important to the development of Europe.'[32] Academic postmodernists would have had reasons to object too since they generally regard Enlightenment rationality as a mixed blessing. But then they really don't add up to a hill of beans in the scheme of things.

Liberal and secular politicians, many with the lawyers' limited historical consciousness, decided to omit a religion that made a major contribution to the dignity and sacred identity of autonomous individuals regardless of their ethnic origins, as the greatness of one God paradoxically lessened human dependence. Its transcendental focus has set bounds to what the powerful could not or, more importantly, *should not* do by providing moral exemplars of good kingship and evil tyranny. The eleventh-century investiture contest between emperor Henry IV and pope Gregory VII contributed to the evolution of a separate civil society beyond the ambit of the state. In the seventeenth century, Jesuit theologians developed theories of resistance to tyrants, up to and including justifications for tyrannicide in times when this was not academic. In the absence of other forms of welfare, Christianity has provided charity to the needy for several centuries, a constant from St Francis of Assisi to the Salvation Army and the Samaritans. As the British socialist politician, Roy Hattersley, pointedly asked, when have committed rationalists ever operated soup-kitchens, hotlines for the suicidal or hostels for crack addicts? Europe also consists of what one might call 'cultural Christians' – a term more routinely used by 'cultural Jews' who have abandoned their religious faith. Entire swathes of European art, literature and music, including Raphael and Rubens, Bach and Handel, Messiaen and Roualt, are incomprehensible without knowledge of Christian sacred culture.

Attempts to airbrush Christianity out of the historical record are as intellectually dishonest as Stalin's photographic conversion of those he had murdered into a bush, lake or jacket.

The unwanted intrusion of Islamic terrorism into Europe's major cities, already depressingly accustomed to the mindless murderousness of Basques and Irish, and the dawning realisation that among Islamic minorities there are those whose primary allegiance is to a foreign religion or, worse, to international terrorists, and who have conceived a murderous alienation from their parents' or grandparents' adoptive homeland, has had immediate consequences beyond elaborating more anti-terrorist legislation. The Germans have introduced citizenship tests whose questions – opponents argue – are tougher than those posed to contestants on the local equivalent of 'Who wants to be a millionaire?' In the Netherlands, where in the wake of the slaughtering of Theodor van Gogh the shock has been greatest, the immigration authorities have produced a video to convey to immigrants the quintessence of 'Dutchness'. This consists of snippets from the life of William of Orange, tulips and windmills, naked sunbathers and a gay wedding. The Netherlands' security minister, the former prison official Rita Verdonk, is currently trying to outlaw the wearing of the burqa, a 'crime' already subject to a £100 fine in parts of Belgium. The imagination reels as it tries to conjure up what an equivalent British video might offer: Elizabeth I, roses and castles, the generous frontage of the celebrity Jordan, and nightly scenes of drunken anarchy on the streets of Cardiff or Nottingham that would embarrass Bosch or Breughel.[33]

Inevitably, perhaps, the allegiance militant religious minorities display towards their religion has led to questioning of both liberalism and the theology of multiculturalism, public doctrines that have come into conflict with one another. Liberal notions of equal human rights have collided with the lesser rights that some minorities accord to women, not to speak of their absolute denial of rights to gay people. Liberals have failed to persuade vociferous religious minorities that their own culture of universal human rights is not a recipe for decadence, or something that these minorities take a cynically instrumental view of. Interestingly, the first words uttered by a man captured after an attempted suicide bombing in London, as he appeared naked with his hands up on the balcony of a block of flats, were 'I know my rights,' a thought eagerly greeted by armies of British human rights lawyers whose blinkered and self-righteous indifference to the primary right of people not to be blown up is truly a sign of decadence. Unfortunately, the lawyers over-generously represented in our legislatures simply parrot this way of thinking, a development that may contribute to mass alienation from our political system.

Nor did liberal multiculturalists, who imagined a Herderian riot of diverse flowers living gaily in a huge garden, take adequate notice of the fact that one aggressive minority would seek to create cultural no-go areas, in which mosques and madrassas in what amount to ghettos of the mind, would be followed by calls for a separate Islamic banking system, sharia law or, more outrageously, an Islamic parliament. The spectacle of Yorkshire as a northern outpost of 'Eurabia' is not confined to the imagination of Harvard and Stanford historian Niall Ferguson. Western societies have tolerated various devils and pests in their midst, providing them with 'people carriers' as part of their welfare package, despite their vocal calls for our destruction. Belatedly but rightly most European governments are throwing such individuals out, but they will have to be much more vigilant about whom they let in, ensuring, for example, that imams speak European languages, and are educated in Western values vis-à-vis homosexuals or women.

It might reasonably be objected that Western societies have long made accommodations with, for example, Orthodox Jews, who do not want to work on their sabbath, or Sikhs, who wish not to discommode their hair by wearing motorcycle crash helmets. London's Soho Chinese community have their own street signs and telephone kiosks adorned with Mandarin, to the delight of those who visit there. But that is not part of a campaign of territorial exclusion or self-assertion. The demand for banks that refuse to charge interest represents an extreme version of Islam, which sits ill with Egyptian banks that routinely charge moderate rates of interest or indeed with learned theological opinion in that highly Islamic country. In other words, the call for sharia-conforming banks is part of a strategy for expanding Islam's space within the host country in conscious rejection of any notion of accommodation, as is intimidation against churches marooned in Islamic-dominated areas.[34] Similarly, the wearing of head-scarves in schools has become an act of provocation, exploited by the militants who encourage schoolgirls in this choice of fashion.

In addition to emphasising 'rights', multiculturalism asserts specific group grievances, whether they concern the Irish potato famine, colonialism, slavery or the Jewish Holocaust, to take the better-known examples. Few dare to highlight the fact that, for example, in the 1840s the British government maintained the navy, collected duties on barrels of brandy, but entirely lacked the administrative apparatus to do anything about a large-scale famine. What has a twenty-year-old Spanish person got to do with the deaths of native Amerindians in the sixteenth century for which his or her government has apologised? Why are countries that fought Nazism for five years being made to feel guilty about the fate of the Jews for which Germany and those who collaborated with it were solely responsible? Encouragingly, the Roman Catholic Church is beginning to baulk at demands for apologies for the Crusades – which were a Christian response to Islamic aggression.

In addition to being often unsympathetic to the victimhood of other victims (as the British Council of Muslims recently demonstrated in its churlish avoidance of Holocaust Day and which some US Jews displayed in their equally churlish response to gypsy victims of Nazi persecution who were excluded from the Holocaust Memorial), such aggrieved groups imagine that a self-identifying cause construed as moral entitlement trumps any collective obligations, largely because, in this case and others in the West, the dominant majority tends to have a history as victor rather than victim and hence cannot mount the same emotionally based assault on popular consciousness. The effect is rather like arguing the moral case of the war in Iraq with the mother of a deceased soldier, a lost cause since motherhood and victimhood combined are mythically powerful. While minorities send out a clear and indignant moral message, the majority is thoroughly confused by its residual Christianity, a liberalism that simultaneously transmits 'tough' as well as 'soft' signals, and a public culture where the 'freedoms' achieved in the 1960s have degenerated into addictions and obsessions suggestive of dependency, not least freakish obsessions with deviant sex, food, housing or life lived through material acquisition. It is a grim spectacle.[35]

Surveying the early-twentieth-century European cultural landscape, there are signs everywhere that the creeds that became hegemonic in the sixty years after the Second World War are in desuetude. The end of the Cold War and the emergence of an Islamist terrorist threat have opened up chasms within the Atlantic community that saw off Hitler and the successors of Josef Stalin. Support for the US-led coalition in Iraq can determine (as it has already done) the fate of European governments, as witnessed by the fall of Spanish president José María Aznar, a man of courage and dignity, and the longevity (until 2005) of chancellor Gerhard Schröder, as the Spanish and German left played their anti-

American cards in a climate rendered almost insane by the re-election of George W. Bush. The complication that Israel seems to represent to harmonious relations between Islam and a Western world of which Israel seems an increasingly tenuous part has resulted in antisemitism – if that is what criticism of Israel is seen to be – becoming as characteristic of the European left as it once was of the right, although some would argue that that poison has lain within 'anti-Zionism' all along. Bitter quarrels have erupted among American Jewish intellectuals, because the allegedly antisemitic view that Israel is complicating US foreign policy is as rife among their gentile colleagues in the US as it is in a Europe which some American Jews hysterically claim is synonymous with that malady. But there are less parochial concerns than what animates New York intellectuals, whether academic or public, whose prodigious wordage in the *New York Review of Books* or *New Republic* passes most people by.

Terrorism has further major consequences. The shadow of Islam has made the expansion of Europe to include Turkey, a secular state with a Muslim majority, an urgent desideratum, further evidence that the tight Franco-German axis around which 'Old Europe' operated is in terminal decline. Turkey has been a reliable member of NATO for decades, and unlike Belgium or Holland is one of the few countries in modern Europe capable of fighting a war. Potential accession states will soon number Morocco, which is already in preliminary negotiations to accede, and in all likelihood Algeria, which from the 1840s was an integral part of France. But there are also important changes in the realm of public ideology that would have seemed unthinkable some years ago when the old anti-Fascist slur of 'racism' routinely silenced all debate.

When Trevor Phillips, the black British and Labour-supporting chairman of the British Commission for Racial Equality pronounced that multiculturalism is inherently divisive, people became alert, even if others have been saying the same thing for several decades. Evidently, the messenger was more important than the message, even if the message has yet to percolate down to the politically correct denizens of local government, the lowest rung – in every sense – of modern government, who in Britain occupy their time (and spend other people's money) with trying to convert Christmas into 'winter lights' or some other silliness that neither Hindus, Jews, Muslims nor Sikhs actually want.[36] The chief rabbi, Sir Jonathan Sacks, the most impressive religious leader in the Kingdom, has similarly argued that Britain needs to develop the ethos of a country house to replace the anomie of a commercial travellers' cheap motel, a metaphor intended to stimulate greater awareness of what all citizens have in common. Elsewhere, changes in climate are coming thick and furious. A certain liberal-left desperation is evident when it is claimed that a suddenly deliberalised Holland was probably never very liberal at all, an assertion difficult to square with the hashish cafés and the canal-side window displays of prostitutes. The war and insurgency in Iraq (and the 'war on terror') have sent shockwaves through liberal ranks, causing bitter divisions between so-called 'tough' liberals like Michael Ignatieff and Christopher Hitchens and those apparently less concerned with whether Iraqis and Afghans should enjoy the same rights as themselves. The erstwhile left is bitterly divided over such issues as torture. News of this novel trend has yet to reach celebrity actors, film-makers and playwrights, who are stuck in the infantile Noam Chomsky cum Harold Pinter view of the world that, in the latter case, has been endorsed with the Nobel Prize. Nor has it penetrated the walls of the 'Left university', which will probably be the last redoubt of Western multiculturalism, just as it has long been the sunset home of Marxism and its derivatives.[37] The voices of militant rationalism and scientistic stridency have become shriller, with Darwinism's high prophet, the zoologist Richard Dawkins, behaving like the hotter sort of seventeenth-century English Protestant in his zeal to mock the faith of people who believe in miracles. In hospitals and research laboratories some scientists push hard against the boundaries of what many lay people regard as decent or seemly. Even in Britain – where the Churches are otherwise preternaturally obsessed with homosexual clergy – when both the Anglican and Roman Catholic archbishops expressed unease about the easy availability of abortion, people listened.[38] Some still gleefully anticipate the onset of the totally secular society, a viewpoint I recently heard confidently expressed by the Spanish minister of religion, despite contrary evidence for the enormous vitality of religions in the US and around much of the rest of the world. Islam is resurgent, but so too is Hinduism in India, with even the Chinese Communist authorities forced to take religion seriously, which means persecuting the Catholic Church and adherents of Falang Gong. Thanks to the John Paul II generation, Christianity is vital in much of formerly Communist eastern Europe, including Russia, where Orthodoxy is experiencing a renaissance that has even spread to western Europe. The young Poles and Ukrainians who come to work in southern and western Europe are conspicuous by a dignified religious faith that makes it easy for them to integrate. So too are many migrants from

Africa, whose vibrant Churches in poorer parts of big European cities are filled with ladies in vivid hats and men in suits and ties, not to speak of the flow of migrants into Spain from Central and Latin America. Although opinion surveys routinely announce the demise of Christianity, there seem to be plenty of Christians in the upper reaches of banks, broadcasting industry and newspapers, while Tony Blair's cabinet includes at least one member of Opus Dei as well as several practising Anglicans.

The Christian Churches of Europe present a confused picture in the early twenty-first century. Some things are obvious enough, but they may require brief recapitulation. How long can they continue to regard Churches in the New World, let alone the Third World, as appendages to an old continent where Christianity appears to some to be in decline? It is striking, to say the least, how few European Christians are prepared to speak publicly about their co-religionists in the US, whom liberal opinion routinely caricatures and blames for America's venturesome foreign policy. I have yet to hear a single European cleric ask whether in fact this is true. The Roman Catholic Church seems unsure whether to make a virtue of its authority and traditional structures – which may consolidate its core believers – or to make judicious compromises with a society that it finds alien, hostile and vapid. Under the present pontiff, Benedict XVI, the goal seems to be to fall back on the commitment of increasingly isolated groups. However, the Churches are not immune to demographic facts. In most European countries, an ageing ministry and a fall in the number of vocations disadvantage the priesthood, whose members are also underpaid and overworked in relation to the herculean kindnesses they perform. Importing young clergy from the Third World, as the Church does in France, will not correct this problem in the long term. On the other hand, courses in Christianity, such as Alpha and Faithworks, which are tailored to a modern pressurised lifestyle, are doing booming business – cynics say as dating agencies – as are the nation's psychotherapists, the secularised alternative to the religion Sigmund Freud regarded as an illusion. But how many of these people are retained by a parish church? It may be a distortion of the media, but the Churches seem to put extraordinary passion into discussing sex, a subject that many lay people find less than compelling. If this is partly a reflection of the homosexualisation of the clergy, this will presumably have grave implications for heterosexuals wishing to pursue a vocation who may not feel comfortable in what are tantamount to gay covens, familiar enough from other walks of life.[39] Perhaps they might like to address other themes once in a while to animate the interest of thinking people within the vast pool of cultural Christians?

For example, there is a palpable fatigue with the culture of living through shopping, or rather with unprecedented levels of credit-card debt, and some chains of superfluous shops face ruination. Perhaps the huge suburban shopping malls will join the cathedrals as part of the heritage industry. They might also wish to explore the reasons for the self-destruction of so many young people, whether those who kill themselves with drugs, or who just degrade themselves through other forms of nihilism. Surely this is to identify real phenomena that the Churches might wish to address, in connection with the wider existential boredom that bedevils modern Western mankind? They may also wish to tackle the disturbing social implications of the 'multicultural society', rather than passing over hard inter-cultural questions in the interests of inter-faith 'dialogue'. In whose interest is it that certain vociferous minorities are protected from public criticism through special legislation, while various self-appointed cultural commissars seek to eradicate Europe's historic religious traditions? This goes beyond the annual rote expressions of alarm about the commercialisation of Christmas, or attempts to supplant it with 'the holidays' or 'winterval'. What is the position of the Churches on the prospect of legal dualism or federalism, in which religious minorities are allowed to practise an alternative law to that in the rest of the land concerned?

What do the Churches have to say about the worrying surrender of sovereignty, not to a federal Europe – although that is also a matter of deep concern – but to the self-appointed 'moderate' leaders of so-called communities, a deal brokered to contain violent people within these minorities? Talk of 'Eurabia' is alarmist, but the example of 'community restorative justice' in Northern Ireland indicates how entire communities can be delivered into the hands of extremely suspect so-called leaders, whose agenda is modest compared to those wishing to restore the medieval Caliphate to most of Spain. What will be the role of Christianity – Europe's historic faith and the culture most people are born into – in relation to the civil religions which anxious governments are actively exploring in every state in Europe as a means of re-creating social harmony now that the post-war welfare state consensus no longer seems sufficient to perform this unifying task? Why is the United States more successful in absorbing immigrants – including Muslim immigrants –

than much of Europe? Is it a matter of greater space and social mobility? Does the absence of a welfare state diminish the opportunities for resentments about how the cake is shared out? Does the strict separation of Church and state mitigate the overwhelmingly and sometimes stridently Christian nature of the US?

Optimists, including all those who still subscribe to the multicultural 'project', will conclude this book with the thought that in the past Europe has successfully accommodated other minority faiths, and that judicious adjustments – the appointment of 'head-scarf mediators', the provision of Muslim cemeteries, and the licensing of humane forms of animal slaughtering, (all contentious in many parts of Europe) and even limited legal dualism or federalism, will allow us all to josh along in the fullness of time. Optimistic secularists may feel that the challenge represented by Europe's fifteen million Muslims may give a final impetus to the separation of Church and state, winding up various anachronistic anomalies, from the Church of England to the Lutheran Establishments of Scandinavia.[40] By contrast, pessimists may object that these measures represent the thin end of the wedge, to be followed by further demands, and that such innovations as separate legal systems are inherently divisive. Demographic factors alone will result in the grim prospect of 'Eurabia' if only to ensure a workforce to support the large over-hang of pensioners of my own generation and beyond. No measures will appease Europe's Islamist radicals whose primary loyalties are to the free-floating mercenary army symbolised by Al-Qaeda, whose solidarities and values have been forged on battlefields stretching from the Balkans, via the Caucasus to Iraq and Afghanistan. On the whole, I conclude this book as an optimist, although certainly not of the Panglossian variety, since the increasingly sharp definition of what is at stake is itself surely part of the solution.

NOTES

Chapter 1: 'Distress of Nations and
Perplexity': Europe after the Great War

1 Annette Becker, *War and Faith.*
   The Religious Imagination in France,
   1914–1930 (Oxford 1998) pp. 146ff.
2 A. L. Rowse, *A Cornish Childhood*
   (London 1974) pp. 255–6
3 On Kipling see David Gilmour, *The*
   Long Recessional. The Imperial Life of
   Rudyard Kipling (London 2002)
   pp. 248ff. The most allusively
   comprehensive account of grief is Jay
   Winter, *Sites of Memory, Sites of*
   Mourning. The Great War in European
   Cultural History (Cambridge 1995)
4 Georges Rouault, *Miserere* (Paris n.d.)
5 See Paul Fussell, *The Great War and*
   Modern Memory (Oxford 1975)
   pp. 187ff.
6 Alex King, *Memorials of the Great War*
   in Britain. The Symbolism and Politics of
   Remembrance (Oxford 1998)
7 George L. Mosse, *The Nationalization*
   of the Masses. Political Symbolism and
   Mass Movements in Germany from the
   Napoleonic Wars through the Third
   Reich (Ithaca 1975) pp. 68–72
8 George L. Mosse, 'Towards a General
   Theory of Fascism' in his *The Fascist*
   Revolution (New York 1999) p. 15
9 Konrad Heiden, *Der Fuehrer,* trans.
   Ralph Manheim (London 1999)
   pp. 121–2 citing Lieutenant Gerhard
   Rossbach
10 George L. Mosse, *Fallen Soldiers.*
   Reshaping the Memory of the World
   Wars (Oxford 1990) p. 165; Nigel Jones,
   A Brief History of the Birth of the Nazis
   (New York 2004) replaces earlier works
   on the Freikorps

11 Horst Möller, *Europa zwischen den*
   Weltkriegen (Munich 1998) p. 122
12 Allan Bullock 'The Double Image' in
   Malcolm Bradbury and James
   McFarlane (eds), *Modernism. A Guide*
   to European Literature 1890–1930
   (London 1976) pp. 58ff.
13 Henri Barbusse, *Under Fire,* trans.
   Robin Buss (London 2003) pp. 296–319
   for the quotations
14 Edward Timms, *Karl Kraus, Apocalyptic*
   Satirist. Culture and Catastrophe in
   Habsburg Vienna (New Haven 1986) is
   the best biography
15 For an excellent discussion of Kraus's
   views on the wartime press see Niall
   Ferguson, *The Pity of War* (London
   1998) ch. 8 pp. 212ff.
16 Karl Kraus, *In Those Great Times.*
   A Kraus Reader, ed. Harry Zohn
   (Manchester 1984) pp. 82–3
17 Karl Kraus, *Die letzten Tage der*
   Menschheit. Tragödie in fünf Akten mit
   Vorspiel und Epilog (Frankfurt am Main
   1986) act 1, scene 27–8, pp. 190–1
   [hereafter cited as *Die letzten Tage*]
18 Kraus, *Die letzten Tage* act 2, scene 15,
   p. 355
19 J. N. Figgis, *Civilisation at the*
   Crossroads (London 1913) p. 95
20 Peter Ackroyd, *T. S. Eliot* (London
   1984) p. 128
21 For a routine debunking of the poem
   see Christopher Hitchens, 'A Breath of
   Dust', *Atlantic Monthly* (2005) 296,
   pp. 142–8
22 For a good knockabout assault on
   modern irrationalism see Francis

Wheen, *How Mumbo-Jumbo Conquered the World* (London 2004)

23 T. S. Eliot 'The Dry Salvages', *Four Quartets* in *Collected Poems 1909–1962* (London 1963) p. 212

24 J. V. Langmead Casserley, *The Retreat from Christianity in the Modern World* (London 1952) pp. 65–6

25 See the classic study by Karl-Dietrich Bracher, *The Age of Ideologies. A History of Political Thought in the Twentieth Century* (London 1984) especially pp. 26ff.

26 Richard J. Evans, *The Coming of the Third Reich* (London 2004). On the multiple factual and interpretative inadequacies of this gargantuan enterprise see the reviews by such German authorities as Heinrich-August Winkler in *Der Spiegel* or Klaus Hildebrand in the *Frankfurter Allgemeine Zeitung*

27 Emile Durkheim, *The Elementary Forms of Religious Life* (Oxford 2001) pp. 157–8

28 Karl Löwith, *My Life in Germany before and after 1933. A Report* (London 1994) p. 15

29 Ibid. p. 63

30 A truth evident to Thomas Mann in his 1938 essay 'A Brother' about Hitler

31 Oskar Jaszi, *Magyariens Schuld, Ungarns Sühne. Revolution und Gegenrevolution in Ungarn* (Munich 1923) pp. 69–70

32 See Ian Kershaw, *The Hitler Myth. Image and Reality in the Third Reich* (Oxford 1987) pp. 18–19

33 Robert P. Ericksen, *Theologians under Hitler* (New Haven 1985) p. 84

34 Klaus Schreiner, '"Wann kommt der Retter Deutschlands?" Formen und Funktionen von politischen Messianismus in der Weimarer Republik', *Saeculum* (1998) 49, pp. 125–6; I am grateful to James Campbell SJ for elucidating these oracles

35 Löwith, *My Life in Germany* p. 18

36 Ulrich Linse, *Barfüssige Propheten. Erlöser der zwanziger Jahre* (Berlin 1983) pp. 33ff.

37 Sebastian Haffner, *Defying Hitler. A Memoir* (London 2002) pp. 51–2

38 Linse, *Barfüssige Propheten* pp. 156ff.

39 Armin Mohler, *Die konservative Revolution in Deutschland 1918–1932. Ein Handbuch* (Darmstadt 1994) p. 138

40 D. H. Lawrence, 'A Letter from Germany' in *Selected Essays* (London 1950) pp. 175–9; see also Harry T. Moore, *The Priest of Love. A Life of D. H. Lawrence* (London 1974) p. 387

41 Hermann Hesse 'The Longing of our Time for a Worldview' in Anton Kaes, Martin Jay and Edward Dimendberg (eds), *The Weimar Republic Sourcebook* (Berkeley 1994) nr 141, p. 366

42 Elke Fröhlich (ed.), *Die Tagebücher von Joseph Goebbels* (Munich 2005) 1/II December 1925–May 1928, p.112, entry for 24 July 1926

43 Christoph Bry, *Der Hitler-Putsch. Berichte und Kommentare eines Deutschland-Korrespondenten (1922–1924) für das 'Argentinische Tag-und Wochenblatt* ed. Martin Gregor-Dellin (Nördlingen 1987) 22 November 1922, pp. 64–6

44 Christoph Bry, *Verkappte Religionen. Kritik des kollektiven Wahns* (Nördlingen 1988) p. 129

45 Ibid. p. 241

46 Adrian Hastings, *A History of English Christianity 1920–2000* (London 2001) p. 174

47 John Kent, *William Temple* (Cambridge 1992) p. 125

48 Edward Norman, *Church and Society in England 1770–1970. A Historical Study* (Oxford 1976) p. 340

49 For the general background see Martin Conway, *Catholic Politics in Europe 1918–1945* (London 1997) pp. 30ff.

50 Kurt Nowak, *Geschichte des Christentums in Deutschland. Religion, Politik und Gesellschaft vom Ende der Aufklärung bis zur Mitte des 20. Jahrhunderts* (Munich 1995) p. 208

51 For the details see Rudolf Morsey, '1918–1933' in Winfried Becker, Günter Buchstab, Anselm Doering-Manteuffel and Rudolf Morsey (eds), *Lexikon der Christlichen Demokratie in Deutschland* (Paderborn 2002) pp. 36–7

Chapter 2: The Totalitarian
Political Religions

1 Ronald Clark, *The Life of Bertrand Russell* (London 1995) p. 380
2 Bertrand Russell, *The Autobiography of Bertrand Russell* (Boston 1968) p. 149; Ray Monk, *Bertrand Russell. The Spirit of Solitude 1872–1921* (New York 1996) p. 581; Caroline Moorehead, *Bertrand Russell. A Life* (London 1992) pp. 312ff; and Alan Ryan, *Bertrand Russell. A Political Life* (New York 1988) pp. 81ff.
3 Bertrand Russell, *The Practice and Theory of Bolshevism* (London 1920) pp. 15–17
4 Philip Boobbyer, *S. L.Frank. The Life and Work of a Russian Philosopher 1877–1950* (Athens, OH 1995) pp. 65–7
5 Robert Service, *Lenin. A Biography* (London 2000) p. 64
6 Nicolas Berdyaev, *The Russian Revolution* (Ann Arbor 1966) p. 58
7 Adam Ulam, The Bolsheviks (Cambridge, Mass. 1965) pp. 205–6
8 Alexandr Blok, 'The Twelve', *Selected Poems*, trans. Jon Stallworthy and Peter France (Manchester 2000) p. 110
9 See the excellent discussion of 'God-building' in Arthur Jay Klinghoffer, *Red Apocalypse. The Religious Evolution of Soviet Communism* (Lanham 1996) pp. 49–51
10 Richard Pipes, *Russia under the Bolshevik Regime 1919–1924* (London 1994) p. 344
11 For examples see William B. Husband, 'Soviet Atheism and Russian Orthodox Strategies of Resistance 1917–1932', *Journal of Modern History* (1998) 70, pp. 87ff. for examples
12 Etienne Fouilloux, 'Erschütterungen (1912–1939)' in Jean-Marie Mayeur (ed.), *Erster und Zweiter Weltkrieg. Demokratien und Totalitäre Systeme (1914–1958). Die Geschichte des Christentums* (Freiburg 1992) vol. 12, p. 932 reproduces extracts from Tikhon's letter
13 Jonathan Daly, 'Storming the Last Citadel'. The Bolshevik Assault on the Church, 1922' in Vladimir Brovkin (ed.), *The Bolsheviks in Russian Society.*

The Revolution and the Civil Wars (New Haven 1997) p. 243
14 The text of the letter dated 19 March 1922 is reproduced as Document 94 in Richard Pipes (ed.), *The Unknown Lenin. From the Secret Archive* (New Haven 1996) pp. 152–5
15 See especially Dimitry Pospielovsky, *The Russian Church under the Soviet Regime 1917–1982* (New York 1984) vol. 1, pp. 43ff.
16 Robert Conquest, *Religion in the USSR* (New York 1968) pp. 20–1
17 See Lynne Viola, 'The Peasant Nightmare. Visions of the Apocalypse in the Soviet Countryside', *Journal of Modern History* (1990) 62, especially pp. 759ff.
18 Sarah Davies, *Popular Opinion in Stalin's Russia. Terror, Propaganda and Dissent 1934–1941* (Cambridge 1997) pp. 79–80
19 William C. Fletcher, *The Russian Orthodox Church Underground 1917–1970* (Oxford 1971)
20 For examples see René Fülöp-Miller, *The Mind and Face of Bolshevism* (New York 1929) pp. 186–8, and Richard Stites, *Revolutionary Dreams. Utopian Vision and Experimental Life in the Russian Revolution* (Oxford 1989) p. 108
21 David Powell, *Antireligious Propaganda in the Soviet Union. A Study in Mass Persuasion* (Cambridge, Mass. 1975) p. 36
22 Daniel Peris, *Storming the Heavens. The Soviet League of the Militant Godless* (Ithaca 1998)
23 See James C. Scott, *Seeing Like a State. How Certain Schemes to Improve the Human Condition have Failed* (New Haven 1998) pp. 193ff.
24 Orlando Figes and Boris Kolonitskii, *Interpreting the Russian Revolution. The Language and Symbols of 1917* (New Haven 1999) p. 40 and more generally for the influence of 1789 on February and October 1917
25 Ibid. pp. 59–60
26 Jennifer McDowell, 'Soviet Civil Ceremonies', *Journal for the Scientific Study of Religion* (1974) 13, p. 267
27 Stites, *Revolutionary Dreams* pp. 84–92

28 Lynne Atwood and Catriona Kelly, 'Programmes for Identity. The "New Man" and the "New Woman" ', in Catriona Kelly and David Shepherd (eds), *Constructing Russian Culture in the Age of Revolution 1881–1940* (Oxford 1998) p. 269

29 Robert C. Tucker, *Stalin as Revolutionary 1879–1929* (London 1973) p. 58

30 Robert Conquest, *Stalin. Breaker of Nations* (London 1991) p. 110; on Stalin's time at the Tiflis seminary see Adam B. Ulam, *Stalin. The Man and his Era* (Boston 1989) pp. 22ff.

31 Robert Service, *A History of Twentieth-Century Russia* (London 1997) p. 153

32 Susan Buck-Morss, *Dreamworld and Catastrophe. The Passing of Mass Utopia in East and West* (Cambridge, Mass. 2000) p. 73

33 See especially Christel Lane, *The Rites of Rulers. Ritual in Industrial Society – the Soviet Case* (Cambridge 1981) pp. 210–16

34 Orlando Figes, *A People's Tragedy. The Russian Revolution 1891–1924* (London 1996), pp. 804ff.; Dimitri Volkogonov, *Lenin. Life and Legacy* (London 1991), pp. 435ff.; Service, *Lenin. A Biography*, pp. 481ff.

35 Didier Misiedlak, 'Religion and Political Culture in the Thought of Mussolini', *Totalitarian Movements and Political Religions* (2005) 6, pp. 395–406, is useful

36 See the important discussion by Jacob L. Talmon, *Myth of the Nation and Vision of Revolution. Ideological Polarization in the Twentieth Century* (New Brunswick 1991) pp. 490–5

37 Martin Clark, *Modern Italy 1871–1982* (London 1990) p. 215

38 R.J.B. Bosworth, *The Italian Dictatorship* (London 1998) p. 41

39 Emilio Gentile, *The Sacralization of Politics in Fascist Italy* (Cambridge, Mass. 1996) p. 20

40 Benito Mussolini, 'Discorso di Pesaro' in E. Susmel and D. Susmel (eds), *Opera omnia di Benito Mussolini* (Florence 1956) 22, p. 197

41 George Mosse, 'The Poet and the Exercise of Political Power' in his *Masses and Man. Nationalist and Fascist Perceptions of Reality* (Detroit 1980)

42 Ernst Nolte, *Three Faces of Fascism* (New York 1966) p. 188

43 See the excellent discussion in Emilio Gentile, 'Fascism in Power. The Totalitarian Experiment' in Adrian Lyttleton (ed.), *Liberal and Fascist Italy* (Oxford 2002) pp. 146–8

44 Henry Spencer, 'The Mussolini Regime' in Guy Stanton Ford (ed.), *Dictatorship in the Modern World* (New York 1935) p. 102

45 For palingenesis see the numerous intemperate effusions of Roger Griffin

46 See Mabel Berezin, *Making the Fascist Self. The Political Culture of Interwar Italy* (Ithaca 1997) pp. 63–5

47 Gentile, *The Sacralization of Politics* p. 35

48 Jens Petersen, 'Die Entstehung des totalitarismusbegriffs in Italien' in Manfred Funke (ed.), *Ein Studien-Reader zur Herrschaftsanalyse moderner Diktaturen* (Düsseldorf 1978) p. 123

49 Gentile, *The Sacralization of Politics* p. 62

50 'Doctrine of Fascism' in Michael Oakeshott (ed.), *The Social and Political Doctrines of Contemporary Europe* (Cambridge 1939) pp. 164–78

51 Berezin, *Making the Fascist Self* p. 191

52 See Ian Kershaw, *The Hitler Myth. Image and Reality in the Third Reich* (Oxford 1987)

53 On these complex tendencies see Richard A. Webster, *The Cross and the Fasces. Christian Democracy and Fascism in Italy* (Stanford 1960) pp. 23–5

54 John Pollard, 'Italy' in Tom Buchanan and Martin Conway (eds), *Political Catholicism in Europe 1918–1965* (Oxford 1996) pp. 78–9

55 Elisa Carrillo, *Alcide de Gasperi. The Long Apprenticeship* (Notre Dame 1965) p. 62

56 Webster, *The Cross and the Fasces* p. 75

57 John Pollard, *The Vatican and Italian Fascism 1929–32. A Study in Conflict* (Cambridge 1985) pp. 27–8

58 For this interaction see Renato Moro, 'Religion and Politics in the Time of

Secularization. The Sacralisation of Politics and Politicisation of Religion', *Totalitarian Movements and Political Religions* (2005) 6, especially pp. 80–3

59 Davies, *Popular Opinion in Stalin's Russia* p. 172

60 As suggested by Robert H. McNeal, *Stalin. Man and Ruler* (London 1988) p. 152

61 Tucker, *Stalin as Revolutionary* pp. 477–81

62 Ibid. pp. 481–2

63 Robert C. Tucker, 'The Rise of Stalin's Personality Cult', *American Historical Review* (1979) 84, pp. 347–66

64 Davies, *Popular Opinion in Stalin's Russia* pp. 150–1

65 Boris Souvarine, *Stalin* (New York 1939) p. 662

66 Mikhail Heller and Aleksandr Nekrich, *Utopia in Power. The History of the Soviet Union from 1917 to the Present* (London 1985) p. 282

67 Lane, *Rites of Rulers* p. 217 citing a poem in a 1938 children's magazine

68 Robert C. Tucker, 'Does Big Brother Really Exist?' in Irving Howe (ed.), *1984 Revisited. Totalitarianism in our Century* (New York 1983) pp. 100–2

69 Robert C. Tucker, 'Lenin's Bolshevism as a Culture in the Making' in Abbott Gleason, Peter Kenez and Richard Stites (eds), *Bolshevik Culture. Experiment and Order in the Russian Revolution* (Bloomington 1985) pp. 25–38

70 See Stephen Kotkin, *Magnetic Mountain. Stalinism as a Civilisation* (Berkeley 1995) especially pp. 292–3

71 Victor Kravchenko, *I Chose Freedom. The Personal and Political Life of a Soviet Official* (London 1947) p. 55

72 Service, *A History of Twentieth-Century Russia* p. 140

73 Klaus-Georg Riegel, 'Rituals of Confession Within Communities of Virtuosi: An Interpretation of the Stalinist Criticism and Self-Criticism in the Perspective of Max Weber's Sociology of Religion', *Totalitarian Movements and Political Religions* (2000) 1, p. 31

74 Kravchenko, *I Chose Freedom* pp. 132–47

75 Arkady Vaksberg, *Stalin's Prosecutor.*

The Life of Andrei Vyshinsky (New York 1990) pp. 42–6

76 Ibid. p. 115

77 Vaksberg, *Stalin's Prosecutor* p. 123

78 Robert Conquest, *The Great Terror* (London 1968) pp. 162–3

79 Ibid. pp. 230–53

80 Figes and Kolonitskii, *Interpreting the Russian Revolution* p. 185

81 Klaus-Georg Riegel, 'Marxism–Leninism as a Political Religion', *Totalitarian Movements and Political Religions* (2005) 6, p. 107

82 Fülöp-Miller, *The Mind and Face of Bolshevism* p. 3

83 See Victoria E. Bonnell, *Iconography of Power. Soviet Political Posters under Lenin and Stalin* (Berkeley 1997) especially pp. 187ff.

84 Frank Manuel and Fritzie Manuel, *Utopian Thought in the Western World* (Cambridge, Mass. 1979) pp. 271–9

85 Kotkin, *Magnetic Mountain* for these remarks

86 Ibid. p. 205

87 Christopher Read, 'Values, Substitutes and Institutions. The Cultural Dimension of the Bolshevik Dictatorship' in Brovkin (ed.), *The Bolsheviks in Russian Society* p. 315

88 For this see Lewis Siegelbaum, *Stakhanovism and the Politics of Productivity in the USSR, 1935–1941* (Cambridge 1988) pp. 63ff.

89 Ibid. p. 230

90 For an excellent discussion of these themes see Katerina Clark, 'Utopian Anthropology as a Context for Stalinist Literature' in Robert C. Tucker (ed.), *Stalinism. Essays in Historical Interpretation* (New Brunswick 1999) pp. 180–98

91 Mikhail Heller, *Cogs in the Wheel. The Formation of Soviet Man* (New York 1988) p. 177

92 The painting is reproduced in Igor Golomstock, *Totalitarian Art* (New York 1990) p. 210

93 Lisa A. Kirschenbaum, *Small Comrades. Revolutionizing Childhood in Soviet Russia, 1917–1932* (New York 2001) p. 158; for details on Morozov see Conquest, *The Great Terror* pp. 668–9, and Jan Feldman, 'New

Thinking about the "New Man".
Developments in Soviet Moral
Theory', *Studies in Soviet Thought*
(1989) 38, pp. 147–63 for the fifty years
it took for the USSR to ditch both
Morozov and Stakhanov as moral
icons. Most exhaustively see Catriona
Kelly, *Comrade Pavlik. The Rise and
Fall of a Soviet Boy Hero* (London
2005)

94 Jochen Hellbeck, 'Fashioning the
Stalinist Soul. The Diary of Stepan
Podlubnyi (1931–1939)', *Jahrbücher
für Geschichte Osteuropas* (1996) 44,
pp. 344ff. for all citations from the
diary

95 Helmut Heiber, *Goebbels. A Biography*
(New York 1972) p. 15

96 See Uriel Tal, ' "Political Faith" of
Nazism Prior to the Holocaust' in his
*Religion, Politics and Ideology in the
Third Reich* (London 2004) p. 28, and
Michael Rissmann, *Hitlers Gott*
(Zurich 2001) p. 41

97 H. R. Trevor-Roper (introduced by)
Hitler's Table Talk 1941–1944 (Oxford
1988) hereafter *HTT*. *HTT*
27 February 1942, p. 342

98 Adolf Hitler *Mein Kampf*, trans. Ralph
Manheim (London 1959) p. 417
[hereafter cited as *HMK*]

99 *HMK* p. 393

100 This important point is made by
Philippe Burrin, 'Die politischen
Religionen: Das Mythologisch-
Symbolische in einer säkularisierten
Welt' in Michael Ley and Julius H.
Schoeps (eds), *Der Nationalsozialismus
als politische Religion* (Bodenheim bei
Mainz 1997) pp. 181–2. Steigmann-
Gall's 'discovery' of the alternative of
'religious politics' in his 'Was National
Socialism a Political Religion or a
Religious Politics?' in Michael Geyer
and Hartmut Lehmann (eds), *Religion
und Nation. Nation und Religion.
Beiträge zu einer unbewältigten
Geschichte* (Göttingen 2004) pp. 386ff.
constructs a very windy road to a very
modest conclusion by the device of
reversing two words used by other
scholars

101 Richard Steigmann-Gall, *The Holy
Reich. Nazi Conceptions of*

Christianity, 1919–1945 (Cambridge
2003) p. 63

102 Claus-Ekkehard Bärsch, *Die politische
Religion des Nationalsozialismus*
(Munich 1998) p. 288

103 See Max Domarus (ed.), *Hitler.
Speeches and Proclamations* (London
1992) 2, p. 1146 for the relevant speech
in September 1938 which constituted
a coded warning to Himmler and
Rosenberg

104 Roger Griffin, *The Nature of Fascism*
(London 1991) p. 32 misunderstands
these events

105 See Claudia Witte, 'Artur Dinter' in
B. Danckwortt, T. Querg and Claudia
Schöningh (eds), *Historische
Rassismusforschung* (Hamburg 1995)
pp. 113–51

106 Steigmann-Gall, *The Holy Reich*
pp. 60–1. Steigmann-Gall's book fails
to address larger debates about
secularisation and surrogacy

107 The recent spate of anti-Catholic
literature summarised in Goldhagen's
*A Moral Reckoning. The Role of The
Catholic Church in the Holocaust and
its Unfulfilled Duty of Repair* (New
York 2002) fails to grapple with any of
these issues and hence amounts to an
offensive caricature of the complex
realities of a bi-confessional society,
not to speak of international
Catholicism. Whereas Hitler attacked
the Roman Catholic Church for being
philosemitic, there is curiously no
record of his acknowledging the
antisemitism that such critics as
Goldhagen claim it was allegedly
permeated with

108 *HTT* 13 December 1941, p. 144

109 *HTT* 20–21 February 1942, pp. 322–3,
and 28 April 1942, p. 445 for his
observatory at Linz. Rissmann, *Hitlers
Gott* makes too much of this in a book
that does not tell us much about
Hitler's God at all

110 *HTT* 20–21 February 1942, p. 325
where he says he was 'liberated from
the superstition that the priests used
to teach' at the age of fourteen

111 See the unjustly neglected Detlev
Grieswelle's *Propaganda der
Friedlosigkeit – Eine Studie zu Hitlers*

Rhetorik 1920–1933 (Stuttgart 1972) pp. 56–7

112 HTT 9 April 1942, p. 419

113 HTT 27 February 1942, p. 343

114 HTT 13 December 1941, p. 143

115 HTT 20–21 February 1942, pp. 322–4

116 HTT 20–21 February 1942, p. 322

117 HTT 13 December 1941, p. 145

118 HTT 7 April 1942, p. 410

119 HTT 4 July 1942, p. 533

120 HTT 11 August 1942, pp. 625–6

121 Domarus (ed.), Hitler. Speeches and Proclamations 2, p. 908

122 Peter Adam, The Arts of the Third Reich (London 1992) p. 172 for the painting

123 On Hitler's exploitation of his 'authentic experiences', where apart from the embroidered truth of his war service he fabricated a period as a 'construction worker', see J. P. Stern, Hitler. The Führer and the People (London 1975) pp. 23ff.

124 Ian Kershaw, The Hitler Myth p. 30

125 Cited by Ralph Georg Reuth, Goebbels (London 1993) p. 54

126 Domarus (ed.), Hitler. Speeches and Proclamations 2, p. 836

127 Ibid. p. 833

128 See especially Uriel Tal, 'Political Faith of Nazism' p.35 and his 'Structures of German "Political Theology" in the Nazi Era' in his Religion, Politics and Ideology in the Third Reich; and Hartmut Lehmann, ' "God our Old Ally". The Chosen People Theme in Late Nineteenth- and Early Twentieth-Century Nationalism' in W. Hutchinson and H. Lehmann (eds), Many are Chosen (Minneapolis 1994) pp. 85ff.

129 Robert P. Ericksen, Theologians under Hitler (New Haven 1985) p. 103

130 Uriel Tal, 'On Modern Lutheranism and the Jews' in his Religion, Politics and Ideology pp. 192ff.

131 The classic study of these issues is Kurt Nowak, 'Euthanasie' und Sterilisierung im 'Dritten Reich'. Die Konfrontation der evangelischen und katholischen Kirche mit dem 'Gesetz zur Verhütung erbkranken Nachwuchses' und der 'Euthanasie'-Aktion (Göttingen 1978) p. 92

132 Sabine Schleiermacher, 'Der Centralausschuss für die Innere Mission und die Eugenik am Vorabend des "Dritten Reiches" ' in T. Strohm and J. Thierfelder (eds), Diakonie im 'Dritten Reich'. Neuere Ergebnisse zeitgeschichtliche Forschung (Heidelberg 1990) pp. 60–77. And Michael Burleigh, 'Between Enthusiasm, Compliance and Protest. The Churches, Eugenics and the Nazi "Euthanasia" Programmes' in Burleigh, Ethics and Extermination. Reflections on Nazi Genocide (Cambridge 1997) pp. 130–141

133 See above all Maurice Oleander, The Languages of Paradise. Race, Religion, and Philology in the Nineteenth Century (Cambridge, Mass. 1992) and the older Leon Poliakov, The Aryan Myth. A History of Racist and Nationalist Ideas in Europe (New York 1971)

134 HMK p. 348

135 HMK p. 263

136 HMK p. 270

137 HMK p. 265

138 HMK p. 278

139 Robert S. Wistrich, 'The Last Testament of Sigmund Freud', Leo Baeck Institute Year Book (2004) 49, pp. 99–104

140 See the discussion in Michael Burleigh, The Third Reich. A New History (London 2000) pp. 219ff.

141 Steigmann-Gall, The Holy Reich p. 48 quoting lectures Althaus gave in 1932

142 Kurt Meier, Kreuz und Hakenkreuz. Die evangelische Kirche im Dritte Reich (Munich 1992) p. 37

143 Hans Kohn, 'Communist and Fascist Dictatorship. A Comparative Study' in Ford (ed.), Dictatorship in the Modern World p. 149

144 Joshua Podro, Nuremberg. The Unholy City (London 1937)

145 Victor Klemperer, The Language of the Third Reich. LTI–Lingua Tertii Imperii. A Philologist's Notebook (London 1999) p. 34

146 Hans-Ulrich Thamer, 'Faszination und Manipulation. Die Nürnberger Reichsparteitage der NSDAP' in Uwe Schultz (ed.), Das Fest.

Kulturgeschichte von der Antike bis zur Gegenwart (Munich 1988) pp. 353ff.

147 George Mosse, *The Nationalization of the Masses* (Ithaca 1975) p. 206

148 Domarus (ed.), *Hitler. Speeches and Proclamations* 2, pp. 727–8

149 See especially Sabine Behrenbeck, *Der Kult um die toten Helden. Nationalsozialistische Mythen, Riten und Symbole* (Vierow 1996) pp. 299ff.

150 Nicholas Goodrick-Clarke, *The Occult Roots of Nazism. Secret Aryan Cults and their Influence on Nazi Ideology* (London 1985) pp. 177ff.

151 Michael Wildt, 'The Spirit of the Reich Main Security Office (RSHA)', *Totalitarian Movements and Political Religions* (2005) 6, p. 347

152 Waldemar Gurian, *Bolshevism. Theory and Practice* (London 1932) pp. 226–7

153 Heinz Hürten (ed.), *Deutsche Briefe 1934–1938. Ein Blatt der katholischen Emigration* (Mainz 1969)

154 Eric Voegelin, *Autobiographical Reflections* pp. 46–7

155 Eric Voegelin, *The Political Religions. The Collected Works of Eric Voegelin* [hereafter *CWEV*] (Columbia, Missouri 2000) 5; see also *The History of the Race Idea. From Ray to Carus. CWEV* (Columbia, Missouri 1998) 3 and *Race and State. CWEV* (Columbia, Missouri 1997) 2; see also Eric Voegelin, *Autobiographical Reflections*, ed. Ellis Sandoz (Baton Rouge 1989) for the biographical details

156 Franz Borkenau, *The Totalitarian Enemy* (London 1940) pp. 130 and 140–1. On Borkenau see Birgit Lange-Enzmann, *Franz Borkenau als politischer Denker* (Berlin 1996). See also William D. Jones, *The Lost Debate. German Socialist Intellectuals and Totalitarianism* (Urbana 1999)

157 Raymond Aron, *The Dawn of Universal History. Selected Essays from a Witness to the Twentieth Century* (New York 2002) pp. 177–223; the best biography of Aron remains Robert Colquhoun, *Raymond Aron. The Philosopher in History* (London 1986)

Chapter 3: The Churches in the Age of Dictators

1 See Norman Sherry, *The Life of Graham Greene* (London 1989) 1, especially pp. 653ff., and Graham Greene, *The Lawless Roads* (London 1939) p. 182

2 Greene, *The Lawless Roads* pp. 100–1

3 Pius XI, *Acerba Animi*, 29 September 1932

4 Anthony Rhodes, *The Power of Rome in the Twentieth Century*, vol. 2: *The Vatican in the Age of Dictators 1922–1945* (London 1973) p. 101

5 On Calles see Enrique Krauze, *Mexico. Biography of Power. A History of Modern Mexico 1810–1996* (New York 1997) pp. 404ff.

6 Sherry, *The Life of Graham Greene* 1, pp. 612-13

7 Frances Lannon, *Privilege, Persecution, and Prophecy. The Catholic Church in Spain 1875–1975* (Oxford 1987) p. 181

8 Rhodes, *The Power of Rome* 2, pp. 121–2

9 For examples see Mary Vincent's excellent *Catholicism in the Second Republic. Religion and Politics in Salamanca, 1930–1936* (Oxford 1996) pp. 184–8

10 Pius XI, *Dilectissima nobis*, 3 June 1933, pius-xi/encyclicals

11 For these details see Frances Lannon, 'The Church's Crusade against the Republic' in P. Preston (ed.), *Revolution and War in Spain 1931-1939* (London 1984) pp. 48–53

12 Stanley Payne, *Spanish Catholicism. An Historical Overview* (Madison, Wis. 1984) p. 160

13 Vincent, *Catholicism in the Second Spanish Republic* pp. 217–20

14 See especially Mary Vincent, 'Spain' in Tom Buchanan and Martin Conway (eds), *Political Catholicism in Europe 1918–1965* (Oxford 1996) pp. 117–18

15 Paul Preston, *A Concise History of the Spanish Civil War* (London 1996) p. 47

16 Ibid. p. 43

17 José M. Sánchez, *The Spanish Civil War as a Religious Tragedy* (Notre Dame 1987) p. 9

18 Franz Borkenau, *The Spanish Cockpit* (London 1937) pp. 112–13

19 Cedric Salter, *Try-Out in Spain* (New York 1943) pp. 19–20

20 Sánchez, *The Spanish Civil War as a Religious Tragedy* p. 17

21 Hugh Thomas, *The Spanish Civil War* (London third edition 1990) p. 271

22 Sánchez,*The Spanish Civil War as a Religious Tragedy* pp. 47–8 for these statistics

23 Stanley Payne, *The Franco Regime 1936–1975* (London 2000) p. 199

24 Payne, *Spanish Catholicism* p. 175

25 Paul Preston, 'The Discreet Charm of a Dictator. Francisco Franco' in his *Comrades. Portraits from the Spanish Civil War* (London 1999) p. 53

26 Payne, *The Franco Regime* p. 206

27 See especially Stanley Payne, *A History of Fascism 1914-45* (London 1995) pp. 262–7

28 On this distinction see the classic analysis by Karl Dietrich Bracher, 'Authoritarismus und Totalitarismus' in his *Wendezeiten der Geschichte. Historische-politische Essays* (Stuttgart 1992) pp. 145–72

29 Charles Delzell (ed.), *Mediterranean Fascism 1919–1945* (New York 1970) p. 332

30 See Tom Gallagher, 'Portugal' in Buchanan and Conway (eds), *Political Catholicism in Europe* pp. 129ff.

31 Tom Gallagher, 'Portugal' in Martin Blinkhorn (ed.), *Fascists and Conservatives* (London 1990) p. 165

32 Payne, *A History of Fascism* p. 249

33 See Klaus-Jörg Siegfried, *Universalismus und Faschismus. Das Gesellschaftsbild Othmar Spann* (Vienna 1974) pp. 144–7

34 Johannes Messner, *Dollfuss. An Austrian Patriot* (London 1935) p. 99

35 Alfred Pfoser and Gerhard Renner, '"Ein Toter fuehrt uns an!" Anmerkungen zur kulturellen Situation im Austrofaschismus' in Emmerich Talos and Wolfgang Neugebauer (eds), *'Austrofaschismus' – Beiträge über Politik: Ökonomie und kultur 1934–1938* (Vienna 1984) pp. 238–42

36 Frank Coppa, 'Two Popes and the Holocaust' in John K. Roth and Elizabeth Maxwell (eds), *Remembering for the Future. The Holocaust in the Age of Genocide* (Basingstoke 2001) 2, p. 399

37 Erika Wienzierl, 'Austria. Church, State, Politics' p. 20

38 Frank Coppa, 'The Vatican and the Dictators' in Richard Wolff and Jörg Hoensch (eds), *Catholics, the State, and the European Radical Right 1919–1945* (New York 1987) p. 211

39 The text of the memorandum can be found in Charles R. Gallagher, 'Pacelli Documents', *America. The National Catholic Weekly* (2005) 192, pp. 1–4. See also Charles R. Gallagher, 'Personal, Private Views', *America* (2003) 189, pp. 1–5

40 Dermot Keogh, *The Vatican, the Bishops and Irish Politics 1919–39* (Cambridge 1986) p. 204

41 On these developments see F. S. L. Lyons, *Ireland since the Famine* (London 1963) p. 495

42 See the fine discussion in R. F. Foster, *Modern Ireland 1600–1972* (Oxford 1988) pp. 516ff.

43 Dermot Keogh and Finín O'Driscoll, 'Ireland' in Buchanan and Conway (eds), *Political Catholicism in Europe* pp. 279ff.

44 John H. Whyte, *Church and State in Modern Ireland 1923–1979* (Totowa, NJ 1980) p. 28

45 Ibid. pp. 52–6

46 Stella Alexander 'Croatia. The Catholic Church and Clergy, 1919–1945' in Wolff and Hoensch (eds), *Catholics, the State, and the European Radical Right* pp. 31ff.

47 Martin, Conway, 'Belgium' in Buchanan and Conway (eds), *Political Catholicism in Europe* pp. 201–3

48 For the above see John Hellman's excellent *Emmanuel Mounier and the New Catholic Left 1930–1950* (Toronto 1981)

49 Robert Paxton, 'France. The Church, the Republic' in Wolff and Hoensch (eds), *Catholics, The State and the European Radical Right* p. 78

50 Douglas Lane Patey, *The Life of Evelyn Waugh* (Oxford 1998) pp. 146–7

51 Marcus Tanner, *Ireland's Holy Wars. The Struggle for a Nation's Soul 1500–2000* (New Haven 2001) pp. 307–9

52 Frederic Hartweg, 'Der französische Protestantismus und die Kirchen in

Deutschland' in Gerhard Besier (ed.), *Zwischen 'nationaler Revolution' und militärischer Aggression* (Munich 2001) p. 240

53 Eamon Duffy, *Saints and Sinners. A History of the Popes* (New Haven 2001) p. 338

54 Robert Speaight, *Georges Bernanos. A Study of the Man and the Writer* (London 1974) pp. 156ff.

55 Malcolm Scott, *Mauriac. The Politics of a Novelist* (Edinburgh 1980) pp. 70–6

56 Bernard Doering, *Jacques Maritain and the French Catholic Intellectuals* (Notre Dame 1983) especially pp. 85–125

57 On this see Thomas Brechenmacher, 'Pope Pius XI, Eugenio Pacelli, and the Persecution of the Jews in Nazi Germany 1933–1939. New Sources from the Vatican Archives', *German Historical Institute Bulletin* (2005) 27, pp. 22–3

58 Martin Conway, *Catholic Politics in Europe 1918–1945* (London 1997) p. 41

59 On this see Philip Jenkins, *The New Anti-Catholicism. The Last Acceptable Prejudice* (Oxford 2003)

60 On this see the discussion in Ronald Rychlak, *Righteous Gentiles* (Dallas 2005) pp. 26–8

61 Philip Hughes, *Pope Pius the Eleventh* (London 1937) pp. 68ff.

62 See Kenneth Whitehead, 'The Pope Pius XII Controversy', *Political Science Reviewer* (2002) 31, p. 349 citing William Teeling, *Pope Pius XI and World Affairs* (New York 1937) p. 67

63 Lord Clonmore, *Pope Pius XI and World Peace* (London 1938) p. 54

64 On this see Ronald Rychlak, 'Goldhagen v. Pius XII', *First Things* (2002) pp. 42–3, and Emma Fattorini, *Germania e Santa Sede. Le nunziatura di Pacelli tra la Grande Guerra e la Repubblica di Weimar* (Bologna 1992) p. 116

65 Gerhard Besier, *Der Heilige Stuhl und Hitler-Deutschland. Die Faszination des Totalitären* (Munich 2004) especially pp. 79–92

66 D. A. Binchy, *Church and State in Fascist Italy* (Oxford 1970) p. 333

67 Pius XI, *Quadragesimo anno*

68 John Pollard,The *Vatican and Italian Fascism 1929–32. A study in conflict* (Cambridge 1985) pp. 136–7

69 Pius XI, *Non abbiamo bisogno* quotations from clauses 12 and 44

70 For the details see Ludwig Volk, *Der bayerische Episkopat und der Nationalsozialismus 1930-1934* (Mainz 1966) pp. 14–19

71 Archivio Nunziatura Monaco, protocollo nr 28961, Busta 396, Fascicolo 7, Foglio 6r–7v and 'Pacelli denounces the Nazis', *Inside the Vatican* (March 2003) pp. 30–1 for an English translation of the document

72 Zsolt Aradi, *Pius XI. The Pope and the Man* (New York 1958) p. 222

73 Sir Ivone Kirkpatrick, *The Inner Circle. Memoirs* (London 1959) p. 47

74 Ernst Hanisch, *Die Ideologie des Politischen Katholizismus in Österreich 1918–1938* (Vienna 1977) p. 30

75 Besier, *Der Heilige Stuhl* pp. 140–5

76 On this see David Bates, 'Legitimität and Légalité. Political Theology and Democratic Thought in an Age of World War' in Michael Geyer and Hartmut Lehmann (eds), *Religion und Nation. Nation und Religion* (Göttingen 2004) pp. 435ff.

77 Giovanni Sale, *Hitler, la Santa Sede e gli Ebrei* (Milan 2003) p. 107

78 Konrad Repgen, 'Über die Entstehung der Reichskonkordats-Offerte im Frühjahr 1933 und die Bedeutung des Reichskonkordats', *Vierteljahreshefte für Zeitgeschichte* (1978) 26, pp. 499ff. And his 'Zur vatikanischen Strategie beim Reichskonkordat', *Vierteljahreshefte für Zeitgeschichte* (1983) 31, pp. 512–15

79 See Michael Feldkamp, *Pius XII und Deutschland* (Göttingen 2000) p. 95

80 See the classic account by Karl-Dietrich Bracher, *Die Auflösung der Weimarer Republik* (Villingen 1971 fifth edition)

81 Ludwig Volk, 'Die Fuldaer Bischofskonferenz von Hitlers Machtergreifung bis zur Enzyklika "Mit brennender Sorge"', in Dieter Albrecht (ed.), *Katholische Kirche im Dritten Reich* (Mainz 1976) p. 40

82 John Jay Hughes, 'The Pope's "Pact with Hitler". Betrayal or Self-Defense?', *Journal of Church and State* (1975) 17, p. 70

83 Besier, *Der Heilige Stuhl* pp. 183ff.
84 John Conway, *The Nazi Persecution of the Churches 1933–1945* (London New York 1968) pp. 25–6
85 Repgen, 'Zur vatikanischen Strategie beim Reichskonkordat', documentary appendix pp. 530–5
86 Kirkpatrick, *The Inner Circle* p. 48
87 'Vatican told Nuncio to Forgo Praise of Hitler', Zenit.org 1 May 2003 referring to an interview with Matteo Napolitano citing Entry 604, p.o. fascicle 113 in the Vatican archives
88 Robert P. Ericksen, *Theologians under Hitler. Gerhard Kittel, Paul Althaus, and Emanuel Hirsch* (New Haven 1985)
89 Robert A. Krieg, *Catholic Theologians in Nazi Germany* (New York 2004) pp. 146ff.
90 Michael B. Lukens, 'Joseph Lortz and a Catholic Accommodation with National Socialism' in Robert P. Ericksen and Susannah Heschel (eds), *Betrayal. German Churches and the Holocaust* (Minneapolis 1999) pp. 162–3
91 Lothar Groppe, 'The Church and the Jews in the Third Reich', *Fidelity* (1983) 1, p. 20
92 Fritz Gerlich, *Der Kommunismus als Lehre vom Tausendjährigen Reich* (Munich 1920)
93 On Gerlich see Erwein Freiherr von Aretin, *Fritz Michael Gerlich. Prophet und Martyrer* (Munich 1983) and Rudolf Morsey 'Fritz Gerlich (1883–1934)' p. 35
94 William Doino, 'Edith Stein's Letter', *Inside the Vatican* (March 2003) pp. 22–7 contains this exchange in facsimile
95 Kurt Nowak, *'Euthanasie' und Sterilisierung im 'Dritten Reich'. Die Konfrontation der evangelischen und katholischen Kirche mit dem 'Gesetz zur Verhütung erbkranken Nachwuchses' und der 'Euthanasie'* (Göttingen 1978) p. 112
96 Krieg, *Catholic Theologians in Nazi Germany* pp. 49–50. Beth A. Griech-Polelle's *Bishop von Galen. German Catholicism and National Socialism* (New Haven 2002) p. 69 omits Pacelli's censuring of the two academics. Her references to 'Dr Victor Brandt' on p. 72, conflating Viktor Brack, who was not a doctor of any description, with Professor Karl Rudolf Brandt, Hitler's accident emergency surgeon (both organisers of the T-4 euthanasia programme), do not inspire much confidence either
97 Gerhard Besier, *Die Kirchen und das Dritte Reich. Spaltungen und Abwehrkämpfe 1934–1937* (Munich 2001) pp.210–11
98 Feldkamp, *Pius XII und Deutschland* p. 105
99 Guenther Lewy, *The Catholic Church and Nazi Germany* (London 1964) p. 133
100 Anon, *The Persecution of the Catholic Church in the Third Reich* (London 1940) pp. 426–7
101 Ibid. p. 105
102 See Wolfgang Dierker, *Himmlers Glaubenskrieger* (Paderborn 2002) p. 151
103 Besier, *Die Kirchen und das Dritte Reich* p. 204
104 Anon, *The Nazi Persecution of the Catholic Church* (London 1939) p. 268
105 Charles R. Gallagher, 'Personal, Private Views', *America. The National Catholic Weekly* (2003) 189, p. 2
106 Lewy, *The Catholic Church and Nazi Germany* p. 125
107 Brechenmacher, 'Pope Pius XI' p. 36
108 See Hubert Wolf, 'Pius XI und die "Zeitirrtümer" ', *Vierteljahreshefte für Zeitgeschichte* (2005) 53, pp. 12ff.
109 Peter Godman, *Hitler and the Vatican* (New York 2004) pp. 194–9 for these texts
110 The text of the encyclical can be found in Anon (ed.), *The Persecution of the Catholic Church in the Third Reich* pp. 523–7 or under Pius XII Encyclicals on the Vatican website
111 For further examples see Heinz-Albert Raem, *Pius XI und der National-sozialismus. Die Enzyklika 'Mit brennender Sorge' vom 14. März 1937* (Paderborn 1979) pp. 214–16 for numerous examples
112 On reactions to the encyclical see

Giovanni Sale, 'L'enciclica contro il Nazismo,' *Civiltà Cattolica* (2004) 11, pp. 114–27

113 Rhodes, *The Vatican in the Age of the Dictators* p. 208

114 Heinz Hürten, *Deutsche Katholiken 1918–1945* (Paderborn 1996) p. 396

115 Richlak, *Righteous Gentiles* p. 36

116 Dierker, *Himmlers Glaubenskrieger* pp. 224–7

117 See the 'Waldemar Gurian Memorial Issue' in the 1955 *Review of Politics*, the very remarkable political science journal he founded in the US, with personal recollections by among others Hannah Arendt, Hans Kohn and Jacques Maritain

118 Waldemar Gurian, *Bolshevism. Theory and Practice* (London 1932)

119 Heinz Hürten (ed.), *Deutsche Briefe 1934–1938. Ein Blatt der katholischen Emigration* (Mainz 1969) vols 1–2. Vol. 1 nr 52, dated 27 September 1935, p. 593

120 Waldemar Gurian, *Hitler and the Christians* (New York 1936) pp. 57–9

121 'Ist der Nationalsozialismus eine Religion?' in Heinz Hürten (ed.), '*Kulturkampf. Bericht aus dem Dritten Reich. Paris*'. Eichstätter Materialien vol 13 (Regensburg 1988) nr 78, dated 8 August 1939, pp. 251–7

122 Thomas Morgan, *A Reporter at the Papal Court. A Narrative of the Reign of Pius XI* (New York 1937) p. 288

123 Feldkamp, *Pius XII und Deutschland* p. 113

124 R. J. B. Bosworth, *Mussolini's Italy* (London 2005) p. 417

125 Binchy, *Church and State in Fascist Italy* pp. 616–17

126 Coppa, 'Two Popes and the Holocaust' p. 401

127 David Dalin, *The Myth of Hitler's Pope* (Washington DC 2005) pp. 70–3

128 'Pope Pius and the Jews. A Champion of Toleration', *Jewish Chronicle* 17 February 1939 p. 16

129 Richard Steigmann-Gall, *The Holy Reich. Nazi Conceptions of Christianity, 1919–1945* (Cambridge 2003) pp. 67ff.

130 Ernst Christian Helmreich, *The German Churches under Hitler* (Detroit 1979) p. 213

131 Günter Brakelmann, 'Nationalprotestantismus und Nationalsozialismus' in Christian Jansen et al. (eds), *Von der Aufgabe der Freiheit. Politische Verantwortung und bürgerliche Gesellschaft im 19. und 20. Jahrhundert* (Berlin 1995) pp. 337ff.

132 For the details see Shelley Baranowski, 'The 1933 German Protestant Church Elections. Machtpolitik or Accommodation', *Church History* (1980) 49, pp. 298ff.

133 See especially Shelley Baranowski, 'Consent and Dissent. The Confessing Church and Conservative Opposition to National Socialism', *Journal of Modern History* (1987) 59, p. 59

134 Victoria Barnett, *For the Soul of the People. Protestant Protest against Hitler* (Oxford 1992) p. 35

135 Ruth Zerner, 'German Protestant Responses to Nazi Persecution of the Jews', in Randolf Braham (ed.), *Perspectives on the Holocaust* (Boston 1983) p. 63

136 Ian Kershaw, *Popular Opinion and Political Dissent in the Third Reich. Bavaria 1933–1945* (Oxford 1983) pp. 164–76

137 John Conway, 'The North American Churches' Reactions to the German Church Struggle' in Besier (ed.), *Zwischen 'nationaler Revolution' und militärischer Aggression* pp. 264–5

138 Adolf Keller, *Church and State on the European Continent* (London 1936) pp. 22–3

139 Frederick Voigt, *Unto Caesar* (London 1938)

140 *Manchester Guardian*, 26 January 1937

141 Adrian Hastings, *A History of English Christianity 1920–2000* (London 2001 fourth edition) p. 321

142 Sir Charles Grant Robertson, *Religion and the Totalitarian State* (London 1937)

143 See the sympathetic study by Owen Chadwick, *Hensley Henson. A Study in the Friction between Church and State* (Oxford 1983) pp. 262–3

144 Herbert Hensley Henson, *Retrospect of an Unimportant Life* (Oxford 1943) 2, p. 413

145 The definitive study of this
memorandum is Martin Greschat
(ed.), *Zwischen Widerspruch und
Widerstand. Texte zur Denkschrift der
Bekennenden Kirche an Hitler (1936)*
(Munich 1987).
146 Robert S. Wistrich 'The Last
Testament of Sigmund Freud', *Leo
Baeck Institute Year Book* (2004) 49,
pp. 100–1
147 'German Martyrs', *Time*, 23 December
1940, p. 38

Chapter 4: Apocalypse 1939–1945

1 See Paul Addison, 'Destiny, History
and Providence. The Religion of
Winston Churchill' in Michael Bentley
(ed.), *Public and Private Doctrine.
Essays in British History Presented to
Maurice Cowling* (Cambridge 1993)
pp. 236ff.
2 Robert Rhodes James (ed.), *Winston S.
Churchill. His Complete Speeches
1897–1963* (8 volumes, London 1974) 6,
p. 6238; on Churchill's oratory see
David Cannadine,'Language.
Churchill as the Voice of Destiny' in
his *In Churchill's Shadow. Confronting
the Past in Modern Britain* (Oxford
2003) pp. 85–113
3 Herbert Hensley Henson, *Retrospect of
an Unimportant Life* (Oxford 1950) 3,
pp. 146–8
4 See Keith Robbins, 'Britain, 1940 and
"Christian Civilization"' in Derek
Beales and Geoffrey Best (eds),
*History, Society and the Churches.
Essays in Honour of Owen Chadwick*
(Cambridge 1985) pp. 279ff.
5 R. Kojecký, *T. S. Eliot's Social Criticism*
(London 1971)
6 Jerzy Klockowski, *A History of Polish
Christianity* (Cambridge 2000) p. 297.
Two hundred and twenty Poles per
thousand were killed in the war, as
opposed to 124 per thousand Russians
and 74 per thousand Germans
7 An exception is obviously José M.
Sánchez, *Pius XII and the Holocaust*
(Washington DC 2002) pp. 137–9
8 See John Pollard, 'The Papacy in Two
World Wars. Benedict XV and Pius
XII Compared', *Totalitarian

Movements and Political Religions
(2001) 2, p. 90
9 As this book is not a history of the
Holocaust, readers should consult the
journals *Yad Vashem Studies* and the
Vierteljahreshefte für Zeitgeschichte for
the recent historiography, and *Civiltà
Cattolica* for relevant material on the
responses of the Catholic Church. For a
comprehensive account of the ongoing
'Pius War' see Joseph Bottum and
David S. Dalin (eds), *The Pius War.
Responses to the Critics of Pius XII*
(Lanham, Md 2004) and especially the
huge bibliography of William Doino
10 For this important point see Robert
Graham, 'Papst Pius XII und seine
Haltung zu den Kriegsmächten' in
Herbert Schambeck (ed.), *Pius XII zum
Gedächtnis* (Berlin 1977) pp. 144–6
11 Pierre Blet et al. (eds), *Actes et
Documents du Saint-Siège* [hereafter
cited as *ADSS*] (Vatican City 1965–81)
1, nrs 18–46, pp. 118–49 for this
correspondence between the Vatican,
its nuncios and the Powers about the
May 1939 peace initiatives
12 *ADSS* 1, nr 113 pp. 230–8
13 Zygmunt Jakubowski, *Pope Pius and
Poland* (New York 1942) pp. 15–16
14 *Inter Arma Caritas. The Vatican Office
of Information for Prisoners of War
Instituted by Pius XII (1939–1947)*
(2 volumes Rome 2004)
15 Peter Hoffmann, 'Roncalli in the
Second World War. Peace Initiatives,
the Greek Famine and the Persecution
of the Jews', *Journal of Ecclesiastical
History* (1989) 40, pp. 77ff.
16 *ADSS* 6, nr 355, pp. 455–7
17 Antonio Gaspari, 'Uncovered
Correspondence of Pius XII', *Inside the
Vatican* (February 2003) pp. 14–16 with
facsimiles of the two letters relaying the
money. Palatucci's policeman nephew
Giovanni would be killed in Dachau
for his role (in conjunction with the
bishop) in rescuing five thousand Jews.
ADSS 8, nr 348, pp. 505–7 for the letter
from the Jews at Ferramonte Tarsia
with its explicit acknowledgement that
the pope 'is not solely the Father and
highest shepherd of Catholics around
the entire world, but at the same time

the fatherly protector and realiser of the humanitarian ideal of humanity as a whole'

18 These examples are taken from *ADSS* 8, nr 275, pp. 430–1; nr 259, pp. 413–15; nr 276, pp. 431–2; and nr 392, pp. 553–5. The quotation is from secretary of state Maglione to Osborne dated 24 November 1941 in *ADSS* 8, nr 207, p. 354. The British refused to relax the naval blockade to enable food to be brought into Greece

19 *ADSS* 1, nr. 213, pp. 315–23

20 'Pope condemns dictators, treaty violators, racism; urges restoring of Poland', *New York Times*, 28 October 1939, front page and p. 9; the quotation from Heinrich Müller is from Saul Friedländer, *Pius XII and the Third Reich* (New York 1966) p. 37 where the document is cited in full

21 *The Pope's Five Peace Points. Address of Pope Pius XII to the Sacred College of Cardinals on Christmas Eve 1939* (London 1940) p. 256 and *Osservatore Romano*, 26–27 December 1939 pp. 1–2; the casualty figures are from Alexander B. Rossino, *Hitler Strikes Poland* (Lawrence, Kansas 2003) p. 234

22 José M. Sánchez, *Pius XII and the Holocaust. Understanding the Controversy* (Washington DC 2002) p. 51

23 August Hlond, *The Persecution of the Catholic Church in German-Occupied Poland* (London 1941) p. 122 for the text of this broadcast to Poles in the US; see also R. M. Macdonald (ed.), *The Pope Speaks* (London 1942) p. 106

24 *ADSS* 3, nr 96, p. 195

25 Harold H. Tittmann Jr, *Inside the Vatican of Pius XII. The Memoirs of an American Diplomat during World War II* (New York 2004) pp. 111–15

26 Klemens von Klemperer, *German Resistance against Hitler. The Search for Allies Abroad 1938–1945* (Oxford 1992) pp. 171ff.

27 For the Foreign Office documents relating to these contacts see Peter Ludlow, 'Papst Pius XII, die britische Regierung und die deutsche Opposition im Winter 1939/40' *Vierteljahreshefte für Zeitgeschichte* (1974) 22, pp. 299–39

28 David Alvarez and Robert Graham, *Nothing Sacred. Nazi Espionage against the Vatican, 1939–1945* (London 1997) pp. 24–33

29 John Conway, *The Nazi Persecution of the Churches 1933–1945* (London 1968) pp. 305–7 for details of protests that would include interventions on behalf of concentration-camp inmates, hostages, Jews expelled from France, starving children in Greece, and refugees from Yugoslavia

30 Friedländer, *Pius XII and the Third Reich* p. 39

31 For a good recent discussion of this see Philippe Burrin, *Ressentiment et apocalypse. Essai sur l'antisémitisme Nazi* (Paris 2004); various essays in Wolfgang Hartwig (ed.), *Utopie und politische Herrschaft im Europa der Zwischenkriegszeit* (Munich 2003) touch on some of these points without offering anything new

32 On these policies see Michael Burleigh, *Death and Deliverance. Euthanasia in Germany 1900–1945* (Cambridge 1994) and Henry Friedländer, *The Origins of Nazi Genocide* (Chapel Hill 1995), although US historians invariably reverse their order of appearance

33 *ADSS* 6, nr 391, pp. 499–500

34 *ADSS* 2, nr 33, pp. 102–3

35 For a facsimile of the sermon see Joachim Kuropka, *Clemens August Graf von Galen. Sein Leben und Wirken in Bildern und Dokumenten* (Cloppenburg 1992) pp. 217ff.

36 *ADSS* 2, nr 76, pp. 230–1, 30 September 1941

37 *ADSS* 2, nr 84, p. 257, Pius XII to Konrad Gröber dated 1 March 1942

38 Charles Pichon, *The Vatican and its Role in World Affairs* (New York 1950) p. 158

39 'Pope is Emphatic about Just Peace. Jews' rights defended', *New York Times*, 14 March 1940

40 John Conway, 'The Meeting between Pope Pius XII and Ribbentrop', *Canadian Journal of History* (1968) 1, p. 107 and *ADSS* 1, nrs 254–9 for the Vatican record of these meetings

41 *Osservatore Romano*, 12 May 1940 pp. 13–14

42 *ADSS* 1, nr 312, p. 453

43 Owen Chadwick, *Britain and the Vatican during the Second World War* (Cambridge 1986) p. 112

44 Ronald J. Rychlak, *Hitler, the War, and the Pope* (Columbus, Miss. 2000) p. 140

45 *ADSS* 1, nr 374, pp. 508–9; see also Owen Chadwick, 'The Papacy and World War II', *Journal of Ecclesiastical History* (1967) 18, p. 76

46 Jonathan Steinberg, *All or Nothing. The Axis and the Holocaust 1941–1943* (London 1990) pp. 17ff.

47 Robert Graham, *The Vatican and Communism during World War II. What Really Happened?* (San Francisco 1996) p. 91

48 Robert Conquest, *Religion in the USSR* (New York 1968) p. 34

49 Graham, *The Vatican and Communism during World War II* pp. 38–40

50 Dimitry Pospielovsky, *The Russian Church under the Soviet Regime 1917–1982* (New York 1984) 1, p. 200

51 Alexander Dallin, *German Rule in Occupied Russia 1941–1945. A Study of Occupation Politics* (London 1981) p. 474

52 Xavier de Montclos, *Les Chrétiens face au Nazisme et au Stalinisme. L'épreuve totalitaire, 1939–1945* (Paris 1983) p. 93

53 John Cornwell, *Hitler's Pope. The Secret History of Pius II* (London 1999) p. 264

54 Graham, *The Vatican and Communism during World War II* pp. 124–7

55 W. D. Halls, *Politics, Society and Christianity in Vichy France* (Oxford 1995) pp. 37–8

56 Michèle Cointet, *L'Église sous Vichy* (Paris 1998) p. 25

57 Antonio Costa Pinto, 'Le Portugal. L'état nouveau de Salazar' in Jean-Pierre Azéma and François Bédarida (eds), *Vichy et les Français* (Paris 1996) pp. 674ff. is good on these comparative issues

58 Renée Bédarida, *Les Catholiques dans la guerre 1939–1945* (Paris 1998) p. 53

59 Julian Jackson, *France. The Dark Years 1940–1944* (Oxford 2000) p. 364

60 Susan Zuccotti, *The Holocaust, the French, and the Jews* (Lincoln, Nebr. 1993) p. 54

61 Bédarida, *Les Catholiques dans la guerre* p. 173

62 Michael Marrus, 'French Churches and the Persecution of Jews in France, 1940–1944' in Otto Dov Kulka and Paul R. Mendes-Flohr (eds), *Judaism and Christianity under the Impact of National Socialism* (Jerusalem 1987) p. 311

63 Halls, *Politics, Society and Christianity in Vichy France* p. 109

64 Henri de Lubac, *Christian Resistance to Anti-Semitism: Memories from 1940–1944* (San Francisco 1990) pp. 66–70 for both texts

65 Cointet, *L'Église sous Vichy* pp. 203–5

66 Pierre Blet et al. (eds), *Pius XII and the Second World War. According to the Archives of the Vatican* (New York 1999) pp. 232–3

67 Jacques Adler, 'The "Sin of Omission"? Radio Vatican and the Anti-Nazi Struggle 1940–1942', *Australian Journal of Politics and History* (2004) 50, pp. 396–406

68 Renée Bédarida, *Pierre Chaillet. Témoin de la résistance spirituelle* (Paris 1988)

68 See especially Renée Bédarida, *Les Armes de l'esprit. Témoinage chrétien, 1941–1944* (Paris 1977)

70 *The Times* dated 11 September 1942 cited by Martin Gilbert, *The Righteous. The Unsung Heroes of the Holocaust* (New York 2003) p. 263

71 Cointet, *L'Église sous Vichy* pp. 265ff.

72 Bob Moore, *Victims and Survivors. The Nazi Persecution of the Jews in the Netherlands 1940–1945* (London 1997) p. 59

73 *Tablet*, 29 August 1942, p. 103

74 Yitshak Arad, 'Stalin and the Soviet Leadership. Responses to the Holocaust' in John K. Roth and Elisabeth Maxwell (eds), *Remembering for the Future. The Holocaust in an Age of Genocide* (Basingstoke 2001) 1, pp. 355ff. The relative neglect of Soviet responses to the Holocaust seems curious when compared with the vast literatures devoted to neutrals and Western democracies

75 This is especially true of Friedländer, *Pius XII and the Third Reich* pp. 117–24

which does not address the four-month
interval between US receipt of the
Riegner Telegram and the Allied
response

76 See Gerhart M. Riegner, 'Riegner
Telegram' in Walter Laqueur (ed.),
The Holocaust Encyclopedia (New
Haven 2001) pp. 562–7

77 *ADSS* 8, nr 300, p. 455

78 *ADSS* 8, nr 298, p. 453, and nr 301,
p. 456

79 *ADSS* 8, nr 303, p. 458

80 The aide-mémoire is published in
Friedländer, *Pius XII and the Third
Reich* pp. 104–10

81 *ADSS* 8, nr. 342, p. 501

82 Chadwick, *Britain and the Vatican
during the Second World War*
pp. 208–11

83 *ADSS* 3, part ii, nr 406, pp. 636ff.

84 Tittmann, *Inside the Vatican of Pius XII*
pp. 118ff.

85 Thomas Moloney, *Westminster,
Whitehall and the Vatican. The Role of
Cardinal Hinsley 1935–1943* (London
1985) p. 174

86 *ADSS* 8, nr 496, pp. 669–70, and
Carlo Falconi, *The Silence of Pius XII*
(Boston 1970) p. 170. For a reasoned
discussion of Pius see Meir Michaelis,
*Mussolini and the Jews. German–Italian
Relations and the Jewish Question
in Italy 1922–1945* (Oxford 1978)
pp. 372–7

87 Anthony Rhodes, *The Power of Rome in
the Twentieth Century*, vol. 2: *The
Vatican in the Age of Dictators 1922–1945*
(London 1973) p. 288

88 The most illuminating study is Martin
Blinkhorn (ed.), *Fascists and
Conservatives* (London 1990)

89 David Dalin, *The Myth of Hitler's Pope*
(Washington DC 2005) pp. 131ff.

90 For a good overview see John S.
Conway, 'The Vatican, Germany and
the Holocaust' in Peter C. Kent and
John F. Pollard (eds), *Papal Diplomacy
in the Modern Age* (Westport 1994)
pp. 105–20

91 *ADSS* 1, nr 210, p. 313, 26 September
1939

92 Montclos, *Les Chrétiens face au Nazisme
et au Stalinisme* p. 142

93 *ADSS* 8, nr 199, pp. 345–7

94 Livia Rothkirchen, 'The Vatican and
the "Jewish Problem" in Slovakia',
Yad Vashem Studies (1967) 6, p. 39

95 Livia Rothkirchen, 'Czechoslovakia'
in David S. Wyman (ed.), *The World
Reacts to the Holocaust* (Baltimore
1996) p. 170

96 Rothkirchen, 'The Vatican and the
"Jewish Problem" ' p. 48

97 Blet et al. (eds), *Pius XII and the
Second World War* pp. 177–8

98 *ADSS* 8, nr 426, p. 598

99 Marcus Tanner, *Croatia. A Nation
Forged in War* (New Haven 1997)
pp. 141ff. for these events

100 *ADSS* 9, nr 130, especially annexe II
Stepinac to Maglione dated 24 May
1943, pp. 222–224, listing the regime's
beneficial measures

101 Hubert Butler, *Independent Spirit.
Essays* (New York 1996) contains
several essays on Croatia, the Church
and Ireland

102 Montclos, *Les Chrétiens face au
Nazisme et au Stalinisme* p. 156

103 Tittmann, *Inside the Vatican of Pius
XII* pp. 50–1

104 Rhodes, *The Power of Rome* 2, p. 326

105 Menachem Shelah, 'The Catholic
Church in Croatia, the Vatican and
the Murder of the Croatian Jews' in
his *Remembering for the Future*
(Oxford 1988) 1, p. 270

106 Stella Alexander, *The Triple Myth.
A Life of Archbishop Alojzije Stepinac*
(New York 1987) p. 81

107 Ibid. p. 85

108 As conceded by the otherwise critical
Menachem Shelah, 'The Catholic
Church in Croatia' p. 332

109 Rhodes, *The Power of Rome* 2,
pp. 330–1

110 Shelah, 'The Catholic Church in
Croatia' p. 334

111 *ADSS* 9, nr 130, annexe iii, pp. 224–9

112 The most interesting book on
Romanian Fascism is Radu Ioanid,
*The Sword of the Archangel. Fascist
Ideology in Romania* (Boulder 1990)

113 Payne, *A History of Fascism 1919–1945*
(London 1998) pp. 392ff.

114 Raul Hilberg, *Die Vernichtung der
europäischen Juden* (Frankfurt am
Main 1990) 2, p. 817

115 Jean Ancel, 'The "Christian" Regime of Romania and the Jews, 1940–1942', *Holocaust and Genocide Studies* (1993) 7, p.21

116 *ADSS* 5, nr 93, pp. 240–1, letter dated 20 September 1941; see also Graham, *The Vatican and Communism during World War II* pp. 38–9

117 Radu Ioanid, *The Holocaust in Romania* (Chicago 2000) p. 246 gives the nuncio his due

118 *ADSS* 8, nr 531, p. 702, nuncio Bernardini to cardinal Maglione dated 29 October 1942

119 *ADSS* 9, nr 52, p. 128, nuncio Cassulo to cardinal Maglione dated 14 February 1943

120 'Declaration' by Alexandre Safran reprinted in Jean Ancel (ed.), *Documents Concerning the Fate of Romanian Jewry during the Holocaust* (New York 1986) 8, pp. 599–601

121 E.g. *ADSS* 9, nr 219, pp. 330–2 Cassulo to Maglione detailing his interventions with Radu Lecca and the concrete improvements he had extracted from the Romanian authorities on behalf of the Jews

122 *ADSS* 10, nr 211, annexes i and ii, pp. 291–2; see also Theodore Lavi, 'The Vatican's Endeavors on Behalf of Rumanian Jewry during the Second World War', *Yad Vashem Studies* (1963) 5, pp. 405–18

123 Hilberg, *Die Vernichtung der europäischen Juden* pp. 794–5

124 Hoffmann, 'Roncalli in the Second World War', see also Michael Bar-Zohar, *Beyond Hitler's Grasp. The Heroic Rescue of Bulgaria's Jews* (Holbrook 1998)

125 Tzvetan Todorov, *The Fragility of Goodness. Why Bulgaria's Jews Survived the Holocaust* (London 1999) p. 128

126 Rychlak, *Hitler, the War and the Pope* pp. 202–4

127 John Caroll-Abbing, *But for the Grace of God* (Rome 1965) p. 56

128 The most balanced account is by Owen Chadwick, 'Weizsäcker, the Vatican and the Jews of Rome', *Journal of Ecclesiastical History* (1977) 28, pp. 179ff.

129 Susan Zuccotti, *The Italians and the Holocaust. Persecution, Rescue and Survival* (London 1987) pp. 132–4 is considerably more balanced in approach than her later *Under his Very Windows. The Vatican and the Holocaust in Italy* (New Haven 2000). On the latter see Ronald Rychlak, 'Comments on Susan Zuccotti's *Under his Very Windows*', *Journal of Modern Italian Studies* (2002) 7, pp. 218ff.

130 Rychlak, *Hitler, the War and the Pope* p. 217

131 *ADSS* 10, nr 117, p. 191, and nr 133 p. 206

132 Jenö Levai, *Hungarian Jewry and the Papacy. Pope Pius XII did Not Remain Silent* (London 1968) pp. 21–2

133 A. C. F. Beales, *The Pope and the Jews* (London 1945) pp. 35–6

134 Hoffmann, 'Roncalli in the Second World War' p. 86

135 *ADSS* 10, nr 270, p. 357, and nr 273, p. 359

136 For numerous examples of Christian religious houses helping Jews see Eugene Levai, *Black Book on the Martyrdom of Hungarian Jewry* and Levai, *Hungarian Jewry and the Papacy* pp. 44–5 (this latter title uses the anglicised version of the Hungarian author's name cited above in note 132)

137 For these statistics see Martin Gilbert, *The Righteous* p. 405

138 István Deák, 'The Pope, the Nazis and the Jews', *New York Review of Books*, 23 March 2000, pp. 44–9. It is to be sincerely hoped that people like Deák will devote as much space to reviewing the large number of recent books defending Pius XII as they do to those attacking him, although one is not optimistic that the *New York Review of Books* will facilitate this on such a generous scale

Chapter 5: Resistance, Christian Democracy and the Cold War

1 G. Quazza, 'The Politics of the Italian Resistance' in S.J. Woolf (ed.), *The Rebirth of Italy 1943–1950* (London 1972) p. 28

2 See the penetrating discussion in Hans Maier, 'Christliche Widerstand. Der Fall des Dritten Reiches' in his *Politische Religionen. Die totalitären Regime und das Christentum* (Freiburg im Breisgau 1995) pp. 62ff. See also the still useful Luigi Sturzo, 'The Right to Rebel' in his *Politics and Morality. Essays in Christian Democracy* (London 1938) pp. 195–212. Sturzo was exiled for his rebellion against the Church's accommodations with Fascism

3 Renée Bédarida, *Les Catholiques dans la guerre 1939–1945* (Paris 1998) p. 126

4 Georges Bernanos, *Plea for Liberty. Letters to the English, the Americans, the Europeans* (New York 1944)

5 See Harry Roderick Kedward, *Resistance in Vichy France* (Oxford 1978) pp. 28ff.

6 W. D. Halls, *Politics, Society and Christianity in Vichy France* (Oxford 1995) pp. 204–5

7 François and Renée Bédarida (eds), *La Résistance spirituelle 1941–1944. Les Cahiers clandestins du témoignage chrétien* (Paris 2001) pp. 117–56 and 213–21 for the texts

8 A. Shennan, *Rethinking France. Plans for Renewal 1940–1946* (Oxford 1989)

9 Maurice Larkin, *France since the Popular Front. Government and People 1936–1996* (Oxford 1997) p. 109

10 Michael Kelly, 'Catholics and Communism in Liberation France, 1944–47' in Frank Tallett and Nicholas Atkins (eds), *Religion, Society and Politics in France since 1789* (London 1991) pp. 187ff.

11 See the excellent discussion in R. E. M. Irving, *Christian Democracy in France* (London 1973) pp. 57–65

12 Georges Bidault, *Resistance. The Political Autobiography of Georges Bidault* (London 1965) pp. 15–17

13 Jean-Marie Mayeur, *Des partis catholiques à la Démocratie chrétienne xix–xx siècles* (Paris 1980) p. 173

14 For the above see James McMillan, 'France' in Tom Buchanan and Martin Conway (eds), *Political Catholicism in Europe 1918–1965* (Oxford 1996) pp. 59ff. and Irving, *Christian Democracy in France* (London 1973)

15 Quazza, 'The Politics of the Italian Resistance' p. 17

16 Richard A. Webster, *The Cross and the Fasces. Christian Democracy and Fascism in Italy* (Stanford 1960) p. 165

17 Peter Hebblethwaite, *Paul VI. The First Modern Pope* (London 1993) p. 194

18 Paul Ginsborg, *A History of Contemporary Italy 1943–1980* (London 1990) pp. 16–17

19 F. Chabod, *L'Italia contemporanea 1918–1948* (Turin 1961) p. 125

20 James Edward Miller, *The United States and Italy 1940–1950. The Politics and Diplomacy of Stabilization* (Chapel Hill 1986)

21 Hebblethwaite, *Paul VI* p. 193

22 Ennio di Nolfo, *Vaticano e Stati Uniti 1939–1952* (Milan 1978)

23 Anthony Rhodes, *The Power of Rome in the Twentieth Century*, vol. 3: *The Vatican in the Age of the Cold War 1945–1980* (London 1992) pp. 144ff.

24 Mayeur, *Des partis catholiques* p. 177

25 John Pollard, 'Italy' in Buchanan and Conway (eds), *Political Catholicism in Europe 1918–1965* p. 87

26 R. E. M. Irving, *The Christian Democratic Parties of Western Europe* (London 1979) p. 58

27 Elisa Carrillo, *Alcide de Gasperi. The Long Apprenticeship* (Notre Dame 1965) p. 84

28 Mario Einaudi and François Goguel, *Christian Democracy in Italy and France* (Notre Dame 1952) p. 31

29 Francesco Traniello, 'Political Catholicism, Catholic Organization, and Catholic Laity in the Reconstruction Years' in Frank Coppa and Margherita Repetto-Alaia (eds), *The Formation of the Italian Republic* (New York 1989) p. 45

30 Martin Clark, *Modern Italy 1871–1982* (London 1984) p. 317

31 Norman Kogon, *A Political History of Italy. The Postwar Years* (New York 1983) p. 29

32 Peter Hebblethwaite, 'Pope Pius XII. Chaplain of the Atlantic Alliance?' in Christopher Duggan and Christopher Wagstaff (eds), *Italy in the Cold War. Politics, Culture and Society 1948–58* (Oxford 1995) p. 73

33 For this see Hans Woller, 'Die Entscheidungswahlen vom April 1948' in Woller (ed.), *Italien und die Grossmächte 1943-1949* (Munich 1988) pp. 85–6

34 Hebblethwaite, 'Pope Pius XII' p. 72

35 Rhodes, *The Power of Rome* 3, p. 163

36 Piero Bevilacqua, 'Custom' in Omar Calabrese (ed.), *Modern Italy. Images and History of a National Identity* (Milan 1984) 3, p. 192

37 Schwering cited by John L. Allen, *Cardinal Ratzinger. The Vatican's Enforcer of the Faith* (London 2000) p. 30

38 Frederic Spotts, *The Churches and Politics in Germany* (Middetown, Conn. 1973)

39 Hans-Peter Schwarz, *Adenauer*, vol. 1: *Der Aufstieg 1876–1952* (Munich 1994) pp. 440–1

40 Dennis L. Bark and David R. Gress, *A History of West Germany* (Oxford 1989) 1, p. 148

41 See the important study by Jean Solchany, 'Vom antimodernismus zum antitotalitarismus. Konservative Interpretationen des national-sozialismus in Deutschland 1945–1949', *Vierteljahreshefte für Zeitgeschichte* (1996) 44, pp. 373–94

42 Rainer Bendel, Lydia Bendel-Maidl and Andreas Goldschmidt, 'Vergangenheitsbewältigung in theologischen Schriften Joseph Bernharts, Romano Guardinis und Alois Winklhofers', *Kirchliche Zeitgeschichte* (2000) 13, pp. 138–77

43 Schwarz, *Adenauer* 1, p. 514

44 For the full text see Konrad Repgen's thoughtful discussion in his 'Die Erfahrung des Dritten Reiches und das Selbstverstandnis der deutschen Katholiken nach 1945' in Victor Conzemius, Martin Greschat and Hermann Kocher (eds), *Die Zeit nach 1945 als Thema kirchlicher Zeitgeschichte* (Göttingen 1988) pp. 127–79

45 John Conway, 'How Shall the Nations Repent? The Stuttgart Declaration of Guilt, October 1945', *Journal of Ecclesiastical History* (1987) 38, p. 621; see also Gerhard Besier and Gerhard Sauter, *Wie Christen ihre Schuld bekennen. Die Stuttgarter Erklärung 1945* (Göttingen 1985)

46 Spotts, *The Churches and Politics in Germany* p. 304

47 Geoffrey Pridham, *Christian Democracy in Western Germany* (London 1977) p. 23

48 For the foundation of the CDU/CSU see Noel D. Cary, *The Path to Christian Democracy. German Catholics and the Party System from Windthorst to Adenauer* (Cambridge, Mass. 1996) pp. 147ff.

49 Spotts, *The Churches and Politics in Germany* p. 172

50 For an excellent discussion of Christian Democracy see Kees van Kersbergen, 'The Distinctiveness of Christian Democracy' in David Hanley (ed.), *Christian Democracy in Europe. A Comparative Perspective* (London 1994) pp. 31–47

51 Schwarz, *Adenauer* 1, p. 773

52 Heinrich-August Winkler, *Der lange Weg nach Westen. Deutsche Geschichte vom 'Dritten Reich' bis zur Wieder-vereinigung* (Munich 2000) 2, pp. 147–51

53 Paul Preston, *Franco* (London 1993) pp. 329–30

54 Juan Linz, 'Staat und Kirche in Spanien' in Martin Greschat and Jochen-Christoph Kaiser (eds), *Christentum und Demokratie im 20. Jahrhundert* (Stuttgart 1992) p. 66

55 Frances Lannon, *Privilege, Persecution, and Prophecy. The Catholic Church in Spain 1875–1975* (Oxford 1987) pp. 215–16

56 Stanley Payne, *Spanish Catholicism. An Historical Overview* (Madison, Wis. 1984) p. 184

57 Stanley Payne, *The Franco Regime 1936–1975* (London 1987) pp. 382–3

Chapter 6: The Road to Unfreedom: The Imposition of Communism after 1945

1 Robert Conquest, an eyewitness to these events in Bulgaria, is characteristically clear-minded about matters of fact and chronology; see his *Reflections on a Ravaged Century* (London 1999) pp. 153–9. For a benign

interpretation of the imposition of 'people's democracy' upon eastern Europe see most obviously Eric Hobsbawm, *The Age of Extremes. The Short Twentieth Century 1914–1991* (London 1994) p. 238 who claims that only after the Italian 1948 elections did the Communists 'follow suit' by 'eliminating' non-Communists from coalition governments of the 'people's democracies'. In fact, the chronology is completely the reverse. Mark Mazower, *Dark Continent. Europe's Twentieth Century* (London 1998) pp. 253ff. gives a similarly selective account of how the Communists actually achieved power. To take one example, from many, Mazower tells readers that the Hungarian Smallholders Party won the 1945 elections to 'illustrate' the alleged 'hesitancies' of Soviet strategy; he does not inform us that two years later the Party's general secretary was abducted by the NKVD while walking home and disappeared, or that prime minister Ferenc Nagy was told to stay abroad, lest anything happen to his infant son at home, a threat that ensured Nagy's resignation. It is clearly not good form in some circles to mention the sordid nature of Communist rule

2 Milovan Djilas, *Tito. The Story from the Inside* (London 1981) p. 176

3 See Maria Schmidt, 'Ungarns Gesellschaft in der Revolution und im Freiheitskampf von 1956', *Kirchliche Zeitgeschichte* (2004) 17, pp. 102–3 which is a sober discussion by a leading expert on Communist Terror based at the former terror headquarters in Budapest

4 Jozsef Cardinal Mindszenty, *Memoirs* (London 1974) p. 31

5 Mazower, *Dark Continent* pp. 255–6 implies such a universal retreat, citing the example of Hungary and general troop statistics; but see R. J. Crompton, *Eastern Europe in the Twentieth Century* (London 1994) p. 244 for the February 1947 peace treaty that specifically sanctioned the Red Army's continued presence

6 John Micgiel, '"Bandits and Reactionaries". The Suppression of

Opposition in Poland' in Norman Nymark and Leonid Gibianskii (eds), *The Establishment of Communist Regimes in Eastern Europe, 1944–1949* (Westview, Conn. 1997) pp. 97–8

7 Stephane Courtois et al., *The Black Book of Communism. Crimes. Terror. Repression* (Cambridge, Mass. 1999) p. 408. Interestingly, this international bestseller by several leading experts was rejected by a British publishing industry that is otherwise so eager to publish anything about the Nazis

8 Schmidt, 'Ungarns Gesellschaft' pp. 102–3

9 For examples see Jerzy Holzer, *Der Kommunismus in Europa. Politische Bewegung und Herrschaftssystem* (Frankfurt am Main 1998) p. 87

10 On this important theme see George Schöpflin, *Politics in Eastern Europe 1945–1992* (Oxford 1993) pp. 68–9

11 Bradley Abrams, 'The Politics of Retribution. The Trial of Josef Tiso in the Czechoslovak Environment' in István Deák, Jan T. Gross and Tony Judt (eds), *The Politics of Retribution in Europe. World War II and its Aftermath* (Princeton 2000) pp. 261–3

12 Anthony Rhodes, *The Power of Rome in the Twentieth Century*, vol. 3: *The Vatican in the Age of the Cold War 1945–1980* (London 1992) p. 65

13 Marcus Tanner, *Croatia. A Nation Forged in War* (New Haven 1997) p. 179

14 Stella Alexander, *The Triple Myth. A Life of Archbishop Alojzije Stepinac* (Boulder 1987) pp. 126–7

15 For the details of the US response see Peter C. Kent, *The Lonely Cold War of Pius XII* (Montreal 2002) pp. 168–73

16 József Fuisz, 'Beitrag der Religionsgemeinschaften zum Aufstand 1956', *Kirchliche Zeitgeschichte* (2000) 17, p. 117

17 Kent, *The Lonely Cold War of Pope Pius XII* p. 113

18 Mindszenty, *Memoirs* pp. 10–11

19 Rhodes, *The Power of Rome* 3, p. 30

20 Mindszenty, *Memoirs* p. 81

21 Jan Siedlarz, *Kirche und Staat im kommunistischen Polen 1945–1989* (Paderborn 1996) pp. 51ff.

22 Adam Michnik, *The Church and the Left* (Chicago 1993) p. 61
23 Michael C. Steinlauf, 'Poland' in David S. Wyman (ed.), *The World Reacts to the Holocaust* (Baltimore 1996) pp. 110–14 seems fair-minded on these complex issues
24 George Weigel, *The Final Revolution. The Resistance Church and the Collapse of Communism* (New York 1992) pp. 107ff. contains an astute appreciation of Wyszyński
25 Michnik, *The Church and the Left* pp. 61–2
26 The accord is reproduced in Siedlarz, *Kirche und Staat* pp. 77–9
27 Kent, *The Lonely Cold War of Pius XII* pp. 174–6

Chapter 7: Time of the Toy Trumpets

1 Philip Larkin, *Collected Poems* (London 2003) p. 58
2 Callum Brown, *The Death of Christian Britain. Understanding Secularization 1800–2000* (London 2001) pp. 172–3
3 Adrian Hastings, *A History of English Christianity 1920–2000* (London 2001 fourth edition) pp. 454–5
4 Hugh McLeod, 'The Sixties. Writing the Religious History of a Crucial Decade', *Kirchliche Zeitgeschichte* (2001) 14, pp. 40–1
5 Hastings, *A History of English Christianity* pp. 551–2; see also Steve Bruce, *God is Dead. Secularization in the West* (Oxford 2002) pp. 66ff., and Bruce, *Religion in Modern Britain* (Oxford 1995) pp. 32–44 for various statistical tables
6 Larkin, *Collected Poems* p. 146
7 On this see the brilliant discussion by Karl-Dietrich Bracher, *The Age of Ideologies. A History of Political Thought in the Twentieth Century* (London 1984) p. 213
8 As noted by Robert Conquest, *Reflections on a Ravaged Century* (London 1999). Conquest spent four years fighting Hitler; witnessed the Communist takeover in Bulgaria as a foreign service officer, and then spent decades researching the reality of the Soviet Union
9 Dominic Sandbrook, *Never Had it So Good. A History of Britain from Suez to the Beatles 1956–1963* (London 2005) p. xii
10 G. I. T. Machin, *Churches and Social Issues in Twentieth-Century Britain* (Oxford 1998) pp. 187–8
11 Gerald Parsons, 'Between Law and Licence. Christianity, morality and "permissiveness" ' in Parsons (ed.), *The Growth of Religious Diversity. Britain from 1945* (London 1994) p. 243
12 Philip Allen, 'A Young Home Secretary' in Andrew Adonis and Keith Thomas (eds), *Roy Jenkins. A Retrospective* (Oxford 2004) pp. 64ff. Allen was the permanent under-secretary of state in the Home Office at the time
13 Roy Jenkins, *A Life at the Centre* (London 1991)
14 Graham Dale, *God's Politicians. The Christian Contribution to 100 Years of Labour* (London 2000) p. 176
15 Arthur Marwick, *The Sixties. Cultural Revolution in Britain, France, Italy and the United States c. 1958–c. 1974* (Oxford 1998) p. 265
16 Jenny Pollock, 'Landmark Social Housing Left to Crumble', *Guardian*, 28 May 2001 gives the general flavour of life in Sheffield's Park Hill flats; anyone driving into the southbound Blackwall Tunnel can see the Goldfinger Tower looming to the right
17 Eric Hobsbawm, *The Age of Extremes. The Short Twentieth Century 1914–1991* (London 1994) p. 330
18 Cynthia Lennon, *John* (London 2005)
19 'Religious belief "falling faster than church attendance" ', *Daily Telegraph*, 20 August 2005 reporting on a new study of 10,500 households conducted by the University of Manchester. Reasonably enough, an Anglican spokesman objected that parents were not the sole means for the transmission of religious faith, the Alpha courses having been successful in recruiting young middle-class people
20 Brown, *Death of Christian Britain* pp. 176–80
21 David Lodge, *How Far Can You Go?* (London 1980) p. 133

22 Sandbrook, *Never Had it So Good* pp. 245ff.

23 Bruce, *God is Dead* pp. 192–4

24 As courageously explained by Trevor Phillips, the Chairman of the Commission for Racial Equality in Britain (September 2005)

25 See John Wolffe, 'How Many Ways to God? Christians and Religious Pluralism' in Parsons (ed.), *The Growth of Religious Diversity* pp. 31–3

26 See the interesting discussion by John Wolffe, 'The Religions of the Silent Majority' in Parsons (ed.), *Growth of Religious Diversity* pp. 318ff.

27 Bruce, *God is Dead* p. 80

28 Bruce, *Religion in Modern Britain* p. 95 for the beginnings of these cults in the late 1960s and early 1970s

29 Bruce, *God is Dead* p. 81

30 Tom Stransky, 'The Secretariat for Promoting Christian Unity' in Adrian Hastings (ed.), *Modern Catholicism. Vatican II and After* (London 1991) pp. 182–3

31 There is a vast literature on Vatican II. For our purposes see the discussion in Hubert Jedin (ed.), *History of the Church*, vol. 10: *The Modern Age* (London 1981) pp. 96–150

32 See John McDade, 'Catholic Theology in the Post-Conciliar Period' in Hastings (ed.), *Modern Catholicism* pp. 422ff. which is knowledgeable and fair-minded

33 Frank Coppa, *The Modern Papacy since 1789* (London 1998) p. 223

34 Peter Hebblethwaite, *Pope Paul VI. The First Modern Pope* (London 1993) p. 380

35 Ibid. p. 437

36 Mary Vincent, 'Spain' in Tom Buchanan and Martin Conway (eds), *Political Catholicism in Europe 1918–1965* (Oxford 1996) p. 125. Vincent makes the telling point that 90 per cent of Spanish bishops had been ordained before 1936 and that most were over seventy-five years of age. The clergy were the youngest in the world

37 Norman B. Cooper, *Catholicism and the Franco Regime* (London 1975) p. 27

38 Stanley Payne, *Spanish Catholicism. An Historical Overview* (Madison, Wis. 1984) pp. 189–91

39 Cooper, *Catholicism and the Franco Regime* pp. 36–7

40 Stanley Payne, *The Franco Regime 1936–1975* (London 1987) pp. 588–90

41 See the important discussion in Juan Linz, 'Staat und Kirche in Spanien' in Martin Greschat and Jochen-Christoph Kaiser (eds), *Christentum und Demokratie im 20. Jahrhundert* (Stuttgart 1992) pp. 80–1

42 Olivier Compagnon, 'Latein Amerika' in Jean-Marie Mayeur (ed.), *Die Geschichte des Christentums, Religion, Politik, Kultur*, vol. 13: *Krisen und Erneuerung (1958–2000)* (Freiburg 2002) p. 507

43 There is a fair-minded discussion of Ratzinger's point of view in John L. Allen's *Cardinal Ratzinger, The Vatican's Enforcer of the Faith* (London 2000) pp. 131ff.

44 This discussion owes much to Christopher Rowland (ed.), *The Cambridge Companion to Liberation Theology* (Cambridge 1999)

Chapter 8: 'The Curse of Ulster': The Northern Ireland Troubles c. 1968–2005

1 Toby Harnden, *'Bandit Country'. The IRA & South Armagh* (London 1999) pp. 239ff., and Kevin Myers, 'An Irishman's Diary', *Irish Independent*, 21 September 2005. I am indebted to Professor Desmond King of Nuffield College Oxford, a Quaker from the Irish Republic, for this and other references to Irish newspapers which I do not see. He is not responsible for my views on them. Also Thomas Hennessey, *A History of Northern Ireland 1920–1996* (London 1997) pp. 82ff.

2 Alvin Jackson, *Ireland 1798–1998* (Oxford 1999) p. 388

3 Jeremy Warner 'Outlook', *Independent*, 18 October 2005

4 On this see the excellent Hubert Butler, 'The Artukovich File' in his *Independent Spirit. Essays* (New York 1995) pp. 465ff.

5 On this see Frank Furedi, 'The Age of Unreason', *Spectator*, 19 November 2005 pp. 40–2. It is worth recalling that Furedi was a founder-guru of the Revolutionary Communist Party

6 Marianne Elliott, *The Catholics of Ulster. A History* (London 2000) pp. 442ff.

7 See Liam Clarke and Kathryn Johnston, *Martin McGuiness. From Guns to Government* (Edinburgh 2003)

8 David McKittrick and David McVea, *Making Sense of the Troubles* (London 2000) p. 25

9 Margaret Thatcher, *The Downing Street Years* (London 1993) p. 385

10 Nigel Lawson, *The View from No. 11. Memoirs of a Tory Radical* (London 1992) pp. 669–70

11 Martin Dillon, *God and the Gun. The Church and Irish Terrorism* (London 1997) p. 140

12 For a succinct statement by (English-based) academic republicans of why this was not a religious conflict see John McGarry and Brendan O'Leary's textbook *Explaining Northern Ireland. Broken Images* (Oxford 1995) pp. 171–213

13 A. T. Q. Stewart, *The Narrow Ground. Aspects of Ulster 1609–1969* (London 1977) is a truly brilliant example of how local history can elucidate the big issues

14 On this see Steve Bruce, *God Save Ulster! The Religion and Politics of Paisleyism* (Oxford 1986) pp. 7ff.

15 Stewart, *The Narrow Ground* pp. 113–22 in particular

16 David Officer, '"For God and Ulster". The Ulstermen on the Somme' in Ian McBride (ed.), *History and Memory in Modern Ireland* (Cambridge 2001) pp. 160–83

17 Ruth Dudley Edwards, *The Faithful Tribe. An Intimate Portrait of the Loyal Institutions* (London 2000) p. 196

18 Marcus Tanner, *Ireland's Holy Wars. The Struggle for a Nation's Soul 1500–2000* (New Haven 2001) p. 324

19 Roy Foster, *Modern Ireland 1600–1972* (London 1989) p. 584

20 See Dudley Edwards in her comprehensive and vivid history of the Orange Order, *The Faithful Tribe* p. 339

21 Elliott, *The Catholics of Ulster* p. 391

22 For a good discussion of this see Bruce, *God Save Ulster!* pp. 90–1

23 See Paul Bew, Peter Gibbon and Henry Patterson, *Northern Ireland 1921–1994. Political Forces and Social Classes* (London 1995) pp. 116–17

24 See David McKittrick et al., *Lost Lives. The Stories of the Men, Women and Children Who Died as a Result of the Northern Ireland Troubles* (Edinburgh 2004). This is probably the most worthwhile book ever written about Northern Ireland

25 For this important connection see Richard English, *Armed Struggle. The History of the IRA* (London 2003) pp. 90ff.

26 McKittrick and McVea, *Making Sense of the Troubles* pp. 40–1

27 Conor Cruise O'Brien in the *Irish Independent*, 8 June 1991

28 Conor Cruise O'Brien, *States of Ireland* (New York 1972) p. 205

29 Ed Maloney, *A Secret History of the IRA* (London 2002) p. 77

30 Sean O'Callaghan, *The Informer* (London 1998) p. 242

31 Dillon, *God and the Gun* p. 129

32 English, *Armed Struggle* p. 123

33 His dubious role is ably discussed in Clarke and Johnston, *Martin McGuiness* pp. 67–80

34 '£85m for Bloody Sunday lawyers', *Daily Mail*, 3 December 2005 p. 50

35 For these statistics see McKittrick et al., *Lost Lives* p. 1534

36 Clarke and Johnston, *Martin McGuiness* pp. 203–4

37 See Dean Godson, *Himself Alone. David Trimble and the Ordeal of Unionism* (London 2004) p. 826

38 McKittrick et al., *Lost Lives* p. 229

39 Ibid. pp. 301–4

40 Ibid. p. 800

41 Roy Mason, *Paying the Price* (London 1999) pp. 164–5

42 Dillon, *God and the Gun* p. 174

43 O'Callaghan, *The Informer* pp. 110–11

44 Dillon, *God and the Gun* pp. 122–3

45 Mason, *Paying the Price* p. 211

46 Anthony Howard, *Basil Hume. The Monk Cardinal* (London 2005) p. 325

47 Godson, *Himself Alone* p. 165. Godson's book is a masterly account of the most recent phase of the 'peace process'

48 Paul Routledge, *John Hume. A Biography* (London 1997) pp. 207ff.

49 Ibid. pp. 230–1
50 Chris Ryder and Vincent Kearney,
    *Drumcree. The Orange Order's Last
    Stand* (London 2002) gives the worm's-
    eye view in great detail
51 Ibid. p. 503
52 I owe this insight to Dean Godson,
    'You'll never guess who's to blame
    for 7/7', *The Times*, 13 December 2005
    p. 16

Chapter 9: 'We Want God, We Want
God': The Churches and the Collapse of
European Marxist–Leninism 1970–1990

1 Michael Simmons, *The Reluctant
   President. A Political Life of Vaclav
   Havel* (London 1991) pp. 119ff.
2 Tony Judt, *Postwar. A History of Europe
   since 1945* (London 2005)
3 Lech Walesa, *A Path of Hope. An
   Autobiography* (London 1987) p. 96
4 Norman Davies, *Heart of Europe. A
   Short History of Poland* (Oxford 1986)
   p. 11
5 See Leszek Kołakowski's obituary of
   John Paul II in the *Independent*, 4 April
   2005 pp. 34–5
6 For biographical information on John
   Paul II see George Weigel's brilliant
   *Witness to Hope. The Biography of John
   Paul II* (New York 1999) and the many
   obituaries published in April 2005
7 Sabrina P. Ramet, *Nihil Obstat.
   Religion, Politics, and Social Change in
   East-Central Europe and Russia*
   (Durham, NC 1998) p. 133
8 Timothy Garton Ash, *The Polish
   Revolution. Solidarity 1980–1982*
   (London 1983) p. 23
9 See the fascinating study by Jan Kubik,
   *The Power of Symbols against the
   Symbols of Power. The Rise of Solidarity
   and the Fall of State Socialism in Poland*
   (University Park, Pa 1994) pp. 38ff.
10 John Lewis Gaddis, *The Cold War*
   (London 2005) is notably fair-minded
   about Reagan, Thatcher and John Paul
   II and the West's victory in the Cold
   War
11 Ibid. p. 221
12 George Weigel, *The Final Revolution.
   The Resistance Church and the Collapse
   of Communism* (Oxford 1992) p. 51

13 The most conspicuous example being
   Eric Hobsbawm, *The Age of Ideologies*
   (London 1994) pp. 460ff.
14 See above all John Clark and Aaron
   Wildavsky, *The Moral Collapse of
   Communism. Poland as a Cautionary
   Tale* (San Francisco 1990)
15 Walesa, *A Path of Hope* pp. 60ff.
16 Clark and Wildavsky, *The Moral
   Collapse of Communism* p. 117
17 Weigel, *The Final Revolution* p. 131
18 Gaddis, *The Cold War* p. 221
19 Weigel, *Witness to Hope* pp. 404–7
20 Roger Boyes, *The Naked President.
   A Political Life of Lech Walesa* (London
   1994) pp. 136–7
21 Jonathan Luxmoore and Jolanta
   Babiuch, *The Vatican & the Red Flag.
   The Struggle for the Soul of Eastern
   Europe* (London 1999) p. 259
22 John Follain, 'Was Kremlin behind
   plot to kill the Pope?', *Sunday Times*,
   3 April 2005 p. 16
23 Ramet, *Nihil Obstat* p. 55
24 Robert F. Goeckel, 'Der Weg der
   Kirchen in der DDR' in Günther
   Heydemann and Lothar Kettenacker
   (eds), *Kirchen in der Diktatur. Drittes
   Reich und SED-Staat* (Göttingen 1993)
   pp. 158–9
25 Roger Engelmann, 'Der Volksaufstand
   vom 17 Juni 1953', *Kirchliche
   Zeitgeschichte* (2002) 17, pp. 44–62
26 Wayne C. Bartee, *A Time to Speak Out.
   The Leipzig Citizen Protests and the Fall
   of East Germany* (Westport 2000) p. 52
27 Heinrich-August Winkler, *Der lange
   Weg nach Westen. Deutsche Geschichte
   vom 'Dritten Reich' bis zur
   Wiedervereinigung* (Munich 2002) 2,
   pp. 156ff.
28 Alan L. Nothnagle, *Building the East
   German Myth. Historical Mythology and
   Youth Propaganda in the German
   Democratic Republic 1945–1989* (Ann
   Arbor 1999) p. 105
29 Gerhard Besier, *Der SED-Staat und die
   Kirche. Der Weg in die Anpassung*
   (Munich 1993) pp. 301–11
30 Gerhard Besier, *Der SED-Staat und die
   Kirche 1969–1990. Die Vision vom
   'Dritten Weg'* (Frankfurt am Main 1995)
   pp. 511ff.
31 Christian Joppke, *East German*

*Dissidents and the Revolution of 1989.
Social Movement in a Leninist Regime*
(London 1995) pp. 88–9

32 Bartee, *A Time to Speak Out* pp. 110ff.

33 Gerhard Besier and Stephan Wolf (eds),
*Pfarrer, Christen und Katholiken. Das
Ministerium für Staatssicherheit der
ehemaligen DDR und die Kirchen*
(Neukirchen-Vluyn 1992)

34 Winkler, *Der lange Weg* 2, p. 494

*Chapter 10: Cubes, Domes and Death
Cults: Europe after 9/11*

1 Joseph Conrad, *The Secret Agent*
(London 1907, edition cited 1963)
p. 269

2 On this architectural angle see Roger
Scruton's important *The West and the
Rest. Globalization and the Terrorist
Threat* (London 2002) p. 101.

3 See Brian Moynihan's outstanding
report 'Hardline Holland' in the
Sunday Times, 27 February 2005
pp. 34–42

4 See *The 9/11 Commission Report* (New
York 2004) pp. 1–14 for details of the
hijackings, and Bob Woodward, *Bush
at War* (New York 2002) and Richard
A. Clarke, *Against All Enemies. Inside
America's War on Terror* (New York
2004) for details of the official
responses. My own responses to 9/11 are
in 'The Age of Anxiety', *Sunday Times*,
22 September 2001, News Review pp.
1–2 and in reviews for the same paper,
the *Evening Standard* and the *Literary
Review* of many books on Islamist
terrorism and the war in Iraq

5 John L. Esposito, *Unholy War. Terror in
the Name of Islam* (Oxford 2002) p. 92
for the cases of Farag Foda and Naguib
Mahfuz. Western television is, of
course, more than prepared to
broadcast programmes suggesting that
these threats are in the mind, e.g. Adam
Curtis, *The Power of Nightmares*, BBC2
(2005)

6 Francis Harris and Anton La Guardia,
'Al Qa'eda in rift over murder of
Muslims', *Daily Telegraph*, 8 October
2005 p. 17 reporting Pentagon
intercepts of a lengthy letter from
Ayman al-Zawahiri concerned about

al-Zarqawi's murderous campaign
against the Shia

7 Jason Burke, *Al-Qaeda. The True Story
of Radical Islam* (London 2003) p. 15
disaggregates a number of groups that
governments deliberately conflate into
Al Qaeda

8 This is the subject of an important
book by Ian Buruma and Avishai
Margalit, *Occidentalism. A Short History
of Anti-Westernism* (London 2004)

9 Ibid. p. 115

10 As ably discussed by Burke, *Al-Qaeda*
pp. 54–5

11 From a large literature see 'Memories of
Sayyid Qutb. An Interview with John
Calvert', *Worldpress.org* (September
2005); John Calvert, 'Sayyid Qutb in
America', *Newsletter 7* (March 2001)
of the International Institute for the
Study of Islam in the Modern World;
and John Calvert 'Sayyid Qutb. The
Face of the Modern Islamist',
www.blissstreetjournal.com/
sayyid_qutb.htm

12 Fred Halliday, 'Saudi Arabia 1997.
A Family Business in Trouble' in his
Nation and Religion in the Middle East
(London 2000) pp. 169ff.

13 Bruce Lawrence (ed.), *Messages to the
World. The Statements of Osama bin
Laden* (London 2005) pp. 4–15, 'The
Betrayal of Palestine' dated 29
December 1994

14 See Chris Mackey and Greg Miller,
*The Interrogators. Inside the Secret War
against Al Qaeda* (London 2004) p. 322
for the story of Prisoner 237

15 John Simpson, *A Mad World, My
Masters. Tales from a Traveller's Life*
(London 2000) pp. 82–4

16 The poem is available under the entry
for Theo van Gogh at
www.crimelibrary.com. I recently
filmed at this location

17 That being the subtext of the Atman
Foundation round table in October
2005 where this line was articulated by
various senior Spanish and Moroccan
politicians. The presence of Tariq
Ramadan, a known Al Qaeda apologist,
ensured that the conference was
boycotted by the PPE and Israel. Only
one participant, a British Muslim,

Imam Sayyid of Brighton, explicitly condemned Islamist terrorism, although I managed to condemn the recent nuclear threats of Iran's president Machmud Achmadinedschad via Iran's ambassador to Madrid

18 For a good discussion of this see Patrick Sookhdeo, 'Will London Burn Too?', *Spectator*, 12 November 2005 p. 16

19 As reported by Melanie Phillips, 'This Lethal Moral Madness', *Daily Mail*, 14 July 2005 p. 15

20 See the informed and unhysterical book by Peter Coleman, *The Liberal Conspiracy. The Congress for Cultural Freedom and the Struggle for the Mind of Postwar Europe* (New York 1989)

21 As brilliantly portrayed in Michel Houellebecq's *Platform* (London 2003)

22 Buruma and Margalit, *Occidentalism* especially pp. 27–9

23 'The fundamental laws of this country', *Daily Telegraph*, 14 July 2005 p. 29 sets out four points that would separate the law-abiding from 'irreconcilables'

24 The most interesting work on US relations with Europe is Niall Ferguson's *Colossus. The Rise and Fall of the American Empire* (London 2004) and Timothy Garton Ash, *History of the Present. Essays, Sketches and Despatches from Europe in the 1990s* (London 1999)

25 Peter Berger, 'Religion and the West', *National Interest* (Summer 2005) p. 113

26 Stephen L. Carter, *The Culture of Disbelief* (New York 1993) pp. 97–8

27 The most impressive book on US conservatism is John Micklethwait and Adrian Wooldridge, *The Right Nation. Why America is Different* (London 2004) which has a fine discussion of think-tanks

28 As volunteered by BBC Radio 3's presenter Philip Dodds in the context of his being invited by the (Roman Catholic) director-general of the BBC Mark Thompson to report on how the corporation's news and current affairs programmes handle religion

29 Simon Jenkins, 'An election dominated by holy rows?', *The Times*, 23 March 2005 p. 19

30 Daniel Hannan, 'Accidental Hero', *Spectator*, 13 November 2004 pp. 21–2

31 Marcel Gauchet, *The Disenchantment of the World. A Political History of Religion* (Princeton 1997) p. 163

32 Celia Bromley-Martin, 'Being Honest about Europe's Roots', *Inside the Vatican* (2003) 11, pp. 10–11

33 'Dutch plan culture and language test for immigrants', *Daily Telegraph*, 5 February 2005 p. 14

34 Charles Moore, 'Islam is not an exotic addition to the English country garden', *Daily Telegraph*, 21 August 2004

35 Channel 4 and Channel 5 in Britain seem to be as obsessed with sexual deviancy as the BBC is obsessed with that old standby of the imaginatively challenged, the Nazis.

36 Ruth Dudley Edwards, 'So Who Will Stop the PC Zealots?', *Daily Mail*, 12 November 2005 pp. 16–17

37 See the important article by James Pierson in the *Weekly Standard*, 3 October 2005

38 See Brian Appleyard's outstanding essay 'Beyond Belief?', *Sunday Times*, 27 March 2005, News Review pp. 1–2

39 A theme bravely explored by John Cornwell, *Breaking the Faith. The Pope, the People and the Fate of Catholicism* (London 2001) pp. 166ff.

40 Jytte Klausen, *The Islamic Challenge. Politics and Religion in Western Europe* (Oxford 2005)

PICTURE CREDITS

1 November 1920. King George V unveiling the cenotaph in Whitehall.
 © Getty Images
 Georges Rouault (French, 1887–1958) *Les Ruines Elles-Memes Ont Peri*
 (*Miserere* No. 34), 1926, etching, aquatint on paper, 65.7 × 50.2 cm.
 © Mackenzie Art Gallery, University of Regina Collection.
 Louis Christian Prediger Haeusser, (1881–1927) © Ullsteinbild/AKG
 Images.

2–3 1st February 1922. Pope Pius XI (Ambrogio Damiano Achille Ratti,
 1857–1939) © Topical Press Agency/Getty Images.
 Group from a Mockery Procession. A scene from the Christmas Festival
 of the Godless.
 Project for a Communist Mass Festival Dedicated to the 'Visible God' the
 Machine. Drawing by M. Dobnynski.
 Cathedral of Christ the Saviour, Moscow, USSR, early 1930s. © Sovfoto.

4–5 'Los Nacionales' Ministerio de Propaganda. The Hoover Institution
 Archives, Stanford University.
 20 August 1936. A loyalist firing squad seen taking aim at a statue of Christ
 on the Cerro de los Angeles, outside Madrid. © AP Photo.
 July, 1936. The Corpse of a Carmelite nun on public display in Barcelona,
 Spain during the Spanish Civil War. © Hulton Deutsch/Corbis.

6–7 Libertas, Shield and Difendetemi.
 'Salva I Tuoi Figli'. Christian Democrat poster for the election campaign
 of 1948.
 Nightime projections in Piazza del Duomo in Milan for the celebration
 of 28 October 1933.
 The homeless and refugees take shelter in makeshift dormitories at
 Castelgandolfo.

8 Gemalde von Hermann. *Im Anfang war das Wort* (*In the Beginning was
 the Word*) © Ullsteinbild/AKG Images.

8 *The Preaching of the Antichrist*, detail of Christ and the Devil, from the
  Chapel of the Madonna di San Brizio, 1499–1504 (fresco), Signorelli,
  Luca (*c*.1450–1523)
  Duomo, Orvieto, Umbria, Italy / The Bridgeman Art Library.

9 Catholic Fashions, Clerical modes for Summer.
  Vichy 1940–1944. La Legion des Combattants.
  Pope Pius XII, talking with children during the 80th birthday party they
  gave for him at the Vatican. © James Whitmore/Time Life
  Pictures/Getty Images.

10–11 Citizen marchers carrying posters of Joseph Stalin at a May Day parade
  in Bucharest, Romania c 1950s. © Sovfoto.
  Anti-Communist Poster Illustration: the transition of the Militant
  Godless to the continuous working week. The Hoover Institution
  Archives, Stanford University.
  18 March 1980. Archbishop Oscar Romero. © Christian Poveda/Corbis.
  22 August 1980. Workers participate in religious confession at the
  entrance to the Lenin shipyard in Gdansk. © Alain
  Keler/Sygma/Corbis.

12–13 *That Was The Week That Was*, Henry Morgan, Elliott Reid, David Frost,
  Nancy Ames, 1964–65. © Everett Collection (EVT)/Rex Features.
  May 1968. A demonstrator hurls a stone to riot police officers in Paris.
  © AP Photo.
  25 March 1989. Beatle John Lennon and his wife Yoko Ono, hold a
  bed-in for peace, Amsterdam. © AP Photo.
  October 1990. Pallbearers at the funeral of Dessie Grew. © Pacemaker
  Press.
  Christological Mural, Belfast, Northern Ireland.
  20 March 1988. Father Alex Reid tends to a British Soldier who was
  mobbed following an IRA funeral at the Milltown Cemetery, Belfast,
  Northern Ireland. © TDY/Rex Features.

14–15 11 September 2001. The South Tower of the World Trade Centre bursts
  into flames after being stuck by hijacked United Airlines Flight 175.
  © Sean Adair/Reuters/Corbis.
  5 August 2003. Captured Taliban soldiers and members of Al Qaeda,
  Sheberghan, Afghanistan. © Sipa Press/Rex Features.
  2 November 2004. The covered body of Theo van Gogh. © Eran
  Oppenheimer/Empics

16 28 October 2005. Riots in the Clichy-Sous-Bois area of suburban Paris,
  following the death of two teenage boys. © Jean-Michel
  Turpin/Corbis.

SELECT BIBLIOGRAPHY

Titles marked with an asterisk have been especially useful to this work, although the author has not used this scheme in cases where books of lasting creative value are concerned whose importance has already been highlighted in the body of the work.

Adam, Peter *The Arts of the Third Reich* (London 1992)

Adonis, Andrew and Thomas, Keith (eds) *Roy Jenkins. A Retrospective* (Oxford 2004)

Albrecht, Dieter (ed.) *Katholische Kirche im Dritten Reich* (Mainz 1976)

Alexander, Stella *The Triple Myth. A Life of Archbishop Alojzije Stepinac* (New York 1987)*

Allen, John L. *Cardinal Ratzinger. The Vatican's Enforcer of the Faith* (London 2000)

—— *Opus Dei. Secrets and Power inside the Catholic Church* (London 2005)

Alvarez, David and Graham, Robert *Nothing Sacred. Nazi Espionage against the Vatican, 1939–1945* (London 1997)

Ancel, Jean (ed.) *Documents Concerning the Fate of Romanian Jewry during the Holocaust* (New York 1986) 8 volumes

—— 'The "Christian" Regime of Romania and the Jews, 1940–1942' *Holocaust and Genocide Studies* (1993) 7 pp. 14–29

Anon (ed.) *The Persecution of the Churches in the Third Reich* (London 1939)*

Anon (ed.) *The Pope's Five Peace Points. Address of Pope Pius XII to the Sacred College of Cardinals on Christmas Eve 1939* (London 1939)

Aretin, Erwein Freiherr von *Fritz Michael Gerlich. Prophet und Martyrer* (Munich 1983)*

Ash, Timothy Garton *The Polish Revolution. Solidarity 1980–1982* (London 1983)

Azéma, Jean-Pierre and Bédarida, François (eds) *Vichy et les Français* (Paris 1996)

Baranowski, Shelley 'The 1933 German Protestant Church Elections. Machtpolitik or Accommodation' *Church History* (1980) 49, pp. 298–315

Bark, Dennis L. and Gress, David R. *A History of West Germany* (Oxford 1989) 2 volumes

Bärsch, Claus-Ekkehard *Die politische Religion des Nationalsozialismus. Die religiöse Dimension der NS-Ideologie in den Schriften von Dietrich Eckart, Joseph Goebbels, Alfred Rosenberg und Adolf Hitler* (Munich 1988)

Bartee, Wayne C. *A Time to Speak Out. The Leipzig Citizen Protests and the Fall of East Germany* (Westport 2000)*

Baum, Gregory (ed.) *The Twentieth Century. A Theological Overview* (New York 1999)

Beales, A. C. F. *The Pope and the Jews. The Struggle of the Catholic Church against Anti-Semitism during the War* (London 1945)

Beales, Derek and Best, Geoffrey (eds) *History, Society and the Churches. Essays in Honour of Owen Chadwick* (Cambridge 1985)

Becker, Winfried, Buchstab, Günther, Doering-Manteuffel, Anselm and Morsey, Rudolf (eds) *Lexikon der Christlichen Demokratie in Deutschland* (Paderborn 2002)*

—— 'Papst Pius XII. Und sein "Schweigen" über den Holocaust' *Kirchliche Zeitgeschichte* (2005) 1840–67*

Bédarida, Renée *Les Armes de l'esprit. Témoignage chrétien, 1941–1944* (Paris 1977)*

—— *Pierre Chaillet. Témoin de la résistance spirituelle* (Paris 1988)

—— *Les Catholiques dans la guerre 1939–1945* (Paris 1998)*

Behrenbeck, Sabine *Der Kult um die toten Helden. Nationalsozialistische Mythen, Riten und Symbole* (Vierow 1996)*

Benjamin, Daniel and Simon, Steven *Age of Sacred Terror. Radical Islam's War against America* (New York 2002)

Bennassar, Bartolomé *La Guerre d'Espagne et ses lendemains* (Paris 2004)*

Bentley, Michael (ed.) *Public and Private Doctrine. Essays in British History Presented to Maurice Cowling* (Cambridge 1993)

Berdyaev, Nicolas *The Origin of Russian Communism* (London 1937)*

Berezin, Mabel *Making the Fascist Self. The Political Culture of Interwar Italy* (Ithaca 1997)

Bergen, Doris *Twisted Cross. The German Christian Movement in the Third Reich* (Chapel Hill 1996)*

Berger, Peter (ed.) *The Desecularization of the World* (Washington, DC 1997)

Bernanos, Georges *Plea for Liberty. Letters to the English, the Americans, the Europeans* (New York 1944)

Besançon, Alain *The Rise of the Gulag. The Intellectual Origins of Leninism* (New York 1981)

———— *The Falsification of the Good* (London 1996)*

Besier, Gerhard *Der SED-Staat und die Kirche. Der Weg in die Anpassung* (Munich 1993)

———— *Der SED-Staat und die Kirche 1969–1990. Die Vision vom 'Dritten Weg'* (Frankfurt am Main 1995)

———— *Der SED-Staat und die Kirche. Höhenflug und Absturz* (Frankfurt am Main 1995)

———— *Die Kirchen und das Dritte Reich. Spaltungen und Abwehrkämpfe 1934–1937* (Munich 2001)

———— (ed.) *Zwischen 'nationaler Revolution' und militärischer Aggression. Transformationen in Kirche und Gesellschaft während der konsolidierten NS-Gewaltherrschaft 1934–1939* (Munich 2001)

———— *Der Heilige Stuhl und Hitler-Deutschland. Die Faszination des Totalitären* (Munich 2004)*

Bew, Paul, Gibbon, Peter and Patterson, Henry *Northern Ireland 1921–1994. Political Forces and Social Classes* (London 1995)*

Bidault, Georges, *Resistance. The Political Autobiography of Georges Bidault* (London 1965)

Binchy, D. A. *Church and State in Fascist Italy* (Oxford 1970)

Blet, Pierre, Martini, Angelo, Graham, Robert A. and Schneider, Burkhart (eds) *Actes et documents du Saint Siège relatifs à la guerre mondiale* (Vatican City 1965–81) volumes 1–11

———— *Pius XII and the Second World War. According to the Archives of the Vatican* (New York 1999)*

Blok, Alexandr *Selected Poems* (Manchester 2000)

Boobbyer, Philip *S. L. Frank. The Life and Work of a Russian Philosopher 1877–1950* (Athens, OH 1995)

Bonnell, Victoria E. *Iconography of Power. Soviet Political Posters under Lenin and Stalin* (Berkeley 1997)

Borkenau, Franz *The Spanish Cockpit* (London 1937)

———— *The Totalitarian Enemy* (London 1940)

Bosworth, R. J. B. *Mussolini's Italy* (London 2005)

Bottum, Joseph and Dalin, David (eds) *The Pius War. Responses to the Critics of Pius XII* (Lexington 2005)*

Boyes, Roger *The Naked President. A Political Life of Lech Walesa* (London 1994)

Bracher, Karl Dietrich *The German Dictatorship* (London 1970)

———— *The Age of Ideologies. A History of Political Thought in the Twentieth Century* (London 1984)*

———— *Wendezeiten der Geschichte. Historisch-politische Essays* (Stuttgart 1992)*

—— *Geschichte als Erfahrung. Betrachtungen zum 20. Jahrhundert* (Stuttgart 2001)*

Bradford, Richard *First Boredom, Then Fear. The Life of Philip Larkin* (London 2005)

Brakelmann, Günther 'Nationalprotestantismus und Nationalsozialismus' in Christian Jansen et al. (eds) *Von der Aufgabe der Freiheit. Verantwortung und bürgerliche Gesellschaft im 19. und 20. Jahrhundert* (Berlin 1995)

Brechenmacher, Thomas 'Pope Pius XI, Eugenio Pacelli, and the Persecution of the Jews' *German Historical Institute London Bulletin* (2005) 27, pp. 17–44*

Brovkin, Vladimir (ed.) *The Bolsheviks in Russian Society. The Revolution and the Civil Wars* (New Haven 1997)

Brown, Callum *The Death of Christian Britain: Understanding Secularization 1800–2000* (London 2001)*

Bruce, Steve *God Save Ulster! The Religion and Politics of Paisleyism* (Oxford 1986)

—— *Religion in Modern Britain* (Oxford 1995)

—— *Religion in the Modern World. From Cathedrals to Cults* (Oxford 1996)*

—— *God is Dead. Secularization in the West* (Oxford 2002)*

—— *Politics and Religion* (Cambridge 2005)*

Bry, Carl Christian *Verkappte Religionen. Kritik des kollektiven Wahns* (Nördlingen 1988)*

—— *Der Hitler-Putsch. Berichte und Kommentare eines Deutschland-Korrespondenten (1922–1924) für das 'Argentinische Tag- und Wochenblatt'* (Nördlingen 1987)

Buchanan, Tom and Conway, Martin (eds) *Political Catholicism in Europe 1918–1965* (Oxford 1996)*

Buck-Morris, Susan *Dream World and Catastrophe. The Passing of Mass Utopia in East and West* (Cambridge, Mass. 2000)

Burleigh, Michael *Death and Deliverance. Euthanasia in Germany 1900–1945* (Cambridge 1994)

—— *Ethics and Extermination. Reflections on Nazi Genocide* (Cambridge 1997)

—— *The Third Reich. A New History* (London 2000)

—— 'Religion and Social Evil. The Cardinal Basil Hume Memorial Lectures' *Totalitarian Movements and Political Religions* (2002) 3, pp. 1–60

—— *Earthly Powers. Religion and Politics in Europe from the French Revolution to the Great War* (London 2005)

Burke, Jason *Al-Qaeda. The True Story of Radical Islam* (London 2003)*

Burrin, Philippe 'Political Religion: The Relevance of a Concept' *History & Memory* (1997) 9, pp. 321–52

—— *Ressentiment et apocalypse. Essai sur l'antisémitisme Nazi* (Paris 2004)

Buruma, Ian and Margalit, Avishai *Occidentalism. A Short History of Anti-Westernism* (London 2004)*

Butler, Hubert *Independent Spirit. Essays* (New York 1996)*

Calabrese, Omar (ed.) *Modern Italy. Images and History of a National Identity* (Milan 1983) volumes 1–4

Camus, Albert, *Between Hell and Reason. Essays from the Resistance Newspaper Combat, 1944–1947* selected and trans. Alexandre de Gramont (Hanover 1991)

Cannadine, David *In Churchill's Shadow. Confronting the Past in Modern Britain* (Oxford 2003)

Carey, John *What Good are the Arts?* (London 2005)

Carrillo, Elisa *Alcide de Gasperi. The Long Apprenticeship* (Notre Dame 1965)

Cary, Noel D. *The Path to Christian Democracy. German Catholics and the Party System from Windthorst to Adenauer* (Cambridge, Mass. 1996)

Chadwick, Owen 'Weizsäcker, the Vatican and the Jews of Rome' *Journal of Ecclesiastical History* (1977) 28, pp 179–199

—— *Hensley Henson. A Study in the Friction between Church and State* (Oxford 1983)*

—— *Britain and the Vatican during the Second World War* (Cambridge 1986)*

—— *The Christian Church in the Cold War* (London 1992)

Chamberlain, Lesley *The Philosophy Steamer. Lenin and the Exile of the Intelligentsia* (London 2006)

Clark, John and Wildavsky, Aaron *The Moral Collapse of Communism. Poland as a Cautionary Tale* (San Francisco 1990)

Clarke, Liam and Johnston, Kathryn *Martin McGuiness. From Guns to Government* (Edinburgh 2003)*

Clarke, Richard A. *Against All Enemies. Inside America's War on Terror* (New York 2004)

Clonmore, Lord *Pope Pius XI and World Peace* (London 1938)

Cobban, Alfred *Dictatorship. Its History and Theory* (London 1939)*

Cohn, Norman *The Pursuit of the Millennium. Revolutionary Millenarians and Mystical Anarchists of the Middle Ages* (Oxford 1970)*

Cointet, Michèle *L'Église sous Vichy 1940–1945. La repentance en question* (Paris 1998)

Conquest, Robert *Religion in the USSR* (New York 1968)

—— *The Great Terror* (London 1968)*

—— *Tyrants and Typewriters. Communiqués from the Struggle for Truth* (Lexington, Mass. 1989)*

—— *Stalin. Breaker of Nations* (London 1991)

—— *Reflections on a Ravaged Century* (London 1999)

—— *The Dragons of Expectation. Reality and Delusion in the Course of History* (London 2005)

Conrad, Joseph, *The Secret Agent* (originally London 1907, this edition 1963)

Conway, John *The Nazi Persecution of the Churches 1933–1945* (New York 1968)

—— 'The Meeting between Pope Pius XII and Ribbentrop' *Historical Papers of the Canadian Historical Association* (1968) 1, pp. 215–27

—— 'The Churches, the Slovak State and the Jews 1939–1945' *Slavonic and East European Review* (1974) 52, pp. 85–112

—— 'The Vatican and the Holocaust: A Reappraisal' *Miscellanea Historiae Ecclesiasticae* (1984) 9, pp. 475–89

—— 'The Vatican, Germany and the Holocaust' in P. Kent and J. Pollard (eds) *Papal Diplomacy in the Modern Age* (Westport 1994)

Conway, Martin *Catholic Politics in Europe 1918–1945* (London 1997)

Cooper, Barry *New Political Religions, or an Analysis of Modern Terrorism* (Columbia, Miss. 2004)*

Cooper, Norman B. *Catholicism and the Franco Regime* (London 1975)

Coppa, Frank *The Modern Papacy since 1789* (London 1998)

Coppa, Frank and Repetto-Alaia, Margherita (eds) *The Formation of the Italian Republic* (New York 1993)

Cornwell, John *Hitler's Pope. The Secret History of Pius XII* (London 1999)

Crampton, R. J. *Eastern Europe in the Twentieth Century* (London 1994)

Dalin, David G. 'Pius XII and the Jews' *Weekly Standard* (26 February 2001) pp. 31–9

—— *The Myth of Hitler's Pope. How Pope Pius XII Rescued Jews from the Nazis* (Washington DC 2005)

Dallin, Alexander *German Rule in Occupied Russia 1941–1945. A Study of Occupation Politics* (London 1981)

Daniels, Anthony *Utopias Elsewhere* (New York 1991)*

Davies, Norman *God's Playground: A History of Poland* (Oxford 1981) 2 volumes*

—— *Heart of Europe. A Short History of Poland* (Oxford 1984)*

Davies, Sarah *Popular Opinion in Stalin's Russia. Terror, Propaganda and Dissent 1934–1941* (Cambridge 1997)

Dawson, Christopher *Religion and the Modern State* (London 1936)*

Deák, István, Gross, Jan T. and Judt, Tony (eds) *The Politics of Retribution in Europe. World War II and its Aftermath* (Princeton 2000)

De Rosa, Gabriele *Luigi Sturzo* (Turin 1977)

Dierker, Wolfgang *Himmlers Glaubenskrieger. Die Sicherheitsdienst der SS und seine Religionspolitik 1933–1941* (Paderborn 2002)

Dillon, Martin *God and the Gun. The Church and Irish Terrorism* (London 1997)

Dionne, E. J. and Diiulio, John J. (eds) *What's God got to do with the American Experiment?* (Washington DC 2000)

Doering, Bernard *Jacques Maritain and the French Catholic Intellectuals* (Notre Dame 1983)

Domarus, Max (ed.) *Hitler. Speeches and Proclamations* (London 1995–2004) 4 volumes

Doosry, Yasmin 'Die sakrale Dimension des Reichsparteitagsgeländes in Nürnberg' in Richard Ferber, (ed.) *Politische Religion, religiöse Politik* (Würzburg 1997) pp. 205–24

Duffy, Eamon *Saints and Sinners. A History of the Popes* (New Haven 2001)*

Duggan, Christopher and Wagstaff, Christopher (eds) *Italy and the Cold War. Politics, Culture and Society 1948–58* (Oxford 1995)

Dunn, Dennis J. *The Catholic Church and the Soviet Government 1939–1949* (New York 1977)

Edwards, Ruth Dudley *The Faithful Tribe. An Intimate Portrait of the Loyal Institutions* (London 1999)*

—— 'A Liberal Dose of Stupidity' *FT Magazine* (2 April 2005) pp. 16–19

Eliot, Thomas Stearns *Collected Poems 1909–1962* (London 1963)

—— *After Strange Gods* (London 1934)

Elliott, Marianne *The Catholics of Ulster. A History* (London 2000)

English, Richard *Armed Struggle. The History of the IRA* (London 2003)

Ericksen, Robert P. *Theologians under Hitler* (New Haven 1985)*

Ericksen, Robert P. and Heschel, Susannah (eds) *Betrayal. German Churches and the Holocaust* (Minneapolis 1999)

Esposito, John L. *Unholy War. Terror in the Name of Islam* (Oxford 2002)

Falasca-Zamponi, Simonetta *Fascist Spectacle. The Aesthetics of Power in Mussolini's Italy* (Berkeley 1997)

Falconi, Carlo *The Popes in the Twentieth Century. From Pius XII to John XXIII* (London 1967)

Feldkamp, Michael 'Eugenio Pacelli: The German Years' *Inside the Vatican* (2002) pp. 40–5

—— *Pius XII und Deutschland* (Göttingen 2000)*

Feldman, Jan 'New Thinking about the "New Man". Developments in Soviet Moral Theory' *Studies in Soviet Thought* (1989) 38, pp. 147–63

Fenn, Richard K. (ed.) *The Blackwell Companion to the Sociology of Religion* (Oxford 2001)

—— *Beyond Idols. The Shape of a Secular Society* (Oxford 2001)

Ferguson, Niall *The Pity of War* (London 1998)

—— *Colossus. The Rise and Fall of the American Empire* (London 2004)*

Figgis, J. N. *Civilisation at the Crossroads* (London 1912)

Fletcher, William C. *The Russian Orthodox Church Underground 1917–1970* (Oxford 1971)

Ford, Guy Stanton (ed.) *Dictatorship in the Modern World* (New York 1938)*

Fouilloux, Étienne 'Église Catholique et Seconde Guerre Mondiale' *Vingtième Siècle. Revue d'Histoire* (2002) 73, pp. 111–24

Freud, Sigmund *The Future of an Illusion* (New York 1961)

Friedländer, Saul *Pius XII and the Third Reich. A Documentation* (New York 1966)

Fuisz, József 'Der Beitrag der Religionsgemeinschaften zum Ungarnaufstand 1956' *Kirchliche Zeitgeschichte* (2004) 17, pp. 113–32

Fülöp-Miller, René *The Mind and Face of Bolshevism* (New York 1929)*

Fussell, Paul *The Great War and Modern Memory* (Oxford 1975)

Gambetta, Diego (ed.) *Making Sense of Suicide Missions* (Oxford 2005)

Gasperi, Maria Romana de *De Gasperi. Ritratto di uno statista* (Milan 1964)

Gentile, Emilio *The Sacralization of Politics in Fascist Italy* (Cambridge, Mass. 1996)

—— *Le religioni della politica. Fra democrazie e totalitarismo* (Rome 2001)*

—— 'Political Religion: A Concept and its Critics' *Totalitarian Movements and Political Religions* (2005) 6, pp. 19–29

Gerlich, Fritz *Der Kommunismus als Lehre vom Tausendjährigen Reich* (Munich 1920)*

Geyer, Michael and Lehmann, Hartmut (eds) *Religion und Nation. Nation und Religion. Beiträge zu einer unbewältigten Geschichte* (Göttingen 2004)

Gilbert, Martin *The Righteous. The Unsung Heroes of the Holocaust* (New York 2003)

Gildea, Robert *Marianne in Chains. In Search of the German Occupation 1940–45* (London 2002)

Ginsborg, Paul *A History of Contemporary Italy 1943–1980* (London 1990)

Gleason, Abbott *Totalitarianism. The Inner History of the Cold War* (New York 1995)

Gleason, Abbott, Kenez, Peter, and Stites, Richard (eds) *Bolshevik Culture. Experiment and Order in the Russian Revolution* (Bloomington 1985)

Glucksmann, André *Dostoievski en Manhattan* (Madrid 2002)

Godman, Peter *Hitler and the Vatican* (New York 2004)

Godson, Dean *Himself Alone. David Trimble and the Ordeal of Unionism* (London 2004)*

Goldinger, Walter and Binder, Dieter *Geschichte der Republik Österreich 1918–1938* (Vienna 1992)

Golomstock, Igor *Totalitarian Art* (New York 1990)*

Goodrick-Clarke, Nicholas *The Occult Roots of Nazism. Secret Aryan Cults and their Influence on Nazi Ideology* (London 1985)

Gotto, Klaus and Repgen, Konrad (eds) *Die Katholiken und das Dritte Reich* (Mainz 1980)

Graham, Robert 'The "Right to Kill" in the Third Reich. Prelude to Genocide' *Catholic Historical Review* (1976) 62, pp. 56–76

—— *The Vatican and Communism during World War II. What Really Happened?* (San Francisco 1990)

Gray, John *Al Qaeda and What it Means to be Modern* (London 2003)

Greene, Graham *The Lawless Roads* (London 1939)

Gregor, James A. *The Faces of Janus. Marxism and Fascism in the Twentieth Century* (New Haven 2000)

Greschat, Martin (ed.) *Zwischen Widerspruch und Widerstand. Texte zur Denkschrift der Bekennenden Kirche an Hitler 1936* (Munich 1987)

Greschat, Martin and Kaiser, Jochen-Christoph (eds) *Christentum und Demokratie im 20. Jahrhundert* (Stuttgart 1992)*

Grieswelle, Detlev *Propaganda der Friedlosigkeit – Eine Studie zu Hitlers Rhetorik 1920–1933* (Stuttgart 1972)

Gurian, Waldemar *Bolshevism. Theory and Practice* (London 1932)

—— *Hitler and the Christians* (New York 1936)*

Haffner, Sebastian *Defying Hitler. A Memoir* (London 2002)

Halliday, Fred *Islam and the Myth of Confrontation* (London 1996)

—— *Nation and Religion in the Middle East* (London 2000)

Halls, W. D. *Politics, Society and Christianity in Vichy France* (Oxford 1995)

Hanisch, Ernst *Die Ideologie des Politischen Katholizismus in Österreich 1918–1938* (Vienna 1977)

Hanley, David (ed.) *Christian Democracy in Europe. A Comparative Perspective* (London 1994)

Hardtwig, Wolfgang 'Political Religions in Modern Germany. Reflections on Nationalism, Socialism and National Socialism' *Bulletin of the German Historical Institute, Washington D.C.* (2001) 28, pp. 3–27

Harnden, Toby *'Bandit Country'. The IRA & South Armagh* (London 1999)

Hastings, Adrian (ed.) *Modern Catholicism. Vatican II and After* (London 1991)

—— *A History of English Christianity 1920–2000* (London 2001 fourth edition)*

Hebblethwaite, Peter *Paul VI. The First Modern Pope* (London 1993)

Heiber, Helmut *Goebbels. A Biography* (New York 1972)

Heiden, Konrad *The Fuehrer* trans. Ralph Manheim (originally 1944, London 1999)

Heineman, Kenneth J. *God is a Conservative. Religion, Politics and Morality in Contemporary America* (New York 1998)

Hellbeck, Jochen 'Fashioning the Stalinist Soul. The Diary of Stepan Podlubnyi (1931–1939)' *Jahrbücher für Geschichte Osteuropas* (1996) 44, pp. 344–373

Heller, Mikhail *Cogs in the Wheel. The Formation of Soviet Man* (New York 1988)*

Heller, Mikhail and Nekrich, Aleksandr *Utopia in Power. The History of the Soviet Union from 1917 to the Present* (London 1985)

Hellman, John *Emmanuel Mounier and the New Catholic Left 1930–1950* (Toronto 1981)

Helmreich, Ernst Christian *The German Churches under Hitler* (Detroit 1979)

Henson, Herbert Hensley *Retrospect on an Unimportant Life* (Oxford 1943) 3 volumes

Heydemann, Günther and Kettenacker, Lothar (eds) *Kirchen in der Diktatur. Drittes Reich und SED-Staat* (Göttingen 1993)

Hitler, Adolf *Mein Kampf* trans. Ralph Manheim (London 1969)

Hlond, August *The Persecution of the Catholic Church in German-Occupied Poland* (London 1941)

Hobsbawm, Eric *The Age of Extremes. The Short Twentieth Century 1914–1991* (London 1994)

Hockerts, Hans *Die Sittlichkeitsprozesse gegen katholische Ordensangehörige und Priester 1936/1937* (Mainz 1971)

Hoffmann, Peter 'Roncalli in the Second World War' *Journal of Ecclesiastical History* (1989) 40, pp. 74–99

Hollander, Paul *Political Pilgrims. Travels of Western Intellectuals to the Soviet Union, China and Cuba* (New York 1981)*

Hoser, Paul 'Hitler und die katholische Kirche' *Vierteljahreshefte für Zeitgeschichte* (1994) 42, pp. 473–92

Howard, Anthony *Basil Hume. The Monk Cardinal* (London 2005)

Howe, Irving (ed.) *1984 Revisited. Totalitarianism in our Century* (New York 1983)

Hughes, John Jay 'The Pope's "Pact with Hitler". Betrayal or Self Defense?' *Journal of Church and State* (1975) 17, pp. 63–80

Hughes, Philip *Pope Pius the Eleventh* (London 1937)

Huntington, Samuel P. *The Clash of Civilizations and the Remaking of World Order* (London 1997)

Hürten, Heinz (ed.) *Deutsche Briefe 1934–1938. Ein Blatt der katholischen Emigration* (Mainz 1969)

—— Waldemar Gurian. Ein Zeuge der Krise unserer Welt in der ersten Hälfte des 20 Jahrhunderts (Mainz 1972)

—— (ed.) 'Kulturkampf. Bericht aus dem Dritten Reich. Paris'. Eichstätter Materialien (Regensburg 1988)

—— Deutsche Katholiken 1918–1945 (Paderborn 1996)

Husband, William B. 'Soviet Atheism and Russian Orthodox Strategies of Resistance 1917–1932' Journal of Modern History (1998) 70, pp. 74–107

Huttner, Markus Totalitarismus und säkulare Religionen (Bonn 1999)

Inglis, Tom, Mach, Zdisław and Mazanek, Rafał (eds) Religion and Politics. East–West Contrasts from Contemporary Europe (Dublin 2000)

Ioanid, Radu The Sword of the Archangel. Fascist Ideology in Romania (Boulder 1990)

Irving, R. E. M. Christian Democracy in France (London 1973)

—— The Christian Democratic Parties of Western Europe (London 1979)

James, Clive The Meaning of Recognition. New Essays 2001–2005 (London 2005)*

James, Robert Rhodes (ed.) Winston S. Churchill. His Complete Speeches 1897–1963 (London 1974) 8 volumes

Jenkins, Philip The New Anti-Catholicism. The Last Acceptable Prejudice (Oxford 2003)

Jenkins, Roy A Life at the Centre (London 1991)

Johnson, Paul Modern Times. A History of the World from the 1920s to the 1990s (London 1992)*

Joppke, Christian East German Dissidents and the Revolution of 1989. Social Movement in a Leninist Regime (London 1995)

Judt, Tony Postwar. A History of Europe since 1945 (London 2005)

Katz, David S. The Occult Tradition (London 2005)

Kean, Thomas H. and others The 9/11 Commission Report. Final Report of the National Commission on Terrorist Attacks upon the United States (New York 2004)*

Kedward, Harry Roderick, Resistance in Vichy France. A Study in Ideas and Motivation in the Southern Zone 1940–1942 (Oxford 1978)

Keller, Adolf Religion and the Modern State (London 1934)*

Kelly, Catriona Comrade Pavlik. The Rise and Fall of a Soviet Boy Hero (London 2005)

Kelly, Catriona and Shepherd, David (eds) Constructing Russian Culture in the Age of Revolution 1881–1940 (Oxford 1998)

Kent, John William Temple (Cambridge 1992)

Kent, Peter C. *The Pope and the Duce. The International Impact of the Lateran Agreements* (London 1983)

—— 'The Vatican and the Spanish Civil War' *European History Quarterly* (1986) pp. 441–64

—— *The Lonely Cold War of Pius XII* (Montreal 2002)

Keogh, Dermot *Ireland and the Vatican. The Politics and Diplomacy of Church–State Relations 1922–1960* (Cork 1995)

Kepel, Gilles *The Revenge of God. The Resurgence of Islam, Christianity and Judaism in the Modern World* (Pennsylvania 1995)

Kershaw, Ian *The Hitler Myth. Image and Reality in the Third Reich* (Oxford 1987)

Kertzer, David I. *The Popes against the Jews. The Vatican's Role in the Rise of Modern Antisemitism* (New York 2001)

Kirkpatrick, Ivone *Inner Circle. Memoirs* (London 1959)

Kirschenbaum, Lisa *Small Comrades. Revolutionizing Childhood in Soviet Russia, 1917–1932* (New York 2001)

Klausen, Jytte *The Islamic Challenge. Politics and Religion in Western Europe* (Oxford 2005)

Klemperer, Klemens von *Ignaz Seipel. Christian Statesman in a Time of Crisis* (Princeton 1972)

—— *German Resistance to Hitler. The Search for Allies Abroad 1938–1945* (Oxford 1992)*

Klemperer, Victor *The Language of the Third Reich. LTI–Lingua Tertii Imperii. A Philologist's Notebook* (London 1999)

Klinghoffer, Arthur Jay *Red Apocalypse. The Religious Evolution of Soviet Communism* (Lanham 1996)*

Klockowski, Jerzy *A History of Polish Christianity* (Cambridge 2000)

Kojecký, R. *T. S. Eliot's Social Criticism* (London 1971)

Kolonitskii, Boris and Figes, Orlando *Interpreting the Russian Revolution. The Language and Symbols of 1917* (New Haven 1999)

Kotkin, Stephen *Magnetic Mountain. Stalinism as a Civilisation* (Berkeley 1995)*

Krauze, Enrique *Mexico. Biography of Power* (New York 1997)

Kravchenko, Victor *I Chose Freedom. The Personal and Political Life of a Soviet Official* (London 1947)*

Kroll, Franz-Lothar *Utopie als Ideologie: Geschichtsdenken und politisches Handeln im Dritten Reich* (Paderborn 1998)

Künzlen, Gottfried *Der neue Mensch* (Frankfurt am Main 1997)*

Kuropka, Joachim *Clemens August Graf von Galen. Sein Leben und Wirken in Bildern und Dokumenten* (Cloppenburg 1992)

Lane, Christel *The Rites of Rulers. Ritual in Industrial Society – the Soviet Case* (Cambridge 1981)

Lannon, Frances *Privilege, Persecution, and Prophecy. The Catholic Church in Spain 1875–1975* (Oxford 1987)

Lapomarda, Vincent *The Jesuits and the Third Reich* (1989)

Larkin, Maurice *France since the Popular Front. Government and People 1936–1986* (Oxford 1988)

Larkin, Philip *Collected Poems* (London 2003)

Lavi, Theodore 'The Vatican's Endeavours on Behalf of Rumanian Jewry during the Second World War' *Yad Vashem Studies* (1963) 5, pp. 405–18

Lawrence, David Herbert *The Plumed Serpent* (London 1926)

—— *Selected Essays* (London 1950)

Lawson, Nigel *The View from No. 11. Memoirs of a Tory Radical* (London 1992)

Lehmann, Hartmut *Säkulisierung. Der europäische Sonderweg in Sachen Religion* (Göttingen 2004)

Levai, Jenö *Hungarian Jewry and the Papacy. Pope Pius XII did Not Remain Silent* (London 1968)*

Lewy, Guenther *The Catholic Church and Nazi Germany* (New York 1965)

Ley, Michael and Schoeps, Julius H. (eds) *Der Nationalsozialismus als politische Religion* (Bodenheim bei Mainz 1997)

Lilla, Mark *The Reckless Mind. Intellectuals in Politics* (New York 2001)

Linse, Ulrich *Barfüssige Propheten. Erlöser der zwanziger Jahre* (Berlin 1983)*

—— *Geisterseher und Wunderwirker. Heilsuche im Industriezeitalter* (Frankfurt am Main 1996)

Longley, Clifford *Chosen People. The Big Idea that Shapes England and America* (London 2002)

Löwith, Karl *My Life in Germany before and after 1933. A Report* (London 1994)*

Luxmoore, Jonathan and Babiuch, Jolanta *The Vatican and the Red Flag. The Struggle for the Soul of Eastern Europe* (London 1999)

Lubac, Henri de *Christian Resistance to Anti-Semitism. Memories from 1940–1944* (San Francisco 1990)*

Lübbe, Hermann 'Tugendterror – Höhere Moral als Quelle politischer Gewalt' *Totalitarismus und Demokratie* (2004) 1, pp. 203–17*

Ludlow, Peter 'Papst Pius XII, die britische Regierung und die deutsche Opposition im Winter 1939/40' *Vierteljahreshefte für Zeitgeschichte* (1977) 22, pp. 299–341*

McBride, Ian (ed.) *History and Memory in Modern Ireland* (Cambridge 2001)

McDermott, Terry *Perfect Soldiers. The 9/11 Hijackers* (New York 2005)

McDowell, Jennifer 'Soviet Civil Ceremonies' *Journal for the Scientific Study of Religion* (1974) 13, pp. 20–35

McElroy, John Harmon *American Beliefs. What Keeps a Big Country and a Diverse People United* (Chicago 1999)*

McGarry, John and O'Leary, Brendan *Explaining Northern Ireland. Broken Images* (Oxford 1995)

McKittrick, David (ed.) *Lost Lives. The Stories of the Men, Women and Children Who Died as a Result of the Northern Ireland Troubles* (Edinburgh 2004)*

McKittrick, David and McVea, Sean *Making Sense of the Troubles* (London 2000)*

Machin, G. I. T., *Churches and Social Issues in Twentieth-Century Britain* (Oxford 1998)*

Mackey, Chris *The Interrogator's War. Inside the Secret War against Al Qaeda* (London 2004)

Maier, Hans *Christlicher Widerstand im Dritten Reich* (Hamburg 1994)

—— *Eine Kultur oder viele? Politische Essays* (Stuttgart 1995)*

—— *Politische Religionen. Die totalitären Regime und das Christentum* (Freiburg im Breissau 1995)*

—— (ed.) *Totalitarismus und politische Religionen* (Paderborn 1996–2003) 3 volumes* [Routledge are publishing an English translation]

—— (ed.) *Wege in die Gewalt. Die modernen politischen Religionen* (Frankfurt am Main 2000)*

—— *Das Dopplegesicht des Religiösen. Religion-Gewalt-Politik* (Freiburg 2004)*

Manuel, Frank and Manuel, Fritzie *Utopian Thought in the Western World* (Cambridge, Mass. 1979)

Marrus, Michael and Paxton, Robert *Vichy France and the Jews* (New York 1983)

Martin, David and Mullen, Peter (eds) *Unholy Warfare. The Church and the Bomb* (Oxford 1983)

Martin, William *With God on our Side. The Rise of the Religious Right in America* (New York 1996)

Marty, Martin E. *Modern American Religion* (Chicago 1986–96) 3 volumes*

Marwick, Arthur *The Sixties. Cultural Revolution in Britain, France, Italy, and the United States c. 1958–c. 1974* (Oxford 1998)

Mason, Roy *Paying the Price* (London 1999)

Mayeur, Jean-Marie *Des partis catholiques à la démocratie chrétienne xix–xx siècles* (Paris 1980)*

Messner, Johannes *Dollfuss. An Austrian Patriot* (London 1934)

Micklethwait, John and Wooldridge, Adrian *The Right Nation. Why America is Different* (London 2004)*

Miccoli, Giovanni *I dilemmi e i silenzio di Pio XII* (Milan 2000)

Michnik, Adam *The Church and the Left* (Chicago 1993)*

Möller, Horst *Europa zwischen den Weltkriegen* (Munich 1998)

Moloney, Ed *A Secret History of the IRA* (London 2002)

Montclos, Xavier de, *Les Chrétiens face au Nazisme et au Stalinisme. L'épreuve totalitaire, 1939–1945* (Paris 1983)*

Moore, Bob *Victims and Survivors. The Nazi Persecution of the Jews in the Netherlands 1940–1945* (London 1997)

Morgan, Thomas *A Reporter at the Papal Court. A Narrative of the Reign of Pius XI* (New York 1937)

Moro, Renato 'Religion and Politics in the Time of Secularisation' *Totalitarian Movements and Political Religions* (2005) 6, pp. 71–86

Mosse, George Lachmann *Masses and Man. Nationalist and Fascist Perceptions of Reality* (Detroit 1980)

—— *The Crisis of German Ideology. Intellectual Origins of the Third Reich* (New York 1981)

—— *The Fascist Revolution. Toward a General Theory of Fascism* (New York 1999)

Musiedlek, Didier 'Religion and Political Culture in the Thought of Mussolini' *Totalitarian Movements and Political Religions* (2005) 6, pp. 395–406

Niebuhr, Reinhold 'The Religion of Communism' *Atlantic Monthly* (1931) 147 pp. 12–24

Norman, Edward R. *Church and Society in England 1770-1970. A Historical Study* (Oxford 1976)*

—— *Secularisation* (London 2002)

Nothnagle, Alan L. *Building the East German Myth. Historical Mythology and Youth Propaganda in the German Democratic Republic 1945–1989* (Ann Arbor 1999)

Nowak, Kurt *'Euthanasie' und Sterilisierung im 'Dritten Reich'. Die Konfrontation der evangelischen und katholischen Kirche mit dem 'Gesetz zur Verhütung erbkranken Nachwuchses' und der 'Euthanasie'-Aktion* (Göttingen 1978)

—— *Geschichte des Christentums in Deutschland: Religion, Politik und Gesellschaft vom Ende der Aufklärung bis zur Mitte des 20. Jahrhunderts* (Munich 1995)

O'Brien, Conor Cruise *States of Ireland* (New York 1972)*
—— *God Land. Reflections on Religion and Nationalism* (Cambridge, Mass. 1988)
—— *Ancestral Voices. Religion and Nationalism in Ireland* (Dublin 1994)*
Oakeshott, Michael (ed.) *The Social and Political Doctrines of Contemporary Europe* (Cambridge 1939)
O'Callaghan, Sean *The Informer* (London 1998)*
Ortega y Gasset, José *The Revolt of the Masses* (New York 1957)
Overy, Richard *The Dictators. Hitler's Germany, Stalin's Russia* (London 2004)

Pacelli, Eugenio *Gesammelte Reden* ed. Ludwig Kass, (Berlin 1930)
Parsons, Gerald (ed.) *The Growth of Religious Diversity. Britain from 1945* (London 1993–4) 2 volumes
Patey, Douglas Lane *The Life of Evelyn Waugh* (Oxford 1998)
Pattie, Charles, Seyd, Patrick and Whiteley, Paul *Citizenship in Britain. Values, Participation and Democracy* (Cambridge 2004)
Payne, Stanley *Spanish Catholicism. An Historical Overview* (Madison, Wis. 1984)*
—— *The Franco Regime 1936-1975* (London 2000)
—— *The Spanish Civil War, the Soviet Union, and Communism* (New Haven 2004)
—— 'On the Heuristic Value of the Concept of Political Religion and its Application' *Totalitarian Movements and Political Religions* (2005) 6, pp. 163–74*
Peris, Daniel *Storming the Heavens. The Soviet League of the Militant Godless* (Ithaca 1998)
Phayer, Michael *The Catholic Church and the Holocaust 1930–1965* (Bloomington 2000)
Pichon, Charles *The Vatican in World Affairs* (New York 1950)
Pipes, Richard *Russia under the Bolshevik Regime 1919–1924* (London 1994)
—— (ed.) *The Unknown Lenin. From the Secret Archives* (New Haven 1996)
Podro, Joshua *Nuremberg. The Unholy City* (London 1937)
Poewe, Karla *New Religions and the Nazis* (Abingdon 2006)
Pois, Robert A. *National Socialism and the Religion of Nature* (London 1986)
Poliakov, Leon *The Aryan Myth. A History of Racist and Nationalist Ideas in Europe* (New York 1971)
Pollard, John *The Vatican and Italian Fascism 1929–32. A Study in Conflict* (Cambridge 1985)*
—— *The Unknown Pope: Benedict XV (1914–1922) and the Pursuit of Peace* (London 1999)

—— 'The Papacy in Two World Wars: Benedict XV and Pius XII Compared'
Totalitarian Movements and Political Religions (2001) 2, pp. 83–96

Pospielovsky, Dimitry *The Russian Church under the Soviet Regime 1917–1982*
(New York 1984) 2 volumes

Powell, David *Antireligious Propaganda in the Soviet Union. A Study in Mass
Persuasion* (Cambridge, Mass. 1975)

Preston, Paul (ed.) *Revolution and War in Spain 1931–1939* (London 1984)

—— *Franco* (London 1995)

—— *A Concise History of the Spanish Civil War* (London 1996)

—— *Comrades. Portraits from the Spanish Civil War* (London 1999)

Pridham, Geoffrey *Christian Democracy in Western Germany. The CDU/CSU
in Government and Opposition 1945–1976* (London 1977)

Raem, Heinz-Albert *Pius XI und der Nationalsozialismus. Die Enzyklika 'Mit
brennender Sorge' vom 14. März 1937* (Paderborn 1979)

Ratzinger, Joseph (now Benedict XVI) and Pera, Marcello *Without Roots. The
West, Relativism, Christianity, Islam* (New York 2006)*

Rémond, Réne *Les Crises du catholicisme en France dans les années trente* (Paris
1996)

Repgen, Konrad 'Über die Entstehung der Reichskonkordat-Offerte im
Frühjahr 1933 und die Bedeutung des Reichskonkordats' *Vierteljahreshefte
für Zeitgeschichte* (1978) 26, pp. 499–533*

—— 'Zur vatikanischen Strategie beim Reichskonkordat' *Vierteljahreshefte
für Zeitgeschichte* (1983) 31, pp. 506–35

—— 'German Catholicism and the Jews: 1933–1945' in Otto Dov Kulka and
Paul R. Mendes-Flohr (eds) *Judaism and Christianity under the Impact of
National Socialism* (Jerusalem 1987)

Reuth, Ralph Georg *Goebbels* (London 1993)

Revel, Jean-François *The Totalitarian Temptation* (New York 1977)*

Rhodes, Anthony *The Power of Rome in the Twentieth Century*, vol. 2: *The
Vatican in the Age of the Dictators 1922–1945* (London 1973)*

—— *The Power of Rome in the Twentieth Century*, vol. 1: *The Vatican in the
Age of Liberal Democracies 1870–1922* (London 1983)*

—— *The Power of Rome in the Twentieth Century*, vol. 3: *The Vatican in the
Age of the Cold War 1945–1980* (London 1992)

Rhodes, James M. *The Hitler Movement. A Modern Millenarian Revolution*
(Stanford 1980)

Riegel, Klaus-Georg 'Marxism as a Political Religion' *Totalitarian Movements
and Political Religions* (2005) 6, pp. 97–126*

Rissmann, Michael *Hitlers Gott. Vorsehungsglaube und Sendungsbewusstsein
des deutschen Diktators* (Zurich 2001)

Rittner, Carol and Roth, John (eds) *Pope Pius XII and the Holocaust* (London 2002)

Robertson, Charles Grant *Religion and the Totalitarian State* (London 1937)

Rogers, David *Politics, Prayer and Parliament* (London 2000)

Rogger, Hans and Weber, Eugen (eds) *The European Right. A Historical Profile* (Berkeley 1966)

Roth, John and Maxwell, Elizabeth (eds) *Remembering for the Future. The Holocaust in the Age of Genocide* (Basingstoke 2001) 3 volumes

Rothkirchen, Livia 'Vatican Policy and the "Jewish Problem" in Independent Slovakia' *Yad Vashem Studies* (1963) 5, pp. 405–18

Routledge, Paul *John Hume. A Biography* (London 1997)

Rowland, Christopher (ed.) *The Cambridge Companion to Liberation Theology* (Cambridge 1999)

Russell, Bertrand *The Practice and Theory of Bolshevism* (London 1920)

Rychlak, Ronald J. *Hitler, the War and the Pope* (Columbus, Miss. 2000)

—— 'The 1933 Concordat between Germany and the Vatican' *The Digest. National Italian Bar Association Law Journal* (2001) 9, pp. 23–47

—— *Righteous Gentiles. How Pius XII and the Catholic Church Saved Half a Million Jews from the Nazis* (Dallas, Texas 2005)*

Sale, Giovanni *Hitler, la Santa Sede e gli Ebrei* (Milan 2003)*

—— 'L'enciclica contro il Nazismo' *Civiltà Cattolica* (2004) 11, pp. 114–27*

Salter, Cedric *Try-Out in Spain* (New York 1943)

Sánchez, José M. *The Spanish Civil War as a Religious Tragedy* (Notre Dame 1987)

—— *Pius XII and the Holocaust. Understanding the Controversy* (Washington DC 2002)

Sandbrook, Dominic *Never had it So Good. A History of Britain from Suez to the Beatles 1956–1963* (London 2005)*

Sartori, Giovanni *La sociedad multiétnica. Pluralismo, multiculturalismo y extranjeros* (Madrid 2001)*

Schambeck, Herbert (ed.) *Pius XII zum Gedächtnis* (Berlin 1977)

Schmidt, Maria 'Ungarns Gesellschaft in der Revolution und im Freiheitskampf von 1956' *Kirchliche Zeitgeschichte* (2004) 17, pp. 100–12

Schmitt, Carl *Political Theology. Four Chapters on the Concept of Sovereignty* (Cambridge, Mass. 1988)

—— *The Concept of the Political* (Chicago 1996)

Schöpflin, George *Politics in Eastern Europe 1945–1992* (London 1993)

Schreiner, Klaus '"Wann kommt der Retter Deutschlands?" Formen und Funktionen von politischen Messianismus in der Weimarer Republik' *Saeculum* (1998) 49, pp. 107–60

Schulz, Uwe (ed.) *Das Fest. Kulturgeschichte von der Antike bis zur Gegenwart* (Munich 1988)

Schwartz, Michael 'Konfessionelle Milieu und Weimarer Eugenik' *HZ* (1995) 261, pp. 403–448

Schwarz, Hans-Peter *Adenauer* (Munich 1994) 2 volumes*

Scott, James C. *Seeing Like a State. How Certain Schemes to Improve the Human Condition have Failed* (New Haven 1998)

Scott, Malcolm *Mauriac. The Politics of a Novelist* (Edinburgh 1980)

Scruton, Roger *Modern Culture* (London 1998)*

—— *The West and the Rest. Globalization and the Terrorist Threat* (London 2002)*

—— *Gentle Regrets. Thoughts from a Life* (London 2005)

Service, Robert *Lenin. A Biography* (London 2000)

Shapiro, James *Oberammergau. The Troubling Story of the World's Most Famous Passion Play* (London 2000)

Shelah, Menachem 'The Catholic Church in Croatia, the Vatican and the Murder of the Croatian Jews' *Holocaust and Genocide Studies* (1989) 4, pp. 323–39

Sherry, Norman *The Life of Graham Greene* (London 1989–2004) 3 volumes

Siedlarz, Jan *Kirche und Staat im kommunistischen Polen 1945–1989* (Paderborn 1996)*

Siegelbaum, Lewis *Stakhanovism and the Politics of Productivity in the USSR, 1935–1941 (Cambridge 1988)*

Simmons, Michael *The Reluctant President. A Political Life of Vaclav Havel* (London 1991)

Souvarine, Boris *Stalin* (New York 1939)

Speaight, Robert *Georges Bernanos. A Study of the Man and the Writer* (London 1974)

Spotts, Frederic *The Churches and Politics in Germany* (Middletown, Conn. 1973)*

Stasiewski, Bernhard 'Die Kirchenpolitik der Nationalsozialismus im Warthegau' *Vierteljahreshefte für Zeitgeschichte* (1959) 1, pp. 46–74

—— (ed.) *Akten deutscher Bischöfe über die Lage der Kirche 1933 bis 1945* (Mainz 1968–76) volumes 1–3

Stehlin, Stewart *Weimar and the Vatican: German–Vatican Diplomatic Relations in the Interwar Years* (Princeton 1983)

Steigmann-Gall, Richard *The Holy Reich. Nazi Conceptions of Christianity, 1919–1945* (Cambridge 2003)

Steinberg, Jonathan *All or Nothing. The Axis and the Holocaust 1941–1943* (London 1990)

Stern, Fritz *The Politics of Cultural Despair* (Berkeley 1961)

—— *Dreams and Delusions. The Drama of German History* (New Haven 1999)*

—— *Das feine Schweigen. Historische Essays* (Munich 1999)

Stern, J. P. *Hitler. The Führer and the People* (London 1975)

Stewart, A. T. Q. *The Narrow Ground. Aspects of Ulster 1609–1969* (London 1977)*

Stites, Richard *Revolutionary Dreams. Utopian Visions and Experimental Life in the Russian Revolution* (Oxford 1989)

Stuart Hughes, H, *The Obstructed Path. French Social Thought in the Years of Desperation* (New York 1966)

Sturzo, Don Luigi *Politics and Morality. Essays in Christian Democracy* (London 1938)

Tal, Uriel *Religion, Politics and Ideology in the Third Reich. Selected Essays* (London 2004)

Talmon, Jacob L. *Myth of the Nation and Vision of Revolution. Ideological Polarization in the Twentieth Century* (New Brunswick 1991)

Talos, Emmerich and Neugebauer, Wolfgang (eds) *'Austrofaschismus' – Beiträge über Politik, Ökonomie und Kultur 1934–1938* (Vienna 1984)

Tanner, Marcus *Croatia. A Nation Forged in War* (New Haven 1997)

Taylor, Peter, *Provos. The IRA & Sinn Fein (London 1997)*

—— *Loyalists* (London 1999)

—— *Brits. The War against the IRA (London 2001)*

Thomas, Hugh *The Spanish Civil War* (London third edition 1990)*

Timms, Edward *Karl Kraus. Apocalyptic Satirist* (New Haven 1986–2005) 2 volumes*

Tittmann, Harold H. *Inside the Vatican of Pius XII. The Memoir of an American Diplomat during World War II* (New York 2004)

Todorov, Tzvetan *The Fragility of Goodness. Why Bulgaria's Jews Survived the Holocaust* (London 1999)

Tucker, Robert C. *Stalin as Revolutionary 1879–1929* (London 1973)

—— (ed.) *Stalinism. Essays in Historical Interpretation* (New Brunswick 1999)

Ulam, Adam *The Bolsheviks* (Cambridge, Mass. 1965)

—— *Stalin. The Man and his Era* (Boston 1989)

Urban, Mark *Big Boys' Rules. The Secret Struggle against the IRA (London 1992)*

Vaksberg, Arkady *Stalin's Prosecutor. The Life of Andrei Vyshinsky* (New York 1990)

Vincent, Mary *Catholicism in the Second Republic. Religion and Politics in Salamanca 1930–1936* (Oxford 1996)*

Voegelin, Eric *The Political Religions* (originally 1938), *The collected Works of Eric Voegelin*, volume 5, Manfred Henningsen (ed.) (Columbia, Missouri 2000) volumes 1–34

—— *Autobiographical Reflections* Ellis Sandoz (ed.) (Baton Rouge 1989)

Voigt, Frederick *Unto Caesar* (London 1938)

Volk, Ludwig (ed.) *Akten Kardinal Michael von Faulhaber 1917–1945* (Mainz 1975–8) volumes 1–3

Vondong, Klaus *Magie und Manipulation: Ideologischer Kult und politische Religion des Nationalsozialismus* (Göttingen 1971)

—— *Die Apokalypse in Deutschland* (Munich 1988)

Walesa, Lech *A Path of Hope. An Autobiography* (London 1987)*

Ward, Keith *The Case for Religion* (Oxford 2004)*

Webster, Richard A. *The Cross and the Fasces. Christian Democracy and Fascism in Italy* (Stanford 1960)

Weigel, George *The Final Revolution. The Resistance Church and the Collapse of Communism* (New York 1992)

—— *Witness to Hope. The Biography of John Paul II* (New York 1999)*

—— *The Cube and the Cathedral* (New York 2005)*

Weinzierl, Erika and Skalnik, Kurt *Österreich 1918–1938* (Graz 1983)

Wheen, Francis *How Mumbo-Jumbo Conquered the World* (London 2004)

Whitehead, Kenneth D. 'The Pope Pius XII Controversy' *Political Science Reviewer* (2002) 31, pp. 283–387

Wildt, Michael 'The Spirit of the Reich Main Security Office (RSHA)' *Totalitarian Movements and Political Religions* (2005) 6, pp. 333–49

Wilkinson, James D. *The Intellectual Resistance in Europe* (Cambridge, Mass. 1981)

Willis, Frank Roy *Italy Chooses Europe* (Oxford 1971)

Wippermann, Wolfgang *Totalitarismustheorien* (Darmstadt 1997)

Wistrich, Robert S. 'The Last Testament of Sigmund Freud' *Leo Baeck Institute Year Book* (2004) 49, pp. 87–104

Wolf, Hubert 'Pius XI und die "Zeitirrtümer" ' *Vierteljahreshefte für Zeitgeschichte* (2005) 53, pp. 1–42

Wolff, Richard and Hoensch, Jörg (eds) *Catholics, the State, and the European Radical Right 1919–1945* (New York 1987)

Woller, Hans (ed.) *Italien und die Grossmächte 1943–1949* (Munich 1988)

Woolf, S. J. (ed.) *The Rebirth of Italy 1943–1950* (London 1972)

Zuccotti, Susan *The Italians and the Holocaust. Persecution, Rescue and Survival* (London 1987)

—— *The Holocaust, the French, and the Jews* (New York 1993)

—— *Under His Very Windows. The Vatican and the Holocaust in Italy* (New Haven 2000)

INDEX

Hungary – *cont.*
  assault on religious schools by
    Communists 331, 332–3
  deportation of Jews and efforts by
    Church to stop 280–2
  rigging of 1947 elections by
    Communists 332
  struggle between Communist regime
    and Church 329–35

immigrants/immigration 358, 470–1
Imperial War Graves Commission 2
Independence Party (Hungary) 332
Index on Censorship (magazine) 422
Inner Mission 106, 180, 205
Innitzer, cardinal Theodor 149–50
internet 469
IRA (Irish Republican Party) 151, 373, 376,
    385, 390, 391–3, 399–400, 402
  ceasefires 409
  Maze hunger strikes 402
  split within 391
  violence of and bombing campaigns
    392–3, 394, 395–8, 405–6, 407–8, 410,
    411, 474
Iran 456
Iran-Iraq war 463
Iranian Revolution (1979) 462–3
Iraq 458, 471
Iraq war 478, 480
Ireland xiii
  Constitution (1937) 152
  culture of 374–5
  and European Economic Community
    375–6
  and Great War 381
  influence of the Church 152
  and Second World War 374
  sentimentality over 373
  and Spanish Civil War 157
  see also Irish Free State; Irish Republic;
    Northern Ireland
Ireland Act (1949) 381
Irish Free State 150–2, 381, 382–3
Irish National Liberation Army (INLA)
    392, 395
Irish Northern Aid Committee
    (NORAID) 373

Irish Republic 386, 391, 413
Irish Republican Army *see* IRA
Iron Guard (was League of the Archangel
    Michael) 270, 271
Irujo, Manuel de 137
Islam 359, 364, 456, 477, 480
Islamic Salvation Front 459
Islamist terrorism xiv, xv, 456–67, 468–9,
    471–2, 476, 478, 479, 483
  catalogue of atrocities 458, 459–60
  ideological indebtedness to West 471–2
  see also Al Qaeda; bin Laden, Osma; 9/11
Israel 458, 479
Italy 8, 291–9
  attempt to protects Jews by Pius XII
    221–2, 278, 279–80, 282–3
  and Catholic Church 292
  Catholicism and politics 33–4, 65–7,
    294, 296, 298
  Christian Democrats in 65, 293–6, 297,
    298, 299
  and Communism 296–7
  Concordat (1929) 68, 70, 166
  defence of Catholic interests by
    Christian Democrats 293–4
  emergence of anti-Fascist movement
    291–2
  forces competing for dominance in
    post-war 292
  and Great War 5, 66
  Jews in 278–80
  peasant unrest 296
  resistance movement 284
  struggle against Communism 296–9
  see also Fascist Italy
Izotov, Nikita 89

Jackie 355
Jacobins xi, 64
Jacques, father 249
jahiliyya 461
Janiszewski, Jerzy 431
Jarrah, Ziad 451
Jaruzelski, general Wojciech 431, 433,
    435
Jaszi, Oskar 18
Jehovah's Witnesses 346
Jenkins, Roy 351–2

suicide bombers 458
Sully 158
Sunday Schools 346
Sword of the Spirit movement 215–16
Szálasi, Ferenc 281, 324
Szefir, Gyula 331
Szeptycki, metropolitan 255
Sztójay, Döme 280

Taliban 466
Ta'Morley, Eoin 395
Tanzania
  bombing of US embassy (1998) 458, 466
Taracón, Vincente Enrique y 366
Tardini 231–2, 255, 293, 339
Tartary 28
Taylor, Myron 93, 223, 255
Tedeschini, Federico 127
Teitgen, Pierre-Henri 289
television 348–9, 353
  'Clean Up TV' campaign 350–1
Temple, William 31–2
Tennenberg memorial (East Prussia) 6
Tere Nouvelle 154
terrorist assaults 456–60, see also 9/11;
  Islamist terrorism
Thälmann, Ernst 'Teddy' 441
Thatcher, Margaret 378, 402, 404, 421–2, 423
Théas, Pierre Marie 248
Thomas, Hugh 134
Tikhon, patriarch 43, 44, 46, 236
Tiso, Dr Jozef 258–9, 260, 261, 262, 341–2
Tisserant, cardinal 238, 268
Tito, marshal 328, 329
Tittmann, Harold 224
Togliatti, Palmiro 291–2, 295, 296
Tomášek, cardinal Frantisek 420
totalitarian regimes
  as political religions 117–22
Transnistria 273, 275
Treoltsch, Ernst 36
Trimble, David 408, 410, 411
Triumph of the Will (film) 113
Troost, Paul Ludwig 115
Trotsky, Leon 43, 44, 53, 83, 89
  Literature and Revolution 89
Truman, Harry 316

Tucholsky, Kurt 17
Tucker, Robert 75
Tuka, Vojtech 259, 260, 261
al-Turabi, Hassan 464
Turatti, Augusto 60
Turkey 479
Tygodnik Powszechny 340, 427
Tynan, Kenneth 427
Tyne Cot Memorial 2

Ukraine 236, 237, 326, 327
Ulbricht, Walter 311, 439, 442, 443
Ulster see Northern Ireland
Ulster Defence Regiment 391, 398
Ulster Volunteer Force (UVF) 388, 390, 397
Uniate Catholics
  assault on by Stalin 326–8
Union Générale des Juifs de France (UGIF) 243
Unione Nazionale (Italy) 96
Unionists (Northern Ireland) 380, 381, 383, 384–5, 388, 390, 405, 407, 410, 411
Unità, L' 335
United Evangelical Lutheran Church 436
United Nations 364
United States 53, 358, 473
  civil rights movement 474
  and Communism in Italy 297
  condemnation of Nazism by Christians 208
  and Franco 317
  and Northern Ireland 373, 408, 409, 412
  and post-war Italy 292–3, 297
  reaction to Nazi persecution of Jews 253
  religion in 273
  and Vatican 293
Unknown Warrior 4
Urban, Jerzy 434
Ustashe (Insurgency-Croatian Revolutionary Organisation) 262–4, 265–6, 329

Val, Merry del 69
Valeri, nuncio 245
Vallat, Xavier 157, 241, 243
Valle de los Caídos 366
van Gogh, Theo xv, 456–7, 459, 476

Wörmann, Ernst 227
Wright, Billy 'King Rat' 409
Wurm, bishop Theophil 204, 208, 210,
    211, 228, 302
Wyszyński, bishop Stefan 338, 340, 341,
    421

Yaroslavsky, Emelyan 49
Young Men's Association 181–2
Young Pioneers 91, 92
youth culture
  in 1960s 354–5

Yugoslavia 153, 262–3, 320, 328–9
Yule Festival 116

Zapiaín, bishop António Pildain 316
al-Zarqawi, Abu Musab 458–9, 463,
    466
al-Zawahari, Ayman 460, 463
Zerapha, Georges 155
Zimbabwe 356
Zinoviev, Grigori 53, 80, 82
Zolli, chief rabbi Israel 280
Zweig, Arnold 7